COMPOSITION AND ORIGIN OF COMETARY MATERIALS

Cover illustration: Picture of Come Hale-Bopp taken on March 21, 1997 with the Schmidt 92/67 telescope by Giovanni Costa, Sergio Dalle Ave and Claudio Lissandrini, Osservatorio Astronomico di Padova, Sede di Asiago.

Space Science Series of ISSI
Volume 8

The International Space Science Institute is organized as a foundation under Swiss law. It is funded through recurrent contributions from the European Space Agency, the Swiss Confederation, the Swiss National Science Foundation, and the Canton of Bern. For more information, see the homepage at http://www.issi.unibe.ch/

The titles published in this series are listed at the end of this volume.

COMPOSITION AND ORIGIN OF COMETARY MATERIALS

Proceedings of an ISSI Workshop,
14–18 September 1998, Bern, Switzerland

Edited by

K. ALTWEGG
Physikalisches Institut, Universität Bern, CH-3012 Bern, Switzerland

P. EHRENFREUND
Leiden Observatory, 2300 RA Leiden, The Netherlands

J. GEISS
International Space Science Institute, CH-3012, Bern, Switzerland

W. F. HUEBNER
Southwest Research Institute, San Antonio, TX 78228-0510, USA

Reprinted from *Space Science Reviews*, Volume 90, Nos. 1–2, 1999

KLUWER ACADEMIC PUBLISHERS
DORDRECHT / BOSTON / LONDON

Space Sciences Series of ISSI

A C.I.P. Catalogue record for this book is available from the Library of Congress.

ISBN 0-7923-6154-7

Published by Kluwer Academic Publishers,
P.O. Box 17, 3300 AA Dordrecht, The Netherlands

Sold and distributed in North, Central and South America
by Kluwer Academic Publishers,
101 Philip Drive, Norwell, MA 02061, U.S.A.

In all other countries, sold and distributed
by Kluwer Academic Publishers,
P.O. Box 322, 3300 AH Dordrecht, The Netherlands

Printed on acid-free paper

All Rights Reserved
© 1999 Kluwer Academic Publishers
No part of the material protected by this copyright notice may be reproduced or
utilized in any form or by any means, electronic or mechanical,
including photocopying, recording or by any information storage and
retrieval system, without written permission from the copyright owner.

Printed in the Netherlands

TABLE OF CONTENTS

Foreword   xi

I: GASES AND GRAINS IN THE COMA

K. ALTWEGG, H. BALSIGER and J. GEISS / Composition of the Volatile Material in Halley's Coma from *In Situ* Measurements   3

J. CROVISIER and D. BOCKELÉE-MORVAN / Remote Observations of the Composition of Cometary Volatiles   19

R. MEIER and T. OWEN / Cometary Deuterium   33

P. EBERHARDT / Comet Halley's Gas Composition and Extended Sources: Results from the Neutral Mass Spectrometer on Giotto   45

M.C. FESTOU / On the Existence of Distributed Sources in Comet Comae   53

A. LI and J.M. GREENBERG / The Distributed CO in Comet Halley   69

E. KÜHRT / H_2O Activity of Comet Hale-Bopp   75

G. CREMONESE / Hale-Bopp and Its Sodium Tails   83

E.K. JESSBERGER / Rocky Cometary Particulates: Their Elemental, Isotopic and Mineralogical Ingredients   91

M.S. HANNER / The Silicate Material in Comets   99

M.N. FOMENKOVA / On the Organic Refractory Component of Cometary Dust   109

II: FROM COMA ABUNDANCES TO NUCLEUS COMPOSITION

W.F. HUEBNER and J. BENKHOFF / From Coma Abundances to Nucleus Composition   117

A. ENZIAN / On the Prediction of CO Outgassing from Comets Hale-Bopp and Wirtanen   131

J. BENKHOFF / On the Flux of Water and Minor Volatiles from the Surface of Comet Nuclei   141

J.M. GREENBERG and A. LI / Morphological Structure and Chemical Composition of Cometary Nuclei and Dust   149

A.C. LEVASSEUR-REGOURD / Polarization of Light Scattered by Cometary Dust Particles: Observations and Tentative Interpretations   163

D. PRIALNIK and M. PODOLAK / Changes in the Structure of Comet Nuclei due to Radioactive Heating   169

III: ORIGIN OF COMETARY MATERIALS

G. WINNEWISSER and C. KRAMER / Spectroscopy Between the Stars   181

W.M. IRVINE / The Composition of Interstellar Molecular Clouds   203

L.J. ALLAMANDOLA, M.P. BERNSTEIN, S.A. SANDFORD and R.L. WALKER / Evolution of Interstellar Ices   219

P. EHRENFREUND / An ISO View on Interstellar and Cometary Ice Chemistry   233

B. FEGLEY Jr. / Chemical and Physical Processing of Presolar Materials in the Solar Nebula and the Implications for Preservation of Presolar Materials in Comets   239

J. GEISS, K. ALTWEGG, H. BALSIGER and S. GRAF / Rare Atoms, Molecules and Radicals in the Coma of P/Halley   253

G. STRAZZULLA / Ion Irradiation and the Origin of Cometary Materials   269–274

J.F. KERRIDGE / Formation and Processing of Organics in the Early Solar System   275

K. LODDERS and R. OSBORNE / Perspectives on the Comet-Asteroid-Meteorite Link   289

IV: CRITICAL MEASUREMENTS FOR THE FUTURE

P.R. WEISSMAN / Diversity of Comets: Formation Zones and Dynamical Paths   301

G. SCHWEHM and R. SCHULZ / Rosetta Goes to Comet Wirtanen   313

R. SCHULZ and G. SCHWEHM / Coma Composition and Evolution of Rosetta Target Comet 46P/Wirtanen   321

W.T. HUNTRESS Jr. / Missions to Comets and Asteroids   329

L. COLANGELI, V. MENNELLA, J.R. BRUCATO, P. PALUMBO and A. ROTUNDI / Characterization of Cosmic Materials in the Laboratory   341

S.A. STERN / Studies of Comets in the Ultraviolet: The Past and the Future — 355

A.J. BALL, H.U. KELLER and R. SCHULZ / Critical Questions and Future Measurements – Collated Views of the Workshop Participants — 363

V: SUMMARY AND INDEXES

K. ALTWEGG, P. EHRENFREUND, J. GEISS, W.F. HUEBNER and A.C. LEVASSEUR-REGOURD / Cometary Materials: Progress Toward Understanding the Composition in the Outer Solar Nebula — 373

Subject Index — 391

Comet Index — 398

Molecule Indexes — 399

Abbreviations — 408

Author Index — 411

List of Participants — 412

Composition and Origin of Cometary Materials
ISSI Workshop, September 14 - 18, 1998, Bern, Switzerland

Group Photographs

1. J.F. Kerridge
2. G. Strazzulla
3. H. Rickman
4. S.A. Stern
5. P. Eberhardt
6. W.T. Huntress Jr.
7. J. Geiss
8. S. Wenger
9. D. Prialnik
10. R. Meier
11. J. Crovisier
12. E.K. Jessberger
13. W. Benz
14. S. Verani
15. L. Colangeli
16. B. Fegley Jr.
17. K. Lodders
18. H.U. Keller
19. G. Cremonese
20. M.J. Mumma
21. W.F. Huebner
22. L.J. Allamandola
23. C. Barbieri
24. J.F. Crifo
25. P. Ehrenfreund
26. P.R. Weissman
27. A. Enzian
28. A.J. Ball
29. R. Schulz
30. S. Graf
31. L.M. Lara
32. J. Benkhoff
33. E. Kührt
34. G. Winnewisser
35. J.M. Greenberg
36. K. Altwegg
37. M.S. Hanner
38. W.M. Irvine
39. M.C. Festou
40. U. Pfander
41. M.N. Fomenkova
42. G. Nusser Jiang
43. A.C. Levasseur-Regourd

FOREWORD

The current volume, the eighth in the "Space Sciences Series of ISSI" (International Space Science Institute) presents the proceedings of the workshop on "Composition and Origin of Cometary Materials", which was held at ISSI in Bern on September 14-18, 1998.

Following the advice of three ISSI Working Groups that had studied several aspects of comet research, ISSI appointed four convenors, Kathrin Altwegg (University of Bern), Pascale Ehrenfreund (University of Leiden), Johannes Geiss (ISSI), and Hans Rickman (Astronomical Observatory Uppsala), to set up the programme and to nominate participants.

The topic of this workshop is highly interdisciplinary in nature, and consequently scientists from different fields, including optical and radio astronomers, chemists, experimental physicists, as well as theorists and modellers of cometary and interstellar processes, participated.

The data discussed at the workshop were mainly those obtained in 1986 of the coma of Comet Halley by the Giotto spacecraft of ESA and the Vega spacecraft of the Soviet Union and those more recently obtained by remote sensing of the comae of the bright Comets Hyakutake (C/1996 B2) and Hale-Bopp (C/1995 O1). In addition, new laboratory data and model results were presented and discussed.

Comparison of data on interstellar and cometary grains confirmed earlier ideas on their origin and evolution: Tiny refractory grains are produced in stellar envelopes and explosions and transported into the interstellar medium. When a dense Interstellar Molecular Cloud (IMC) forms, these grains serve as nuclei for the condensation of gas-phase molecules and radicals. During recycling of such cloud materials through the diffuse interstellar medium, the mantles of the grains are processed by UV and cosmic radiation. It is thought that the organic material (also called CHON) found in the coma of Halley's Comet is the result of such processing. Cycling between IMCs and the diffuse interstellar medium may have occurred several times until 4.6×10^9 years ago. At that time a molecular cloud collapsed to form protostellar discs — among them the solar nebula. During the collapse phase and in the solar nebula some evaporation, molecular alteration, and recondensation occurred, but it appears that the CHON material found in comets has preserved most of its interstellar identity.

New results presented at the workshop confirmed that at least some constituents of comet ices are also of interstellar origin: Two types of species are considered as specific indicators for IMC chemistry: deuterated molecules and certain radicals. In IMCs, deuterium is highly enriched in condensable molecules. Similarly, D is enriched (above the proto-solar abundance) by an order of magnitude in cometary water and by two orders of magnitude in the hydrogen-cyanide of Comet Hale-Bopp. Such strong enrichments result from ion-molecule reactions at very low temperatures. The D/H ratios measured in cometary water and cometary hydrogen are similar to values found in the hot cores of IMCs. This suggests that significant fractions of H_2O and HCN molecules of Comets Halley and Hale-Bopp were synthesised in the ancient IMC from which the solar system formed 4.6×10^9 years ago. Some CH_2 and C_4H, which were detected in the inner coma of Comet Halley, appear to be primary species that have been vaporised from the ice of the nucleus. If so, these radicals are likely to be of IMC origin, and their survival places important constraints on the thermal history of comet ices.

This volume is a collection of the papers resulting from the invited and contributed presentations, each of which was reviewed by an independent referee. Furthermore, it contains an article by the four editors and Ann-Chantal Levasseur-Regourd that summarises and integrates the various views presented at the workshop. The article concludes with a list of recommendations. At the end, the reader will find a list of abbreviations as well as a subject index and an index of comets. Furthermore, we have included an index of molecules, put together by Stephan Graf and Ursula Pfander, which should be useful for the reader. Comet missions mentioned in this volume are those that had been approved at the time of the workshop. This list may change even before this volume is published.

We wish to express our sincere thanks to all those who have made this volume possible. First of all, we should like to thank the authors for their contributions and the reviewers for their critical and timely reports, which have significantly contributed to the quality of this volume. It is a pleasure to thank the directorate and the staff of ISSI for their support in organising this workshop, in particular D. Taylor, X. Schneider, G. Nusser Jiang, U. Pfander, and S. Wenger. We thank Stephan Graf for designing the chemical index and his help with the manuscript, Paul Wild for his advice concerning the Comet Index, and Diane Taylor for correcting the English in some of the papers. The editors could not have published this volume at the foreseen date without the support of the editorial assistant Ursula Pfander. Her help to authors and editors in producing camera-ready texts, tables, and figures, and her preparing the camera-ready version of the volume including the indexes is very much appreciated.

September 1999
K. Altwegg, P. Ehrenfreund, J. Geiss, W. F. Huebner

I: GASES AND GRAINS IN THE COMA

COMPOSITION OF THE VOLATILE MATERIAL IN HALLEY'S COMA FROM IN SITU MEASUREMENTS

K. ALTWEGG and H. BALSIGER
Physikalisches Institut, University of Bern, Sidlerstr. 5, CH-3012 Bern, Switzerland

J. GEISS
International Space Science Institute, Hallerstr. 6, CH-3012 Bern, Switzerland

Abstract. The investigation of the volatile material in the coma of comets is a key to understanding the origin of cometary material, the physical and chemical conditions in the early solar system, the process of comet formation, and the changes that comets have undergone during the last 4.6 billion years. So far, in situ investigations of the volatile constituents have been confined to a single comet, namely P/Halley in 1986. Although, the Giotto mission gave only a few hours of data from the coma, it has yielded a surprising amount of new data and has advanced cometary science by a large step. In the present article the most important results of the measurements of the volatile material of Halley's comet are summarized and an overview of the identified molecules is given. Furthermore, a list of identified radicals and unstable molecules is presented for the first time. At least one of the radicals, namely CH_2, seems to be present as such in the cometary ice.

As an outlook to the future we present a list of open questions concerning cometary volatiles and a short preview on the next generation of mass spectrometers that are being built for the International Rosetta Mission to explore the coma of Comet Wirtanen.

Keywords: P/Halley, Volatiles, Radicals

1. Introduction

The European Giotto mission, the Russian Vega missions and the Japanese Sakigake and Suisei missions to Comet P/Halley, the Giotto flyby at comet Grigg-Skjellerup and the NASA ICE flyby at comet Giacobini-Zinner have so far been the only spaceprobes coming close enough to a comet to probe the coma in situ. For several years to come, Giotto will remain the only mission carrying mass spectrometers suitable for composition measurements of neutrals and ions. During the last several years the remote sensing of cometary comae in a broad range of wavelengths has made a big step forward concerning sensitivity, spatial and spectral resolution, and number of identified molecules. This has advanced our knowledge about the composition of comae and the diversity of comets. The in situ measurements have provided information not available through remote sensing. The density profiles of individual molecules and ions as a function of the distance from the nucleus determined by Giotto Mass Spectrometers have been used to study the chemistry in the coma of Comet P/Halley. From this, parent molecules were identified and information on the composition of the nucleus was deduced, which is the main goal of cometary coma composition measurements. Molecules that are not accessible to remote observations due to low emissivity or due to interference of the emissions from more abundant species may be detected through mass spectrometry. On the other hand, the mass resolution of mass spectrometers is limited and species having almost the same mass (e.g., CO and N_2) could so far not be separated. In addition, the number of comets where in situ measurements will be made will always be very limited. Remote sensing and in situ observations therefore are complementary.

The present paper gives an overview of the results deduced from in situ measurements in the last 12 years including some new results on minor species. A summary of the results achieved with remote sensing is given by Crovisier and Bockelée-Morvan (1999). The present paper also highlights some open questions that may only be answered in the future. One of the upcoming missions that could answer these questions is the international Rosetta mission to Comet P/Wirtanen. We therefore give a short introduction to ROSINA, the instrument on the Rosetta mission that is dedicated to mass spectrometry of the volatile material.

2. Experimental

The Giotto payload contained three magnetic mass spectrometers for analyzing the volatile material of the coma and its interaction with the solar wind: the Neutral Mass Spectrometer (NMS) and the Ion Mass Spectrometer (IMS) with its two sensors HERS (High Energy Range Spectrometer) and HIS (High Intensity Mass Spectrometer). For details on the NMS see Krankowsky *et al.* (1986); for details on the IMS see Balsiger *et al.* (1987). The NMS instrument included also an energy analyzer that could be used as mass spectrometer, at least inside the contact surface where the temperatures were low. Another energy analyzer (PICCA) for positive ions was part of the Rème Plasma Analyzer experiment (RPA) and had limited mass analyzing capabilities (Rème *et al.*, 1986). In Table I the performance of the different Giotto sensors are summarized.

For most sensors, except the IMS-HIS, there are only data for the inbound leg of the Giotto trajectory available. Inbound, HIS got usable data up to a distance 1300 km from the nucleus and, on the outbound leg (although somewhat deteriorated due to the wobbling of the spacecraft), from 1600 km on outwards.

The interpretation of the data was not straightforward. This is one of the reasons why even 12 years after the flyby the neutral and the ion data have not been fully published. Interferences from different molecules on the same mass peaks are hard to resolve with the limited mass resolution and mass range of the sensors. With refined chemical models it is now possible to get results from ion data even for minor peaks and for ion masses where several molecules can contribute. Unfortunately for many of the species there are no or only inaccurate reaction rates

TABLE I
Performance of the Giotto Sensors for Volatile Material

Sensor	Species	Mass Range	Distance (km) [a]
NMS mass analyzer	Ions and neutrals	12 - 36	2000 [b]
NMS energy analyzer	Ions and Neutrals	12 - 50	2000 [b]
HERS (IMS)	Ions	1-4 / 12- 32	60000 [b,c]
HIS (IMS)	Ions	12-57	inbound and outbound
PICCA (RPA)	Ions	12 - 100	10000 [b,d]

[a] Closest distance from comet up to where usable data have been collected.
[b] NMS, PICCA and IMS-HERS were damaged close to the nucleus.
[c] HERS was optimized for solar wind-coma interaction. Therefore the field of view was not in the direction of the spaceprobe-comet line close to the comet.
[d] PICCA was saturated close to the nucleus.

available, be it for photoionization, photodissociation or other chemical reactions that are needed to deduce the abundances of the appropriate parent molecules from the measured ion densities. Therefore, some of these rates have to be estimated, which leads to uncertainties in the deduced abundances. Details on data analysis, chemical models used, and references for chemical reaction rates can be found, for example, in Schmidt *et al.* (1988), Altwegg *et al.* (1993), Häberli *et al.* (1995), Meier *et al.* (1994), Huebner *et al.* (1987), and Anicich (1993).

3. Results: Highlights of Abundance Measurements in P/Halley's Coma

Fig. 1 shows a composite ion mass spectrum of the IMS-HIS sensor (Altwegg *et al.*, 1993) and the RPA-PICCA sensor (Mitchell *et al.*, 1986). There are data peaks at almost all mass numbers within the mass range from 12 amu/e to 100 amu/e. The combined results of all mass spectrometers on board Giotto have yielded the identification of a large number of parent and daughter molecules, radicals, and ions. This has provided insight into the coma chemistry and physics, thus enabling conclusions to be drawn about the nucleus composition. The work of the numerous people who have contributed to the data analysis is herewith acknowledged. It is not possible to discuss all the Giotto results in the framework of this paper. The most important or surprising ones are highlighted here.

3.1. Elemental Abundances in the Cometary Coma

It has been shown that the oxygen/silicon and the carbon/silicon ratio in Comet Halley is very close to the solar abundances (Fig. 2, Geiss, 1988).

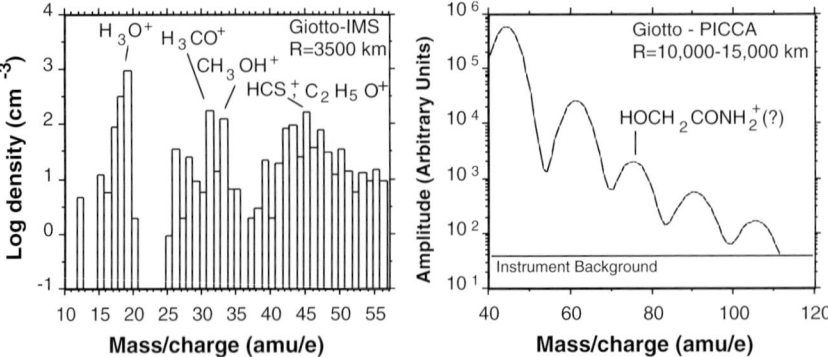

Figure 1. Combined mass spectrum from the IMS-HIS sensor (Altwegg et al., 1993) and the RPA-PICCA sensor (Mitchell et al., 1986) at Comet P/Halley. The HIS data are at 3500 km from the nucleus, the PICCA data at approx. 10000 km from the nucleus.

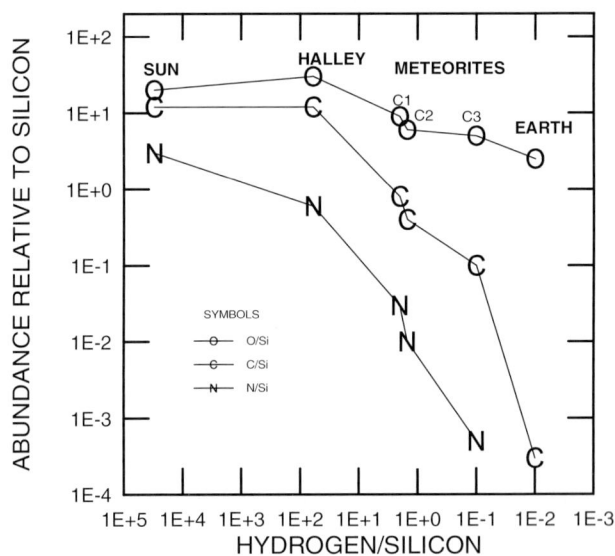

Figure 2. Elemental abundances relative to silicon for several bodies of the solar system (Geiss, 1988).

"Before Giotto" C1-chondrites were the most pristine solar system material accessible for analysis, and most of the elemental average abundances in the solar system were derived from these objects. Comet Halley is clearly more pristine than C1-chondrites, because we find in the comet the full complement of carbon and oxygen. Only nitrogen seems to be somewhat deficient. Comets therefore represent the material in the solar system that has experienced the least change during its history. The deficiency of nitrogen may well be due to the fact that part of the

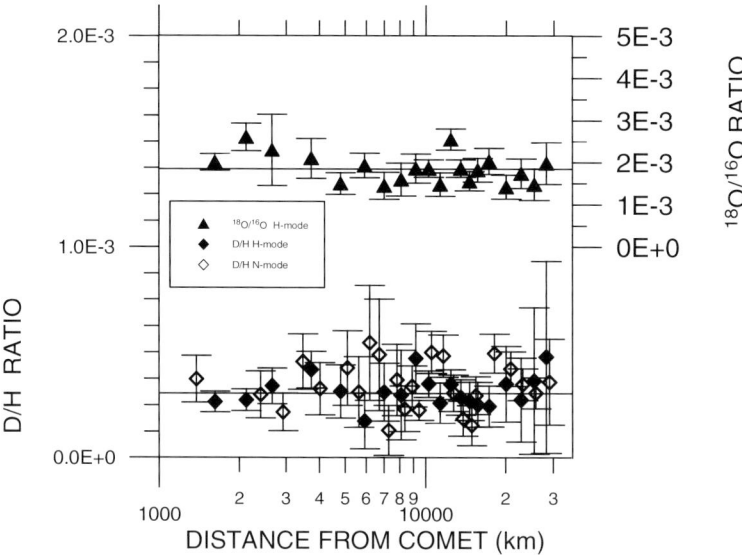

Figure 3. D/H and $^{18}O/^{16}O$ in the hydronium ion of Comet Halley. Data from the IMS-HIS instrument.

nitrogen was present as N_2, which is hard to condense or which may have been lost since the formation of comets. For a discussion on the nitrogen abundance in Comet Halley see Geiss *et al.* (1999).

3.2. Isotopic Ratios in the Volatile Material Released by Comet Halley

Isotopic ratios are excellent tracers for the origin of materials. In Comet P/Halley the deuterium/hydrogen ratio has been determined in the hydronium ion and a value for the D/H ratio in cometary water of $(3.08 \pm 0.3) \times 10^{-4}$ has been determined (Balsiger *et al.*, 1995; Eberhardt *et al.*, 1995). The results from the HIS-IMS sensor are shown in Fig. 3.

The Giotto measurements have been confirmed also in different comets by remote sensing measurements. For a comparison see Meier *et al.* (1999). The cometary D/H ratios are about twice the value as for sea water on Earth, and it is enriched by about a factor of 15 relative to D/H in the protosolar cloud (Geiss and Gloeckler, 1998). Such a high deuterium enrichment can only be achieved by chemical reactions at very low temperatures. Since reactions between neutrals do not proceed, it is generally agreed that the high deuterium abundances must have been produced by ion-molecule reactions. (cf. Geiss and Reeves, 1981).

These reaction rates are rather slow. Thus, for obtaining a high D/H ratio one either needs a very long time or a very high ion density. Long reaction times are available in interstellar molecular clouds and, indeed, high deuterium enrichments are observed in a variety of heavier molecules in such clouds (cf. Irvine, 1999;

Winnewisser and Kramer, 1999). Thus, the presently most popular hypothesis to account for the deuterium enrichment in cometary water is an origin in the dark molecular cloud from which our solar nebula emerged.

However, this hypothesis has been questioned. The other option, the high ion density, could probably have been achieved by X-rays in the early protoplanetary disc. More work is needed to distinguish between the two possibilities.

So far water has been the only cometary molecule for which the D/H ratio has been determined from in situ measurements. For methanol an upper limit of 1% for the D/H ratio has been given (Eberhardt et al., 1994).

$^{18}O / ^{16}O$ has also been determined in cometary water (Balsiger et al., 1995); (Eberhardt et al., 1995). The value is compatible to the solar system value. The only other isotope so far determined is ^{34}S (Altwegg, 1996) in the sulfur ion. In this case the value is also compatible with the solar system value. Due to the large interference between ^{13}C and CH it is very difficult to determine the carbon isotopic ratio. But the data seem to be compatible with a value of $^{12}C / ^{13}C = 90$ (Altwegg et al., 1994), i.e., the solar system value.

3.3. MOLECULAR ABUNDANCES

Table II shows a list of clearly identified molecules, unsaturated molecules and radicals from Giotto measurements. Table III shows "probably" identified molecules. These are measurements that suffer from interference in the mass peak from other possible molecules or parent molecules that are deduced only from their daughter molecules, e.g. OCS, which is beyond the mass range of the mass spectrometers, but where we see an extended source of sulfur ions with a lifetime matching the lifetime of OCS. Table IV is a list of molecules where we can only give upper limits because their densities are either too small or again because of an interference with other possible molecules. From these tables it is evident that the composition of P/Halley's coma is very complex containing a surprisingly large number of organic molecules. Meanwhile the remote sensing capabilities have improved considerably and there exists a large number of molecules which have been identified in different comets.

3.3.1. *Oxidized vs. Reduced Hydrocarbons*

Although the solar nebula was very rich in atomic hydrogen, the number of oxidized hydrocarbons found in Comet P/Halley is surprisingly large and the ones in the reduced form relatively small. Most of the carbon in the volatile material is in CO, CO_2, and H_2CO. The abundance of CH_4 is $< 0.8\ \%$ of water, whereas the CO and CO_2 are the two most abundant molecules besides water.

TABLE II

Observed molecules in the coma of Comet Halley

Species	Source strength	Species measured	Remarks	Reference
H_2O	100 %	Neutrals		Krankowsky et al., 1986
CO	17 %	Neutrals	extended source related to dust?	Eberhardt et al., 1987
H_2CO	3.8 %	Ions H_3CO^+	practically no nucleus source	Geiss et al., 1991; Meier et al., 1993
CO_2	3.5 %	Neutrals		Krankowsky et al., 1986
NH_3	1.5 %	Ions NH_3^+, NH_4^+	optical values are partly lower (≤ 1 %)	Allen et al., 1987; Meier et al., 1994
CH_3OH	1.25% / 1.7%	Ions $CH_3OH_2^+$ / CH_3OH^+		Altwegg, 1996; Eberhardt et al., 1995
CH_2	0.27%	Ions CH_3^+	radical, quasi nucleus source	Altwegg et al., 1994
H_2S	.15% / .4 %	Ions H_3S^+		Altwegg, 1996; Eberhardt et al., 1995
C_2H_2	0.3%	Neutrals		Reber, 1997
C_2H_4	0.3%	Neutrals		Reber, 1997
C_2H_6	0.4%	Neutrals		Reber, 1997
HCN,	0.1%	Ions H_2CN^+		Geiss et al., 1991
CH_3CN	0.14%	CH_4CN^+		Geiss et al., 1999

3.3.2. *Nitrogen Compounds*

We give only an upper limit for N_2 which is very hard to identify in mass spectrometry because of the overlap with CO. NH_3 is the most abundant nitrogen-bearing compound in the volatile material although other nitrogen bearing molecules, as for example XCN compounds (X=H, CH_3, etc.), may have a significant abundance (Geiss et al., 1999).

3.3.3. *Sulfur Compounds*

Several sulfur-bearing molecules have been detected, namely H_2S, OCS and CS_2 (Altwegg, 1996). This suggests that Halley has condensed, besides oxygen and carbon, also its full complement of sulfur. About 20 % of the cometary sulfur may be in the volatile material. Therefore, the value for the sulfur abundance in the solar

TABLE III
Probably observed molecules in the coma of Comet Halley

Species	Source strength	Remarks	Reference
N_2	≤1.5%	in situ, ions	Balsiger et al., 1993
	0.02%	optical	Wyckoff et al., 1991
CS_2, CS	0.2%	optical	Feldman et al., 1987
	0.1%	in situ	Altwegg, 1996
OCS	0.2%	Ions S^+	Altwegg, 1996
CH_3CHO [a]	0.5%	in situ	present work
Acetaldehyde			
C_3H_2	unspecified	in situ	Korth et al., 1989
	0.1 %	in situ	present work
C_2H_5CN	0.028%	in situ	Geiss et al., 1999

[a] Isomers included

TABLE IV
Upper limits for molecules in the coma of Comet Halley

Species	Upper Limit	Remarks	References
NO	≤0.5%	in situ, IMS	Geiss et al., 1991
Nitric oxide			
O_2	≤0.5%	in situ, IMS	Leger et al., 1998
CH_4	≤0.8%	in situ, IMS	Altwegg et al., 1994
Methane			
SO	≤0.5%	in situ, IMS	present work
Sulfur monoxide			
CH_3NH_2	≤0.15%	in situ, IMS	Geiss et al., 1999
Methylamine			
HC_3N	≤0.04 %	in situ, IMS	Geiss et al., 1999

system has to be revised upward because so far sulfur has been determined from C1-chondrites which do not contain volatile sulfur compounds because they did not condense or have been lost.

3.3.4. *Unsaturated Molecules and Radicals*

The most prominent peak in the ion mass range below the water group is mass/charge = 15 amu/e (Altwegg *et al.*, 1994). The density of these ions decreases much faster with distance from the nucleus than H_3O^+. The only molecule that fits the distance profile for this ion mass is CH_2 which protonates with hydronium ions very easily to form CH_3^+. Modeling shows that CH_2 has either to be the daughter product of a very fast decaying molecule or to be present as a radical directly in the ice. There is no parent molecule for CH_2 known with a lifetime short enough to explain the data. Positive evidence of radicals in the ice would be very interesting, giving information about the physical and chemical conditions that prevailed during comet formation. With the presence of CH_2 as a parent molecule and the identified NH_3, the abundance and density profiles of the whole mass/charge group 12-16 amu/e (CH_n and NH_n) can be explained including the optically observed CH^+. The CH^+ which has been observed for a long time in comets (Arpigny *et al.*, 1987) is (at least close to the nucleus) partly a daughter product of CH_2. There remain some discrepancies for distances very close to the comet for carbon (mass 12 and 13 amu). From the IMS-HIS data we found evidence for other unsaturated molecules and radicals in the coma of Comet Halley (see Table V). Most of them have a high proton affinity and therefore react with the hydronium ion to form the protonated ion, which can be detected by an ion mass spectrometer. Because it is extremely difficult to obtain chemical reaction rates for most of these molecules and radicals, the error limits for the abundances are rather large; nevertheless, the species are positively present in the coma. Except for CH_2, which we believe is incorporated as a radical in the ice, the other radicals could still be daughter products from heavier molecules. For the abundance of CO, which is also an unsaturated molecule, see section 3.4.

3.4. EXTENDED SOURCES

The distance profiles of most species are compatible with the release of the parent molecule directly from the ice in the nucleus. The density profile shows approx. a $1/r^2$ dependence. However, for CO and H_2CO (Eberhardt *et al.*, 1987; Meier *et al.*, 1993) this is clearly not the case. The profile of CO is not compatible with a release of all of the CO directly from the nucleus. To fit the distance profile a large portion of the CO has to be released slowly, for example from dust. There are discussions that this phenomenon could also be explained by temporal features like a jet or a region on the comet which is much richer in CO than the neighboring regions. But it seems unlikely that only CO and not the water production rate is affected by such a phenomenon. On the other hand, there are calculations showing

TABLE V
Abundance of Radicals / Unsaturated Molecules

Ion Mass (amu/e)	Parent Molecule	Abundance rel. to water (‰)	Quality
15	CH_2	2.7	a
26	C_2H	0.1-0.6	c[*]
28	CO	35	a[**]
39	C_3H_2	0.5-1.2	b[***]
41	C_2O	0.2-0.5	b[***]

Quality:
a Abundances of other contributing molecules cannot be assessed
b Contributions of other molecules possible, but minor or not very probable
c Major contribution of other molecules

[*] Contribution of C_2H_2 from neutral data of the NMS (Reber, 1997)
[**] Eberhardt et al., 1987
[***] Reaction rates uncertain

that it is not possible to store this amount of CO in comet dust. The discussion is ongoing (see also the paper by Festou, 1999).

3.5. CHEMICAL ION PILE-UP

One of the most puzzling facts in the ion spectra at the time of the Giotto encounter was the fact that the highest ion density is not close to the nucleus but around 10000 km from the nucleus on the Giotto trajectory (Fig. 4).

The upper panel shows the hydronium ion density as a function of distance; the lower panel shows mass/charge=32 amu/e. The pile-up has been explained by Häberli et al. (1995) and Häberli et al. (1996) as being the result of an increase in electron temperature and therefore a decrease in the recombination rate of the hydronium ion. Also here it was suggested that this may only be a temporal or spatial feature (Eberhardt et al., 1995). The same pile-up is, however, also found in the HIS outbound data at 8000 km from the nucleus on the Giotto trajectory which is fully compatible with the inbound data and excludes a spatial variation. Meanwhile the feature has also been detected in Comets Hale-Bopp and Hyakutake (Bouchez et al., 1998) and it has been stationary over several days. We conclude therefore that the ion pile-up is not a temporal variation of outgassing but an inherent phenomenon of the coma.

In the mass/charge 32 amu/e profile, the density increases sharply right at the contact surface. This is due to the long-lived sulfur ion which does not recombine.

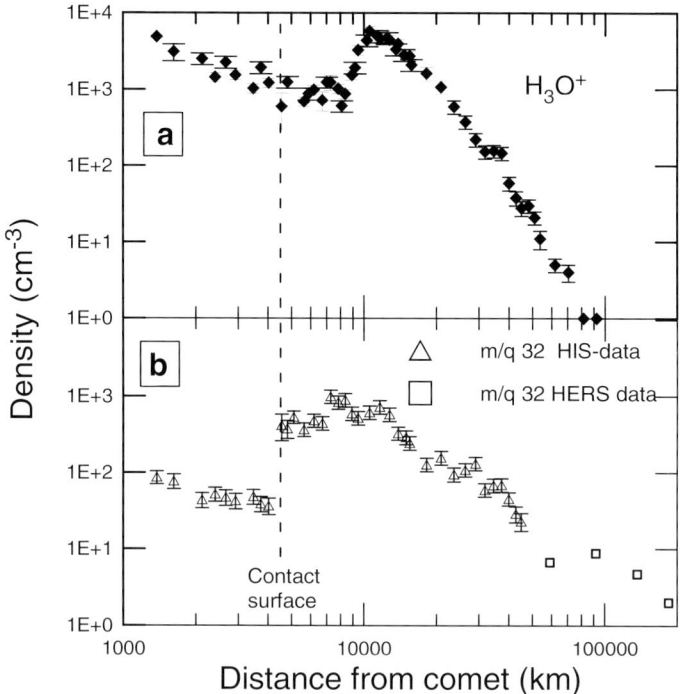

Figure 4. Density profiles of hydronium ions and mass/charge 32 amu/e ions as a function of distance from the nucleus. Data are from IMS-HIS. The peak around 10000 km is called the chemical pile-up and is due to an increase in the electron temperature and therefore a decrease in recombination (Altwegg, 1996).

All sulfur ions which are created on the sunward side of the nucleus outside of the contact surface are carried with the solar wind back towards the contact surface which leads to this sharp density increase at the boundary.

4. Open Questions

Remote sensing of cometary coma composition and in situ measurements have led to a new understanding of the complex processes involved in the formation and evolution of comets; they are contributing to the understanding of the nucleus composition and finally to the history of the solar system material in general. However, in the last few years many new questions have arisen that need to be solved to get more insight into the formation of the solar system, the processing of cometary materials, and finally the origin of our Earth. Below is a list of measurements still to be made and the associated topics that could benefit from it. The list is almost certainly incomplete and will evolve with time.

Elemental abundances:
- Nitrogen abundance: Physical and chemical conditions during comet formation;
- Noble gases: Processing of comets

Isotopic abundances:
- D/H in heavy organic molecules: Origin of material
- Other isotopes in different molecules (C, O etc.): Origin of material

Molecular abundances:
- Heavy organic molecules: Origin of material; processing of material prior to incorporation in comets
- Reduced vs. oxidized molecules: Chemical and physical conditions during molecule formation; origin of material
- Series of molecules, e.g. $C_n H_m$: Origin of material; processing of material prior to incorporation in comets
- O_2, O_3: Origin of terrestrial oxygen
- Radicals : Physical and chemical conditions during comet formation; processing of comets

Physical and chemical processes:
- Extended sources: Composition of dust;
- Molecular abundances as function of heliospheric distance: Nucleus composition, and processing of nucleus
- Molecular abundance differences in jets: Homogeneity of nucleus composition; spatial and temporal differences
- Abundance differences between Oort cloud comets and Kuiper belt comets: Physical and chemical conditions in the different comet forming regions; chemistry in the solar nebula and sub-nebulae

5. ROSINA on Rosetta

Several missions are being planned that address mainly cometary dust or the nucleus directly (Huntress, 1999). The only mission that will conduct a thorough measurement of the volatile component of a coma is Rosetta, where two instruments are dedicated to the volatile material: ROSINA (Rosetta Orbiter Sensor for Ions and Neutral Analysis) and Berenice. Rosetta will be launched in 2003 and will arrive at Comet Wirtanen in 2011. For a detailed description of this mission, see Schwehm *et al.* (1999).

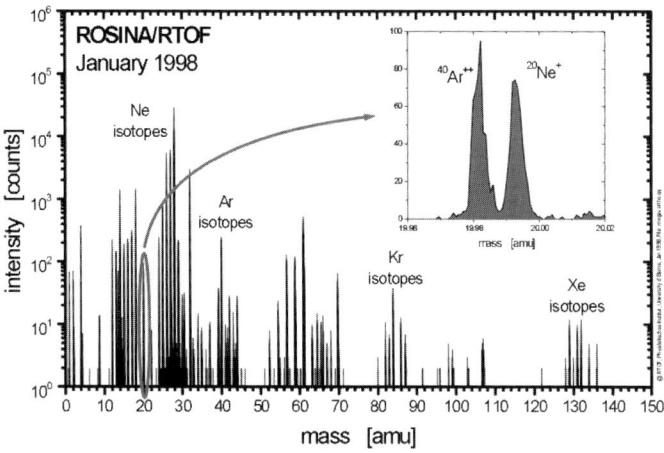

Figure 5. Mass spectrum from the RTOF lab prototype.

TABLE VI
Scientific capabilities of ROSINA/Rosetta

	Mass Range [amu]	Mass Resolution m/Δ m (at 1%)	Sensitivity Gas [A/Torr]	Dynamic Range
DFMS	12-100	3000	10^{-5}	10^{10}
RTOF	1-$\geq$300	$\geq$500	10^{-3}	$10^6/10^8$

Berenice will concentrate on certain isotopes (O, D) in several molecules whereas ROSINA will investigate the elemental, molecular and isotopic abundances of the volatiles over a very large mass range. In Table VI the characteristics of ROSINA are given. It is not possible to achieve with one sensor all scientific goals, i.e. a very high mass resolution combined with a very broad mass range and a very high sensitivity. ROSINA therefore will include two mass spectrometers, namely DFMS (Double Focusing Magnetic Spectrometer) and RTOF (Reflectron Time Of Flight spectrometer). They will be launched under high vacuum with a cover that will only be released in space. The principles of the sensors are given in Balsiger *et al.* (1997). The sensors have already been built as lab prototypes with the final ion optics geometry.

Fig. 5 gives a spectrum from the RTOF sensor. RTOF has a high mass range (1 to $\geq$ 500 amu) combined with a high sensitivity. Fig.6 shows a high mass resolution

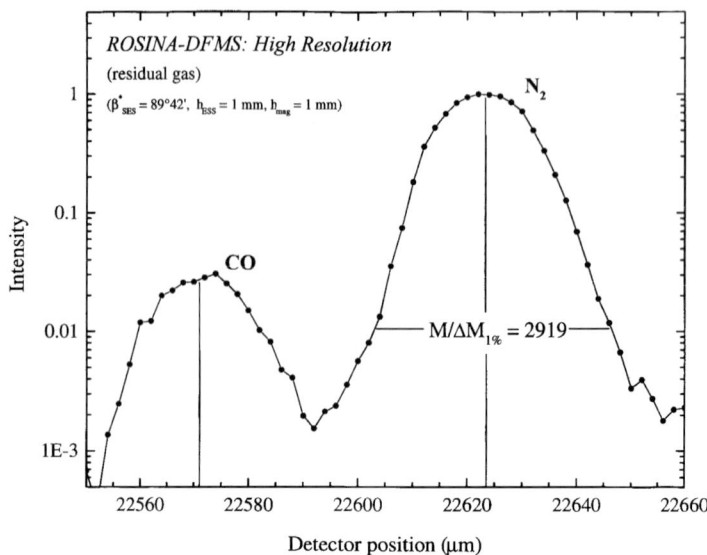

Figure 6. Mass spectrum from the DFMS prototype

spectrum taken with the DFMS. N_2 and CO can clearly be separated, even if the abundance of N_2 is less than 1% of CO.

Rosetta will meet the comet near its aphelion and will fly in close vicinity of the comet up to perihelion. ROSINA will be one of the instruments giving interesting and important data from the very first onset of cometary activity to the active phase at perihelion. It will be possible to follow the evolution of the coma, to measure differences in the coma composition as a function of heliocentric distance, to identify heavy organic molecules and get isotope rations for many elements in different molecules. With this mission we hope to answer many of the above questions, although Rosetta will go only to one single comet. Comet Wirtanen seems to be a Kuiper belt object and may therefore be rather different from, for example, Halley, which is an Oort cloud comet. With the help of remote sensing, eventually also the differences between comets can be assessed.

References

Altwegg, K., Balsiger, H., Geiss, J., Goldstein, R., Ip, W.-H., Meier, A., Neugebauer, M., Rosenbauer, H., and Shelley, E.: 1993, 'The ion population between 1300 km and 230 000 km in the coma of comet P/Halley', *Astron. Astrophys.* **279**, pp. 260-266.

Altwegg, K., Balsiger, H., Geiss, J.: 1994,'Abundance and origin of the CH_n^+ ions in the coma of comet Halley', *Astron. Astrophys.* **290**, pp. 318-323.

Altwegg, K.: 1996, 'Sulfur in the coma of comet Halley from in situ measurements', Habilitationsschrift, University of Bern.

Allen, M., Delitsky, M., Huntress, W., Yung, Y., Ip, W.-H., Schwenn, R., Rosenbauer, H., Shelley, E., Geiss, J.: 1987,' Evidence for methane and ammonia in the coma of comet P/Halley', *Astron. Astrophys.* **187**, pp. 502-512.

Anicich, V. G.: 1993, 'Evaluated bimolecular ion-molecule gas phase kinetics of positive ions for use in modeling the chemistry of planetary atmospheres, cometary comae, and interstellar clouds',*J. Phys. Chem. Ref. Data* **22** (**6**), p 1496.

Arpigny, C., Zeippen, C.J., Klutz, M., Magain, P., Hutsemekers, D.: 1987, 'On the interpretation of the CH cometary spectrum', in *Symposium on the Diversity and Similarity of Comets*, edited by E.J. Rolfe and B. Battrick, **ESA SP-278**, pp. 607-612.

Balsiger, H., Altwegg, K., Bühler, F., Geiss, J., Ghielmetti, A.G., Goldstein, B.E., Goldstein, R., Huntress, W.T., Ip, W.-H, Lazarus, A.J., Meier, A., Neugebauer, M., Rettenmund, U., Rosenbauer, H., Schwenn, R., Sharp R.D., Shelley, E.G., Ungstrup, E. and Young, D.T.: 1986,'Ion composition and dynamics at comet Halley', *Nature* **321**, pp.330-336.

Balsiger, H., Altwegg, K., Benson, J., Bühler, F., Fischer, J., Geiss, J. Goldstein, B.E., Goldstein, R., Hemmerich, P., Kulzer, G., Lazarus, A.J., Meier, A., Neugebauer, M., Rettenmund, U., Rosenbauer, K. Säger, T Sanders., R. Schwenn, E. G. Shelley, D. Simpson, D., and Young, D.T.: 1987,'The ion mass spectrometer on Giotto', *J. Phys. E: Sci. Instrum.* **20**, p 759.

Balsiger, H., Altwegg, K., Geiss, J.: 1995,'D/H and $^{18}O/^{16}O$ - ratio in the hydronium ion and in neutral water from in situ ion measurements in comet Halley', *J. Geophys. Res.* **100**, pp. 5827-5834.

Balsiger, H., Altwegg, K., Arijs, E., Bertaux, J.-L., Berthelier, J.-J., Bochsler, P., Carignan, G.R., Eberhardt, P., Fisk, L.A., Fuselier, S.A., Ghielmetti, A.G., Gliem, F., Gombosi, T.I., Kopp, E., Korth, A., Livi, S., Mazelle, C., Rème, H., Sauvaud, J.A., Shelley, E.G., Waite, J.H., Wilken, B., Woch, J., Wollnik, H., Wurz, P., Young, D.T.: 1997, 'Rosetta Orbiter Spectrometer for Ion and Neutral Analysis ROSINA', *Adv. Space.Res.* **21** , pp. 1527-1535.

Bouchez, A.H., Brown, M.E., Spinrad, H., Misch, A.: 1998, 'Observations of the ion pile-up in Comets Hale-Bopp and Hyakutake', *Icarus*, in press.

Crovisier, J., and Bockelée-Morvan, D.:1999, 'Remote observations of the composition of cometary volatiles', *Space Sci. Rev.*, this volume.

Eberhardt, P., Krankowsky, D., Schulte, W., Dolder, U., Lämmerzahl, P., Berthelier, J.-J., Woweries, J., Stubbemann, U., Hodges, R.R., Hoffman, J.H., Illiano, J.M.: 1987, 'The CO and N_2 abundances in comet P/Halley', *Astron. Astrophys.* **187**, p. 481.

Eberhardt, P., Meier, R., Krankowsky, D., Hodges, R.R.: 1994,'Methanol and hydrogen sulfide in comet P/Halley', *Astron. Astrophys.* **288**, pp. 315-329.

Eberhardt, P., Reber, M., Krankowsky, D., Hodges, R.R.: 1995,'The D/H and $^{18}O/^{16}O$ - ratios in water from comet P/Halley', *Astron. Astrophys.* **302**, p. 301.

Eberhardt, P., Krankowsky, D.: 1995, 'The electron temperature in the inner coma of comet P/Halley', *Astron. Astrophys.* **295**, p. 795.

Feldman, P. D., Festou, M.C., A'Hearn, M.F., Arpigny, C., Butterworth, P.S., Cosmovici, C.B., Danks, A.C., Gilmozzi, R., Jackson, W.M., McFadden, L.A., Patriarchi, P., Schleicher, D.G., Dozzi, g., Wallis, M.K., Weaver, H.A., Woods, T.N.: 1987, 'IUE observations of comet P/Halley: evolution of the ultraviolet spectrum between September 1985 and July 1986', *Astron. Astrophys.* **187**, p. 325.

Festou, M.C.: 1999, 'On the existence of distributed sources in comet comae', *Space Sci. Rev.*, this volume.

Geiss, J. and Reeves, H.: 1981, *Astron. Astrophys.* **93**, pp. 189-199.

Geiss, J.: 1988, 'Composition in Halley's comet', *Reviews in modern Astron.* **1**, pp. 1-27.

Geiss, J., Altwegg, K., Anders, E., Balsiger, H., Ip, W.-H., Meier, A., Neugebauer, M., Rosenbauer, H., and Shelley, E.G.: 1991, 'Interpretation of the ion mass spectra in the mass per charge range 25-35 amu/e obtained in the inner coma of Halley's comet by the HIS-sensor of the Giotto IMS experiment', *Astron. Astrophys.* **247**, pp. 226-234.

Geiss, J., Gloeckler, G.: 1998, 'Deuterium and Helium-3 in the Protosolar Cloud', *Space Sciences Series of ISSI* **4**, and *Space Sci. Rev.* **84**, 239-250.

Geiss, J., Altwegg, K., Balsiger, H., and Graf, S.: 1999, 'Rare atoms, molecules and radicals in the coma of P/Halley', *Space Sci. Rev.*, this volume.

Häberli, R., Altwegg, K., Balsiger, H., Geiss, J.: 1995, 'Physics and chemistry of ions in the pile-up region of comet P/Halley', *Astron. Astrophys.* **297**, p. 881.

Häberli, R., Altwegg, K., Balsiger, H., Geiss, J.: 1996, 'Heating of the thermal electrons in the coma of comet P/Halley', *J. Geophys.Res.* **101**, p. 15579.

Huebner, W. F., Keady, J.J., Lyon, S.P.: 1992,'Solar photo rates', *AP&SS* **195**, p. 1.

Huntress, W.: 1999, ' Missions to Comets and Asteroids', *Space Sci. Rev.*, this volume.

Irvine, W.M.: 1999, 'The Composition of Interstellar Molecular Clouds', *Space Sci. Rev.*, this volume.

Krankowsky, D., Lämmerzahl, P., Dörflinger, D., Herrwerth, I., Stubbemann, U., Woweries, J., Eberhardt, P., Dolder, U., Fischer, J., Herrmann, U., Jungck, M., Meier, F., Schulte, W., Berthelier, J.-J., Illiano, J.M., Godefroy, M., Gogly, G., Thèvenet, P., Hoffman, J.H.: 1986, 'The Giotto Neutral Mass Spectrometer', *ESA SP-* **1077**, p. 109.

Korth, A., Marconi, M.L., Mendis, D.A., Krueger, F.R., Richter, A.K., Lin, R.P., Mitchell, D.L., Anderson, K.A., Carlson, C.W., Rème, H., Sauvaud, J.A., d'Uston, C.: 1989, 'Probable Detection of Organic Dust-born Aromatic $C_3H_3^+$ Ions in the Coma of Comet Halley', *Nature* **337**, p. 53.

Korth, A., Krueger, F.R., Mendis, D.A., and Mitchell, D.L.: 1989, 'Organic ions in the coma of comet Halley', in: *Asteroids, comets, meteors III*, edited by C.-I. Lagerkvist, H. Rickman, B. A. Lindblad, and M. Lindgreen, Uppsala Univ. Press, Uppsala.

Léger, A., Ollivier, M., Altwegg, K., Woolf, N.J., 1998:'Is the presence of H_2O and O_3 in an exoplanet a reliable signature of a biological activity?', *Astron. Astrophys.*, in print.

Meier, R., Eberhardt, P., Krankowsky, D., Hodges, R.R.: 1993, 'The extended formaldehyde source in comet P/Halley', *Astron. Astrophys.* **277**, p. 677.

Meier, R., Eberhardt, P., Krankowsky, D., Hodges, R.R.: 1994, 'Ammonia in comet P/Halley', *Astron. Astrophys.* **287**, pp. 268-278.

Meier, R. and Owen, T.C.: 1999, 'Cometary Deuterium', *Space Sci. Rev.*, this volume.

Mitchell, D.L., Lin, R.P., Anderson, K.A., Carlson, C.W., Curtis, D.W., Korth, A., Richter, A.K., Rème, H., Sauvaud, J.A., d'Uston, C., Mendis, D.A.: 1986, 'Derivation of heavy (10-210 amu) ion composition and flow parameters for the Giotto PICCA instrument' *ESA SP-* **250**, p.203.

Reber, M.: 1997, 'Die D/H - und $^{18}O/^{16}O$ -Verhältnisse in Wasser sowie die Formaldehyd-Häufigkeit im Kometen P/Halley - Messungen mit dem Giotto Neutralmassenspektrometer', PhD-Thesis, University of Bern.

Rème, H., Cottin, F., Cros, A., Médale, J.L., Sauvaud, J.A., d'Uston, C., Korth, A., Richter, A.K., Loidl, A., Anderson, K.A., Carlson, C.W., Curtis, D.W., Lin, R.P., Mendis, D.A.: 1986, 'The Giotto RPA-Copernic Plasma Experiment", *ESA SP-* **1077**, p. 33.

Schmidt, H. U., Wegmann, R., Huebner, W.F., and Boice, D.C.: 1988, 'Cometary gas and plasma flow with detailed chemistry', *Computer Phys. Comm.* **49**, pp. 17-59.

Schwehm, G., and Schulz, R.: 1999, 'Rosetta goes to comet Wirtanen', *Space Sci. Rev.*, this volume.

Winnewisser, G., and Kramer, C.: 1999, 'Spectroscopy between the stars', *Space Sci. Rev.*, this volume.

Wyckoff, S., Tegler, S.C., Engel, L.: 1991, 'Ammonia abundances in four comets', *ApJ* **367**, p. 641.

Address for correspondence: Kathrin Altwegg, altwegg@phim.unibe.ch

REMOTE OBSERVATIONS OF THE COMPOSITION OF COMETARY VOLATILES

J. CROVISIER and D. BOCKELÉE-MORVAN
Observatoire de Paris-Meudon, F-92195 Meudon, France

Abstract. The volatile species released in the coma are an important clue to the composition of the cometary nucleus ices. Their identification and the measurement of their abundances is possible by remote sensing. Considerable progress has been made recently using radio and infrared spectroscopy, especially with the observations of the two exceptional comets C/1996 B2 (Hyakutake) and C/1995 O1 (Hale-Bopp).) 24 molecules likely to be parent molecules outgassed from the nucleus have now been identified. Significant upper limits exist for many other species, and the presence of unidentified lines suggests that further species are to be identified. In addition, isotopic varieties have been observed for hydrogen, carbon, nitrogen and sulphur. We will review these results with a special emphasis on the reliability of the identifications and of the molecular production rate determinations. A critical point is to assess whether a given species is a genuine parent molecule outgassed from nuclear ices, or is a secondary product coming from grains or from gas-phase photochemistry. Ground-based spectral imaging, such as radio interferometry, may help resolving this problem.

Keywords: comet, composition, ice, molecule, spectroscopy

1. Introduction

A fundamental goal of cometary science is to determine the composition of comet nucleus ices. In the absence of a direct analysis of this material, one has to analyze the cometary atmosphere formed by volatiles outgassed from the nucleus. This can be done in several steps:
- the identification of molecules present in the coma;
- the determination of their abundances;
- the determination of their origin;
- the determination of the abundance of volatiles in nucleus ices.

We will discuss here how this can be achieved by remote sensing, using ultraviolet, visible, infrared or radio spectroscopy. A complementary method is in situ analysis of the gas by mass spectroscopy — which up to now has only been performed during the space exploration of Comet Halley; it is discussed elsewhere in the proceedings of this workshop (Altwegg *et al.*, 1999; see also Eberhardt, 1999).

We have entered a new era where "parent" molecules can now be directly observed, thanks to new techniques such as radio and infrared spectroscopy. Spectacular results have been recently obtained during the apparitions of the exceptional

TABLE I

Parent molecules in comets: the chronology of first detections.

Year	comet technique	molecules	references a)
1973	C/1973 E1 (Kohoutek)		
	radio	HCN?	[1]
1976	C/1975 V1 (West)		
	UV	CO (also CS)	[2] [3]
1983	C/1983 H1 (IRAS-Araki-Alcock)		
	radio	NH_3?	[4]
	UV	S_2	[5]
1985–86	1P/Halley		
	radio	HCN, H_2CO?	[6] [7] [8] [9]
	IR	H_2O, CO_2	[10] [11]
1990	C/1989 X1 (Austin), C/1990 K1 (Levy)		
	radio	H_2CO, H_2S, CH_3OH	[12] [13]
1996	C/1996 B2 (Hyakutake)		
	radio	NH_3, HNC, CH_3CN,	[14] [15] [16]
		OCS? HNCO?	[17] [18]
	IR	CH_4, C_2H_2, C_2H_6	[19] [20]
1997	C/1995 O1 (Hale-Bopp)		
	radio	OCS, HNCO, HC_3N, SO_2, HCOOH,	
		H_2CS, NH_2CHO, $HCOOCH_3$ (also SO)	[21] [22] [23]

a) References: [1] Huebner et al., 1974; [2] Feldman and Brune, 1976; [3] Smith et al., 1980; [4] Altenhoff et al., 1983; [5] A'Hearn et al., 1983; [6] Despois et al., 1986; [7] Schloerb et al., 1986; [8] Winnberg et al., 1987; [9] Snyder et al., 1989; [10] Mumma et al., 1986; [11] Combes et al., 1986; [12] Bockelée-Morvan et al., 1991; [13] Colom et al., 1992; [14] Palmer et al., 1996; [15] Irvine et al., 1996; [16] Dutrey et al., 1996; [17] Woodney et al., 1997a; [18] Lis et al., 1997; [19] Mumma et al., 1996; [20] Brooke et al., 1996; [21] Lis et al., 1998; [22] Bockelée-Morvan et al., 1998b; [23] Woodney et al., 1998.

comets C/1996 B2 (Hyakutake) and C/1995 O1 (Hale-Bopp): the number of identified parent molecules was multiplied by a factor of almost three. A summary of the chronology of the discovery of "parent" molecules in comets is given in Table I. Our state-of-the-art knowledge of the composition of cometary volatiles is summarized in Table II. Other reviews on this topic were recently given by Bockelée-Morvan (1997); Bockelée-Morvan and Crovisier (1998); Crovisier (1998b), Despois (1998), Eberhardt (1998, 1999), Irvine et al. (2000), Mumma (1997), Rauer (1999).

2. Identification of Molecular Species

Since the first observation of a cometary spectrum (which was made visually by Donati (1864) in comet Tempel 1864 II), cometary spectroscopy and laboratory molecular spectroscopy have been intimately linked. At that time, the cometary spectral features were not identified. Nowadays, although considerable progress has been made, many features in cometary spectra are still unidentified.

A high spectral resolution is of a great help for identification: this is possible for rotationally resolved spectra in the radio and (more recently) infrared domains (Crovisier, 1998a; Despois, 1998).

2.1. The Reliability of Identifications/Detections

To be reliable, the identification or detection of a cometary line should comply with different criteria:
— a good signal-to-noise ratio is required;
— the radio lines should have their shapes and central velocities consistent with those observed for other cometary lines;
— several lines should be observed for confirmation (or the same line should be confirmed in several comets);
— the relative intensities of the observed lines should be in reasonable agreement with predictive models.

These criteria are fulfilled for most of the species listed in Tables I and II. Exceptions are $HCOOCH_3$ and H_2CS which were observed by a single line in a single comet (Bockelée-Morvan et al., 1998b; Woodney et al., 1998).

2.2. Unidentified Lines

There are still many unidentified lines in cometary spectra, suggesting that we may expect further molecular discoveries, and showing that presently available molecular data are still insufficient for the interpretation of these spectra.

In the visible, detailed spectral surveys show a host of unidentified lines (Arpigny, 1995; Brown et al., 1996; Morrison et al., 1997; Wyckoff et al., 1999). They are presumably due to radicals and/or ions, for which laboratory spectral data are needed. Progress in the knowledge of the spectrum of the NH_2 radical recently resulted in the identification of many features in the red region of cometary spectra (Huet et al., 1996).

In the infrared, many unidentified lines have been spotted from the IRTF/CSHELL high-resolution spectra of Comets C/1996 B2 (Hyakutake) and C/1995 O1 (Hale-Bopp) (e.g., Magee-Sauer et al., 1998). They might be due to radicals, since the infrared spectra of simple, stable molecules are well known.

A few unidentified lines have also been noted in the radio domain, but they are observed with a limited signal-to-noise ratio.

2.3. UNIDENTIFIED BANDS

The 3.4 μm emission band is known since the 1986 observations of Comet Halley. Methanol was soon recognized to be a major contributor to this emission (see Bockelée-Morvan *et al.*, 1995, and references therein). Several minor constituents recently identified among those listed in Table II (such as some hydrocarbons and HCOOH) are also contributing. It is highly likely that further species are contributing. Progress on their identification will need a good model of the methanol infrared fluorescence spectrum (there is yet no descriptive model of the relevant infrared bands of methanol), let alone other species. Unfortunately, this spectral region could not be well observed in Comet Hale-Bopp because of the very strong dust continuum.

The presence of PAH's in comets is still an open issue. The identification of phenanthrene was claimed from the near-UV spectra of Comet Halley observed by the TKS instrument on the Vega probes (Moreels *et al.*, 1993). This identification may be questioned, since it was not confirmed in any other comet, and it relies on rather low-resolution spectra. Admittedly, the TKS observations sampled a region quite close to the nucleus of Comet Halley, but this was also the case for the ground-based observations of Comet Hyakutake (e.g., Meier *et al.*, 1998c) which passed close to the Earth. The 3.28 μm emission band, characteristic of aromatics, has definitely been seen, but only in a few comets (see the review by Bockelée-Morvan *et al.*, 1995) and not in Comet Hale-Bopp. The ISO spectra of Comet Hale-Bopp (Crovisier *et al.*, 1997) do not show any of the PAH bands which are so ubiquitous in the interstellar medium. However, the comet was only observed by ISO at $r_h > 2.9$ AU, where PAH's might have not sublimed.

3. From Line or Band Intensities to Column Densities

The fundamental mechanism governing emission by cometary molecules is fluorescence excited by the Sun radiation field. In the inner coma (collisional region), collisions affect the rotational distribution of molecules. This distribution is basically out of equilibrium, between *thermal* and *fluorescence* equilibrium.

To convert *whole band* intensities observed in the infrared, visible or ultraviolet into column densities is rather straightforward: in principle, only the band fluorescence rate (the so-called g-factor) is needed. In most cases, band resonant fluorescence occurs in first approximation (especially for the fundamental bands of vibration in the infrared) and the g-factor can be readily evaluated from the band strength and the solar spectrum. In other cases, the more complex fluorescence cascade must be worked out (e.g., for hot bands of water).

For interpreting single rotational lines (radio) or ro-vibrational lines (infrared), the rotational distribution of the molecule must be known, This requires additional information. It may be provided from the observation of several lines of the

TABLE II

Molecular abundances observed in comets C/1996 B2 (Hyakutake) and C/1995 O1 (Hale-Bopp) at $r_h \approx 1$ AU.

Molecule	a)	b)	C/1996 B2		C/1995 O1		remark
H_2O	IR	5	100	[1]	100	[2] [3]	
CO	R IR UV	> 5	6–30	[1] [4] [5]	20	[2] [6] [7]	extended[c]
CO_2	IR	2			$20^{d)}$	[8]	
CH_4	IR	–	0.7	[1]	0.6	[2]	
C_2H_2	IR	–	≈ 0.5	[9]	0.1	[2]	
C_2H_6	IR	–	0.4	[1]	0.3	[2]	
CH_3OH	R IR	> 5	2	[4]	2	[6] [10]	
H_2CO	R IR	4	0.2–1	[4]	1	[6]	extended[e]
HCOOH	R	–	–		0.06	[11]	
$HCOOCH_3$	R	–	–		0.06	[11]	
NH_3	R IR	1?	0.5	[12]	0.7	[13] [14]	
HCN	R IR	> 5	0.1	[4]	0.25	[2] [6]	
HNCO	R	–	0.07	[4]	0.06	[10] [11]	
HNC	R	–	0.01	[4] [15]	0.04	[6] [16]	extended
CH_3CN	R	–	0.01	[17]	0.02	[6]	
HC_3N	R	–	–		0.02	[10] [11]	
NH_2CHO	R	–	–		0.01	[10] [11]	
H_2S	R	5	0.8	[17]	1.5	[6]	
OCS	R IR	–	0.1	[18]	0.3	[10] [19]	extended?
SO	R	–	–		0.2–0.8	[10] [11] [20]	extended
CS_2	(R) (UV)	> 5	0.1	[17]	0.2	[6]	from $CS^{f)}$
SO_2	R	–	–		0.1	[11] [20]	
H_2CS	R	–	–		0.02	[21]	
S_2	UV	1	0.005	[5]	–		

[a] technique of observation (R: radio);
[b] number of detections in other comets;
[c] abundance given for *nuclear* + *extended* source;
[d] measured at $r_h = 2.9$ AU;
[e] abundance given for *extended* source;
[f] CS_2 abundance has been inferred from its presumed daughter CS.

References:
[1] Mumma et al., 1996; [2] Weaver et al., 1998; [3] Dello Russo et al., 1999; [4] Lis et al., 1997; [5] Weaver et al., 1996; [6] Biver et al., 1998; [7] DiSanti et al., 1999; [8] Crovisier et al., 1997; [9] Brooke et al., 1996; [10] Lis et al., 1998; [11] Bockelée-Morvan et al., 1998b; [12] Palmer et al., 1996; [13] Bird et al., 1997; [14] Bird et al., 1998; [15] Irvine et al., 1996; [16] Irvine et al., 1998; [17] Bockelée-Morvan, 1997; [18] Woodney et al., 1997a; [19] Dello Russo et al., 1998; [20] Kim et al., 1998; [21] Woodney et al., 1997b.

molecule. The radio lines of methanol are particularly suitable for this, since it is often possible to observe several of these lines simultaneously with the same instrumentation (e.g., Biver *et al.*, 1998). Series of ro-vibrational lines have also been observed by ISO or CSHELL/IRTF: for H_2O (Crovisier *et al.*, 1997, 1999; Dello Russo *et al.*, 1999) or CO (DiSanti *et al.*, 1999), HCN, C_2H_2.... But for a single line observation, one has to assume the same rotational temperature as that observed for other species (if possible, at the same moment within a similar field of view).

Another approach is theoretical modelling of rotational excitation. One has to solve for the evolution between thermal equilibrium and fluorescence equilibrium. The kinetic temperature of the gas and its evolution with distance to nucleus must be known, as well as the collisional cross sections (which are in fact very poorly known).

Optical depth effects may be a problem, but only for abundant species (e.g., H_2O) or in productive comets (e.g., HCN radio lines in Comet Hale-Bopp).

4. From Column Densities to Molecular Production Rates

In order to convert column densities into molecular production rates, the knowledge of the density distribution is needed. This is intimately linked to the problem of the origin of the molecules:
- molecule directly sublimed from the nucleus;
- daughter species coming from the photodissociation of a parent;
- molecule coming from a "distributed source" such as dust, or synthesized by chemical reactions in the coma, or produced by still unknown mechanisms.

The simplistic Haser or vectorial models are popularly used in this context. They require the knowledge of the molecular velocities (which can be inferred from radio line shapes) and photodissociation rates (which can be evaluated from laboratory data and the solar spectrum). There is still a need for laboratory data on photodissociation rates. Some of them are just unknown; this is the case for most radicals, but also for the recently discovered species H_2CS, $HCOOCH_3$ and NH_2CHO. It must be noted that those for SO and SO_2 are still debated (discrepant values are reported in the literature; Kim *et al.*, 1998).

Spherical symmetry, which is a fundamental hypothesis in these models, is broken in real cases because of anisotropic outgassing. The resulting errors on production rate evaluations from observations averaged over wide circular fields of view are not very large (typically 10–20% for realistic cases; Biver *et al.*, in preparation). Much more severe could be the errors resulting from the assumption of a parent distribution when the molecule is indeed coming from a distributed source.

The direct mapping of "parent" molecules is now possible. At radio wavelengths, successful observations with spatial resolution of the order of the arcsecond could be made with the BIMA, OVRO and IRAM Plateau-de-Bure interferometers (Wink *et al.*, 1998; Despois 1998, and references therein); with larger fields of view and coarser spatial resolution, "on-the-fly" mapping and multibeam receivers were used (Lovell *et al.*, 1998; Womack *et al.*, 1998). In the infrared, long-slit spectral observations with CSHELL/IRTF could study the extension of molecules such as CO and OCS. Such observations are of a considerable help for finding a solution to these problems.

CO and H_2CO have manifestly distributed sources. Eberhardt (1998 and references therein) argues that the mass spectrometry observations of Giotto in Comet Halley are fully consistent with CO coming from H_2CO (an interesting critical discussion of this interpretation, however, is given by Crifo *et al.*, 1999). This does not seem to be the case for other comets, for which the CO production rate is much larger than that of H_2CO. In Comet Hale-Bopp, 1-D spectral imaging with CSHELL/IRTF (DiSanti *et al.*, 1999; Weaver *et al.*, 1998) has definitely shown the existence of a distributed source when $r < 2$ AU. On the other hand, radio interferometric observations of Comet Hyakutake around closest approach have directly shown a nuclear source of CO (Bockelée-Morvan *et al.*, in preparation). As a matter of fact, both a nuclear and a distributed source seem to coexist, the latter being turned on only at small heliocentric distances (DiSanti *et al.*, 1999). However, the nature of the distributed sources and of possible desorption mechanisms from dust for CO and H_2CO are unknown: there is no quantitative explanation available (Greenberg and Li, 1998).

The case for HCN and CN is now well documented. There is no doubt that HCN is a major parent for the CN radical. Indeed, observations of Comet Hale-Bopp at $r_h > 2.9$ AU indicate that all CN could come from HCN (Rauer *et al.*, 1997). However, these two species have incompatible spatial distribution and production rates at smaller r_h's (Biver *et al.*, in preparation; Festou *et al.*, 1998). An additional source for CN seems to be required (HC_3N and CH_3CN have been identified, but their contribution is nominal). A'Hearn *et al.* (1995) have proposed a contribution from dust, but once again, the desorption mechanism problem is not solved.

The role of chemical reactions to synthesize molecules in the coma is now exemplified by HNC. This species was first detected in Comet Hyakutake (Irvine *et al.*, 1996). In Comet Hale-Bopp the [HNC]/[HCN] ratio was found to increase as the comet approached the Sun (Biver *et al.*, 1997, 1998; Irvine *et al.*, 1998). This strongly suggests that a major part of the cometary HNC is produced by chemical reactions (Rogers and Charnley, 1998; Irvine *et al.*, 1998). Could other minor species, listed in Table II, be formed in this way?

5. From Gas Molecular Production Rates to Ice Molecular Abundances

Because of the different volatilities of cometary ice components, sublimation fractionation occurs. Further complications are due to the evolution of the outer layers of the nucleus during sublimation–recondensation processes. These problems will not be discussed here (they are the subject of several other papers given at the workshop). As a result, the molecular production rates observed in the gas phase *do not* directly reflect the ice molecular abundances. Important clues to this phenomenon were obtained from the observation of the production rate evolutions in Comet Hale-Bopp over $0.9 < r_h < 7$ AU (Biver *et al.*, 1997, 1998; Bockelée-Morvan and Rickman, 1998).

6. Results and Discussion

Our present knowledge of volatile abundances is summarized in Table II. This table must be read and used with the cautions explained in the previous sections. The signal-to-noise ratio of the observations is generally not a problem. Most of the uncertainties in the abundance determinations may be attributed to absolute calibration, modelling uncertainties, and the difficulty to evaluate the water production rates to which abundances are usually normalized. Despite these shortcomings, we believe that for most molecules, the overall error is less (but not much less) than a factor of two. The real problem is for species such as CO or H_2CO which are known (or suspected) to have extended sources. Which fraction is coming from the nucleus or from an extended source, and for this latter, what is its scale length, are very ill-known parameters. In such cases, the errors may be large.

The production rate of water, which appears to dominate the composition of cometary ices, is commonly used to normalize cometary volatile abundances. This is a problem, since most of the time $Q[H_2O]$ is ill known: direct observations of water, which have to be performed in the infrared domain, are rare; indirect determinations (e.g., from the OH radical observed in the radio or UV) have to rely on sophisticated modelling.

Several isotopic species have now been directly observed from their radio lines, as can be seen in the summary given in Table III. The detection of HDO and other species are state-of-the-art observations for which the signal-to-noise ratio is limited. For these observations, single lines are observed and the determination of production rates rely on modelling; the problem of the $Q[H_2O]$ evaluation noted above is also critical. The same enriched [D]/[H] ratio ($\approx 3 \times 10^{-4}$) has been observed in comets 1P/Halley (from in situ mass spectroscopy), C/1996 B2 (Hyakutake) and C/1995 O1 (Hale-Bopp). As discussed by Bockelée-Morvan *et al.* (1998a) and elsewhere in the proceedings of this workshop, the [D]/[H] ratio is a touchstone for formation scenarios of comets. The three comets for which [D]/[H] was measured are presumed to come from the Oort cloud. This ratio will be difficult to measure by remote sensing in Jupiter-family comets, which are much weaker objects. For carbon, nitrogen, and sulphur which are heavier atomic species less sensitive to fractionation during chemical reactions, the isotopic abundances appear to be cosmic (solar).

TABLE III
Isotopic ratios in comets.

Isotopes Molecule	comet	method [a]	cosmic ratio	comet ratio	ref. [b]
[D]/[H]					
H_3O^+	1P/Halley	M	1.5×10^{-5}	$3.08 \pm 0.53 \times 10^{-4}$	[1]
				$3.02 \pm 0.22 \times 10^{-4}$	[2]
H_2O	C/1996 B2	R		$2.9 \pm 1.0 \times 10^{-4}$	[3]
	C/1995 O1	R		$3.3 \pm 0.8 \times 10^{-4}$	[4]
HCN	C/1995 O1	R		$2.3 \pm 0.4 \times 10^{-3}$	[5]
NH	C/1996 B2	V		$< 6 \times 10^{-3}$	[6]
CH	C/1996 B2	V		$< 3 \times 10^{-2}$	[6]
CH_3OH	1P/Halley	M		$< 1 \times 10^{-2}$	[7][c]
[^{18}O]/[^{16}O]					
H_3O^+	1P/Halley	M	2.0×10^{-3}	$1.93 \pm 0.12 \times 10^{-3}$	[1]
				$2.13 \pm 0.18 \times 10^{-3}$	[2]
[^{13}C]/[^{12}C]					
CN	1P/Halley	V	1.1×10^{-2}	$1.05 \pm 0.13 \times 10^{-2}$	[8]
HCN	C/1996 B2	R		$2.9 \pm 1.0 \times 10^{-2}$	[9][d]
	C/1995 O1	R		$1.11 \pm 0.18 \times 10^{-2}$	[10]
				$0.90 \pm 0.09 \times 10^{-2}$	[11]
[^{15}N]/[^{14}N]					
CN	1P/Halley	V	3.6×10^{-3}	$< 3.6 \times 10^{-3}$	[8]
HCN	C/1995 O1	R		$3.1 \pm 0.4 \times 10^{-3}$	[11]
[^{34}S]/[^{32}S]					
atomic S	1P/Halley	M	4.2×10^{-2}	$4.5 \pm 1.0 \times 10^{-2}$	[12]
CS	C/1995 O1	R		$3.7 \pm 0.4 \times 10^{-2}$	[11]

[a] M: mass spectroscopy; R: radio; V: visible;
[b] References: [1] Balsiger et al., 1995; [2] Eberhardt et al., 1995; [3] Bockelée-Morvan et al., 1998a; [4] Meier et al., 1998b; [5] Meier et al., 1998a; [6] Meier et al., 1998c; [7] Eberhardt et al., 1994; [8] Kleine et al., 1995; [9] Lis et al., 1997; [10] Lis et al., 1998; [11] Jewitt et al., 1997; [12] Krankowsky et al., 1986;
[c] The ratio listed here is [deuterated CH_3OH]/[CH_3OH] and not [D]/[H];
[d] This ratio is possibly affected by blending of the $H^{13}CN$ line by a SO_2 line.

The ortho-to-para ratio (OPR) of water is supposed to be conserved during radiative or collisional transitions and to keep memory of the state of water at its formation. It can be estimated from high-resolution infrared spectra such as those obtained some time ago from the KAO and recently by ISO. It has yet been measured in four comets: 1P/Halley, C/1986 P1 (Wilson), C/1995 O1 (Hale-Bopp) and 103P/Hartley 2 (Mumma et al., 1993; Crovisier et al., 1997, 1999). For three of these comets (the "Oort cloud comets" P/Halley and Hale-Bopp and the Jupiter-family comet P/Hartley 2), the OPR was found to be smaller than 3, corresponding to spin temperatures in the range 25–35 K. The meaning of such cold temperatures is still to be understood. Further results can be expected from other species (CH_4, H_2CO, CH_3OH...).

Is there a variation of composition from comet to comet? The sample of comets for which the abundance of even a limited number of parent molecules is known is still extremely small (Table II). However, there are strong suspicions of variations for species such as CO, CH_3OH or H_2S. An important challenge is to investigate a possible correlation between chemical differences and the dynamical classes of the comets, which could betray their different origins and/or thermal histories. From the observations of radicals in almost a hundred comets, A'Hearn et al. (1995) have identified two classes of comets: the first has *typical* abundances while the second is depleted in *carbon-chain* molecules. They suggested that Jupiter-family comets are linked with the latter class. How does this translate into possible differences in the production of parent molecules? Jupiter-family comets seem to be much less productive of CO with $[CO]/[H_2O] \leq 1.5\%$ at $r_h \approx 1$ AU (Weaver et al., 1994; Feldman et al., 1997), but the statistics is still very poor, these weaker comets being much harder to observe. On the other hand, Schulz et al. (1998) questioned the reality of the two taxonomic classes introduced by A'Hearn et al. by pointing out that a heliocentric distance effect could be the reason for the dichotomy, since *carbon-depleted* comets were rather those observed at large heliocentric distances ($r_h > 1.6$ AU), while *typical* comets were observed at $r_h \approx 1$ AU.

How complete is Table II? We believe it could be complete for major species with abundances relative to water of the order of — let us say — roughly 1%. On the other hand, it is obvious that many more minor species are still to be discovered. Stringent upper limits on many species have been obtained in comets Hyakutake and Hale-Bopp that have not yet been exploited nor reported. There are still elusive molecules such as O_2, N_2 or H_2 which, having no strong spectral signatures, cannot be investigated by remote sensing. N_2 abundances have been derived in the past from the claimed detection of the $B-X$ bands of N_2^+; these bands were not reported in recent high-resolution cometary spectra.

References

A'Hearn M.F., Feldman, P.D. and Schleicher, D.J.: 1983, 'The discovery of S_2 in comet IRAS-Aracki-Alcock 1983d', *Astrophys. J.* **274**, L99–L103.

A'Hearn, M.F., Millis, R.L., Schleicher, D.G., Osip, D.J. and Birch, P.V.: 1995, 'The ensemble properties of comets: results from narrowband photometry of 85 comets, 1976–1992', *Icarus* **118**, 223–270.

Altenhoff, W.J., Batrla, W., Huchtmeier, W.K., Schmidt, W.K., Stumpff, P. and Walmsley, C.M.: 1983, 'Radio observations of comet 1983d', *Astron. Astrophys.* **125**, L19–L22.

Altwegg, K., Balsiger, H. and Geiss, J.: 1999, 'Composition of the volatile material in Halley's coma from in situ measurements', *Space Sci. Rev.*, this volume.

Arpigny, C.: 1995, 'Spectra of comets: ultraviolet and optical regions', in *Laboratory and Astronomical High Resolution Spectra*, ed. A.J. Sauval, R. Blomme and N. Grevesse, ASP Conf. Series Vol. **81**, 362–383.

Balsiger, H., Altwegg, K., Geiss, J.: 1995, 'D/H and $^{18}O/^{16}O$ ratio in the hydronium ion and in neutral water from in situ ion measurements in comet Halley', *J. Geophys. Res.* **100**, 5827–5834.

Bird, M.K., Huchtmeier, W.K., Gensheimer, P., Wilson, T.L., Janardhan, P. and Lemme, C.: 1997, 'Radio detection of ammonia in comet Hale-Bopp', *Astron. Astrophys.* **325**, L5–L8.

Bird, M.K., Janardhan, P., Wilson, T.L., Huchtmeier, W.K., Gensheimer, P. and Lemme, C.: 1998, 'K-band radio observations of comet Hale-Bopp: detections of ammonia and (possibly) water', *Earth Moon Planets* (in press).

Biver, N., *et al.*: 1997, 'Evolution of the outgassing of comet Hale-Bopp (C/1995 O1) from radio observations', *Science* **275**, 1915–1918.

Biver, N., *et al.*: 1998, 'Long term evolution of the outgassing of comet Hale-Bopp from radio observations', *Earth Moon Planets* (in press).

Bockelée-Morvan, D., Colom, P., Crovisier, J., Despois, D., Paubert, G.: 1991, 'Microwave detection of hydrogen sulfide and methanol in comet Austin', (1989c$_1$) *Nature* **350**, 318–320.

Bockelée-Morvan, D., Brooke, T.Y. and Crovisier, J.: 1995, 'On the origin of the 3.2- to 3.6-μm emission features in comets', *Icarus* **116**, 18–39.

Bockelée-Morvan, D.: 1997, 'Cometary volatiles : The status after comet C/1996 B2 Hyakutake', in *Molecules in Astrophysics: Probes and Processes, IAU Symp. 178*, ed. E.F. van Dishoeck, Kluwer, 219–235.

Bockelée-Morvan, D. and Crovisier, J.: 1998, 'New results on the composition of comets', in *Planetary Systems: the Long View*, IXth Rencontres de Blois, Éditions Frontières, *in press*.

Bockelée-Morvan, D. and Rickman, H.: 1998, 'Gas production curves and their interpretation', *Earth Moon Planets* (in press).

Bockelée-Morvan, D., *et al.*: 1998a, 'Deuterated water in comet C/1996 B2 (Hyakutake) and its implication for the origins of comets', *Icarus* **133**, 147–162.

Bockelée-Morvan D., *et al.*: 1998b, 'A molecular survey of comet C/1995 O1(Hale-Bopp) at the IRAM telescopes', *Earth Moon Planets* (in press).

Brooke, T.Y., Tokunaga, A.T., Weaver, H.A., Crovisier, J., Bockelée-Morvan, D. and Crips, D.: 1996, 'Detection of acetylene in the infrared spectrum of Hyakutake', *Nature* **383**, 606–608.

Brown, M.E., Bouchez, A.H., Spinrad, A.H. and Johns-Krull, C.M.: 1996, 'A high-resolution catalog of cometary emission lines', *Astron. J.* **112**, 1197–1202.

Colom, P., Crovisier, J., Bockelée-Morvan, D., Despois, D. and Paubert, G.: 1992, 'Formaldehyde in comets: I. Microwave observations of P/Brorsen-Metcalf (1989 X), Austin (1990 V) and Levy (1990 XX)', *Astron. Astrophys.* **264**, 270–281.

Combes, M., *et al.*: 1986, 'Infrared sounding of comet Halley from Vega 1', *Nature* **321**, 266–268.

Crifo, J.-F., Rodionov, A.V., Bockelée-Morvan, D.: 1999, 'The dependence of the circumnuclear coma structure on the properties of the nucleus, III. First modelling of a CO-dominated coma, with applications to comets 46P/Wirtanen and P/Schwachmann-Wachmann 1', *Icarus* **138**, 85–106.

Crovisier, J., Leech, K., Bockelée-Morvan, D., Brooke, T.Y., Hanner, M.S., Altieri, B., Keller, H.U. and Lellouch, E.: 1997, 'The spectrum of comet Hale-Bopp (C/1995 O1) observed with the Infrared Space Observatory at 2.9 AU from the Sun', *Science* **275**, 1904–1907.

Crovisier, J.: 1998a, 'Infrared observations of volatile molecules in comet Hale-Bopp', *Earth Moon Planets* (in press).

Crovisier, J.: 1998b, 'Physics and chemistry of comets: recent results from comets Hyakutake and Hale-Bopp; answers to old questions and new enigmas', *Faraday Discuss.* **109**, 437–452.

Crovisier, J., et al.: 1999, 'ISO spectroscopic observations of short-period comets', in *The Universe as seen by ISO*, ESA **SP-427** (in press).

Dello Russo, N., DiSanti, M.A., Mumma, M.J., Magee-Sauer, K. and Rettig, T.W.: 1998, 'Carbonyl sulfide in comets C/1996 B2 (Hyakutake) and C/1995 O1 (Hale-Bopp): evidence for an extended source in Hale-Bopp', *Icarus* **135**, 377–388.

Dello Russo, N., Mumma, M.J., DiSanti, M.A., Magee-Sauer, K., Novak, R. and Rettig, T.W.: 1999, 'Water production and release in comet C/1995 O1 Hale-Bopp', *Icarus*, in press.

Despois, D., Crovisier, J., Bockelée-Morvan, D., Schraml, J., Forveille, T. and Gérard, E.: 1986, 'Observations of hydrogen cyanide in comet Halley', *Astron. Astrophys.* **160**, L11–L12.

Despois, D.: 1998, 'Radio line observations of molecular and isotopic species, implications on the interstellar origin of cometary ices', *Earth Moon Planets* (in press).

DiSanti, M.A., Mumma, M.J., Dello Russo, N., Magee-Sauer, K., Novak, R. and Rettig, T.W.: 1999, 'Identification of two sources of carbon monoxide in comet Hale-Bopp, , *Nature* **399**, 662-665.

Donati, G.B.: 1864, 'Schreiben des Herrn Professors Donati', *Astron. Nachrichten* **62**, 375–378.

Dutrey, A., et al.: 1996, 'Comet C/1996 B2 (Hyakutake)', *IAUC* No 6364.

Eberhardt, P., Meier, R., Krankowsky, D. and Hodges, R.R.: 1994, 'Methanol and hydrogen sulfide in P/Halley', *Astron. Astrophys.* **288**, 315–329.

Eberhardt, P., Reber, M., Krankowsky, D. and Hodges, R.R.: 1995, 'The D/H and $^{18}O/^{16}O$ ratios in water from comet P/Halley', *Astron. Astrophys.* **302**, 301–316.

Eberhardt, P.: 1998, 'Volatiles, isotopes and origin of comets', in *Asteroids, Comets, Meteors 1996*, COSPAR Colloquium 10 (in press).

Eberhardt, P.: 1999,'Composition of comets: the in situ view', in *Cometary Nuclei in Space and Time, IAU Coll. 168*, ed. M.F. A'Hearn, ASP Conf. Series (in press).

Feldman, P.D. and Brune, W.H.: 1976, 'Carbon production in comet West (1975n)', *Astrophys. J.* **209**, L45–L48.

Feldman, P.D., Festou, M.C., Tozzi, G.P. and Weaver, H.A.: 1997, 'The CO_2/CO abundance ratio in 1P/Halley and several other comets observed by *IUE* and *HST*', *Astrophys. J.* **475**, 829–834.

Festou, M.C., Barale, O., Davidge, T., Stern, S.A., Tozzi, G.P., Womack, M. and Zucconi, J.M.: 1998, 'Tentative identification of the parent of CN radicals in comets: C_2N_2', *BAAS* **30**, 1089.

Greenberg, J.M. and Li, A: 1998, 'From interstellar dust to comets: the extended CO source in comet Halley', *Astron. Astrophys.* **332**, 374–384.

Huebner, W.F., Snyder, L.E. and Buhl, D.: 1974, 'HCN radio emission from comet Kohoutek (1973f)', *Icarus* **23**, 580–585.

Huet, T.R., Hadj Bachir, I., Bolvin, H., Zellagui, A., Destombes, J.L. and Vervloet, M.: 1996, 'NH_2 transitions of cometary interest in the near infrared', *Astron. Astrophys.* **311**, 343–346.

Irvine, W.M., et al.: 1996, 'Spectroscopic evidence for interstellar ices in Comet Hyakutake', *Nature* **383**, 418–420.

Irvine, W.M., Bergin, E.A., Dickens, J.E., Jewitt, D., Lovell, A.J., Matthews, H.E., Schloerb, F.P. and Senay, M.: 1998, 'Chemical processing in the coma as the source of cometary HNC', *Nature* **393**, 547-550.

Irvine, W.M., Schloerb, F.P., Crovisier, J., Fegley, B. and Mumma, M.J.: 2000, 'Comets: a link between interstellar and nebular chemistry', in *Protostars and Planets* **IV**, V. Manning, A. Boss and S. Russell (edts), University of Arizona Press, Tucson (in press).
Jewitt, D.C., Matthews, H.E., Owen, T.C. and Meier, R.: 1997, 'Measurements of $^{12}C/^{13}C$, $^{14}N/^{15}N$, and $^{32}S/^{34}S$ ratios in comet Hale-Bopp', (C/1995 O1) *Science* **278**, 90–93.
Kim, S.J., Bockelée-Morvan, D., Crovisier, J. and Biver, N.: 1998, 'Fluorescence and collisional processes of SO and SO_2 in comet Hale-Bopp', (C/1995 O1) *Earth Moon Planets* (in press).
Kleine, M., Wyckoff, S., Wenhinger, P.A. and Peterson, B.A.: 1995, 'The carbon isotope abundance ratio in comet Halley', *Astrophys. J.* **439**, 1021–1033.
Krankowsky, D., *et al.*: 1986, 'In situ gas and ion composition measurements at Comet Halley', *Nature* **321**, 326–329.
Lis, D.C., Keene, J., Young, K., Phillips, T.G., Bockelée-Morvan, D., Crovisier, J., Schlike, P., Goldsmith, P.F. and Bergin, E.A.: 1997, 'CSO observations of comet C/1996 B2 (Hyakutake)', *Icarus* **130**, 355–372.
Lis, D.C., Mehringer, D., Benford, D., *et al.*: 1998, 'New molecular species in comet c/1995 O1 (Hale-Bopp) observed with the Caltech Submillimeter Observatory', *Earth Moon Planets* (in press).
Lovell, A.J., Schloerb, F.P., Dickens, J.E., De Vries, C.H., Senay, M.C. and Irvine, W.M.: 1998, 'HCO^+ imaging of comet C/Hale-Bopp 1995 O1', *Astrophys. J.* **497**, L117–L120.
Magee-Sauer, K., Mumma, M.J., DiSanti, M.A., Dello Russo, N. and Rettig, T.W.: 1998, 'CSHELL observations of HCN, C_2H_2, and NH_3 in comets C/1995 O1 (Hale-Bopp) and C/1996 B2 (Hyakutake)', *BAAS* **30**, 1064.
Meier, R., Owen, T.C., Jewitt, D.C., Matthews, H.E., Senay, M., Biver, N., Bockelée-Morvan, D., Crovisier, J. and Gautier, D.: 1998a, 'Deuterium in comet C/1995 O1 (Hale-Bopp): Detection of DCN', *Science* **279**, 1707–1710.
Meier, R., Owen, T.C., Matthews, H.E., Jewitt, D.C., Bockelée-Morvan, D., Biver, N., Crovisier, J. and Gautier, D.: 1998b, 'A determination of HDO/H_2O in comet C/1995 O1 (Hale-Bopp), *Science* **279**, 842–844.
Meier, R., Wellnitz, D., Kim, S.J. and A'Hearn, M.F.: 1998c 'The NH and CH bands of comet C/1996 B2 (Hyakutake)', *Icarus* **136**, 268–279.
Moreels, J., Clairemidi, J., Hermine, P., Brechignac, P. and Rousselot, P.: 1993, 'Detection of a polycyclic aromatic molecule in P/Halley', *Astron. Astrophys.* **282**, 643–656.
Morrison, N.D., Knauth, D.C., Mullis, C.M. and Lee, W.: 1997, 'High-resolution optical spectra of the head of comet C/1996 B2 (Hyakutake)', *Publ. Astron. Soc. Pacific*, **109**, 676–681.
Mumma, M.J., Weaver, H.A., Larson, H.P., Davis, D.S. and Williams, M.: 1986, 'Detection of water vapor in Halley's comet', *Science* **232**, 1523–1528.
Mumma, M.J., Weissman, P.R. and Stern, S.A.: 1993, 'Comets and the origin of the solar system: Reading the Rosetta stone' in *Protostars and Planets* **III**, E.H. Levy and J.I. Lunine (edts), University of Arizona Press, Tucson, 1177–1252.
Mumma, M.J., DiSanti, M.A., Dello Russo, N., Fomenkova, M., Magee-Sauer, K., Kaminski, C.D. and Xie, D.X.: 1996, 'Detection of abundant ethane and methane, along with carbon monoxide and water, in comet C/1996 B2 (Hyakutake): evidence for interstellar origin', *Science* **272**, 1310–1314.
Mumma, M.J.: 1997, 'Organic volatiles in comets: their relation to interstellar ices and solar nebula material', in *From Stardust to Planetesimals*, Y.J. Pendleton and A.G.G.M. Tielens eds, ASP Conf. Series **122**, 369–398.
Palmer, P., Wootten, A., Butler, B., Bockelée-Morvan D., Crovisier, J., Despois, D., Yeomans, D.K.: 1996, 'Comet Hyakutake: First secure detection of ammonia in a comet' *BAAS* **28**, 927–928.
Rauer, H., Arpigny, C., Boehnhardt, H., Colas, F., Crovisier, J., Jorda, L., Küppers, M., Manfroid, J., Rembor, K. and Thomas, N.: 1997, 'Optical observations of comet Hale-Bopp (C/1995 O1) at large heliocentric distances before perihelion', *Science* **275**, 1909–1912.

Rauer, H.: 1999, 'Remote observations of cometary volatiles and implications on comet nuclei', in *Cometary Nuclei in Space and Time, IAU Coll. 168*, ed. M.F. A'Hearn, ASP Conf. Series (in press).
Rodgers, S.D. & Charnley, S.B.: 1998, 'HNC and HCN in comets', *Astrophys. J.* **501**, L227–L230.
Schloerb, F.P., Kinzel, W.M., Swade, D.A. and Irvine, W.M.: 1986, 'HCN production rates from comet Halley', *Astrophys. J.* **310**, L55–L60.
Schulz, R., Arpigny, C., Manfroid, J., Stüwe, J.A., Tozzi, G.P., Rembor, K., Cremonese, G. and Peschke, S.: 1998, 'Spectral evolution of ROSETTA target comet 46P/Wirtanen', *Astron. Astrophys.* **335**, L46–L49.
Smith, A.M., Stecher, T.P. and Casswell, L.: 1980, 'Production of carbon, sulphur, and CS in comet West', *Astrophys. J.* **242**, 402–410.
Snyder, L.E., Palmer, P. and de Pater, I.: 1989, 'Radio detection of formaldehyde emission from comet Halley', *Astron. J.* **97**, 246–253.
Weaver, H.A., Feldman, P.D., McPhate, J.B., A'Hearn, M.F., Arpigny, C. and Smith, T.E.: 1994, 'Detection of CO Cameron band emission in comet P/Hartley 2 (1991 XV with the *Hubble Space Telescope*', *Astrophys. J.* **422**, 374–380.
Weaver, H.A., Feldman, P.D., McPhate, J.B., A'Hearn, M.F., Arpigny, C., Brandt, J.C. and Randall, C.E.: 1996, 'Ultraviolet spectroscopy and optical imaging of comet Hyakutake (1996 B2) with HST', in *Asteroids, Comets, Meteors 1996*, COSPAR Colloquium 10 (book of abstracts).
Weaver, H.A., Brooke, T.Y., Chin, G., Kim, S.J., Bockelée-Morvan, D. and Davies, J.K.: 1998, 'Infrared spectroscopy of comet Hale-Bopp', *Earth Moon Planets* (in press).
Wink, J., et al.: 1998, 'Evidences for extended sources and temporal modulations in molecular observations of C/1995 O1(Hale-Bopp) at IRAM interferometer', *Earth Moon Planets* (in press).
Winnberg, A., Ekelund, L. and Ekelund, A.: 1987, 'Detection of HCN in comet P/Halley', *Astron. Astrophys.* **172**, 335–341.
Womack, M. Homoch, A., Festou, M.C., Mangum, J., Uhl, W.T. and Stern, S.A.: 1998, 'Maps of HCO^+ emission in C/1995 O1 (Hale-Bopp',) *Earth Moon Planets* (in press).
Woodney, L.M., McMullin, J. and A'Hearn,M.F.: 1997a, 'Detection of OCS in comet Hyakutake (C/1996 B2)', *Planet. Space Sci.* **45**, 717–719.
Woodney, L.M., McMullin, J., A'Hearn,M.F. and Samarasinha, N.: 1997b, *IAU Circ.* 6607
Woodney, L.M., A'Hearn, M.F., Lisse, C., McMullin, J. and Samarasinha, N.: 1998, 'Sulfur chemistry at millimeter wavelengths in C/Hale-Bopp', *Earth Moon Planets* (in press).
Wyckoff, S., Heyd, R.S. and Fox, R.: 1999, 'Unidentified molecular bands in the plasma tail of comet Hyakutake (C/1996 B2)', *Astrophys. J.* **512**, L73–L76.

Address for correspondence: Jacques Crovisier, Observatoire de Paris-Meudon, F-92195 Meudon, France (`crovisie@obspm.fr`)

COMETARY DEUTERIUM

ROLAND MEIER and TOBIAS C. OWEN
Institute for Astronomy, University of Hawaii
2680 Woodlawn Dr.
Honolulu, HI 96822, USA

Abstract. Deuterium fractionations in cometary ices provide important clues to the origin and evolution of comets. Mass spectrometers aboard spaceprobe Giotto revealed the first accurate D/H ratios in the water of Comet 1P/Halley. Ground-based observations of HDO in Comets C/1996 B2 (Hyakutake) and C/1995 O1 (Hale-Bopp), the detection of DCN in Comet Hale-Bopp, and upper limits for several other D-bearing molecules complement our limited sample of D/H measurements. On the basis of this data set all Oort cloud comets seem to exhibit a similar $(D/H)_{H_2O}$ ratio in H_2O, enriched by about a factor of two relative to terrestrial water and approximately one order of magnitude relative to the protosolar value. Oort cloud comets, and by inference also classical short-period comets derived from the Kuiper Belt cannot be the only source for the Earth's oceans. The cometary O/C ratio and dynamical reasons make it difficult to defend an early influx of icy planetesimals from the Jupiter zone to the early Earth. D/H measurements of OH groups in phyllosilicate rich meteorites suggest a mixture of cometary water and water adsorbed from the nebula by the rocky grains that formed the bulk of the Earth may be responsible for the terrestrial D/H. The D/H ratio in cometary HCN is 7 times higher than the value in cometary H_2O. Species-dependent D-fractionations occur at low temperatures and low gas densities *via* ion-molecule or grain-surface reactions and cannot be explained by a pure solar nebula chemistry. It is plausible that cometary volatiles preserved the interstellar D fractionation. The observed D abundances set a lower limit to the formation temperature of (30 ± 10) K. Similar numbers can be derived from the ortho-to-para ratio in cometary water, from the absence of neon in cometary ices and the presence of S_2. Noble gases on Earth and Mars, and the relative abundance of cometary hydrocarbons place the comet formation temperature near 50 K. So far all cometary D/H measurements refer to bulk compositions, and it is conceivable that significant departures from the mean value could occur at the grain-size level. Strong isotope effects as a result of coma chemistry can be excluded for molecules H_2O and HCN. A comparison of the cometary $(D/H)_{H_2O}$ ratio with values found in the atmospheres of the outer planets is consistent with the long-held idea that the gas planets formed around icy cores with a high cometary D/H ratio and subsequently accumulated significant amounts of H_2 from the solar nebula with a low protosolar D/H.

Keywords: Deuterium, Origin, Gas, Composition, Planets, Water, HCN, Interstellar Medium, Comets

1. Introduction

Primordial deuterium (D) was synthesized in the Big Bang during the first few hundred seconds of the radiation era (Wagoner *et al.*, 1967). Since these early days, nucleosynthesis in the interiors of stars has converted a large fraction of the primordial D to ^3He via the reaction $D(p,\gamma)^3He$. Among the stable isotopes,

D has the lowest binding energy per nucleon and can only be formed in extreme environments. No mechanism is known that would produce significant amounts of D during the chemical evolution of the galaxy (Epstein et al., 1976).

Of all the elements, the stable isotopes of hydrogen are the most diverse from each other in terms of their masses and chemical properties. At low temperatures sizeable differences in the zero-point energies of a deuterated molecule and its H-bearing counterpart lead to the well-known strong fractionation effects in kinetically controlled chemical reactions. D/H determinations provide a sensitive tool to explore formation temperatures. They can help constrain the origin of cometary material and trace cometary ices to the various reservoirs of our solar system.

2. Observations

A'Hearn et al. (1985) were probably the first to attempt a quantitative measurement of the D/H ratio in a comet. They targeted OH, a photodissociation product of the most abundant cometary volatile, H_2O. They searched for strong deuterium lines in the $A-X$ 0–0 UV-band near 3064 Å using high resolution spectra from the International Ultraviolet Explorer (IUE). A'Hearn et al.'s study did not reveal any positive detections but resulted in upper limits of $(D/H)_{OH} < 0.008$, 0.007, 0.006, 0.004, and 0.006 for Comets C/1978 T1 (Seargent), C/1979 Y1 (Bradfield), 2P/Encke, C/1982 M1 (Austin), and C/1983 H1 (IRAS-Araki-Alcock), respectively. Employing the same instrument and technique, Schleicher and A'Hearn (1986) derived a more stringent upper limit of $< 4 \times 10^{-4}$ for Comet 1P/Halley and 8×10^{-3} for 21P/Giacobini-Zinner. Recently Meier et al. (1998c) reported upper limits for two new hydrides, NH and CH. They analyzed echelle spectra of Comet C/1996 B2 (Hyakutake) (Table I).

Two independent mass spectrometers aboard the spaceprobe Giotto recorded mass-resolved ion spectra of H_3O^+ with a dynamic range large enough to detect the rare isotopes at masses 20^+ and 21^+. From these data two independent groups inferred $(D/H)_{H_2O}$ ratios for Comet 1P/Halley. The results are summarized in Table I together with upper limits for the D/H in H_2CO and CH_3OH (CDH_2OH and CH_3OD averaged) set by these same experiments. The limits were derived from protonated ions of the corresponding molecules assuming that the $^{13}C/^{12}C$ ratio is solar.

A new generation of powerful sub-mm telescopes and the apparition of two exceptionally bright comets finally led to the first ground-based detections of cometary D. Initial attempts at 225.897 GHz (HDO $3_{12}-2_{21}$) were performed by Crovisier et al. (1993) and yielded only upper limits of 0.011 and 0.007 for Comets C/1989 X1 (Austin) and C/1990 K1 (Levy). Bockelée-Morvan et al. (1998) announced the first positive detection of the HDO $1_{01}-0_{00}$ transition at 464.925 GHz in Comet C/1996 B2 (Hyakutake) and an upper limit of 0.01 for $(D/H)_{HCN}$ (Table I). One year later Meier et al. (1998a, b) reported on the double-detection of HDO

TABLE I

D/H ratios in cometary molecules including upper limits for undetected species. Listed are D/H values and not ratios of deuterated/non-deuterated species. These two ratios are different for molecules with multiple H atoms (*e.g.*, D/H in water is $0.5[HDO]/[H_2O]$). The averaged value for Comet 1P/Halley is the average of Balsiger *et al.*'s (1995) result, corrected by +11% to account for the D/H fractionation between H_3O^+ and H_2O, and Eberhardt *et al.*'s (1995) measurement. NH_3 is most likely the parent molecule of NH, but the D/H ratio may be different in these two molecules depending on the branching ratios of NH_2D to NH and ND. Theoretically expected D/H values for ion-molecule reactions at an equilibrium temperature of $T = 30$ K are also listed using Millar *et al.*'s (1989) Table 2 with the "old" branching ratios. Millar *et al.* adopted two different sets of branching ratios for dissociative electron recombination ("old" and "new" ratios). Recent measurements of the branching ratio for dissociative electron recombination of H_3O^+ by (Williams *et al.*, 1996) suggest that the "old" branching ratios are more realistic, but more laboratory work is clearly needed.

Molecule	Comet	D/H ratio ($\pm 1\sigma$)	Theory	Reference
H_2O	1P/Halley	$(3.08 \pm {}^{+0.19}_{-0.26}) \times 10^{-4}$	0.0005	Balsiger *et al.* (1995)
	1P/Halley	$(3.02 \pm 0.07) \times 10^{-4}$		Eberhardt *et al.* (1995)
	1P/Halley (average)	$(3.16 \pm 0.11) \times 10^{-4}$		Eberhardt *et al.* (1995)
	C/1996 B2 (Hyakutake)	$(2.9 \pm 1.0) \times 10^{-4}$		Bockelée-Morvan *et al.* (1998)
	C/1995 O1 (Hale-Bopp)	$(3.3 \pm 0.8) \times 10^{-4}$		Meier *et al.* (1998a)
HCN	C/1995 O1 (Hale-Bopp)	$(2.3 \pm 0.4) \times 10^{-3}$	0.0017	Meier *et al.* (1998b)
H_2CO	1P/Halley	$< 2 \times 10^{-2}$	0.0050	Balsiger *et al.* (1995)
CH_3OH	1P/Halley	$< \sim 1 \times 10^{-2}$	0.0013	Eberhardt *et al.* (1994)
NH (NH_3)	C/1996 B2 (Hyakutake)	$< 6 \times 10^{-3}$	0.0002	Meier *et al.* (1998c)
CH (?)	C/1996 B2 (Hyakutake)	$< 3 \times 10^{-2}$	0.001-0.01	Meier *et al.* (1998c)

and DCN in Comet C/1995 O1 (Hale-Bopp). As opposed to HDO, where water measurements with other instruments had to be cited for the D/H ratio, the DCN 5–4 transition at 362.046 GHz could be directly compared with spectra of HCN and the optically thin $H^{13}CN$ line using the same telescope and same receiver (Table I). Bockelée-Morvan *et al.* (private commun.) discovered another serendipitous $3_{12} - 2_{21}$ HDO line in one of their spectra of Comet Hale-Bopp, and Blake *et al.* (private commun.) recently claimed interferometric detections of both HDO and DCN in the same comet.

3. The D/H Ratio in Water

In the interstellar medium (ISM) and in the accretion disk that evolved from the collapsing natal cloud, molecular hydrogen dominated the bulk composition in the volatile phase. All D enrichments occurred at the expense of this large and inexhaustible hydrogen reservoir which served as a buffer for $(D/H)_{H_2}$. In the ISM, gas densities are low ($\sim 10^4$ cm^{-3}), time-scales long (10^6-10^8 y), and temperatures range from ~ 10 K in dark interstellar clouds up to $\sim 100-150$ K near star forming regions ("hot cores"). Ion-molecule reactions and gas-grain interactions dominate the chemistry and lead to high D fractionations in all minor species other than H_2. In most parts of the solar nebula, however, the high gas density and the concurrent absence of an efficient ionization source prevented the build-up of a significant ion population.

It has been speculated that X-ray emissions from the early sun during a T-Tauri phase may have led to much higher ionization rates than previously assumed. If this holds true, ion-molecule reactions in the early solar nebula could have affected the deuterium chemistry at larger heliocentric distances despite the short lifetime of the accretion disk. Aikawa *et al.* (1997) believe that a minimum-mass solar nebula surrounded by weak-lined T Tauri stars can be irradiated by cosmic rays down to $R \geq 10$ AU. Ions in this area then start to deplete CO and N_2 and build up CO_2, NH_3 and HCN. Another consequence of Aikawa *et al.*'s model are strong variations in the molecular abundances as a function of the distance to the central star. Recently this view was challenged by Willacy *et al.* (1998). These authors developed a time-dependent, 1-dimensional model that included gas phase reactions, mantle chemistry and gas-surface interactions. They found ion densities were negligibly small in the protoplanetary disk. Over much of the disk gas is absorbed by the dust and keeps its interstellar composition with little heliocentric variability, even if ionization by cosmic ray hits is important in the outer regions of the solar nebula, and even if decay of ^{26}Al acts as a strong ionization source in the inner regions of the protoplanetary disk, where the optical depth is too high to allow X-rays to penetrate the nebula deeply.

Neutral-neutral reactions can only exchange D with nebular H_2 if temperatures rise above ~ 300 K (Grinspoon and Lewis, 1988). At lower temperatures the equilibrium is never reached within the lifetime of the solar system, much less within the 10^5-10^6 years of the accretion phase. Even in a slowly cooling solar nebula, D enrichments remain modest and do not exceed a factor of ~ 3 (Lécluse and Robert, 1994). In addition, at high temperatures, nebular chemistry tends to set the D/H ratio of all species back close to the low protosolar value in H_2. Summarizing, it is exceedingly difficult to produce highly enriched species in the solar nebula (Geiss and Reeves, 1981).

To date we have HDO measurements from three different comets, for all of which dynamical considerations favor an origin in the Oort cloud. Based on our limited sample, Oort cloud comets appear to have similar $(D/H)_{H_2O}$ ratios, roughly a factor of 2 larger than in terrestrial water. Obviously radial mixing in the early solar nebula was not very thorough. Not all of the nebular gas was cycled through the chemically active inner zone, where D exchange with nebular H_2 decreased the $(D/H)_{H_2O}$ and erased all species-dependent D fractionations. We do not yet have an HDO determination in a short-period, Jupiter family comet. Those formed in the Kuiper Belt outside the orbits of Uranus and Neptune, i.e. outside the source region for Oort cloud comets (Duncan et al., 1987), where a substantial admixture of nebular H_2O with low D/H seems even less likely. Although we must await an HDO measurement in a classical short-period comet to be certain, the high D abundance in the three comets where it has been measured implies that today's comets cannot be the only source for the Earth's oceans.

Jupiter formed at the so-called "snow-line", where most of the water vapor from the inner solar nebula condensed to form ice. Delsemme (1998) pointed out that as Jupiter must have scattered most of the planetesimals in its vicinity out of the solar system, some struck the inner planets, and only a small fraction of these 100 m sized bodies reached the Oort cloud. He speculates these planetesimals may have had a lower D/H ratio than Oort cloud comets and delivered much of the Earth's water. Unfortunately this hypothesis is hard to test. Owen and Bar-Nun (1995) also postulated that icy planetesimals from the Jupiter zone (called Type I) could have contributed water to Earth. However, they pointed out that such planetesimals would also bring in carbon and nitrogen, producing a value of O/C that is half the value found on the Earth's surface today. Hence these authors suggested that either some subsequent loss of carbon occurred, or an additional source of water is required. Mixing cometary water with water adsorbed from the warm, inner solar nebula by the rocky grains that formed the bulk of the Earth is a plausible scenario to explain the origin and the $(D/H)_{SMOW}$ ratio in terrestrial water, as water vapor in the inner solar nebula could have a $D/H < 1 \times 10^{-4}$ (Lécluse and Robert, 1994; Owen, 1997; Meier et al., 1998a). The average D/H ratio in carbonaceous chondrites is consistent with the SMOW value. These meteorites may represent some of the building blocks of the Earth and would then have contributed to the subsequent outgassing of water (Deloule et al., 1997). Deloule et al. (1997) interpreted large departures from the mean D abundance at the grain-size level, but not at the chondrule or whole rock scale as evidence for a coexisting nebular and interstellar component in these meteorites.

4. Implications on the Origin and Evolution of Cometary Material

The D/H ratio in cometary volatiles varies dramatically from molecule to molecule. In H_2O and HCN we find enrichments of at least an order of magnitude above the protosolar value of $(2.6 \pm 0.7) \times 10^{-5}$, derived from the DH/H_2 ratio in the Jovian atmosphere (Mahaffy et al., 1998), or $(3.02 \pm 0.17) \times 10^{-5}$, if the $^3He/^4He$ ratio in the solar wind is used (Gautier and Morel, 1997; Geiss, 1993). The value of D/H in cometary HCN is 7 times the value in H_2O. Such enhanced and species-dependent deuterium abundances cannot easily be produced in the environment of the solar nebula. However, they are quite common in interstellar clouds, where ion-molecule reactions at low temperatures dictate the chemistry. This allows us to conclude that cometary material may still retain a memory of the ISM. A sublimation-recondensation process during the formation of the solar nebula (Lunine et al., 1991) does not necessarily change the D/H ratio. Therefore even though the icy grains incorporated in a comet nucleus may no longer be in their pristine state, their interstellar D signature could still exist. If depletions in the most volatile species are neglected, the resemblance of the chemical (volatile) inventory in comets to the ISM has always been striking. The discovery of abundant saturated hydrocarbons further supports the scenario of preserved ISM chemistry, or at least a kinetic formation mechanism (Mumma et al., 1996).

On the basis of a simple ion-molecule model in the gas phase (Millar et al., 1989), Meier et al. (1998b) inferred an effective formation temperature from the D/H in HCN and H_2O. Interestingly the temperature does not appear to depend on the choice of the molecule. Gas-grain surface reactions can lead to similar, species-dependent D enrichments, but typically they occur at higher temperatures (Brown et al., 1989). Unfortunately they are also more difficult to predict quantitatively. But whatever role gas-grain interactions may have played, pure ion-molecule calculations will at least set a lower bound to T. Hence, we conclude the effective temperature in the natal cloud could not have been below (30 ± 10) K. This limit makes it hard to defend a scenario in which comets were assembled directly in dark clouds of the ISM at temperatures near 10–15 K. Much more promising are hot cores, where temperatures are higher and D fractionations less extreme. The derived temperature is somewhat lower but still consistent with the average temperature in the early solar system at the Uranus-Neptune distance.

There is mounting evidence that $T \approx 30$ K is a characteristic temperature for Comet Hale-Bopp, and possibly comets in general. In the water of Comet Hale-Bopp Crovisier et al. (1997) found an ortho-to-para ratio (OPR) of (2.45 ± 0.10). This ratio is lower than the canonical equilibrium value of 3 and corresponds to a temperature of ~ 25 K. A slightly higher temperature of 29 K was derived by Mumma et al. (1993) from a revised OPR$= 2.5 \pm 0.1$ for Comet 1P/Halley. Neither collisions nor radiative processes can easily transform H_2O from the para to the ortho state or vice versa. Mumma et al. (1993) attributed the OPR of 3.2 ± 0.2 found for Comet C/1986 P1 (Wilson) to radiation damage to the outer layers of this dynamically new comet during its lifetime in the Oort cloud. The exact meaning of OPRs is still the subject of debate, but it is conceivable that OPRs preserve the initial grain formation temperature in the pre-solar interstellar cloud and have not been re-equilibrated since that time. Almost 15 years ago S_2 was first discovered in Comet IRAS-Araki-Alcock. It was then re-observed in Hyakutake and is now believed to be a regular constituent in the coma of comets. If S_2 is indeed a true parent molecule and is not produced from a yet to be identified more complex species, temperatures no higher than $\sim$30 K are allowed for S_2 to remain in the matrix, or to prevent S_2 from reacting with other species (A'Hearn and Feldman, 1985).

On the other hand, Owen and Bar-Nun (1995) have argued that if the heavy noble gases on Earth and Mars were delivered by comets, a formation temperature of (50 ± 15) K would be most suitable for trapping these gases in the cometary ice grains. Notesco et al. (1997) found a temperature near 65 K to be most compatible with observed abundances of CO/H_2O, CH_4/H_2O and C_2H_6/H_2O in Comet Hyakutake as reported by Mumma et al. (1996). Laboratory experiments by Bar-Nun and colleagues (unpublished) have demonstrated that if cometary ices formed at temperatures greater than 30 K, they should contain $< 3 \times 10^{-3}$ of initial ambient Ne. At $\sim$25 K, this number rises to 10^{-2}. The upper limit of $Ne/O < 0.005$ set by Krasnopolsky et al. (1997) in Hale-Bopp therefore points toward 30 K as a lower limit to the grain formation temperature. All of these temperatures derived from gas trapping in ice must refer to the conditions in the outer solar nebula where infalling interstellar ice grains sublimed and recondensed on cold refractory cores (Lunine et al., 1991). Hence they are not necessarily in conflict with the $\sim$30 K temperature deduced from measurements of OPRs or D/H.

Remote methods of observing comets subtend large areas of typically $10^2 - 10^5$ km around the nucleus and integrate over long columns. By default they can only probe bulk compositions (Eberhardt, 1999). In this context any temperature determination has to be viewed with some caution. Cometary ices may turn out to be highly differentiated at the grain level, and possibly even at the level of planetesimals. On small scales the averaged effective temperature may thus have little in common with the true formation temperatures of individual icy grains. Meteoritic samples exhibit remarkable isotopic anomalies at small scales. Jessberger (1999) discovered extreme variations of the $^{12}C/^{13}C$ ratio in grains of Comet 1P/Halley ranging between 1 and 5000, and to a lesser extent also excursions in the $^{24}Mg/^{25}Mg$ ratio. Such enormous departures from solar system values can neither be explained by nebular or interstellar chemistry nor a solar origin, and must be the result of nucleosynthesis in a diverse set of stars. The evolution of cometary ices was probably decoupled from the history of the refractory material. Yet there is ample reason to believe they, too, may contain a highly heterogeneous collection of icy particles (Greenberg and Li, 1999). A definitive answer to the question, how diverse cometary ices really are on a microscopic scale, must be deferred to future sample return missions.

Gautier (1998) proposed that the currently observed bulk D abundances are the result of an intimate mixture of two discrete phases. The first was a solar nebula component from water vapor that was incorporated into minerals in the form of hydroxyl groups with a D/H ratio somewhere between the terrestrial and the protosolar value. A second, interstellar component came from material in the presolar cloud with $D/H \approx 7 \times 10^{-4}$, similar to the deuterium rich phase in the meteorites Semarkona and Bishunpur (Deloule et al., 1997). Based on an evolutionary model,

Gautier concluded that turbulence produced a transport of volatiles from the inner active zone of the solar nebula to the outer regions and significantly diluted the D/H ratio, even in the Uranus-Neptune region. One difficulty in this model is the fact that transport is so efficient that comets must be created and ejected very rapidly. In the case of a highly viscous disk ($\alpha \geq 0.01$) the time available is less than 10^4 years. Otherwise the D/H ratio for comets would become too low. Perhaps the efficiency of transport has been overestimated in these models. The disk may have been geometrically thin so that vertical transport of energy to the surface of the disk prevailed over radial transport (Pringle, 1981). Obviously one cannot exclude that some H_2O (and other species) with a low nebular D/H may have been carried out to the Uranus-Neptune region. But such material does not appear to have participated in the formation of comets in appreciable amounts. Otherwise, differences in the D/H between HCN and H_2O would have been small or non-existent. An accurate measurement of the D/H ratio in a Jupiter family comet, or perhaps in the distant future, directly in a Kuiper Belt object, could help resolve this issue. Any gradient in the D/H ratio with heliocentric distance would indicate a non-negligible admixture of material that was reprocessed in the solar nebula.

Our entire knowledge about the chemical composition in cometary nuclei is tied to studies of the coma composition. In the past the question has been raised, whether, and how closely, observed D fractionations in the coma actually represent the conditions in the nucleus. At least for the bulk molecule, H_2O, notable isotope effects in the nucleus, at the surface of the comet, during sublimation and expansion, or as a result of coma chemistry can be excluded (Eberhardt *et al.*, 1995). Klinger *et al.* (1989) investigated deuterium enrichments on icy surfaces after heavy outgassing. They found only negligibly small changes in the deuterium abundances of less than 1%. Furthermore, under steady state conditions mass conservation requires that the isotopic composition in the escaping material and on the active surface must be identical (Eberhardt *et al.*, 1995). In the water of comet 1P/Halley the D/H ratio is constant over the cometocentric distance range between 1,500 and 30,000 km (Balsiger *et al.*, 1995). This clearly rules out any significant D fractionation in the coma.

The case for a trace species such as HCN is more problematic. While it is exceedingly difficult to ascertain that no D was exchanged in the interior of the nucleus, we can use the HNC/HCN ratio to estimate upper limits for the D contamination in HCN due to chemical reactions in the coma. In the inner coma the major loss mechanism for HCN is the reaction $HCN + H_3O^+ \rightarrow HCNH^+ + H_2O$. Dissociative electron recombination will eventually neutralize most of the $HCNH^+$ ions forming HCN and HNC at a branching ratio of 0.45 and 0.55 with ejection of one hydrogen atom (Irvine *et al.*, 1998). During the recombination process we assume two extreme cases. None or all of the protonated HCN molecules will return the hydrogen with the low D/H from H_3O^+. At the time of the DCN measurement Irvine *et al.* (1998) obtained HNC/HCN ≈ 0.15. From this measurement upper limits of $< +1.7\%$ and $< -10\%$ for the change in $(D/H)_{HCN}$ due to coma chemistry can be derived. The -10% limit for D depletion is probably of no significance, since D tends to accumulate in the neutral molecule.

5. Comparison with other Deuterium Reservoirs in the Solar System

Icy planetesimals formed the cores of the giant planets Jupiter and Saturn (Hubbard and MacFarlane, 1980). They grew fast and big enough to capture some of the hydrogen reservoir of the solar nebula. Fractionated hydrogen from the icy core did not noticeably contribute to the overall D/H in the vast H_2 atmosphere of Jupiter. The HD/H_2 ratio in the jovian atmosphere thus directly reflects the D/H in the primitive solar nebula (Mahaffy et al., 1998). The same argument works for Saturn (Griffin et al., 1996). For the outer planets, Uranus and Neptune, the situation is more involved. Their gas shells are not as massive as the atmospheres of Jupiter or Saturn, and the contributions from the cores can no longer be neglected (Hubbard and MacFarlane, 1980). Using gravitational moments Podolak et al. (1995) computed the size of the ice cores in Uranus and Neptune. Their model predicts a gas-to-ice mass ratio of 0.16 for Uranus and a range of 0.08–0.21 for Neptune. If ice is predominantly water ice with a cometary D/H ratio, and if the gaseous portion can be represented by H_2 with a protosolar deuterium abundance, the D/H ratio in a perfectly mixed Uranus would be 1.5×10^{-4}, or $(1.3-2.0) \times 10^{-4}$ for Neptune. These numbers are worst case upper limits. Measurements clearly revealed enhanced D/H ratios in Uranus and Neptune (Feuchtgruber et al., 1999). Although the D/H values remain below 10^{-4}, they provide strong evidence for D contamination by the core. The icy part in the cores of the gas planets is probably still fractionated and convection was not strong enough to equilibrate the D abundance in the inner regions with the surrounding H_2 layer (Hubbard et al., 1995).

The atmospheres of the terrestrial planets evolved with time and the preferred escape of H over D shifted the protosolar D/H toward higher values. Titan, the only satellite with a substantial (N_2) atmosphere, probably formed in the subnebula of Saturn and produced its atmosphere by degassing from its constituent ices. As the value of D/H in Titan's methane is approximately 4 times lower than the cometary values described here, cometary impact is not a likely source of Titan's atmosphere (Griffith and Zahnle, 1995; Coustenis et al., 1998). Similarly we can see that the high D/H ratio in comets is irreconcilable with an origin in the subnebulae of the giant planets (Fegley and Prinn, 1989).

Looking to the future, the most important new data will come from investigations of D/H in various molecular species in short period comets. While the faintness of these objects makes them very difficult subjects for observations from Earth, the predictability of their orbital motions makes them the *only* comets to which spacecraft can be reliably targeted. During the next decade we may hope for a new collection of in situ isotope measurements from at least three missions to comets.

References

A'Hearn, M.F., and Feldman, P.D.: 1985, In *Ices in the Solar System* (Klinger, J., Benest, D., and Smoluchowski, R., Eds.), pp. 463–471, D. Reidel Publishing Company, Dordrecht (USA).
A'Hearn, M.F., Schleicher, D.G., and West, R.A.: 1985, *Astrophys. J.* **297**, 826–836.
Aikawa, Y., Umebayashi, T., Nakano, T., and Miyama, S.M.: 1997. *Astrophys. J. Lett.* **486**, L51–L54.
Balsiger, H., Altwegg, K., and Geiss, J.: 1995, *J. Geophys. Res.* **100**, 5827–5834.
Bockelée-Morvan, D., Gautier, D., Lis, D.C., Young, K., Keene, J., Phillips, T., Owen, T., Crovisier, J., Goldsmith, P.F., Bergin, E.A., Despois, D., and Wootten, A.: 1998, *Icarus* **133**, 147–162.
Brown, P.D., and Millar, T.J.: 1989, *Mon. Not. R. Astron. Soc.* **237**, 661–671.
Coustenis, A., Salama, A., Lellouch, E., Encrenaz, Th., de Graaw, Th., Bjoraker, G.L., Samuelson, R.E., Gautier, D., Feuchgruber, H., Kessler, M.F., and Orton, G.S.: 1998, *Bul. Am. Astron. Soc.* **30**, 1060.
Crovisier, J., Bockelée-Morvan, D., Colom, P., Despois, D., and Paubert, G.: 1993, *Astron. Astrophys.* **269**, 527–540.
Crovisier, J., Leech, K., Bockelée-Morvan, D., Brooke, T.Y., Hanner, M.S., Altieri, B., Keller, H.U., and Lellouche, E.: 1997, *Science* **275**, 1904–1907.
Crovisier, J.: 1998, *Bul. Am. Astron. Soc.* **30**, 1059–1060.
Delsemme, A.H.: 1998, *Planet Space Sci.*, in press.
Deloule, E., Doukhan, J.-P., and Robert, F.: 1997, *Proc. Lunar Planet. Sci. Conf. 28th*, 291–292.
Duncan, M., Quinn, T., and Tremaine S.: 1987, *Astron. J.* **94**, 1330–1338.
Eberhardt, P., Meier, R., Krankowsky, D., and Hodges, R.R.: 1994, *Astron. Astrophys.* **288**, 315–329.
Eberhardt P., Reber, M., Krankowsky, D., and Hodges, R.R.: 1995, *Astron. Astrophys.* **302**, 301–316.
Eberhardt, P.: 1999, In *Proc. IAU Colloq. 168, Cometary Nuclei in Space and Time* (A'Hearn, M.F., Ed.), Astron. Soc. Pacific Conf. Ser., in press.
Epstein, R.I., Lattimer, J.M., and Schramm, D.N.: 1976, *Nature* **263**, 198–202.
Fegley, B., Jr., and Prinn, R.G.: 1989, In *The formation and evolution of planetary systems* (Weaver, H.A., and Danly, L., Eds.), pp. 171–211, Cambridge U. Press, Cambridge (UK).
Feuchtgruber, H., Lellouch, E., Bézard, B., Encrenaz, Th., de Graaw, Th., and Davis, G.R.: 1999, *Astron. Astrophys.* **341**, L17–L21.
Gautier D. and Morel, P.: 1997, *Astron. Astrophys.* **323**, L9–L12.
Gautier, D.: 1999, In *Planetary systems: The long view* (Celnikier, L., Ed.), in press, Blois (France).
Geiss, J., and Reeves, H.: 1981, *Astron. Astrophys.* **93**, 189–199.
Geiss, J.: 1993, In *Origin and evolution of the elements* (Prantzos, N., Vangioni-Flam, E., and Cassé, M., Eds.), pp. 87–106, Cambridge Univ. Press, Cambridge (UK).
Greenberg, M., and Li, A.: 1999, *Space Sci. Rev.*, this volume.
Griffith, C.A., and Zahnle, K.: 1995, *J. Geophys. Res.* **100**, 16,907 – 16,922.
Griffin, M.J., et al.: 1996, *Astron. Astrophys.* **315**, L389–L392.
Grinspoon, D.H. and LewisJ.S.: 1988, *Icarus* **72**, 430–436.
Hubbard, W.B., and MacFarlane, J.J.: 1980, *Icarus* **127**, 307–318.
Hubbard, W.B., Podolak, M., and Stevenson, D.J: 1995, In *Neptune and Triton* (Cruikshank, D.P., Ed.), pp. 109–140, U. of Arizona Press, Tucson (USA).
Irvine, W.M., Bergin, E.A., Dickens, J.E., Jewitt, D., Lovell, A.J., Matthews, H.E., Schloerb, F.P., and Senay, M.: 1998, *Nature* **393**, 547–550.
Jessberger, E.: 1999, *Space Sci. Rev.*, this volume.
Klinger, J., Eich, G., Bischoff, A., Joó, F., Kochan, H., Roessler, K., Stichler, W., and Stöeffler, D.: 1989, *Adv. Space Res.* **9**, (3), 123–125.
Krasnopolsky, V.A., Mumma, M.J., Abbott, M., Flynn, B.C., Meech, K.J., Yeomans, D.K., Feldman, P.D., and Cosmovici, C.B.: 1997, *Science* **277**, 1488–1491.
Lécluse, C., and Robert, F.: 1994, *Geochim. Cosmochim. Acta* **58**, 2927–2939.
Lunine, J.I., Engle, S., Riszk, B., and Horanyi, M.: 1991, *Icarus* **94**, 333–344.

Mahaffy, P.R., Donahue, T.M., Atreya, S.K., Owen, T.C., and Niemann, H.B.: 1998. *Space Sci. Rev.* **84**, 251–263.
Meier, R., Owen, T.C., Matthews, H.E., Jewitt, D.C., Bockelée-Morvan, D., Biver, N., Crovisier, J., and Gautier, D.: 1998a, *Science* **279**, 842–844.
Meier, R., Owen, T.C., Jewitt, D.C., Matthews, H.E., Senay, M., Biver, N., Bockelée-Morvan, D., Crovisier, J., and Gautier, D.: 1998b, *Science* **279**, 1707–1710.
Meier, R., Wellnitz, D., Kim, S.J., and A'Hearn, M.F.: 1998c, *Icarus* **136**, 268–279.
Millar, T.J., Bennett, A., and Herbst, E.: 1989, *Astrophys. J.* **340**, 906–920.
Mumma, M.J., Weissman, P.R., and Stern, S.A.: 1993, In *Protostars and planets* **III**, (Levy, E.H, and Lunine, J.I., Eds.), pp. 1177–1252, U. of Arizona Press, Tucson (USA).
Mumma, M.J., DiSanti, M.A., Dello Russo, N., Fomenkova, M., Magee-Sauer, K, Kaminski, C.D., and Xie, D.X.: 1996, *Science* **272**, 1310–1314.
Notesco, G., Laufen, G., and Bar-Nun, A.: 1997, *Icarus* **125**, 471–473.
Owen, T.C., and Bar-Nun, A.: 1995, *Icarus* **116**, 215–226.
Owen, T.C.: 1997, In *From stardust to planetesimals* (Pendleton, Y.J., and Tielens, A.G.G.M., Eds.), pp. 435–450, Astron. Sci. Pacific Public. 122, San Francisco (USA).
Podolak, M., Weizman, A., and Marley, M.: 1995, *Planet. Space Sci.* **43**, 1517–1522.
Pringle, J.E.: 1981, *Ann. Rev. Astron. Astrophys.* **19**, 137–162.
Schleicher, D.G., and A'Hearn, M.F.: 1986, In *New insights in astrophysics. Eight years of UV astronomy with IUE* (Rolfe, E.J., Ed.), pp. 31–33, ESA SP–263, ESA, Paris (France).
Wagoner, R.V., Fowler, W.A., and Hoyle, F.: 1967, *Astrophys. J.* **148**, 3–49.
Willacy, K., Klahr, H.H., Millar, T.J., and Henning, Th.: 1998, *Astron. Astrophys.* **338**, 995–1005.
Williams, T.L., Adams, N.G., Lucia, M.B., Herd, Ch.R., and Geoghegan, M.: 1996. *Mon. Not. R. Astron. Soc.* **282**, 413–420.

Address for correspondence: Tobias Owen, owen@IfA.Hawaii.edu

COMET HALLEY'S GAS COMPOSITION AND EXTENDED SOURCES: RESULTS FROM THE NEUTRAL MASS SPECTROMETER ON GIOTTO

PETER EBERHARDT

Physikalisches Institut, University of Bern, Sidlerstrasse 5, CH-3012 Bern, Switzerland

Abstract. The Neutral Mass Spectrometer on the Giotto spacecraft established that H_2O is the dominant species in Comet Halley's volatiles and determined the abundance of more than 10 parent species. The instrument discovered strong extended H_2CO and CO sources in the coma of Comet Halley. Polymerized H_2CO associated with the cometary dust and evaporating slowly as the monomer is most likely the extended H_2CO source. Photodissociation of the H_2CO into CO fully accounts for the extended CO source.

Keywords: comets, C/Halley, volatiles, extended source

1. Introduction

In this paper a brief review of the results obtained with the Neutral Mass Spectrometer (NMS) on the Giotto spacecraft is presented. The main emphasis will be on the extended H_2CO and CO sources in the coma of Comet Halley discovered by the NMS. The compositional information obtained by the NMS will only be briefly discussed.

2. Instrument Description

On the Giotto spacecraft the Neutral Mass Spectrometer (NMS, Krankowsky *et al.*, 1986a) measured the composition of the cometary gas. The instrument had two sensors, both equipped with a fly-through electron impact ion source. The M-analyser was a double focusing magnetic mass spectrometer, covering the 1 amu/q to 38 amu/q mass range with a mass independent peakwidth of ≈ 0.3 amu/q (FWHM). A focal plane detector allowed simultaneous registration of all masses. The E-analyser was a parallel plate energy analyser, also with a focal plane detector. Because of the high relative velocity of the gas molecules and their low temperature, an energy analysis is essentially equivalent to a mass analysis with limited resolution. In the neutral mode the E-analyser covered the 1 amu/q to 56 amu/q mass range.

The electron bombardement sources used in the NMS create not only ions of the parent molecules but also charged, lighter fragments. When analyzing a gas mixture, this leads to a complex mass spectrum and great care is required to derive a gas composition, especially for minor components present at the percent level and lower.

The Neutral Gas Experiment (NGE) on the two VEGA spacecraft did not provide any useful data on the gas composition (Curtis et al., 1987).

From a mass spectrometrist's point of view the cometary coma corresponds to an oversized ion source which combines photoionization and chemical ionization. Both these are methods used in standard laboratory mass spectrometry for analyzing gas composition. Thus, it is possible to derive the cometary gas composition from the measured composition of cometary ions. In the laboratory such ion sources would be calibrated with known gas mixtures. This is not feasible for the cometary coma. Modeling of the processes in the cometary coma are required to derive the neutral gas composition from the ion abundances.

Three instruments on Giotto were capable of measuring the cometary ion composition: the Ion Mass Spectrometer (IMS), the NMS, and the Positive Ion Cluster Composition Analyser (PICCA). The IMS (Balsiger et al., 1986) had two sensors. For ion composition measurements in the inner coma the High-Intensity-range Spectrometer (HIS) was specially suited. It was a magnetic mass spectrometer covering the 12 amu/q to 56 amu/q mass range with approximately unit mass resolution. By disabling the electron emission in the ion source and adjusting some voltages the NMS could also measure cometary ions. The PICCA instrument was solely an energy analyser (Rème et al., 1986).

3. Gas Composition

The composition of cometary volatiles as determined from the in situ measurements and from remote observations has been reviewed by the author (Eberhardt, 1996). In Table I the gas composition as derived from the in situ measurements by the experiments on the Giotto spacecraft is listed. The data given in Table I are based on neutral gas and ion results obtained by the NMS, except the upper limit for CH_4. The NMS provided the most reliable data and the data evaluation and the modeling used is the most precise one used for the evaluation of the Giotto ion composition data. Compositional data obtained from IMS agree with the values given in Table I, but have higher uncertainties. The abundances determined from the in situ measurements in the coma of Comet Halley fall in the range of parent molecule abundances observed in other comets derived with ground based methods. The only major difference is the H_2CO. The direct H_2CO production from the nucleus is in the range observed for other comets. The strong extended source of H_2CO is an unusual feature for most comets. Only for Hyakutake and Hale-Bopp has the presence of an extended source been reported. It must be pointed out, that with ground based observations it is difficult to distinguish between the direct production from the nucleus and the contribution from an extended source and to derive the exact strength for an extended source. The abundances obtained

TABLE I

Molecular abundances in the gas released from the nucleus of Comet Halley (parent molecules). All results are from experiments on the Giotto spacecraft. The abundances observed for other comets are taken from the review of Eberhardt (1996). a: Krankowsky et al., (1986b); b: Eberhardt et al., (1987); c: Eberhardt (1996); d: Meier et al., (1993); e: Reber et al., (1999); f: Eberhardt et al., (1994); g: Altwegg et al., (1994); h: Meier et al., (1994).

Molecule	Abundance	Remarks	Ref.	Other Comets
H_2O	100		a	100
CO	3.5	Production from nucleus	b,c	1-5
	11	Including extended source	b,c	1-30
CO_2	3		a,b	3*
H_2CO	< 0.4	Production from nucleus	d	0.05-0.6**
	7.5	Including extended source to 25'000km	e	
CH_3OH	1.7		f	0.8-4.1
CH_4	< 1		g	0.7
C_2H_2	≈ 0.3		e	0.3-0.9
C_2H_4	≈ 0.3		e	
C_2H_6	≈ 0.4		e	0.4
NH_3	1.5		h	1
HCN	≈ 0.2	Includes HNC	e	0.03-0.3
H_2S	0.41		f	0.3

*: Comet Levy higher
**: Upper limit for production from nucleus, may include some contribution from extended source

from in situ measurements are often more reliable than from remote observations as the determination of the reference molecule H_2O is, in most cases, not carried out simultaneously with the same method and instrument. Thus, systematic errors inherent to the observations and modeling required for deriving production rates introduce sizable and often underestimated uncertainties.

Remote observations of the two very bright comets Hyakutake and Hale-Bopp have revealed the presence of a number of additional parent molecules so far not recognized unambiguously in the Giotto NMS data. These molecules include (average abundance relative to H_2O in brackets): Carbon Disulfide (CS_2; 0.2 % in Comet Halley, Meier and A'Hearn, 1997); Carbonyl Sulfide (OCS; ≈ 0.3 %, Woodney et al., 1997; Lis et al., 1998); Sulfur Monoxide (SO(SO_2 parent); ≈ 0.5 %, Lis et al., 1998); Sulfur Dimer (S_2; ≈ 0.01 %, see Eberhardt, 1996); Methyl Cyanide (CH_3CN; ≈ 0.02 %, Bockelée-Morvan et al., 1998); Isocyanic Acid (HNCO; ≈ 0.07 %, Lis et al., 1997, 1998); Formamide (NH_2CHO; ≈ 0.01 %, Bockelée-Morvan et al., 1998); Cyanoacetylene (HC_3N; ≈ 0.02 %, Lis et al., 1998).

4. Extended Sources

The in situ measurement of the CO density in the coma of Comet Halley with a spatial resolution of a few 100 km revealed that only about 1/3 of the CO originates directly from the nucleus (Eberhardt et al., 1987, see Fig. 1). The remainder of the CO comes from an extended source located in the innermost 25'000 km of the coma. No suitable parent molecule for this extended source could be identified, and Eberhardt et al. concluded that the CO evaporated from CHON-type dust grains in the coma either as CO or as a short lived parent. Gas release from the dust was also suggested to explain the CN-jets observed in Comet Halley's coma (A'Hearn et al., 1986).

From the radio detection of H_2CO in Comet Halley Snyder et al. (1989) obtained evidence that H_2CO also originates from an extended source. Meier et al. (1993) derived the H_2CO density inside the contact surface from the H_3CO^+ ion density measured by the NMS with a spatial resolution of about 400 km. The resulting $n(R)_{H_2CO}$ profile is not compatible with a nuclear origin of the H_2CO. Based on the gas mass spectra acquired by the NMS the in situ measured $n(R)_{H_2CO}$ profile has now been extended to 25'000 km (Eberhardt et al., 1995; Reber et al., 1999). The extended H_2CO source persists out to a distance of 20'000 km with a distinct local maximum in the source strength $q(R)_{H_2CO}$ from 15'000 km to 19'000 km, coincident with an enhancement of the submicron dust density (McDonnell et al., 1991). The H_2CO photodissociates into CO, and the extended H_2CO source can fully account for the extended CO source. The correlation of the strength of the extended H_2CO source with the submicron dust strongly suggests that the CHON component of the dust is the carrier of the H_2CO in the coma, probably in polymerized form as POM (Huebner, 1987). The POM will directly evaporate as the formaldehyde-monomer. In a comet simulation experiment it was found that H_2CO polymerization occurs on the surface during the sublimation process, especially if NH_3 is present (K. Roessler, private communication). It is, thus, not clear whether the H_2CO in the nucleus is already polymerized or if the transformation to POM occurs in the surface region during the sublimation process. If the polymerization occurs in the surface region, the polymer will act as glue between individual cometary dust grains. Larger aggregates will be formed. This aggregation process will affect predominantly the very fine submicron grains, as these are most abundant by number and have the highest surface to volume ratio. The polymerization process of the cometary H_2CO would thus enhance the formation of a crust on or close to the cometary surface. Strong temporal variations

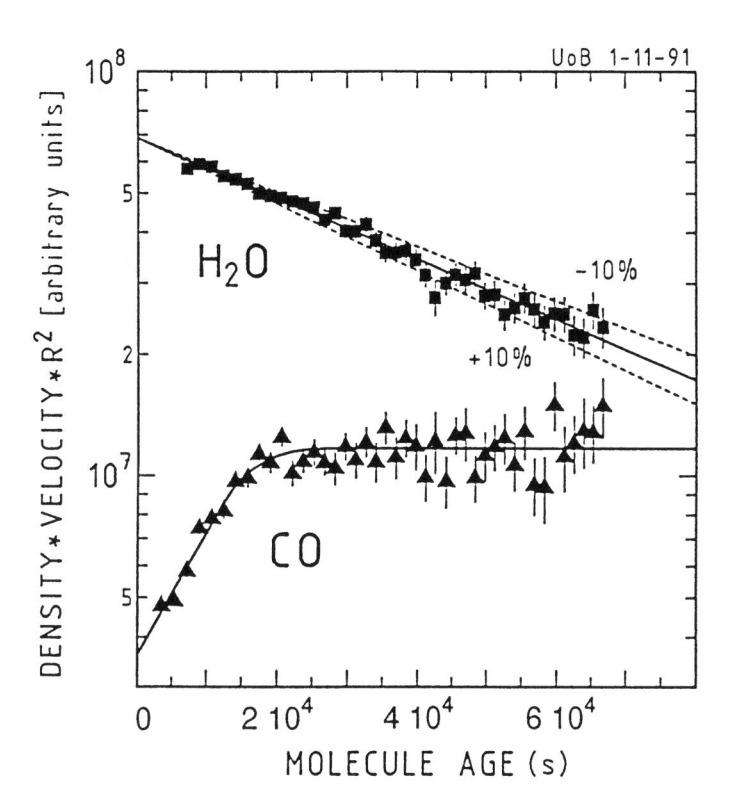

Figure 1. Radial density profiles for H_2O and CO densities obtained from gas mass spectra recorded by the Giotto NMS. To eliminate the geometric and kinematic effects the density multiplied by the expansion velocity and the square of the cometocentric distance are plotted. The gas expansion velocity measured in situ by the E-analyzer of the NMS was used (Lämmerzahl *et al.*, 1987). For the definition and calculation of the molecular age see Meier *et al.* (1993). As the expansion velocity is approximately 1 km/s the molecular age is very approximately equivalent to the cometocentric distance in km. The linear decline of the H_2O profile in this semi-logarithmic plot is due to photodestruction of H_2O. The solid line is a least square fit to the data points and corresponds to a photodestruction rate of $k_{H_2O} \approx 1.7 \; 10^{-5} \; s^{-1}$ (see also Krankowsky and Eberhardt, 1990). This corresponds to $k_{H_2O} \approx 1.4 \; 10^{-5} \; s^{-1}$ at 1 AU. The dotted curves labeled -10% and $+10\%$ are the expected profiles for a 10 % smaller or 10 % larger photodestruction rate. The excellent fit of the data points to a straight line implies that the H_2O production rate was essentially constant during approximately 60'000 s prior to the encounter and that no extended H_2O source is present for distances larger than approximately 5'000 km. The CO profile shows a completely different behaviour. The solid line is an approximation to the data points. It consists of 2 straight line segments approximating the data points with molecular age below 16'000 s ($\approx$ 14'000 km) and above 30'000 s ($\approx$ 28'000 km, line segment drawn horizontally) joined by an eye-ball curve. The increase of the CO profile for distances less than $\approx$ 25'000 km results from an extended CO source in the cometary coma. The essentially constant CO values for distances larger than approximately 28'000 km reflect the long lifetime of CO against photodestruction and the virtual absence of a noticeable local CO production at these distances. Note that the CO data points have been shifted arbitrarily in the y-direction. The CO density includes a small contribution from other molecules which produce a signal on mass 28 amu/q.

in the POM abundance in the coma will be expected. The POM will accumulate in the cometary crust and only when the gas production gets very high will the accumulated POM-rich material be driven off, leading to a temporal enhancement of the H_2CO production in the extended source. Such enhancements will most likely occur around or after perihelion.

The detailed $n(R)_{H_2CO}$ and $q(R)_{H_2CO}$ profiles can only be explained by dust with two different components or precursors. Within the contact surface the strength $q(R)_{H_2CO} \times R^2$ of the extended H_2CO source decreases approximately exponentially with increasing cometocentric distance with a scale length of 2'500 - 3'000 km (Meier *et al.*, 1993). This observation is fully compatible with the H_2CO release being a first order process such as the direct sublimation of H_2CO from polymerized formaldehyde on the dust. At a distance of 15'000 km to 19'000 km, where we observed a local maximum in the source strength $q(R)_{H_2CO} \times R^2$, the carrier with the 2'500 - 3'000 km scale lengths has lost already more than 99 % of the H_2CO. To explain this distant local maximum requires a different type of carrier dust, which releases the H_2CO on a much longer length scale. For the further discussion we will designate these two precursor components of the H_2CO as component I and component II.

The production rate of H_2CO from component I decreases exponentially with cometocentric distance (Meier *et al.*, 1993). This is typical for a first order process depleting a reservoir, similar to the photodestruction of a parent molecule. The most straightforward assumption is that the physics and chemistry of component II is similar to component I but that the scale length is different. The scale length for the release is governed by two factors: the expansion velocity of the dust and the time constant of the H_2CO release from the dust. If we assume, that the fine grained dust is the carrier of the polymerized H_2CO, then the expansion velocity of the dust will be similar or slightly less than the expansion velocity of the gas (similar to Samarasinha and Belton (1994) we assume $v_D = 0.9 \ v_G$). The two polymerized H_2CO components must then have time constants of the order of 4'000 s and 20'000 s for the sublimation as H_2CO. To explain the local maximum in the H_2CO production in the 15'000 km to 19'000 km range the mixing ratio of the two components must be variable with time. Approximately 20'000 s prior to the encounter of the Giotto spacecraft with Comet Halley an enhancement in the production of component II must have occurred. Such a model can explain the observed radial H_2CO and CO density profiles quite well. A possible, but speculative explanation is that one of the precursors is recently formed POM, the other older (presolar) polymers.

The observed two scale length could also be due to two dust components with very different grain size. Micron and submicron sized dust particles, moving essentially with the gas speed and releasing the H_2CO with a time constant of the order of 20,000 s would then be the carrier responsible for the extended H_2CO source at larger distances (component II). One to two orders of magnitude larger dust particles (or aggregates of fine particles) with an expansion velocity of only about 1/5 of the gas speed but the same time constant for the H_2CO release would then be responsible for the extended H_2CO source close to the nucleus, essentially within the contact surface (component I). The mixing ratio of the two dust components would have to be time dependent. The coarser dust could be cometary crust fragments with recently formed POM, the finer dust could be the older (presolar) polymers.

The release of the H_2CO from component II could obey a more complicated law. Boice *et al.* (1990) have discussed the release of polymerized H_2CO molecules, especially $(POM)_5$, from the carrier dust. Subsequent photodestruction then leads to the formaldehyde monomer and finally to the formation of the relatively long lived CO molecules. Boice *et al.* (1990) were able to reproduce the measured CO density profile with such a model. However, all physical constants in the complex POM photochemistry had to be estimated and hence the results are not very constrained. Furthermore, the available laboratory data on the sublimation of para-formaldehyde show, that the sublimation from a formaldehyde polymer occurs mainly as the monomer (see also Meier *et al.*, 1993). Special grain geometries etc. have been proposed as extended sources which show a delayed release (E-berhardt *et al.*, 1987; Samarasinha and Belton, 1994). For instance agglomerates of grains held together by POM as glue would initially have a low release rate of H_2CO. When the agglomerates start to disintegrate because the interstitial glue has partially evaporated, larger surface areas of POM would be exposed and the sublimation rate would increase until the supply of POM is exhausted. However, the initially large aggregates would have a low expansion velocity and the break up of the aggregates would have to occur on a time scale much longer than 20'000 s. Another explanation for a delayed release of the H_2CO could be multilayered grains. The POM-layer on the individual grains could be covered by a layer of material with a very low vapor pressure. Such a protective layer could delay the sublimation of the H_2CO until the grains have reached a larger cometocentric distance. The outermost rim of the POM layer, processed by irradiation with UV or very low energy ions (solar wind like), could form such a protective layer. These models are highly speculative and it seems to the author much more likely, that the spatial distribution of the extended formaldehyde source reflects time dependent mixing of two components with different scale length for the formaldehyde release as discussed above. The variable mixing rate of the two components might even reflect local compositional inhomogeneities on the surface of Halley's nucleus. One of the components could for instance be associated with vents on the surface of the nucleus (see Samarasinha and Belton, 1994).

Acknowledgements

The author would like to thank an anonymous reviewer and Dr. W. Huebner for interesting comments. Research supported by the Swiss National Science Foundation.

References

A'Hearn, M.F., Hoban, S., Birch, P.V., Bowers, C., Martin, R., and Klinglesmith III, D.A.: 1986, *Nature* **324**, 649.
Altwegg, K., Balsiger, H., and Geiss, J.: 1994, *A&A* **290**, 318.
Balsiger, H., Altwegg, K., Bühler, F., Fischer, J., Geiss, J., Meier, A., Rettenmund, U., Rosenbauer, H., Schwenn, R., Benson, J., Hemmerich, P., Säger, K., Kulzer, G., Neugebauer, M., Goldstein, B.E., Goldstein, R., Shelly, E.G., Sanders, T., Simpson, D., Lazarus, A.J., and Young, D.T.: 1986, *ESA-SP* **1077**, 129.
Bockelée-Morvan, D., Wink, J., Despois, D., *et al.*: 1998, *Earth, Moon, & Planets*, in print.
Boice, D.C., Huebner, W.F., Sablik, M.J., and Konno, I.: 1990, *GRL* **17**, 1813.
Curtis, C.C., Fan, C.Y., Hsieh, K.C., Hunten, D.M., Ip, W.-H., Keppler, E., Richter, A.K., Umlauft, G., Afonin, V.V., Dyachkov, A.V., Erö Jr., J., and Somogyi, A.J.: 1987, *A&A* **187**, 360.
Eberhardt, P., Krankowsky, D., Schulte, W., Dolder, U., Lämmerzahl, P., Berthelier, J.J., Woweries, J., Stubbemann, U., Hodges, R.R., Hoffman, J.H., and Illiano, J.M.: 1987, *A&A* **187**, 481.
Eberhardt, P., Meier, R., Krankowsky, D., and Hodges, R.R.: 1994, *A&A* **288**, 315.
Eberhardt, P., Reber, M., Krankowsky, D., and Hodges, R.R.: 1995, *BAAS* **27**, 1143.
Eberhardt, P.: 1996, Proc. 1996 Asteroids, Comets, & Meteor Conference, Versailles, *COSPAR Colloquia Series*, in print.
Huebner, W.F.: 1987, *Science* **237**, 628.
Krankowsky, D., Lämmerzahl, P., Dörflinger, D., Herrwerth, I., Stubbemann, U., Woweries, J., Eberhardt, P., Dolder, U., Fischer, J., Herrmann, U., Hofstetter, H., Jungck, M., Meier, F.O., Schulte, W., Berthelier, J.J., Illiano, J.M., Godefroy, M., Gogly, G., Thévenet, P., Hoffman, J.H., Hodges, R.R., and Wright, W.W.: 1986a, *ESA-SP* **1077**, 109.
Krankowsky, D., Lämmerzahl, P., Herrwerth, I., Woweries, J., Eberhardt, P., Dolder, U., Herrmann, U., Schulte, W., Berthelier, J.J., Illiano, J.M., Hodges, R.R., and Hoffman, J.H.: 1986b, *Nature* **321**, 326.
Krankowsky, D., and Eberhardt, P.: 1990, in *Comet Halley: Investigations, Results, Interpretations, Vol. 1: Organization, Plasma, Gas*, Mason, J.W. (ed.), Ellis Horwood Ltd., New York, p. 273.
Lämmerzahl, P., Krankowsky, D., Hodges, R.R., Stubbemann, U., Woweries, J., Herrwerth, I., Berthelier, J.J., Illiano, J.M., Eberhardt, P., Dolder, U., Schulte, W., and Hoffman, J.H.: 1987, *A&A* **187**, 169.
Lis, D.C., Keene, J., Young, K., Phillips, T.G., Bockelée-Morvan, D., Crovisier, J., Schilke, P., Goldsmith, P.F., and Bergin, E.A.: 1997, *Icarus* **130**, 355.
Lis, D.C., Mehringer, D.M., Benford, D., Gardner, M., Phillips, T.G., Bockelée-Morvan, D., Bivier, N., Colom, P., Crovisier, J., Despois, D., and Rauer, H.: 1998, *Earth, Moon, and Planets*, in print.
McDonnell, J.A.M., Lamy, P.L., and Pankiewicz, G.S.: 1991, in *Comets in the Post-Halley Era 2*, R.L.Newburn Jr., M.Neugebauer, and J.Rahe, Kluwer Acad.Publ., Dordrecht, 1043.
Meier, R., Eberhardt, P., Krankowsky, D., and Hodges, R.R.: 1993, *A&A* **277**, 677.
Meier, R., Eberhardt, P., Krankowsky, D., and Hodges, R.R.: 1994, *A&A* **287**, 268.
Meier, R., and A'Hearn, M.F.: 1997, *Icarus* **125**, 164.
Reber, M., Eberhardt, P., and Krankowsky, D.: 1999, in preparation.
Rème, H., Cotin, F., Cros, A., Médale, J.L., Sauvaud, J.A., d'Uston, C., Korth, A., Richter, A.K., Loidl, A., Anderson, K.A., Carlson, C.W., Curtis, D.W., Lin, R.P., and Mendis, D.A.: 1986, *ESA-SP* **1077**, 33.
Samarasinha, N.H., and Belton, M.J.S.: 1994, *Icarus* **108**, 103.
Snyder, L.E., Palmer, P., and de Pater, I.: 1989, *AJ* **97**, 246.
Woodney, L.M., McMullin, J., and A'Hearn, M.F.: 1997, *Planet.Space Sci.* **45**, 717.

Address for correspondence: Peter Eberhardt, Physikalisches Institut, University of Bern, Sidlerstrasse 5, CH-3012 Bern, Switzerland, eberhardt@phim.unibe.ch

ON THE EXISTENCE OF DISTRIBUTED SOURCES IN COMET COMAE

M.C. FESTOU [*]

Southwest Research Institute, Department of Space Studies, 1050 Walnut St., Suite 426, Boulder, CO 80302, USA

Abstract. The main production process of species occurring in the coma of comets is the photodestruction of molecules initially present in the nucleus ices and non-refractory grains or trapped inside the nucleus "material". Grains can also be a source of molecules in the coma. Chemical reactions may occur between coma species. Consequently, although chances that an abundant coma species has not been detected are now small, the coma composition is certainly quite different from that of the nucleus. Except for the molecules released directly at the nucleus surface, all coma species are produced in an "extended region" or come from "a distributed source". Since the recent literature is rich in reports on observations of molecules and species possibly not initially present in the comet ices or not released at the nucleus, a general discussion of how coma species are stored, processed or produced is presented, based mostly on observational results. What is at stake is the proper modeling of the coma structure, hence an accurate derivation of the nucleus composition from coma observations.

Keywords: Comets, coma, sources of gases

1. Introduction

All molecules, radicals and atoms that we observe in comet comae and tails come from the destruction and / or transformation of molecules that were once stored in the nucleus. We still do not really know in which form. Specialists of the nucleus structure tell us those volatiles are stored as ices that loosely glue grains together and that molecules in gaseous form are likely to be trapped in the icy lattice or in the grains. Until we observed comet C/1995 O1 (Hale-Bopp), even though we knew both from observations (e.g. the activity of distant comets and 2060 Chiron) and from theoretical considerations that water ice could not evaporate much beyond 4–5 AU from the sun, the common paradigm about comet activity was that water ice controlled the production of volatiles at the nucleus surface. The observation in comet Hale-Bopp of numerous species long before water ice started to significantly sublimate, carries a strong message: very volatile species and solid grains find their way towards the surface without the help of subliming water molecules. As important: water ice is not a serious obstacle to their escape into space. A key question is whether such volatile species can also extract icy grains from the inactive icy patches. The consequences of these observational facts

[*] On leave from the Observatoire Midi-Pyrénées, Toulouse, France.

has not yet been fully exploited. However, the concept of water ices having bonds with some molecules is certainly not valid (we knew already that the stoichiometry to have clathrates or hydrates in the nucleus was not correct).

The principal production mechanism in comet comae is the photodestruction of molecules initially residing in the nucleus (see Huebner *et al.*, 1991, for an extensive list of other photolytic processes that may also occur). The possibility that coma neutrals and ions could be produced in chemical reactions involving coma neutrals and ions has been considered first on a theoretical basis, beginning with the pioneering work of Aikin (1974) that predicted the prevalence in the inner coma of the H_3O^+ ion. Daniel Malaise very often publicly commented in the 1970s on how much our coma models were too simple because they were ignoring the possible reshuffling of the coma composition in the first few hundreds of km from the nucleus. *In situ* observations of 1P/Halley in March 1986 in part validated his point of view as some observed species can only be the products of the transformation via chemical reactions of species (or their destruction products) released at the nucleus surface. Yet, the instruments that scrutinized the coma of P/Halley did not probe the first 10^3 km from the nucleus and some short-lived species certainly escaped detection, like the elusive S_2 molecule.

One of the highlights of the exploration of Halley's comet is the discovery that some large molecules existed (see Festou *et al.*, 1993a, for a review of the 1986–1987 campaign). Many works, especially recent ones, indicate that some species are created from a "distributed" parent source. In this paper, we will qualify as a distributed production mechanism any production path that is not *obviously* a photolytic process. Icy grains might have been detected in C/Hale-Bopp and some molecular species have line profiles that are not characteristic of species produced directly at the nucleus. Day after day, there is stronger and stronger evidence that large molecules and / or grains contribute to the production of the species we detect in the coma. Without a clear idea of what the storage / production / processing mechanisms are in comets, there is no reliable evaluation of coma species abundances and, consequently, no accurate determination of the nucleus composition from observations of the coma. Our objective here is to summarize what we know of the connection that exists between atmospheric and nuclear species, without technically addressing modeling issues.

Modeling the spatial distribution of coma species is much more complex than hitherto assumed. However, most data sets offer very limited possibilities when used separately to constrain and / or justify the use of elaborated coma models, as we will illustrate below. Modeling the coma now raises the following key questions:

(i) How are the molecules stored in and then released from the nucleus? What is their spatial distribution and outflow velocity near the nucleus surface?

(ii) Which species are produced in the coma by non-photolytic processes?

(iii) What is the contribution of the distributed sources relative to that of the nucleus sources?

(iv) What are the relative contributions of solid grains and "CHON" particles to the production of molecules present in the coma?

(v) Is the CHON particle abundance correlated with the gas, with the dust production rate?

Below, we will comment on these points but will not deliver definitive answers. The interaction of the solar wind particles with the outer coma will not be considered as it is deemed that the species detected in the EUV and at shorter wavelengths tell us something about solar wind rather than cometary particles. Technical means at our disposal have improved so much during the last 15 years that it would be unfair to comment on interpretations of observations collected a few decades ago and we will restrict ourselves to discussing only the most recent (and so much richer) data sets. Many Comet Hale-Bopp observations have not yet been published and this is why this paper may appear incomplete. However, given the exceptionally high gas production of this comet, the physical conditions prevailing in its coma are radically different from those existing in other comets and we restrict ourselves to quoting / commenting some selected results that apply to most comets.

2. The observations

2.1. Gas Phase Reactions

In comet comae, all observed phenomena are transient in nature. Neutral species move away from the nucleus and leave the coma quite rapidly (lest we forget the small fraction of coma molecules and radicals that fly back towards the inner coma). With outflow velocities of order 1 km s^{-1} and spherical expansion, parent molecule densities rapidly decrease and this reduces both the ion production and the efficiency of all two- and three- body reactions. The crossing time of the "chemical sphere", the region inside which electron recombination and charge exchange reactions are more effective than photolytic processes, is at most a few 10^3 s in very active comets like P/Halley. Therefore, time is of the essence. No reaction having a small rate coefficient or not involving one of the dominant coma species can produce important amounts of reaction products. Charge exchange reactions can occur at larger distances than indicated above. Their main effect is to load the solar wind (SW) with heavy particles, without producing large quantities of new species. They also change the flow of the SW and this induces a loss / redistribution of comet ions. Models that assume that the medium can reach chemical equilibrium rest on an erroneous basis and only the results that are independent of that initial hypothesis are solid.

In 1986, observations of the charged particles and waves in the coma of P/Halley revealed a particularly complex interaction of the solar wind with the coma. This interaction is one of the least documented among the physical phenomena occurring in comet comae, mostly because ions are not abundant and thus not easy to remotely observe. The very existence of an axisymmetric ion tail is the indication that models involving ions ought to include the coupling / interaction between the two charged particle populations. In the neutral magnetic cavity, where SW particles do not penetrate, ignoring this interaction is probably a reasonable zero order approximation. However, outside the contact surface, cometary ions are violently deflected into the tail (see e.g. Schmidt et al., 1988). In very low gas production comets, cometary ions could even reach the nucleus surface (a SW proton will be stopped by a column of H atoms of order a few 10^{14} cm^{-2}, which is obtained at the nucleus when the water production is of order 10^{29} s^{-1}, whereas a water atmosphere will convert all solar protons into H_2O^+ ions and fast H atoms near the 10^9 cm^{-3} level, a density that is obtained at the nucleus surface when the water production is $\sim 10^{27}$ s^{-1}).

Aikin (1974) brought first the idea that the electron recombination of the H_3O^+ ion may constitute an important source of OH radicals (and 2 H atoms). A thorough investigation of the OH emission in comets as observed with the International Ultraviolet Explorer by Budzien et al. (1994) indicated that the small excess of emission that was present at the center of the coma could be due to this phenomenon. (These authors also mentioned the possibility of an increased velocity of the water molecules or prompt emission of OH.) The spatial distribution of such an OH population remains to be modeled, together with that of the electrons created by the photodissociation of the main neutrals. High spatial resolution observations of the inner OH coma of comet Hale-Bopp (Weaver et al., 1999) reveal a brightness peak that is displaced sunward. It is apparent that the production pattern of the water molecules at the nucleus makes it very difficult to separate the various effects responsible for the observed spatial distributions.

Cometary atomic and molecular ions offer a formidable resistance to the penetration of SW particles into the inner coma. Outside the contact surface cometary ions are slowed down through this interaction, increasing the possibilities of electron recombinations. The mass spectra peaks observed deep inside of Halley's comet coma where no coma neutral or ion was known to exist are the best clue that chemical reactions are indeed taking place and producing species that would not otherwise be present in the coma (e.g. at mass 19, the best candidate is H_3O^+). However, none of the abundant neutrals is produced in chemical reactions. A list of the species detected in P/Halley (see e.g. Huebner et al., 1991) shows that exotic ions like $C_3H_3^+$ can also be considered as excellent candidates for such a fate.

The discovery of numerous parent molecules in the coma of recent comets allows one to model the chemistry of the inner coma with unprecedented accuracy. In older studies, only global quantities like the coma content were compared to the observations. Modern measurements should allow one to perform much more quantitative tests of the models as we know the relative abundance of the main parent species and as an extensive list of secondary and tertiary products is available. Very often, even the volume or column densities of these latter are known to us. One of the first results to be obtained is a complete explanation for the presence of the HNC molecule, hence an accurate determination of its production rate (other species of interest comprise SO, HCO, HCO$^+$ and especially the CO_2^+ ion that can lead to the formation of CO molecules in the a $^3\Pi$ state after its dissociative recombination with electrons). Another area worthy of interest is the excitation of comet lines by electron impact.

2.2. GAS JETS

The discovery of CN "jets" in P/Halley by (A'Hearn *et al.*, 1986) —later confirmed by Cosmovici *et al.* (1988) who additionally observed jets of C_2, C_3 and ([O I] + NH$_2$)— was the key work that spread the idea that "distributed sources" could play an important role in comets. Regarding gas jets, one must say that it is hard to tell what they are. In general, a spherically symmetric image is produced *from* the data and then subtracted out of the original image to reveal some *residual* signal. This technique mainly reveals that the coma is not spherically symmetric. In comet Hale-Bopp, the anisotropies are readily visible in the unprocessed images. When the continuum signal anisotropies do not coincide with those of the gas, the expression "gas jet" is usually used, without any justification. As a matter of fact, numerical simulations show that isolated active regions produce an atmosphere that angularly extends over many tens of degrees, something that can hardly be called a jet (see e.g. Keller *et al.*, 1994). The mere existence of numerous active regions at the surface of the nucleus of P/Halley almost prevents the use of the term "gas jets" to describe how gas is angularly distributed near the surface of that comet nucleus. Since only the images obtained by A'Hearn *et al.* (1986) have been the object of a quantitative analysis, it is very difficult to assess the importance of the second source of gas that seems required to explain some observations. Klavetter and A'Hearn (1992) have determined the angular distribution of the CN signal and found that a source function having 2 or 3 maxima at specific ejection angles could fit the data. The CN gas density excesses are still described as gas jets, but are likely to be connected to the transit of active regions in sunlight, as seen in images of the dust distribution near P/Halley's comet surface (see Keller, 1991). The CN and dust coma anisotropies are not spatially correlated and are thus thought to be unrelated.

Few authors have modeled the distribution of gaseous species in the coma with time dependent models. The analysis of the OH and C_2 emissions of comet 2P/Encke by Festou and Barale (1999) seems to be the only one that involves a source of parent molecules that temporally *and* spatially vary (the coma is produced by an active region on a rotating nucleus). In P/Halley, we know that the global activity of the nucleus was highly variable, changing by factors of up to 3 from one day to the next. We also know that this activity was related to the appearance / disappearance of active regions in sunlight. Schleicher *et al.* (1990) successfully modeled the periodic variations of their photometric data using a globally time dependent source of parent molecules. Klavetter and A'Hearn (1994) indicate that

50% of the CN signal in P/Halley must be attributed to a source of parent molecules of unknown nature (molecules or grains) not released at the nucleus. Based on our modeling of the coma, we believe that the observations can be explained by one (or more) active region(s) passing in front of the sun and that invoking a distributed source is not absolutely required (but this remains to be demonstrated and it is not clear whether all source parameters can be retrieved from the data). In the case of comet Hale-Bopp, a very similar situation will be encountered as a powerful active region dominated the nuclear activity (see e.g. Lederer et al., 1999) near perihelion passage, and it was, in particular, responsible for the presence of dust and gas haloes. This latter type of structure and the "jets" observed in P/Halley suggest that the best way to describe them is to model the nucleus output with an angularly and temporally dependent source of nuclear material.

Most pre-Halley interpretations of the spatial distribution of emission by C_3, CH, CN, NH_2, OH and C_2 have required only one parent. A few works only suggest that the C_2 molecule could have more than one precursor molecule but the second parent production rate is never found to dominate the production. For CN, there is almost unanimous agreement between all analyses of the radial profiles and one is forced to conclude that if large molecules or grains contribute to the production of the CN parent, their contribution is small relative to that of the source that comes directly from the nucleus.

One may also consider the problem from another viewpoint by examining the relative production rates of comets. No correlation exists between the dustiness of comets and the CN, C_2, C_3 and NH_2 production rates relative to that of water (85 comets were studied by A'Hearn et al., 1995) and this indicates that solid particles are not important contributors to the production of these species (there are some exceptions to this and we do not know what explains them). This also suggests that if large molecules are always present in comet comae, their abundance has more chances to be correlated with the gas production rates than with those of the dust.

There is still a lot to learn about how molecules leave the nucleus and we must fully understand this aspect of comet aeronomy before we can seriously discuss the nature of other sources of material in the coma.

2.3. THE HALO OF ICY GRAINS

It seems hard to accept the idea that volatiles could reside in the nucleus in the form of ices only and not in the grains. Molecules can be stored in grains in multiple ways and later be released in the coma. It is thus not surprising that icy grains have been searched for in the coma for decades. They have not been found (see Hanner, 1981), even though there are reports of an absorption seen through narrow filters near $3\,\mu$m that could be due to water ice (Hanner, 1984, and references therein). However, it is only with comet Hale-Bopp, observed with modern IR spectrometers from the ground (Davies et al., 1997) and from space by ISO (Lellouch et al., 1998), that more solid evidence was brought about the presence of icy grains in the coma. Theoretical considerations (Hanner, 1981) easily establish that even pure icy grains cannot survive more than a few hundred km from the nucleus in a comet situated at 1 AU from the sun, much less than the usual spatial resolution reached at IR wavelengths. Slightly dirty ices heat up so rapidly that particles would evaporate in minutes or even seconds and would be detectable only when the comet is at least 2–3 AU away from the sun. As a matter of fact, the interpretation of the water line observations obtained near 1 AU never required the assumption of a release of the molecules in an extended region of the coma. As we have seen before, the spatial distribution of the UV emission of OH is almost entirely explainable by a source of water coming directly out of the nucleus. Radio observations of OH are even more interesting in that respect since the velocity of the radicals can be deduced from the observed linewidths. None of the OH radio line observations has required to assume that water was coming out of icy grains. From the work of Hanner (1981) it can be concluded that pure micron-size icy grains are not injected into the coma at 1 AU from the sun and that if icy grains exist in significant amounts in comet comae, since they escape detection, they either have to be large and made of pure water ice or be slightly contaminated by absorbing molecules so that they vaporize rapidly and remain unseen.

The observations of Comet Hale-Bopp at 7 AU pre-perihelion by Davies *et al.* (1997) show very weak absorptions at $\sim 1.5\,\mu$m and $\sim 2.05\,\mu$m. Although the agreement between model and data is only moderate, there is a strong suggestion that water ice has indeed been detected. In the future, the absorption should be searched near 3 μm where this feature should be much stronger. The report by Lellouch *et al.* (1998) is somewhat puzzling. Not only are the ice emissions against a strong continuum spectrum very weak and not well defined but it is not obvious that the reported conclusions are consistent with the direct detection of the water lines by ISO (Crovisier *et al.*, 1997). A cloud of grains containing 2×10^{12} g of water ice within the FOV is consistent with the water production of 3.6×10^{29} molecule s^{-1} reported by Crovisier *et al.* (1997) only if the lifetime of the grains is ~ 2 days and if the grain diameter is 15 μm. Yet, if the two water clouds have similar total water molecule contents, they have very different morphologies, and thus very different column densities. It is not clear that the water lines would be detected if the column density were not very high at the center of the coma, as one would expect from a nuclear source. The two data sets should be interpreted with the same coma model to establish their compatibility. Also, the question of what could drive such large quantities of solid particles would have to be addressed. The detection in P/Halley of solid particles that are rich in C, H, O and N atoms has often been interpreted as an indication that grains similar to those found in the interstellar medium were present in the coma. Such dirty icy grains would necessarily have a rather high absorption coefficient, and their lifetime would be much shorter than the 2 days quoted above.

For the sake of completeness, let us mention the case of the comets that split. The most recent example is comet C/1996 B2 (Hyakutake). Harris *et al.* (1997)

indicate that volatile-rich particles or even icy grains are probably present around the nucleus of the comet. In such a case, the temporary existence of icy grains around the nucleus or in a secondary condensations does not pose any serious theoretical difficulty, except of course for the mechanism that leads to the nucleus breakup.

2.4. GRAIN EVAPORATION AND FRAGMENTATION

Festou et al. (1993b) give a description of the early observations of the Na 5890–96 Å emission that indicate a production from grains is quite possible (see also the fairly complete quotations of Combi et al., 1997, Cremonese et al., 1997, and Wilson et al., 1997). In short, the emission is detected only when the comet is closer than ~ 1.4 AU from the sun, and it ceases when the comet is closer to the sun than ~ 0.5 AU, apparently the signature of temperature-related thresholds. Sodium emission is observed sometimes quite far from the nucleus, even on the sunward side. The long anti-sunward extension is easily explained by the rapid removal of the atoms from their production region by a strong radiation pressure force (the emission rate, the highest of the known comae species, is in the range 3–50 s^{-1}, depending on the heliocentric velocity of the comet). The photoionization rate of Na at 1 AU is $\sim 5.7 \times 10^{-6}$ s^{-1} (average between the theoretical values of Huebner et al., 1992, and Combi et al., 1997); the laboratory value of 1.7×10^{-5} should be ignored, as indicated by the second author), which should prevent the atoms from being seen farther away than $\sim 10^6$ km from the nucleus. However, sodium emission has been observed to extend out to tens of millions of km. A distributed source that replenishes the coma / tail in atoms is thus necessary to explain the tailward observations.

The morphology of the near nucleus Na emission indicates a production at the nucleus or from a short-lived parent (Combi et al., 1997). These latter authors also demonstrate that far from the nucleus, some atoms are likely to be produced by the electron recombination of an ion like NaX^+. Interestingly, this distant source is more powerful than the nuclear source by a factor of 3–5.

Recent observations of Comet Hale-Bopp have allowed us to better understand the nature of two other populations of Na atoms (Cremonese et al., 1997; Wilson et al., 1997). These are a straight and very long Na tail formed of atoms rapidly accelerated by radiation pressure and a Na tail of molecular origin (the emission does not coincide with either the ion or the dust tail). The study of the kinematics of the Na atoms in comet Hale-Bopp by Arpigny et al. (1998) shows that most sodium atoms these authors detected have been released by a nuclear source. However, sunward velocities are not compatible with the deceleration one may expect from radiation pressure and this could imply either a molecular source not related to the grains or collisional transport of the atoms (partial to total cancellation of the action of the radiation pressure force), as suggested by Combi et al. (1997). Emission coincident with the dust tail has also been observed so that there are at least four different populations/sources of Na atoms (nucleus + molecules + ions + grains). Their exact nature and how independent they are is still puzzling. The role of the chemistry in the formation of a molecular species composed of Na has also to be investigated. Last, let us note that the appearance of the Na structures necessarily strongly depends on the observing geometry and on the heliocentric distance of the comet, which probably explains some discrepancies from one report to another. Actually, no observation to date has reported simultaneously on all source mechanisms of Na atoms, making the study of their relationship quite difficult.

The dust-gas coupling region of the coma of comet P/Halley has been investigated by Keller (1991). It appears as a region of strong fragmentation and can be viewed as an extended source of grains and gas. The scale length for the fragmentation is found to be small, of order 20 km, and such a source is not distinguishable in remote observations from the nuclear source because of its coupling with the dominant water molecules. Observations of the spatial distribution of the dust particles at larger distances from the nucleus are not usually interpreted in terms of grain fragmentation but rather as fading particles (e.g. Baum *et al.*, 1992). The problem might be more complex than it seems as, while particles would fragment, the optical properties of the grains and the radiation force acting upon them could vary with distance to the nucleus. The particle trajectories could then differ significantly from those of particles leaving the nucleus and keeping a constant size (the rarely observed tailward striae are a good example that the trajectory of coma grains can be quite elaborate). The question of grain fragmentation is thus far from settled, although the heuristic fragmentation model of Combi (1994) seems quite successful at reproducing the gross dust coma structures (that model cannot address the question of molecule production from grains). Once more, one should emphasize the fact that separating spatial and temporal parameters from a series of images taken at nearly the same time is quite difficult. Comet Hyakutake has ejected many chunks of its nucleus after its passage near the Earth in March 1996 and the presence of volatile-rich particles in the coma is in this case not a surprise.

2.5. LARGE MOLECULES

Since the CN and dust structures that have been found in P/Halley were not spatially correlated, it has been suggested (A'Hearn *et al.*, 1986) that the CHON particles detected *in situ* by Kissel *et al.* (1986) were the unseen source of extra CN radicals and this was supported by the observation that small grains were more abundant far from the nucleus of P/Halley (Vaisberg *et al.*, 1987). Ground-based images of the C_2 emission have been analyzed by Combi and Fink (1997) in terms of two sources of radicals to explain the abnormal variation of the C_2 parent lifetime with heliocentric distance. The results are inconclusive since there are multiple ways to produce a flat emission profile in the inner coma, as observed in the data. In order to improve on this type of study, not only high S/N digital data must be obtained at high spatial resolution, if possible simultaneously on 2-3 different emissions, but a wide interval of heliocentric distances must be tested.

Formaldehyde is ubiquitous in the interstellar medium and is consequently expected to be present in comets, the only solar system objects in which the original solar nebula matter has not been completely processed. It can polymerize and account for some of the large molecules. Indeed, polymerized forms of molecules have been tentatively identified in P/Halley by Korth et al. (1986), Huebner (1987), and Mitchell et al. (1987) from the observation of peaks at masses 45, 61, 75, 91, 105 amu in the PICCA spectra. The peaks are separated by 14 or 16 mass units and suggest polymers containing N and O atoms. In such polymers, the O atoms may be replaced by S atoms to form S-POMs. The H atoms can also be replaced by radicals like NH_2, NH, CN, OH, etc. The bonds lifetime is 10^3 to 10^4 s (Huebner, personal communication), which offers an interesting solution for the origin of the many cometary radicals for which no parent species have yet been found. However, Mitchell et al. (1989) offer an alternate explanation on the basis of RPA2-PICCA resolved mass spectra that show a 2 amu alternation in the 35–50 amu mass range for the dominant ions. They attribute this to a blend of compounds connected to hydrocarbons having a C/O ratio ~ 1.2–1.4.

The claim by Moreels et al. (1994) that a polycyclic aromatic was fluorescing in the coma of P/Halley in the near-UV region is hard to evaluate. The instrument was not calibrated before the flight, even in relative value, and the continuum removal procedure leads to a very uncertain residual. The features seen around 350 nm are at the noise level while the $\sim$ 370 nm feature could be real (but could due to C_3 as well as a PAH).

2.6. THE CASE OF CO AND RELATED MOLECULES

The *in situ* observations of CO (Eberhardt et al., 1987) and H_2CO (Meier et al., 1993) in P/Halley require some CO and H_2CO production in the coma. The production region extends up to at least a few 10^4 km from the nucleus of P/Halley and the nuclear source is found to be about half the total source of CO in the coma. The connection that exists between these two molecules is weak: the destruction time scale of H_2CO and the production time scale of CO are similar, but the production rates do not agree. H_2CO is present in P/Halley at the 4% percent level, with little production in the nucleus (Meier et al., 1993), whereas the total CO production is of order 15–20% that of water, with a nuclear source of $\sim$ 7% (different figures are being quoted by different authors, depending on the model used, the origin of the data and the assumptions made to derive the production rates). Reviews of the problem are found in Krankowsky (1991) and Meier et al. (1993). Consequently, formaldehyde cannot be the distributed source of CO.

Methanol is potentially a candidate for the source of H_2CO but the long scale length of the features attributed to CH_3OH (Combes et al., 1988) and the long theoretical lifetime of methanol suggest that this molecule is not suited for producing formaldehyde. In addition, its main dissociation path does not lead to the formation of the latter (Bockelée-Morvan et al., 1994). The absence of a species *requiring a distributed source* and *produced as abundantly* as CO favors a direct release from grains as the distributed source of CO. These characteristics may explain the observation by DiSanti et al. (1999) of a turn on / turn off of the extended source rather close to the sun. This could also explain the appearance of the CO radio lines, as reported in the next section. The interpretation of the various gas production curves shown by Bockelée-Morvan and Rickman (1999) and spectral line shape variations with heliocentric distance should soon tell us the last word on the origin of CO in comets. The last important point to raise concerning CO is that everything indicates that it is the most abundant species in comets located far from the sun and that the CO line shape suggests a dominant release at the nucleus. If CO were coming primarily from grains, the driver of the grains would have to be found. CO_2 could be the culprit, as this molecule is quite abundant in comets (Feldman et al., 1997). (Note that CO_2 has too long a lifetime to be an important contributor to the CO production.) However, at the present time, given its higher volatility, CO is an excellent candidate for being the first abundant species to leave the nucleus.

3. A Brief Discussion

Most remote sensing works that report on distributed sources rely on the departure of the observed emissions from what would be expected if species were produced from the photolysis of molecules isotropically leaving the nucleus. Each time a departure from the canonical coma spherical symmetry with a 1-2 step photodissociation path is observed, a distribution source is invoked. The review of existing observations we just performed shows that there exist multiple ways to produce anisotropic patterns, and most of them do not require any distributed source. Yet, we have only considered the most likely and important mechanisms at play. For example, in the inner coma, the velocity profile of the parent molecules increases outward to reach outflow velocities in excess of 1 km s^{-1} at 1 AU from the sun (Lämmerzahl et al., 1987). This is associated with a spatial distribution of the parent species steeper than the canonical $1/d^2$ law predicted from spherical expansion. In a comet like Hale-Bopp, the importance of photolytic heating in the first $5\text{--}8 \times 10^4$ km from the nucleus is enormous. Even daughter species velocities should be affected, and this is indeed reflected in numerous radio observations that show that secondary species achieve the outflow velocity of the nuclear molecules in all of the above-mentioned region.

Many inner coma species column density profiles are thought to be too flat to be produced through a single photolytic step. This can result from the existence of multiple photolysed parent molecules or an evaporation of a parent from grains. Among the other mechanisms that can produce the same result, let us mention chemical reactions, prompt emission, collisional quenching of lines arising from metastable states, the Greenstein effect and spatially variable excitation rates (e.g. non-fluorescence equilibrium). Page limit constraints do not allow us to discuss all these points in detail.

As we already mentioned, observations of comet P/Halley have shown us that the source function at the nucleus should be much more complex than is usually assumed. An active region on a rotating nucleus can produce very different coma structures, depending on its location over the surface and the location of the observer relative to the Sun and the comet (studies of the dust coma by Zdenek Sekanina can be taken as a good example of the wide range of possibilities that exist). When coma anisotropies are observed, the very first modeling step should be to adopt a time-dependent source of nucleus material. Adding an angular dependency to the gas output function is the next step to take. It is only if those first two tests have failed that one should consider the possible presence of a distributed source.

In situ observations allow one to access coma parameters more directly than remote observations do (no inversion of the signal required). They have their own limitations, though. The most important is probably that a snapshot at one object is not the ideal means of separating temporal and spatial variabilities. Despite these circumstances, it seems difficult not to accept that CO comes from two sources of similar strengths; one originating at the nucleus while the second could be a release / vaporization from grains as well as the dissociation of a heavy molecule. If the first assumption is valid, it is surprising that only this species exhibits such a behavior. Some observations of comet Hale-Bopp show that molecules usually believed to be released at the nucleus (like HCN and OCS) are indeed also produced in the coma. Since much data still remain to be published, and since the question would require examining the situation of each species in turn, we elect not to discuss this point here.

The "gas jet" source meets many difficulties. We have mentioned an alternate model that needs to be tested. The source of CN radicals in P/Halley was estimated by A'Hearn and collaborators to account for 10–50% of the total CN production: the upper limit appears incompatible with the round structure of the CN isophotes usually seen in comets. (This remark favors large molecules vs. grains for an eventual additional source of CN since we expect them to be redistributed by collisions with the dominant coma species.) As for the other gas jets, the fluxes measured by Cosmovici *et al.* (1988) correspond to total radical masses that greatly surpass the amount of available CHON particles determined by McDonnell *et al.* (1987). Perhaps the image processing techniques used to reveal the presence of jets have prevented an accurate derivation of the abundances. It is still unclear whether the time scales derived for the various observed distributed sources (CO, CN, H_2CO, C_2 ...) are mutually compatible, and, as noted by Krankowsky (1991), the mechanism by which radicals are released in gas jets has still to be clarified.

Among the recent observations that we have not reviewed and that are of utmost importance for the problem we are discussing are the radio observations of molecules (H_2CO, CH_3OH, OCS, CO ...), and that are stable enough to reside in the nucleus or in grains. Many observing programs have been conducted but not a single full analysis of the data is available yet. One of the puzzling results in these observations is that the CO line profile when the comet is far from the sun is strongly displaced sunward, which can be entirely explained by a unique source at the nucleus. Later, when the comet approaches the Sun, the shape of the line evolves towards a more complex structure with peaks that suggest multiple active regions. Although collisional processes could result in similar line profiles because of the large water production of Comet Hale-Bopp, profiles are not all alike and this is perhaps the signature of the presence of multiple sources of the observed species. A recent work by DiSanti *et al.* (1999) also reveals the presence of two CO sources in Comet Hale-Bopp when this latter is near the sun. The strength of the demonstration resides in the comparison with the water molecule profile, this latter being quite different from that of CO. (We note in passing that this rules out the production of water in a halo of icy grains.) The CO production scale is similar to the destruction time scale of the H_2CO molecule but the production rates do not agree by one order of magnitude. This fact also favors a direct release of the CO molecules from grains. Another important result of that work is that the second source would turn on near ~ 1.5 AU, which should be reflected in the radio line profiles.

4. Conclusion

We have stated a few times that models are, in general, too simple, which weakens the conclusions reached in many works. In general, the analyses of observations related to distributed sources (1) do not consider all the possible explanations, and (2) do not establish the uniqueness of the proposed solutions. The main reason that we can identify for such a situation is that most observations are insufficient to arrive at a unique conclusion. Our knowledge of the coma of comets is mature enough to build a new generation of coma models that includes many phenomena that are usually ignored. Numerous observations should be combined at the interpretation level in order to truly constrain the models. There is clearly a future for new investigations of the 2D structure of monochromatic comet emissions. However, for new advances to be made, the data have to be of very high quality, multiple species should be studied in parallel and the dependency on heliocentric distance of the observed phenomena should be monitored. Coma models like those of Combi and his collaborators (see e.g. Combi *et al.*, 1999) combine the hydrodynamics formalism to describe the inner coma and the kinematics approach for the intermediate and outer comae, and their use is recommended, the simpler model of Festou and Barale (1999) goes also in the right direction. One way to progress in this direction is perhaps to make existing models of excitation, production, spatial distribution, ... etc available to all instead of making simplified analyses because such models are hard to develop and validate. The recent advances made in electronic communication could allow us to go in the right direction.

Acknowledgements

I would like to thank an anonymous referee for pointing out a few imprecise or inaccurate statements contained in the original version of this manuscript.

References

A'Hearn, M.F., Hoban, S., Birch, P.V. *et al.*: 1986, *Nature* **324**, pp. 649–651.
A'Hean, M.F., Millis, R.L, Schleicher, D.G. *et al.*: 1995, *Icarus*, **118**, pp. 223–270.
Aikin, A.C.: 1974, *ApJ* **193**, pp. 263–264.
Arpigny, C.A., Rauer, H., Manfroid, J. *et al.*: 1998, *A&A* **334**, pp. L55–L56.
Baum, W.A., Kreidl, T.J., and Schleicher, D.G.: 1992, *AJ* **104**, pp. 1216–1225.
Bockelée-Morvan, D. and Rickman, H.: 1999, to appear in *Earth, Moon Planets*.
Bockelée-Morvan, D., Crovisier, J., Colom, P., and Despois, D.: 1994, *A&A* **287**, pp. 647–665.
Budzien, S.A., Festou, M.C., and Feldman, P.D.: 1994, *Icarus*, **95**, pp. 65–72.
Combes, M., Moroz, V.I., Crovisier, J. *et al.*: 1988, *Icarus*, **76**, pp. 404–436.
Combi, M.F.: 1994, *AJ*, **108**, pp. 304–312.
Combi, M.R., and Fink, U.: 1997, *ApJ* **484**, pp. 879-890.
Combi, M.R., DiSanti, M.A., and Fink, U.: 1997, *Icarus* **130**, pp. 336–354.
Combi, M.R., Cochran, A.L., Cochran, W.D. *et al.*: 1999, *ApJ*, **512**, pp. 961–968.
Cosmovici, C.B., Schwarz, G., Ip, W.-H. and Mack, P.: 1988, *Nature*, **332**, pp. 705–709.
Cremonese, G., Boehnhardt, H., Crovisier, J. *et al.*: 1997, *ApJ* **490**, pp. L199–L202.
Crovisier, J., Leech, K., Bockelée-Morvan *et al.*: 1997, *Science* **275**, pp. 1904–1907.
Davies, J.K., Roush, T.L., Cruishank, D.P. *et al.*: 1997, *Icarus* **127**, pp. 238–245.
DiSanti, M.A., Mumma, M.J., Dello Russo, N. *et al.*: 1999, *Nature*, **399**, pp. 662–665.
Eberhardt, P., Krankowsky, D., Schulte, W. *et al.*: 1987, *A&A* **187**, pp. 481–484.
Feldman, P.D., Budzien, S.A., Festou, M.C. *et al.*: 1992, *Icarus*, **95**, pp. 65–72.
Feldman, P.D., Festou, M.C., Tozzi, G.P., and Weaver, H.A.: 1997, *ApJ* **475**, 829–834.
Festou, M.C., Tozzi, G.P., Smaldone, L.A. *et al.*: 1990, *A&A*, **227**, pp. 609–618.
Festou, M.C., Rickman, H. and West, R.: 1993a, *A&Ap Rev.* **4**, pp. 363–437.
Festou, M.C., Rickman, H. and West, R.: 1993b, *A&Ap Rev.* **5**, pp. 37–163.
Festou, M.C. and Barale, O.: 1999, *A&A*, submitted.
Hanner, M.S.: 1981, *Icarus* **47**, pp. 342–350.
Hanner, M.S.: 1981, *ApJ* **277**, pp. L75–L76.
Harris, W.M., Combi, M.R., Honeycutt, R.K. *et al.*: 1997, *Science*, **277**, pp. 676–681.
Huebner, W.F.: 1987, *Science* **237**, pp. 628–630.
Huebner, W.F., Boice, D.C., Schmidt, H.U. and Wegmann, R.: 1991, in *Comets in the Post-Halley Era*, R.L. Newburn, Jr. *et al.* (eds.), Kluwer Acad. Publ. **2**, pp. 907–936.
Huebner, W.F., Keady, J.J., and Lyon, S.P.: 1992, *Astrophys. Space Sc.* **195**, pp. 1–289.
Keller, H.U.: 1991, in *Physics and Chemistry of Comets*, Huebner, W.F. (Ed.), Springer-Verlag, pp. 13–68.
Keller, H.U., Knollenberg, J., and Markievicz, W.J.: 1994, *Plan. Sp. Sc.* **42**, pp. 367–382.
Kissel, J., Brownlee, D E., Büchler, K. *et al.*: 1986, *Nature* **321**, pp. 336–337.
Klavetter, J.J., and A'Hearn, M.F.: 1992, *Icarus* **95**, pp. 73–85.
Klavetter, J.J., and A'Hearn, M.F.: 1994, *Icarus* **107**, pp. 322–334.
Korth, A., Richter, A.K., Loidl, A. *et al.*: 1986, **321**, pp. 335–336.
Krankowsky, D.: 1991, in *Comets in the post-Halley Era*, Newburn, R.L. *et al.* (eds.), Kluwer Acad. Publ., vol. 2, pp. 855–877.
Lämmerzahl, P., Krankowsky, R., Hodges, R.R. *et al.*: 1987, *A&A* **187**, pp. 169–173.

Lederer, S.M., Campins, H., Osip, D.J., and Schleicher, D.G.: 1999, *PaSS*, in press.
Lellouch, E., Crovisier , J., Lim, T. *et al.*: 1998, *A&A* **339**, pp. L9–L12.
McDonnell, J.A.M., Alexander, W.M., Burton, W.M. *et al.*: 1987, *A&A* **187**, pp. 719–741.
Meier, R., Eberhardt, P., Krankowsky, D., and Hodges, R.R.: 1993, *A&A* **277**, pp. 677–690.
Mitchell, D.L., Lin, R.P., Anderson, K.A. *et al.*: 1987, *Science*, **237**, pp. 626–628.
Mitchell, D.L., Lin, R.P., Anderson, K.A. *et al.*: 1989, *Adv. Space Res.* **9** (2), pp. 35–39.
Moreels, G., Clairemidi, J., Hermine, P. *et al.*: 1994, *A&A* **282**, pp. 643–656.
Schmidt, H.U., Wegman, R., Huebner, W.F. and Boice, D.C.: 1988, *Comp. Phys. Comm.*, **49**, pp. 17–49.
Schleicher, D.G., Millis, R.L., Thompson, D. T. *et al.*: 1990, *AJ* **100**, pp. 896–912.
Vaisberg, O.L., Smirnov, V., Omelchenko, A. *et al.*: 1987, *A&A* **187**, pp. 753–760.
Weaver, H.A., Feldman, P.D., A'Hearn, M.F. *et al.*: 1999, *Icarus* , in press.
Wilson, J.K., Baumgartner, J., and Mendillo, M.: 1997, *GJR Lett.* **25**, 225.

Address for correspondence: Michel C. Festou, Observatoire Midi-Pyrénées, 14 avenue E. Belin, F31400 Toulouse, France (festou@obs-mip.fr)

THE DISTRIBUTED CO IN COMET HALLEY

AIGEN LI[1,2] and J. MAYO GREENBERG[2]
1. *Beijing Astronomical Observatory, Beijing 100012, P.R. China*
2. *Raymond and Beverly Sackler Laboratory for Astrophysics at Leiden Observatory, Sterrewacht Leiden, Postbus 9504, 2300 RA Leiden, The Netherlands*

Abstract. Examination of the spatial distribution of CO intensity of Comet Halley indicates that a large fraction of CO originates from the refractory organic component in the coma, rather than directly from the volatiles in the nucleus. Based on the fluffy aggregate interstellar dust comet model, we have estimated the upper limits of the total amount of CO provided by coma dust. The implications from the comparison of the predicted results with the observed value have been discussed.

Keywords: ISM: dust - Comets: general - Comets: Halley

1. Introduction

The in situ radial profile of the CO intensity, detected by the Neutral Mass Spectrometer (NMS) on the Giotto spacecraft on approach to Comet Halley, multiplied by the square of the radial distance from the nucleus (R), apparently shows an almost linear increase from 10^3 km, the closest distance where the NMS detection was available, up to ~ 13500 km, a distance where the intensity times R^2 reaches its maximum. Thereafter it becomes almost constant (Eberhardt *et al.*, 1987), a fact which was attributed to the presence of an extended source for CO in the coma. Eberhardt *et al.* (1987) suggested that the nucleus rate of production of CO as a parent molecule in Comet Halley was, relative to H_2O, $Q(CO)/Q(H_2O) \leq 0.07$ (a more recent investigation by Greenberg and Li (1998a) estimated $Q(CO)/Q(H_2O) \approx 0.06$) while the total CO production rate was about $Q(CO)/Q(H_2O) \approx 0.15$. Thus the extended source must account for $Q'(CO)/Q(H_2O) \approx 0.09$. Eberhardt *et al.* (1987) speculated that the most likely extended source of the CO was some molecules evaporated from the organic component of the comet dust.

The purpose of this paper is to quantitatively investigate the extended CO source resulting from the relatively volatile fraction of the C=O containing complex organic refractory molecules in the cometary dust. From this one may gain insights on the morphological structure and chemical composition of cometary dust and nuclei.

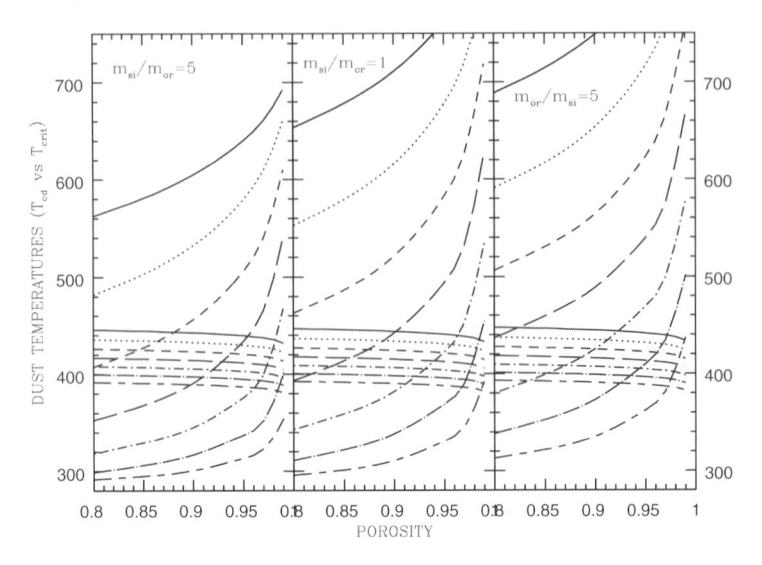

Figure 1. Comparison of the calculated comet dust temperatures (T_{cd}, thick lines) with the estimated critical temperatures (T_{crit}, thin lines) at 1 AU as a function of grain masses and porosities (only for $m_{or}/m_{si} = 1/5$ (left), $m_{or}/m_{si} = 1$ (middle), and $m_{or}/m_{si} = 5$ (right)). The lines in each set correspond to (from top downward for T_{cd}, and upward for T_{crit}) 10^{-12} g, 10^{-11} g, 10^{-10} g, 10^{-9} g, 10^{-8} g, 10^{-7} g, 10^{-6} g.

2. Organic Mantles as a Potential Source for CO

Before the interstellar dust aggregates into cometesimals it has an outer mantle of volatile ices (accreted in the dense molecular cloud phase) with H_2O and CO being the dominant components (Greenberg, 1982; Greenberg and Li, 1999). These outer mantle volatiles constitute the source of the direct nucleus emissions of H_2O and CO and evaporate rapidly after cometary dust comes off the nucleus. The non-volatile organic mantles contain C=O bearing complex molecules and provide the distributed CO in the coma as the dust is heated.

In this work, the chemical characteristics of the laboratory organics – the residues of the photolysis of $H_2O : CO : NH_3 = 5:5:1$ mixtures at 10 K (Briggs et al., 1992) are taken as a first approximation to interstellar organic mantles. Adding up all the possible CO groups in the GCMS analyzed molecules gives a fraction by mass of ~ 0.16 of that component of the organic refractory.

The temperature of the aggregated comet dust is a critical parameter for the determination of the CO emission rate. It can be calculated as a function of dust size, porosity (fluffiness), and the organic fraction (m_{si}/m_{or}) on the basis of the dust energy balance between absorption and emission (see Greenberg and Li, 1998a, 1998b for details). We have considered a wide range of organic mantle thickness, ranging from $m_{or}/m_{si}=1/5$ to $m_{or}/m_{si}=5$. Note that the most likely m_{or}/m_{si} in the nucleus may be ≈ 1, which is consistent with the aggregated interstellar dust model (Greenberg, 1998; Li and Greenberg, 1998) and the *in situ* measurements of Halley (Krueger and Kissel, 1987). For illustrative purpose, in Fig. 1 we present the calculated dust temperatures only for $m_{or}/m_{si}=1/5$, $m_{or}/m_{si}=1$, and $m_{or}/m_{si}=5$. Note that the thinner the organic mantles, the colder the grains are (Greenberg and Hage, 1990).

3. The Total Amount of CO from Dust and its Implications

In order to estimate the total amount of CO which can be provided by comet dust it is necessary to determine the maximum mass (m_{max}^{CO}) of dust which still can evaporate its volatile organic refractory mantle within the scale length. Thus we need to calculate the "critical temperature" of dust with a certain mass and porosity – the lowest temperature at which the dust is still able to do such a job. For convenience, we define a notation m_{dust}^{CO} for the integrated total dust mass up to m_{max}^{CO}, the maximum mass of dust which still can evaporate its volatile organic refractory mantle within the scale length.

The critical temperatures are dependent on the grain size, porosity and its evaporation rate. They are calculated and also presented in Fig. 1 as a function of grain porosity for a variety of grain masses (Greenberg and Li (1998a) for further details).

The total amount of CO contributed by coma dust can be estimated from

$$Q'(CO)/Q(H_2O) = \frac{m_{CO}}{(m_{or})_V} \frac{(m_{or})_V}{m_{or}} \frac{m_{or}}{m_{dust}} \frac{m_{dust}}{m_{gas}} \frac{m_{gas}}{m_{H_2O}}, \quad (1)$$

where $m_{CO}/(m_{or})_V$ is the mass fraction of CO in the volatile part of the organic mantle (≈ 0.16, see Sect. 2); $(m_{or})_V/m_{or}$ is the mass ratio of the volatile organic component to the full whole organic mantle ($\approx 1/2$, see Greenberg and Li, 1998a, 1998b); m_{or}/m_{dust} is the mass fraction of the organic mantle in an individual dust particle (most likely 1/2, Krueger and Kissel 1987, Greenberg, 1998); m_{dust}/m_{gas} is the mass ratio of dust to gas, approximated by the dust to gas production rate; m_{gas}/m_{H_2O} is the mass ratio of gas to water vapor (≈ 1.4, Greenberg (1998)). The dust to gas production rate m_{dust}/m_{gas} (note here m_{dust} is actually m_{dust}^{CO}, the integrated total dust mass up to m_{max}^{CO}; while m_{max}^{CO} can be provided by the critical temperature constraint). For a given dust mass, m_{dust}/m_{gas} can be inferred from the McDonnell size distribution (see Fig. 14 of McDonnell *et al.*, 1991). Then we can obtain $Q'(CO)/Q(H_2O)$. Fig. 2 presents the maximum CO production rate predicted from the comet dust model as a function of porosity. It can be seen that the predicted absolute values are well below the detected number ($Q'(CO)/Q(H_2O) \approx 0.15$). Even for $m_{max}^{CO} = 10^{-6}$ g (for which the particles are certainly too large to evaporate their organic mantle even with a porosity of P = 0.99), $Q'(CO)/Q(H_2O)$ is only $\approx 1.4\%$ ($m_{dust}/m_{gas} \approx 0.25$, McDonnell *et al.*, 1991). In order to fully account for the observed $Q'(CO)/Q(H_2O) \approx 0.09$, a dust to gas production rate

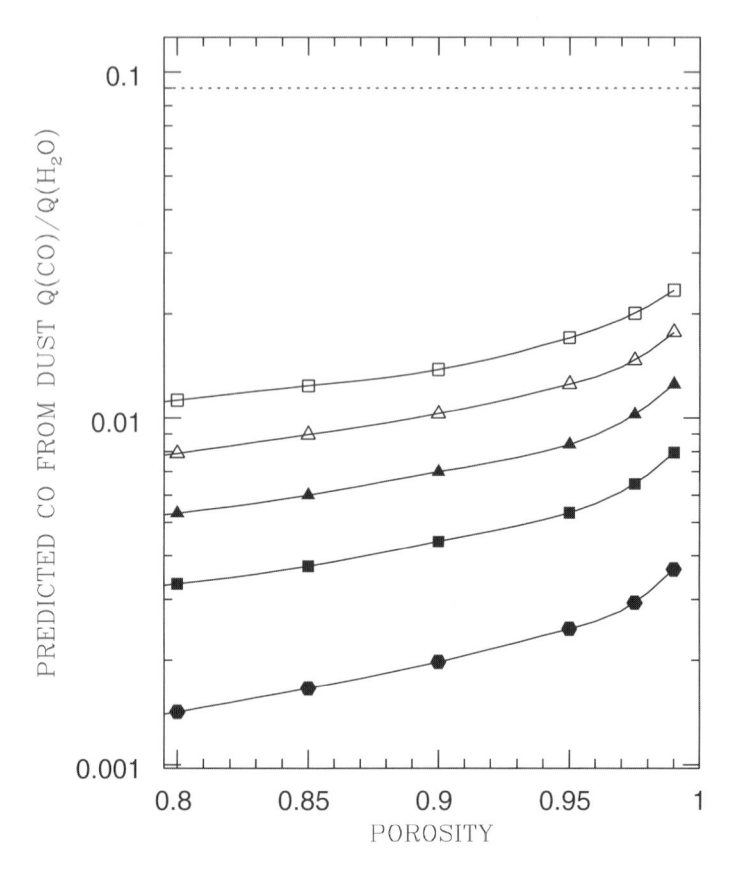

Figure 2. The maximum CO production rate provided by comet dust as a function of grain porosity. Filled hexagons – $m_{or}/m_{si} = 1/5$; filled squares – $m_{or}/m_{si} = 1/2$; filled triangles – $m_{or}/m_{si} = 1$; open triangles – $m_{or}/m_{si} = 2$; open squares – $m_{or}/m_{si} = 5$. The dotted line gives the observed abundance of distributed CO.

as high as 1.6 ($m_{dust}^{CO}/m_{gas} \approx 1.6$) is required. According to the McDonnell size distribution, $m_{dust}^{CO}/m_{gas} = 1.6$ corresponds to a maximum dust size (mass) of $m_{max}^{CO} \approx 1.9 \times 10^{-2}$ g, in other words, all dust grains up to $m = 1.9 \times 10^{-2}$ g should be heated sufficiently to evaporate all of their available CO bearing organics. However, it is obvious that such giants ($m \leq 1.9 \times 10^{-2}$ g) will not be so hot!

The above discussions are limited to $m_{or}/m_{si} = 1$. In the case of thinner organic mantles, as expected and also illustrated in Fig. 2 for $m_{or}/m_{si} = 1/2$ and $m_{or}/m_{si} = 1/5$, the maximum CO production rates are much less than those for $m_{or}/m_{si} = 1$. On the other hand, even if grains have an unrealistically thick mantle as high as $m_{or}/m_{si} = 5$, the maximum CO production rate is still far less than the observed one; e.g., grains with $P = 0.99$ and $m_{or}/m_{si} = 5$ can only contribute $Q'(CO)/Q(H_2O) \approx 0.024$. Results for $m_{or}/m_{si} = 2$ and $m_{or}/m_{si} = 5$ are also presented in Fig. 2.

4. Discussion

It seems that the dust explanation encounters a severe problem in fully accounting for the large number of CO molecules. Other materials such as H_2CO, and C_3O_2, or POM (polyoxymethylene, or polyformaldehyde), CH_3OH, CO_2 etc. (see Greenberg and Li (1998a) and references therein) have been ruled out as a possible source for the distributed CO (Greenberg and Li, 1998a). However, it is true that if the organic mantles were pure POM there would be a large enough CO source *if* all of it could be evaporated and photodissociated. The problem with this is that polyformaldehyde is very weakly absorbing in the visual (as evidenced by its being a white powder) so that the dust would be little warmer than if it were pure silicates and therefore be too cool to evaporate the polyformaldehyde in sufficient quantities to provide its CO fraction as a distributed component.

A likely solution to this problem may be that the dust to gas production rate (m_{dust}/m_{gas}) was underestimated in McDonnell *et al.* (1991) by neglecting efficient dust fragmentation and sublimation in the innermost coma (Greenberg and Li, 1998a). An alternative solution lies in the observed CO abundance itself. The extended CO abundance could have been overestimated due to the crossing of dust jets, the time variation of the comet nucleus activity, and the anisotropic outgassing nature of the nucleus (see e.g. Crifo and Rodionov, 1999).

5. Conclusion

We have quantitatively investigated the extended CO source in Comet Halley. The nonvolatile organic component of cometary dust contains C=O bearing complex molecules and provides the distributed CO in the coma as the dust is heated sufficiently to evaporate. Based on the fluffy aggregate interstellar dust comet model, the upper limits of the total amount of CO provided by dust can be approximately determined as a function of porosity, assuming that the laboratory organics are representative of comet dust organics in terms of the mass fraction containing CO groups. We found that the maximum CO production rate predicted by the comet dust model given the dust/gas observational constraint of the McDonnell comet dust size distribution is an order of magnitude less than the observed abundance of distributed CO. Possible solutions such as the underestimation of the dust to gas ratio, and/or the overestimation of the distributed CO production rate, have been discussed.

Acknowledgements

This work was supported in part by NASA grant NGR 33-018-148 and by a grant from the Netherlands Organization for Space Research (SRON). We thank

Dr. J. Crovisier for useful suggestions. One of us (AL) wishes to thank Leiden University for an AIO fellowship and the World Laboratory for a scholarship. AL also thanks the Nederlandse organisatie voor wetenschappelijk onderzoek (NWO) and the National Science Foundation of China for financial support.

References

Briggs, R., Ertem, G., Ferris, J.P., Greenberg, J.M., *et al.*: 1992, *Origins of Life and Evolution of the Biosphere* **22**, 287.
Crifo, J.F., and Rodionov, A.V.: 1999, *Icarus* **138**, 85.
Eberhardt, P., Krankowsky, D., Schulte, W., *et al.*: 1987, *Astron. Astrophys.* **187**, 481.
Greenberg, J.M.: 1982, in *Comets* (edited by L.L. Wilkening), p. 131. University of Arizona Press, Tuscon.
Greenberg, J.M. and Hage, J.I.: 1990, *Astrophys. J.* **361**, 260.
Greenberg, J.M.: 1998, *Astron. Astrophys.* **330**, 375.
Greenberg, J.M. and Li, A.: 1998a, *Astron. Astrophys.* **332**, 374.
Greenberg, J.M. and Li, A.: 1998b, in *Solar System Ices* (edited by B. Schmitt, C. de Bergh, and M. Festou), Chapter 13, p. 337. Kluwer, Dordrecht.
Greenberg, J.M. and Li, A.: 1999, *Space Sci. Rev.*, this volume.
Krueger, F.R. and Kissel, J.: 1987, *Naturwissenschaften* **74**, 312.
Li, A. and Greenberg, J.M.: 1997, *Astron. Astrophys.* **323**, 566.
McDonnell, J.A.M., Lamy, P.L. and Pankiewicz, G.S.: 1991, in *Comets in the Post-Halley Era* (edited by R.L. Newburn, M. Neugebauer, and J. Rahe), p. 1043. Kluwer, Dordrecht.

Address for correspondence: Aigen Li, agli@astro.Princeton.edu

H_2O-ACTIVITY OF COMET HALE-BOPP

E. KÜHRT

Deutsches Zentrum für Luft- und Raumfahrt, Institut für Weltraumsensorik und Planetenerkundung, Berlin, Germany

Abstract. Due to the outstanding brightness of Comet Hale-Bopp measurements of water production rates were possible over a wide range of heliocentric distances (up to 5 AU). A variety of observing techniques have been used, including radio observations, IR- and UV-measurements. The H_2O-production of a comet is closely connected with the energy balance and the composition of its surface. By comparing measured and calculated rates it is possible to derive properties of the nucleus. The results of this study demonstrate the importance of seasonal effects and show that a low thermal conductivity enhances the water production rate. The observations can be matched with a relatively low, lunar-like thermal conductivity. A lower size limit for the nucleus of Hale-Bopp is derived.

1. Introduction

Direct investigations of the physical and chemical properties of a cometary nucleus require a lander mission as planned with Rosetta and Deep Space 4. However, ground-based observations of the coma together with modelling efforts can also give interesting indications of the nature of comets. The outstanding brightness of Comet Hale-Bopp enabled observers to measure production rates of several volatiles over a wide range of heliocentric distances and to establish an exceptional data base for the analysis of cometary activity and physical properties (for a review see Bockelée-Morvan and Rickman, 1998). This paper investigates the water production of Hale-Bopp. The analysis of more volatile species like CO, CO_2, and others is more complex because they are obviously released below the surface and, therefore, their rates also depend on the orbital history of the comet and on micro-structural properties of the nucleus (Prialnik *et al.*, 1993). Only water molecules probably come directly from the surface (Benkhoff and Huebner, 1995). The production rates of this species offer the simplest way to obtain information about the nucleus.

A one-dimensional thermal model has been applied to calculate the energy balance on Hale-Bopp's surface. The H_2O-activity has been calculated as a function of heliocentric distance by an integration of the local flux over the entire surface of the comet. The influence of heat conduction and of the nucleus' rotational properties is discussed and the results are compared with experimental data published by different groups.

Weaver *et al.* (1997, 1998) analyzed IUE and HST spectra and deduced the water production from OH-data. Crovisier *et al.* (1997, 1998) detected H_2O-molecules directly with the ISO-PHT-S at 2.7 μm. Colom *et al.* (1998) obtained spectra at 18 cm from ground-based radio telescopes and detected OH. Unfortunately, the data reduction is not straightforward but needs additional model assumptions. Colom *et al.* apply the Haser model to determine the water production rates. Because of the high coma density of Hale-Bopp at small heliocentric distances and the resulting extended collisional zone quenching effects must be taken into account. Several uncertainties in models describing collisional quenching can cause a considerable error of the water production rates around perihelion. Furthermore, the Haser model assumes constant molecule velocities but different physical processes can cause deviations from this approximation at all heliocentric distances. The resulting uncertainties restrict the interpretation of coma data.

2. Thermal Model

A one-dimensional thermal model has been applied to calculate the energy transport on Hale-Bopp's surface at 18 latitudes over a complete diurnal cycle. The balance of solar energy input, re-radiation, latent heat of sublimation and heat conduction provides the upper boundary condition of the problem. The strong obliquity of the spin axis, which causes pronounced seasonal effects, and the rotation of the nucleus are included. Unfortunately, there is no observational indication as to the real shape and distribution of activity of Hale-Bopp. Hence the crude assumptions are made that the nucleus is spherical and that the activity is randomly distributed over the surface. Table I summarizes the parameters used in the model.

The investigated infrared emissivities cover the range typical for solar system bodies. The analyzed heat conductivities include low lunar-like values and high values representative for monolithic ice. Figure 1 gives a sketch of the orbital parameters. Because of the strong obliquity of the spin axis there are pronounced seasons on Hale-Bopp. The sketch shows where they begin on the northern hemisphere of the comet.

The heat transport problem and the activity are defined by the set of equations (1) to (5) with the temperature T, the solar constant C_s the Stefan-Boltzmann constant σ, the thermal velocity of the water molecules v_{th}, the cometocentric latitude φ and longitude L. The heliocentric distance, R_h, is obtained from the Kepler equation and the local solar zenith angle, θ, from simple geometric considerations.

Heat diffusion equation:

$$\rho c \frac{\partial T}{\partial t} = K \frac{\partial^2 T}{\partial x^2} \tag{1}$$

Upper boundary condition:

$$\frac{C_S(1-A)}{R_h^2} \max(0, \cos\theta) = \varepsilon \sigma T^4 - K \frac{dT}{dx} + H F_{gas} \tag{2}$$

TABLE I

From measurements:		
Spin period	$P = 11.5$ h	(Jorda et al., 1998)
Obliquity of the spin axis	$i = 86°$	(Jorda et al., 1998)
Angle between ascending node and perihelion	$\Theta = 65°$	(Jorda et al., 1998)
Latent heat of water sublimation	$H = 2660$ kJ kg^{-1}	
Unknown but not very critical for the given problem:		
Density	$\rho = 700$ kg m^{-3}	
Specific heat	$c = 1400$ J kg^{-1} K^{-1}	
Bond albedo	$A = 0.04$	
Unknown and critical:		
Thermal conductivity	$K = 0.001...1$ W K^{-1}m^{-1}	
Emissivity	$\varepsilon = 0.9...1$	
Fit parameter:		
Potentially active surface area	$A_S = 4\pi f R^2$, where R is the radius of the nucleus and f is the fraction of active areas compared to the whole surface	

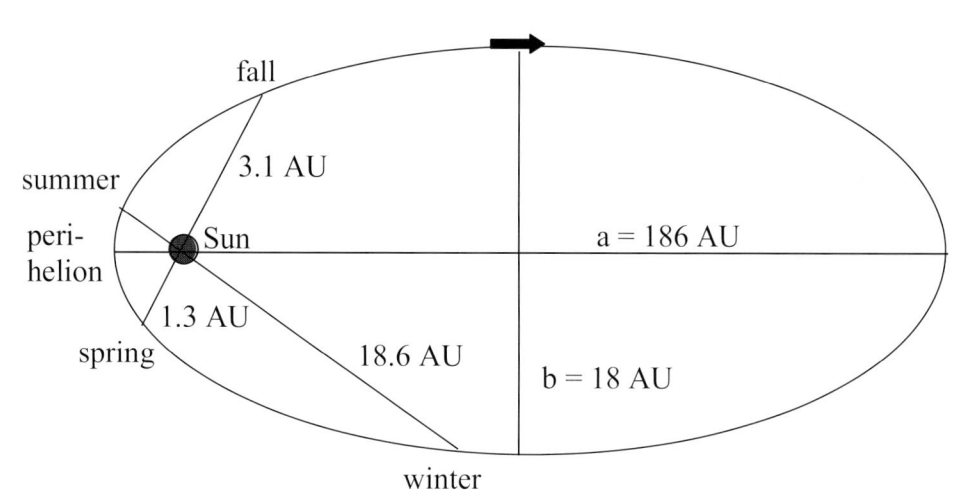

Figure 1. Sketch of the orbital parameters and the seasons on Hale-Bopp

Lower boundary condition:

$$K\frac{dT}{dx} = 0 \qquad (3)$$

Local gas flux (kg m^{-2}s^{-1}):

$$F_{gas}(R_h, \varphi, L) = \frac{2p_v}{\pi v_{th}} \qquad (4)$$

Vapour pressure:

$$p_v(Pa) = 3.5610^{-12}\exp(\frac{6141K}{T}) \qquad (5)$$

Total gas flux (kg s^{-1}):

$$F(R_h) = \frac{As}{4\pi}\int_0^{2\pi}\int_{-\pi/2}^{\pi/2} F_{gas} \cos\varphi d\varphi dL \qquad (6)$$

The coupled set of nonlinear equations (1) to (5) is solved numerically. The lower boundary given by Eq. (3) was set to a depth of some meters (depending on the thermal conductivity) where the temperature variations are strongly damped. The back flux of molecules to the surface caused by collisions in the coma is not considered in Eq. (2) because its net effect in the energy balance is small. To derive the H_2O-activity as a function of the heliocentric distance an integration of the local flux over the entire surface of the comet is performed (Eq. (6)).

3. Results and Discussion

Figures 2 to 4 compare computed water production rates with experimental data published by Colom *et al.* (1998), by Weaver *et al.* (1997; 1998), and by Crovisier *et al.* (1997; 1998). Filled symbols represent the data measured after perihelion. The only free parameter in the model is the size of the active surface area in Eq. (6). It was chosen in such a way that the calculated flux equals that measured at perihelion.

Figure 2 shows how the thermal properties of the surface control the water flux. A lower conductivity produces a higher flux but a lower asymmetry of the curves. At large heliocentric distances the inbound and outbound fluxes differ up to two orders of magnitude. Near perihelion the differences are less pronounced because nearly all solar energy is consumed by sublimation on the surface and the heat flux is of minor importance. Consequently, measurements at large heliocentric distances are needed for shading some light into the thermal behaviour of a cometary nucleus. Beyond 2 AU the curves computed with a high thermal conductivity fail to describe the observations. Unfortunately, only a few data are available after perihelion but

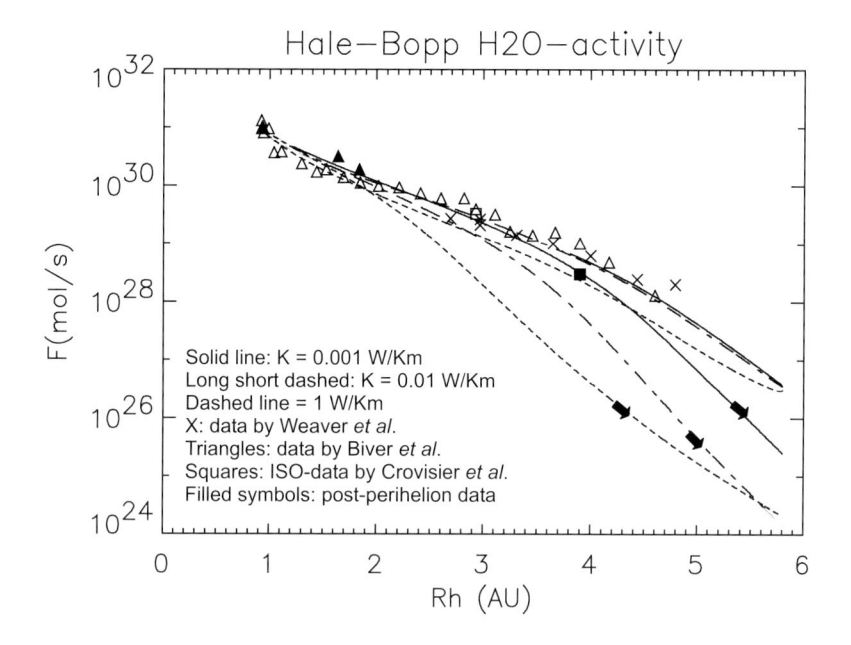

Figure 2. Water production of Hale-Bopp, computed curves with $\varepsilon = 1$

the direct measurement of H_2O with ISO at 3.9 AU should be reliable enough to discriminate between the extremes of thermal conductivity. A conductivity of 1 W K^{-1} m^{-1} (typical for solid ice) seems to be too high to interpret the measurements. A low conductivity of 0.001 W K^{-1} m^{-1} which is typical for porous or granulated matter (e.g. lunar regolith) agrees well with the experimental data. One possible source of deviations found near perihelion are the errors in the data reduction described in the introduction.

The water production also depends on the infrared emissivity of the surface (Fig. 3). A grey body with $\varepsilon = 0.9$ has a higher surface temperature than a black body ($\varepsilon = 1$) and produces more water. However, the differences are too weak to discriminate between both values on the basis of the measurements.

The influence of the obliquity of the spin axis is demonstrated in Fig. 4. At zero obliquity (no seasons) the thermal memory enhances the outbound flux compared to the inbound. However, at a low conductivity of the surface this effect is negligible. The strong seasons that occur at a high obliquity produce a strong asymmetry of the water production with respect to perihelion. Here seasonal effects enhance the flux before perihelion. This result is not of a general nature but depends on the occurrence of the different seasons along the orbit given by the angle between the ascending node and perihelion. At a decreasing thermal conductivity the seasonal

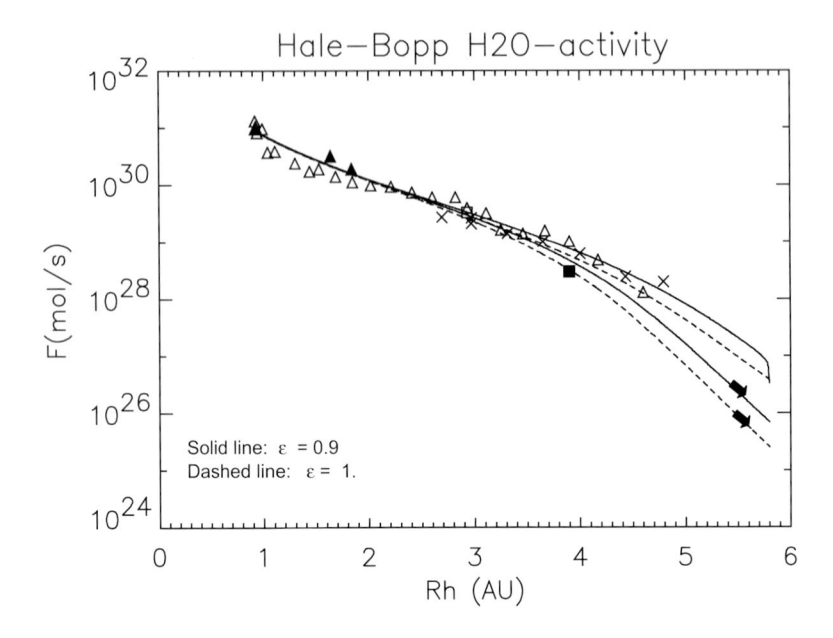

Figure 3. Water production of Hale-Bopp, computed curves with K = 0.001 W/Km, data points as in Fig. 2

effect becomes weaker and vanishes at zero conductivity. Computations with zero obliquity and a high thermal conductivity completely fail to describe the data. This may be the reason why Enzian *et al.* (1998) did not obtain agreement between the modelled and measured H_2O-activity of Hale-Bopp without postulating an extended source.

The physical explanation for the behaviour demonstrated in Figs. 2 to 4 lies in the non-linearity of the energy balance on the surface. Equations (4) and (5) indicate that an exponential law governs the dependence of the sublimation flux on the temperature. This temperature dependence is stronger than that of other terms in Equation (2). The performed computations show that the sublimation flux at a given level of energy input integrated over the surface is higher the stronger the temperature differences over the surface are. Consequently, seasonal effects, long spin periods, a low thermal conductivity and a low emissivity which cause high temperature variations enhance the level of total gas production.

While the analysis is quite complex and the model assumptions are crude, the results indicate that only a low thermal conductivity matches the observed water production. The best fit of the low conductivity curve in Fig. 2 (solid line) was achieved with an active surface area of about 2000 km^2. This area corresponds

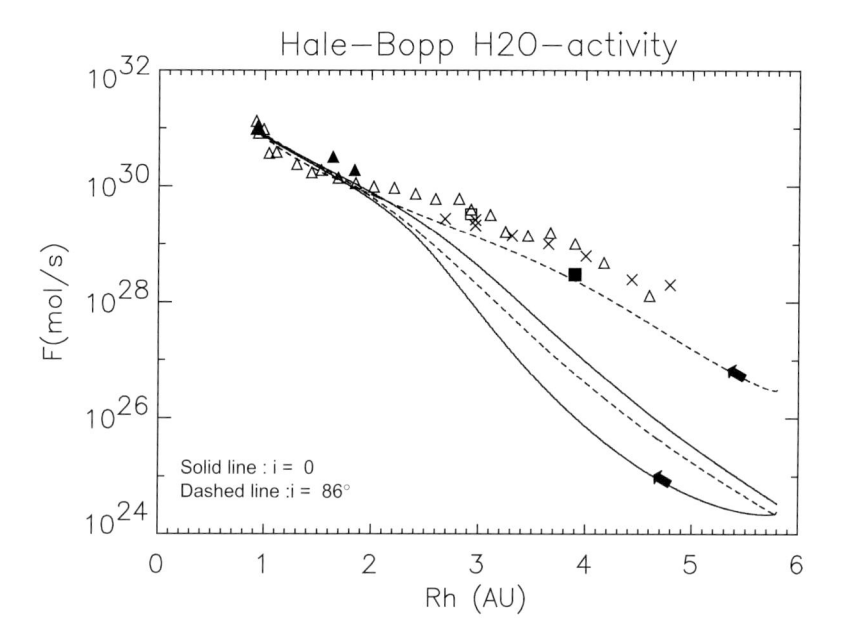

Figure 4. Water production of Hale-Bopp, computed curves with K = 1 W/Km, ε = 1, data points as in Fig. 2

to a sphere of pure ice (f = 1) with a radius of 13 km (see Table I) and gives a lower limit for the size Hale-Bopp. This result is in agreement with the minimum diameter of 27 km derived by Weaver *et al.* (1997) from HST images.

4. Conclusions

- At large heliocentric distances the water production rate of a comet depends strongly on thermal properties of the surface and on the orientation of the spin axis.
- The complex combination of seasons, spin state, heliocentric distance, and thermal properties controls the nonlinear energy balance on the surface and, therefore, the sublimation. Seasonal effects must be taken into account to interpret experimental data adequately.
- Sublimation is extremely sensitive to the surface temperature. With the same energy input the water production is higher the more pronounced the temperature differences over the surface are. Consequently, low thermal conductivity, strong seasons, and slow rotation enhance the total activity.

- Generally, there is a tendency for higher production rates after perihelion as a consequence of the thermal inertia of the cometary material. However, overlapping seasonal effects can reverse this trend. The asymmetry of the production rate with respect to perihelion depends on the thermal conductivity.
- A thermal model with a low heat conductivity of about 0.001 $Wm^{-1} K^{-1}$ fits the experimental data over a wide range of heliocentric distances. A lower limit of 13 km for the nuclear radius of Hale-Bopp has been derived from comparison of model results with measurements.

Acknowledgements

The author would like to express his gratitude to the International Space Science Institute (ISSI) in Bern for the support of this work and to an unknown referee for helpful comments.

References

Benkhoff, J. and Huebner, W.F.: 1995, 'Influence of the Vapor Flux on Temperature, Density, and Abundance Distributions in a Multicomponent, Porous, Icy Body', *Icarus* **114**, 348-354.
Bockelée-Morvan, D. and Rickman, H.: 1998, 'C/1995 O1 (Hale-Bopp) Gas Production Curves and their Interpretation', *Proceedings of the First Intern. Conf. on Comet Hale-Bopp*, Tenerife.
Colom, P., Gérard, E., Crovisier, J., and Bockelée-Morvan, D.: 1998, 'Observations of the OH Radicals in Comet C/1995 O1 (Hale- Bopp) with the Nancy Radio Telescope', *Proceedings of the First Intern. Conf. on Comet Hale-Bopp,* Tenerife.
Crovisier, J., Leech, K., Bockelée-Morvan, D., Brooke, T.Y., Hanner, M.S., Altieri, B., Keller, H.U., and Lellouch, E.: 1997, 'The Spectrum of Comet Hale-Bopp (C/1995 O1) Observed with the Infrared Space Observatory at 2.9 AU from the Sun', *Science* **275**, 1904-1907.
Crovisier, J.: 1998, 'Infrared Observations of Volatile Molecules in Comet Hale-Bopp'' *Proceedings of the First Intern. Conf. on Comet Hale-Bopp* Tenerife.
Enzian, A., Klinger, J., and Cabot, H.: 1998, 'Simulation of the water and carbon monoxide of comet Hale-Bopp using a quasi 3D nucleus model', *Planet. Space Sci.* **46**, 851-858.
Jorda, L., Rembor, K., Lecacheux, J., Colom, P., Colas, F., Frappa, E.,and Lara, L.M.: 1998, 'The Rotational Parameters of Hale-Bopp (C/1995 O1) from Observations of the Dust Jets at Pic du Midi Observatory', *Earth Moon and Planets*, in press.
Prialnik, D., Egozi, U., Bar-Nun, A., Podolak, M., and Greenzweig, Y.: 1993, 'On Pore Size and Fracture in Gas-Laden Comet Nuclei', *Icarus* **106**, 499-507.
Weaver, H.A., Feldman, P.D., A'Hearn, M.F., Arpigny, C., Brandt, J.C., Festou, M.C., Haken, M., McPhate, J.B., Stern, and S.A., Tozzi, G.P.: 1997, 'The Activity and Size of the Nucleus of Comet Hale-Bopp', (C/1995 O1) *Science* **275**, 1900-1904.
Weaver, H.A., Feldman, P.D., A'Hearn, M.F., Arpigny, C., Brandt, J.C, and Stern, S.A.: 1998, 'Hubble Space Telescope Observations of Comet Hale-Bopp', *Proceedings of the First Intern. Conf. on Comet Hale-Bopp*, Tenerife.

Address for correspondence: Ekkehard Kührt, ekkehard.kuehrt@dlr.de

HALE-BOPP AND ITS SODIUM TAILS

G. CREMONESE
Osservatorio Astronomico, vic. Osservatorio 5, 35122 Padova, Italy

Abstract. The sodium emissions have been observed in several new and long-period comets, but only for comet Mrkos 1957d (Nguyen-Huu-Doan, 1960) was a sodium tail detected on a Schmidt plate obtained with a objective prism. Comet Hale-Bopp 1995 O1 offered the first great opportunity to get an image of a long sodium tail. It was more than 3×10^7 km long, defined as a third type of tail, as it was composed only of neutral atoms (Cremonese, 1997a). After the discovery of the sodium tail another team announced it had observed it (Wilson *et al.*, 1998), but it was soon realized they had seen a different sodium tail. The image of Wilson *et al.* (1998) showed a very diffuse sodium tail superimposed on the dust tail, most likely due to the release of sodium atoms from dust particles. It was different from the narrow tail found in the image obtained by the European Hale-Bopp Team and its position angle was 15-20 degrees lower. Spectroscopic observations have been performed on the dust tail, at different beta values, and along the narrow sodium tail showing that the sodium emissions had very different line profiles. The analysis of these profiles will yield important insights into the sources in the inner coma and in the dust tail. This work will report on preliminary analysis of both sodium tails and emphasize the high-resolution spectroscopy performed on the dust tail.

1. Introduction

Sodium atoms may provide an important tool to probe the mechanisms operating in comets because they have the strongest spectrum of any nonvolatile atomic component of these objects. Sodium is characterized by its high efficiency of resonant scattering of solar radiation and can be observed at a lower column density than most other species in the inner planetary system. Understanding the nature of sodium in comets may provide important clues to the relationship between volatile and nonvolatile components of the nucleus and the nuclear formation processes which led to their fractionation. In particular, the analysis of the sources of the two sodium tails observed may allow us to infer the characteristic processes working on the dust grains and on the molecules embedded in the volatile nuclear matrix, described in the following sections.

Sodium has been observed in some comets from the beginning of this century, for example, by Newall (1910) in comet 1910 A, which showed a short yellow tail. The first clear identification of a sodium tail in a comet was reported by Nguyen-Huu-Doan (1960) for comet Mrkos 1957d, in the range of heliocentric distances r=0.57-0.82 AU. He observed a long sodium tail on a plate obtained with an objective prism in front of a small telescope.

In the following years, sodium was observed in other bright comets such as Ikeya-Seki (C/1965f), Bennett (C/1969 Y1), Kohoutek (C/1973 E1), West (C/1976 VI) and Halley (C/1986 III) (Spinrad *et al.*, 1968; Oppenheimer, 1980; Rahe *et al.*, 1976; Combi *et al.*, 1997), but the field of view usually subtended a few 10^5 km, which made it impossible to detect sodium in the tail.

A sodium tail, well separated from the dust and the ion tails, was only discovered on comet Hale-Bopp (Cremonese, 1997a). Other observers then became interested in investigating the production rate and the spatial distribution of sodium in this object (Arpigny *et al.*, 1998; Brown *et al.*, 1998; Rauer *et al.*, 1998; Wilson *et al.*, 1998). A more accurate analysis of the data revealed that Comet Hale-Bopp had two distinct sodium tails with a completely different morphology. In order to distinguish them, Cremonese and Fulle (1999) suggested the two definitions of *narrow sodium tail* and *diffuse sodium tail*. The first feature was characterized by fast atoms coming directly from a source located in the nuclear region along a direction different from the two other tails, while the second feature, completely superimposed on the dust tail, was most likely due to neutral atoms released in situ by dust particles. The study of these features and the related sources might help to understand some key points of mechanisms working in the cometary coma and in the nuclear region that may be important for other atoms and molecules. Similar observations can be applied to other comets, thanks to the easy detection of sodium.

In this work I will describe the different nature of the two sodium tails and mention some hypotheses on the source mechanisms.

2. Analysis and Discussion

The instrumentation utilized to perform wide-field imaging and high-resolution spectroscopy has been described in Cremonese *et al.* (1997b).

The high-resolution spectra we obtained on both sodium tails pointed out a strong difference in the emission profiles, depending on the position from which they were taken. The spectrum along the narrow sodium tail at 0.4 degrees separation (1.6×10^6 km) from the nucleus showed a clear double peak, with the ratio between the two components changing along the slit height (Fig. 1 and 2). The double peak can be explained by two sources working simultaneously in the outer coma, where the narrow sodium tail almost coincided with the dust tail and the diffuse sodium tail. The slower peak (S in the figures) was due to the sodium atoms released locally by the dust particles, while the faster one (F in the figures) was due to the sodium atoms coming from the nuclear region and already accelerated by the fluorescence mechanism. The slower component had a velocity of 6 km s^{-1} and the faster a velocity of 37.7 km s^{-1}, well fitted by the model of Cremonese *et al.* (1997b). The two emission profiles shown in Fig. 1 and 2 were obtained from two slit positions on the same spectrum, separated by 135,000 km. The comparison with another spectrum obtained along the narrow sodium tail, but farther away from

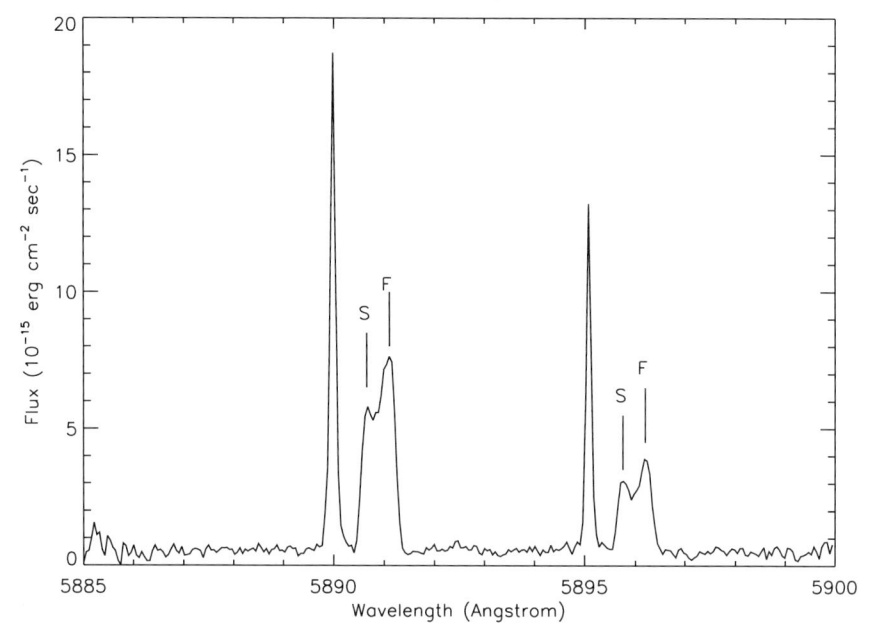

Figure 1. Profile of the sodium D-lines toward the dust tail at 0.4 degrees from the nucleus obtained on April 23.9 UT. The highest peaks are due to the terrestrial emissions. F points at the fast component and S on the slower one. The spectrum has not been corrected by the comet velocity and the y-axis shows the absolute flux in erg cm^{-2} sec^{-1}.

the nucleus, emphasizes their diversity. In the latter case the emission was obtained at 0.89 degree (3.6×10^6 km) separation from the nucleus and the slow component is much less evident, with the peak corresponding to a corrected velocity of 62 km s^{-1}. It means that, far from the nucleus, it is possible to measure only the sodium released in the nuclear region.

Cremonese *et al.* (1997b) used two models to fit the narrow sodium tail and these high-resolution spectra. In both cases they assumed that the gas was produced by the nucleus and/or a near-nucleus source with dimensions much smaller than the tail length. The sodium atoms underwent acceleration in the antisun direction as a result of simple radiation pressure, through photon scattering (fluorescence) in the resonant D-lines transitions. The experimental lifetime for photoionization of 6.17×10^4 s, at a heliocentric distance of 1 AU, assumed in previous studies of the Moon, Mercury, Io, and other comets, did not fit the observed brightness distribution. Such a distribution is consistent with their model after adopting a longer photoionization lifetime of 1.69×10^5 s, as suggested by Huebner *et al.* (1992) following a theoretical recalculation of the cross section. Cremonese *et al.* (1997b) found that the syndyne — an ideal line connecting all the particles sensitive to a same solar radiation pressure force — defined by $\beta = 82 \pm 3$, where β is the ratio of radiation pressure acceleration to solar gravity acceleration, fitted well the narrow sodium tail and the calculation of the mass of any sodium-bearing molecule

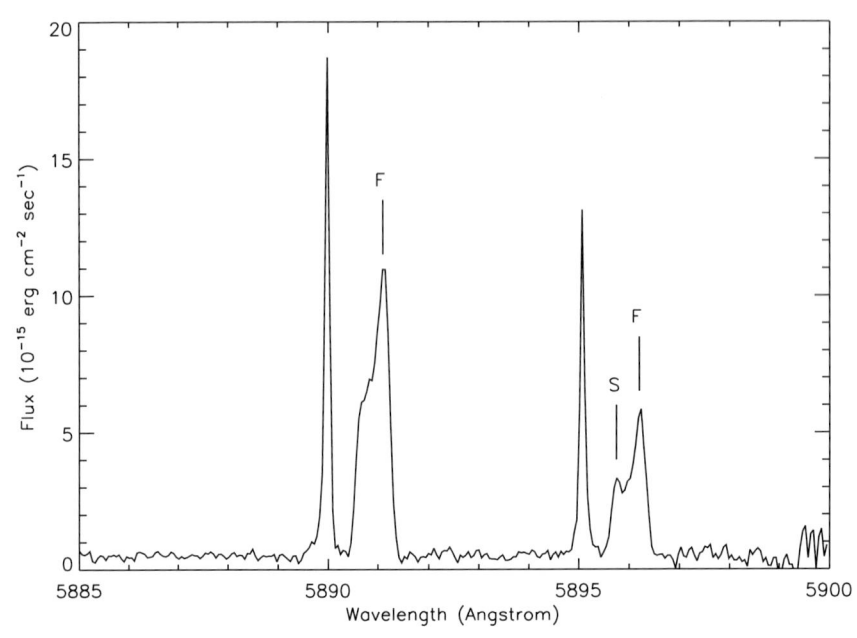

Figure 2. Profile of the sodium D-lines toward the fast sodium tail axis, on the same spectrum as Fig. 1, but at the opposite slit edge.

in the tail, provided $m \leq 22\pm1$ amu. This demonstrates that the narrow sodium tail is composed of neutral sodium atoms alone, as sodium-bearing molecules would otherwise be too massive to be consistent with such a high β ratio. Furthermore, the spectra obtained along the narrow sodium tail did not show any evidence of molecular ion emissions. Then the position angle of the ion tail was about 7-8 degrees lower than the narrow sodium tail. In other words, H_2O^+, NH_2^+ or other molecular ions should follow a different direction. The models adopted by Cremonese *et al.* (1997b) did not reveal the exact nature of the source of the sodium atoms observed in the tail, but spectroscopic observations obtained in the coma by other authors suggested an extended source (Arpigny *et al.*, 1998, Rauer *et al.*, 1998, Brown *et al.*, 1998). Combi *et al.* (1997) hypothesized that the extended source was due to molecular recombination of sodium bearing ions, explaining the morphological similarity between the sodium and H_2O^+ in the near tail in other comets. Arpigny *et al.* (1998) and Brown *et al.* (1998) showed that such a hypothesis did not agree with the observations of the coma, as the sunward extent of the water coma was appreciably less than the distances at which sodium was observed, and the tailward velocity of H_2O^+ was higher than that of the sodium atoms. All these models have been used to explain the source for the narrow tail, while the diffuse tail was due to the in situ release of sodium atoms by dust particles that most likely occurred also in the coma.

In order to study the dust source, we obtained some high-resolution spectra on the dust tail at different distances from the nucleus. We put the slit perpendicular to

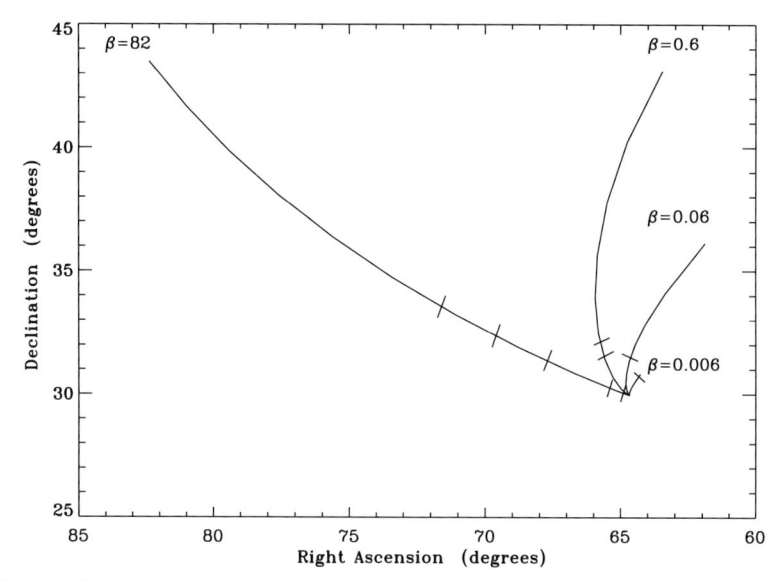

Figure 3. Plot of the syndynes considered for the high-resolution spectroscopy along the dust and narrow sodium (β = 82) tails on April 24.9 UT. The slit positions have been reported on each syndyne.

some specific syndyne and the spectra showed very broad emission lines, without a well defined double component. In this case the sodium emissions are not as strongly red-shifted as along the narrow sodium tail pointing at atoms with much lower velocities. That means the sodium atoms, not yet accelerated, have been released locally from the dust particles. Assuming that the fluorescence mechanism worked even on the atoms released from the dust, the broadening of the profiles is most likely due to the contribution of atoms having different velocities coming from dust particles located around the region observed from the slit. We report in Table I the intensities of these emissions that increase for higher beta values, that is for smaller particles pointing at a dependence on the age of the dust particles. The slit positions on the syndynes are reported in Fig. 3, where β = 82 corresponds to the narrow sodium tail (Cremonese *et al.*, 1997b).

Cremonese *et al.* (1997b) reported a sodium production rate of $Q(Na) \simeq 5 \times 10^{25}$ s^{-1} along the narrow tail, a value very similar to what Wilson *et al.*, 1998) measured on the diffuse tail. The water production rate at this time was of order $\sim 1 \times 10^{31}$ (at r=0.98 AU).

3. Conclusion

It seems clear that the two sodium tails of the comet Hale-Bopp observed were due to two or more different mechanisms, working simultaneously in the coma. The narrow sodium tail was due to the release of neutral atoms in the nuclear region

TABLE I

Sodium D-line intensities obtained on some syndynes on the dust tail on April 24.9 UT.

β	distance from the nucleus (10^6 km)	Intensity (Rayleigh)
0.6	5.2	20
0.6	8.8	8
0.06	4.8	7
0.006	3.6	–

from an extended source that can be associated to a sodium bearing molecule, not yet identified, which may come from the nucleus. Using a Monte Carlo model of sodium dynamics, Brown et al. (1998) suggested that the intensities and velocities in Comet Hale-Bopp can be explained, if 55% of the observed sodium is produced at the nucleus and the remaining 45% in an extended source. They assumed that the sodium is accelerated only by solar radiation pressure.

The release of sodium atoms by dust particles, for instance through an evaporation process (Huebner, 1970), would be responsible for the diffuse sodium tail and the slower component observed in the spectra in the outer coma. Presumably, even the processes proposed for the regolith of Moon and Mercury and their sodium atmospheres can be applicable, including thermal desorption, solar photodesorption and solar wind sputtering (e.g., Sprague et al., 1992).

The modelling of these observations could provide important insights not only into the sources of the sodium tails, but also into processes working in comets. The sodium atom can be used as a tracer of these processes since it is easy to observe. Sodium and the production mechanisms involved could be more common in the Solar System than previously assumed. The discovery of the two sodium tails of comet Hyakutake (Mendillo et al., 1998) testifies to this.

References

Arpigny, C., Rauer, H., Manfroid, J., Hutsemèkers, D., Jehin, E., Crovisier, J., Jorda, L.: 1998, 'Production and kinematics of sodium atoms in the coma of comet Hale-Bopp', *A&A* **334**, L53.

Brown, M.E., Bouches, A.H., Spinrad, H., Misch, A.: 1998, 'Sodium velocities and sources in Hale-Bopp', *Icarus* **134**, 228.

Combi, M. R., Di Santi, M. A., Fink, U.: 1997, 'The spatial distribution of gaseous atomic sodium in the comae of comets: evidence for direct nucleus and extended plasma sources', *Icarus* **130**, 336.

Cremonese, G., and the European Hale-Bopp Team: 1997a, *IAU Circ.* **6631**.

Cremonese, G., Boehnhardt, H., Crovisier, J., Fitzsimmons, A., Fulle, M., Licandro, J., Pollacco, D., Rauer, H., Tozzi, G.P., West, R.M.: 1997b, 'Neutral sodium from comet Hale-Bopp: a third type of tail', *ApJ* **490**, L199.

Cremonese G., Fulle, M.: 1999, 'Sodium in comets', *Earth, Moon and Planets*, in press.

Huebner, W. F.: 1970, 'Dust from cometary nuclei', *A&A* **5**, 286.

Huebner, W. F., Keady, J. J., Lyon, S. P.: 1992, 'Solar photo rates for planetary atmospheres and atmospheric pollutants — index of species — full names and chemical structures', *Ap&SS* **195**, 1.

Mendillo, M., Wilson, J.K., Baumgardner, J., Cremonese, G., Barbieri, C.: 1998, 'Imaging studies of sodium tails in comets', *DPS meeting* **30**, 29.01.

Newall, H.F., 1910, 'The spectrum of the daylight comet 1910 a', *MNRAS* **70**, 459.

Nguyen-Huu-Doan: 1960, 'Le spectre de la comete Mrkos 1957d', *Journal des Observateurs* **43**, 61.

Oppenheimer, M.: 1980, 'Sodium D-line emission in comet West (1975n) and the sodium source in comets', *ApJ* **240**, 923.

Rahe, J., McCracken, C.W., Donn, B.D., 1976, 'Monochromatic and white-light observations of comet Bennett 1969I (1970II)', *A&A Suppl.* **23**, 13.

Rauer, H., Arpigny, C., Manfroid, J., Cremonese, G., Lemme, C.: 1998, 'The spatial sodium distribution in the coma of comet Hale-Bopp (C/1995 O1)', *A&A* **334**, L61.

Spinrad, H., Miner, E.: 1968, 'Sodium velocity fields in comet 1965f', *AJ* **153**, 355.

Sprague, A., Kozlowski, R., Hunten, D., Wells, W., Grosse, F., 1992, 'The sodium and potassium atmosphere of the Moon and its interaction with the surface', *Icarus* **96**, 27.

Wilson, J.K., Mendillo, M., Baumgardner, J., 1998, 'Three tails of comet Hale-Bopp', *Geophys.Res.Letter* **25**, 225.

Address for correspondence: Gabriele Cremonese, Osservatorio Astronomico, vic. Osservatorio 5, 35122 Padova, Italy, cremonese@pd.astro.it

ROCKY COMETARY PARTICULATES: THEIR ELEMENTAL, ISOTOPIC AND MINERALOGICAL INGREDIENTS

ELMAR K. JESSBERGER

Institut für Planetologie, Westfälische Wilhelms-Universität Münster, Wilhelm-Klemm-Strasse 10, D-48149 Münster, Germany

Abstract. The determination of the chemical composition of solid cometary dust particles was one of the prime objectives of the three missions to Comet Halley in 1986. The dust analysis was performed by time-of-flight mass-spectrometry. Within the experimental uncertainty the mean abundances of the rock-forming elements in cometary dust particles are comparable to their abundances in CI-chondrites and in the solar photosphere, i.e. they are cosmic. H, C, and N, on the other hand, in cometary dust are significantly more abundant than in CI-chondrites, approach solar abundances, are to some extent related to O, and reside in an omnipresent refractory organic component dubbed CHON. Element variations between individual dust grains are characterized by correlations of Mg, Si, and O, and to a lesser extent of Fe and S. From particle-to-particle variations of the rock forming elements information on the mineralogy of cometary dust can be obtained. Cluster analysis revealed certain groups that partly match the classifications of stratospheric interplanetary dust particles. About half of Halley's analyzed particles are characterized by anhydrous Fe-poor Mg-silicates, Fe-sulfides, and rarely Fe metal. The Fe-poor Mg-silicates link Halley's dust to that of Hale-Bopp as shown by recent IR observations. No significant deviation from *normal* of the isotopic composition of the elements is unequivocally present with the notable exception carbon: ^{12}C-rich grains with ^{12}C/^{13}C-ratios up to $\approx 5{,}000$ link cometary dust to presolar circumstellar grains identified in certain chondrites.

1. Introduction

Why analyze the composition of cometary particles? The history of the bodies in the solar system is largely governed by their size: large bodies may be or may have been at some time hot inside, may have fractionated into core and mantle or were subjected to various kinds of metamorphic processes (Bischoff, 1998). Small bodies not. Here small bodies are understood as those only that never were fragments of larger ones, but that have most probably always been small. Small planetary bodies are more likely than large ones to have preserved the original states of the material they accreted of. Comets are believed to represent the smallest solar system bodies, but most of them are inaccessible for mankind stored in the Oort cloud. Only if they approach the sun they become visible by means of their developing coma. Then they can be studied by remote sensing techniques and in a one single case yet by in situ experiments. The chemical aspects of the latter is the topic of this chapter.

One would prefer to have an "intact and cold" sample of a comet in the laboratory to analyze the chemical and isotopic composition, to date the material, to study the mineralogy, the physical properties and the structure of its dust and ice, but a mission to this end as yet is precluded by its enormous costs. Thus the desired information rests on what can be extracted from the (very similar) experiments PIA and PUMA-1&2 onboard the missions GIOTTO and VEGA-1&2 to Comet Halley in spring 1986. The instrument that functioned best was PUMA-1 and the resulting data form the basis of the present contribution.

2. The Instrument

The impact-ionization time-of-flight instrument was developed by Jochen Kissel and coworkers (Kissel, 1986a and 1986b). The high velocity of the space craft relative to Halley's coma of 70-80 km/s effectively ionizes the elements of the particles that collide with the Ag-doped Pt target. The ions are accelerated into a time-of-flight mass spectrometer for separation and, after passing a focusing ion-mirror, are detected with a multiplier. To enable the required dynamic range of about five orders of magnitude, the multiplier signals of the dynodes 5, 8, 11, 14, 17, and 20 were logarithmically summed resulting in an overall logarithmic amplifier signal. Since the amplification of a multiplier critically depends on the applied voltage, there remains some uncertainty as to the conversion of the logarithms to real numbers of ions. The resulting mass spectra are recorded in three different modes (Kissel, 1986a). In mode-0 spectra the multiplier output was taken every 66.66 ns resulting in approximately four intensity measures per mass unit. These mode-0 spectra most closely resemble normal laboratory mass spectra. In the higher modes the ion currents were sampled using a peak detector. In the mode-1 spectra, for each maximum appearing at the multiplier an amplitude sample and a sample of a $T^{\wedge 2}$-generator — an electronic signal that is generated by two successive capacitors and that is thought to be proportional to the mass — was taken. A second set was recorded half a mass later as a minimum value. Furthermore, the amplitude was measured every 1.13 μs (the *time sample*) if no other maximum occurred. In mode-2 spectra no minimum value was recorded, and in mode-3 spectra also no minimum values were taken and the time sample only after 8.6 μs. In addition, the analogue channels were stimulated by *test pulses* with preset amplitudes and proper timing.

A very important step in extracting useful information from the mass spectra is their *interpretation* that includes first of all converting the time scale (in mode-0) or the $T^{\wedge 2}$-sample scale (mode-1 to mode-3) into a mass scale for each spectrum. This task is not trivial because of non-linearities and occasional errors in data storage (Schulze *et al.*, 1997). Perhaps the most difficult problem was created by the fact that the data transmission rate from the space craft to Earth was limited such that only about 1% of the spectra were recorded in mode-0. For the majority of the spectra, the modes 1–3, the arrival time of the ions as produced by the $T^{\wedge 2}$-generator induced a serious uncertainty in identifying the masses. This limitation, which is independent from the instrument but was dictated by the mission, euphemistically may be called data *compression*, but effectively is data *elimination*. Certainly one has to keep in mind the year and the technical state-of-the-art of space-qualified equipment when the mission and the experiments were designed, pre-1985. Another problem of yet unknown origin is that the amplitudes of the test pulses unexpectedly vary with distance from the nucleus. At present the irregularities are studied in detail together with a new and complete interpretation of the full data sets from all three experiments.

Nevertheless, a number of independent studies have convincingly demonstrated that indeed useful and in fact exciting information can be extracted from the spectra. From ion intensities the elemental and isotopic abundances can be calculated. In principle, one first would have to calibrate the device, i.e. to determine the number of ions generated upon impact at that high speed from various types of particles. However, there is no accelerator for any kind of macroscopic particles with dimensions in the range of 0.05 to 5 μm to reach 70 or 80 km/s. Thus, the yields were extrapolated from lower speed calibration shots (Kissel and Krueger, 1987a). Consequently the resulting atom abundances are uncertain by a factor of about two. Another complication might arise from the fact that a larger particle that triggers the front-end-channels — these are required to set the start signal for a spectrum — are accompanied by very small particles (Utterback and Kissel, 1990) whose secondary ions then overlay the *normal* spectrum. Such a mix cannot now be disentangled on Earth.

3. Results

The principal results have been reported in quite a number of refereed journals and book chapters (Kissel *et al.*, 1986c, 1986d; Jessberger *et al.*, 1986, 1988, 1989; Langevin *et al.*, 1987; Sagdeev *et al.*, 1987, 1989; Solc *et al.*, 1987; Lawler *et al.*, 1989; Maas *et al.*, 1989; Grün and Jessberger, 1990; Jessberger and Kissel, 1991; Mukhin *et al.*, 1991; Fomenkova *et al.*, 1992, 1994; Lawler and Brownlee, 1992; Fomenkova and Chang, 1994, 1996; Schulze *et al.*, 1997.) They were reviewed most recently by Eberhardt (1998) and by Jessberger and Fomenkova in Sekanina *et al.* (1999). Below I will present an itemized list of the fourteen major findings on which probably all researchers on that topic would agree. In this list I will concentrate on the overall results, on abundances of the rock forming elements and on the state in which they occur. The elements that form the refractory organic component are discussed by Fomenkova (this volume).

1. The cometary particulates are intimate mixtures of two end-member components, dubbed CHON (rich in the elements H, C, N, and O) and ROCK (rich in rock-forming elements as Si, Mg, Fe), respectively.

2. The CHON component is refractory organic material (Jessberger et al., 1986; Kissel and Krueger, 1987b).
3. The ROCK component comprises silicates, metals, oxides, sulfides and others.
4. Both end-member components do not occur as pure components but are intimately mixed down to the finest scale (Lawler and Brownlee, 1992).
5. CHON- and ROCK-rich particulates each comprise about 25% of the particulates while 50% are MIXED. The latter group is defined by the ratio of carbon to any rock forming element between 0.1 and 10 (Fomenkova et al., 1992).
6. The refractory organic CHON-component probably forms rims around ROCK-rich cores, an interpretation of PUMA-1 results (Kissel and Krueger, 1987b) that is in accordance to models of cometary grains by Greenberg (1982).
7. As far as the isotopic information is concerned, no significant (i.e. greater than a factor of two) deviation from the normal composition of those elements was found for which one technically could detect one like Mg and S (Grün and Jessberger, 1990). The notable exception, however, is carbon.
8. An unequivocal detection of isotopically light carbon was reported by Jessberger and Kissel, 1991 : $^{12}C/^{13}C$ ratios as high as 5,000 in a particle composed of almost pure carbon link cometary matter to certain circumstellar graphite grains extracted from carbonaceous chondrites (Zinner et al., 1990; Anders and Zinner, 1993). All $^{12}C/^{13}C$ ratios *below* the normal value of 89 in Jessberger and Kissel (1991) are strict lower limits that result from noise intensities of unknown origin in the spectra (and not from the molecular $^{12}CH^+$-interference as we incorrectly stated there). Two implications are important: (A) There is no single fixed or defined *cometary* carbon isotopic composition although our result is obtained only from one comet, Halley. (B) The presence of the wide range of isotopic compositions of carbon excludes any equilibration processes affecting the carbon carrier during comet formation or later in its history.
9. Particulate masses and densities were estimated to range from 10^{-16} to 10^{-11} g (approximate diameter range 0.02 to 2 μm) with densities from 0.3 to 3 g/cm^3 (Maas et al., 1989).
10. The bulk abundances of the rock-forming elements in Halley's dust are indistinguishable from the solar and chondritic abundances within a factor of two (Jessberger et al., 1988).
11. With the plausible assumption that the whole comet — dust and ice — has approximately solar composition with the exception of hydrogen and nitrogen, we inferred an overall dust/ice ratio of two (Grün and Jessberger, 1990).

The following three items require the assumption that the rock-forming elements in Halley's dust indeed are present in their cosmic abundances because the extraction of mineralogical information would be precluded by the inherent absolute uncertainty of the ion-to-atom yields of a factor of two.

12. Inspecting the variation of the elemental composition from particle to particle some clues as to the mineralogical composition can be drawn: Mg/Fe-ratios display a rather wide range, while that of the Si/Mg ratios is very narrow,

and Mg-rich Fe-poor silicates constitute for sure at least 40%, maybe more than 60% of the particles. The next largest ($\approx$10%) group of particles are iron(+nickel)-sulfides while Fe-oxides play only a very minor rôle (<1%).

13. The presence of unequilibrated high-temperature minerals like Mg-rich silicates and Fe-sulfides (formed above $\sim$600 K) is evidence that equilibration at low temperatures is too slow a process to have affected the cometary dust particles (Schulze et al., 1997).

14. The presence of particles that are rich in refractory elements and resemble Ca, Al-rich inclusions known from chondrites — the earliest formed inner-solar-system materials — or carbonates or sulfates have not been unambiguously identified (Jessberger et al., 1988; Schulze et al., 1997).

Although the above results were always discussed in the framework of what is known on meteorites and on stratospheric dust particles (IDPs) (Jessberger et al., 1998; Sekanina et al., 1999), until recently an effective proof of a direct link of cometary dust and any class of IDPs was not existent. On the contrary, from measurements and modeling the contribution of comets to the dust inventory actually reaching the Earth surface was estimated to be less than 10% only (Kortenkamp and Dermott, 1998). This situation now is dramatically changed with the new results from ISO IR-spectrometry that identify the mineral assemblage in Comet Hale-Bopp dust that resembles closely anhydrous chondritic aggregate IDPs: As Hanner (1999) shows it consists of crystalline Mg-rich, Fe-poor enstatite and forsterite minerals that are formed at high temperatures plus lower-temperature glassy or amorphous grains. Mg-rich, Fe-poor particles in turn dominate Halley's dust although a firm mineral identification from the mass spectra is hampered by the inherent uncertainties. In any case, now there is indeed a link between Halley's dust, Hale-Bopp's dust, IDPs that originate from *many* comets, and through the isotopically light carbon to circumstellar grains. This chain opens the exciting prospect to study cometary particles — certain IDPs — in the laboratory! It is now for the first time that material from the outer solar system is available for mankind. The other side of the coin is, however, that IDPs are very small and thus extremely difficult to analyze, and also that they are subject to contaminating processes during their residence in the stratosphere as has been demonstrated and discussed by Jessberger et al. (1992, 1998) and Arndt et al. (1996a, 1996b).

Acknowledgements

I am most grateful to Jochen Kissel who ingeniously developed the instruments and, moreover, who initiated and proficiently accompanied the scientific evaluation of these exciting data.

References

Anders, E. and Zinner, E.: 1993, 'Interstellar grains in primitive meteorites: diamond, silicon carbide, and graphite', *Meteoritics* **28**, 490-514.

Arndt, P., Bohsung, J., Maetz, M., and Jessberger, E.K.: 1996a, 'The elemental abundances in interplanetary dust particles', *Meteoritics & Planet. Sci.* **31**, 817-833.

Arndt, P., E. K. Jessberger, J. Warren, and M. Zolensky: 1996b, 'Bromine contamination of IDPs during collection', *Meteoritics & Planet. Sci.* **31**, A8.

Bischoff, A.: 1998, 'Aqueous alteration of carbonaceous chondrites: Evidence for preaccretionary alteration — a review', *Meteoritics & Planet. Sci.* **33**, 1113-1122.

Eberhardt, P.: 1998, 'The in situ view', IAU Colloq. 168, Cometary Nuclei in Space and Time, ed. M. A'Hearn, *Astron. Soc. of the Pacific Conf. Series*, in print.

Fomenkova, M., Kerridge, J., Marti, K., and McFadden, L.: 1992, 'Compositional trends in rock-forming elements of comet Halley dust', *Science* **258**, 266-269.

Fomenkova, M. and S. Chang: 1994, 'Carbon in comet Halley dust particles', in *Analysis of Interplanetary Dust* (ed. M. Zolensky *et al.*), pp. 193-202, Amer. Inst. of Phys.

Fomenkova, M., Chang, S., and Mukhin, L.: 1994, 'Carbonaceous components in the comet Halley dust', *Geochim. Cosmochim. Acta* **58**, 4503-4512.

Fomenkova, M.N. and Chang, S.: 1996, 'The link between cometary and interstellar dust', In *The Cosmic Dust Connection* (ed. M. Greenberg), 459-465, Kluwer, Netherlands.

Fomenkova, M.: 1999, *Space Sci. Rev.*, this volume.

Greenberg, J. M.: 1982, 'What are comets made of? A model based on interstellar dust', In *Comets* (ed. L. Wilkening), 131-163, University Arizona Press, Tucson.

Grün, E. and Jessberger, E. K.: 1990, 'Dust', in: *Physics of Comets in the Space Age* (ed. W. Hubner), 113-176, Springer Verlag, Heidelberg.

Hanner, M.S.: 1999, *Space Sci. Rev.*, this volume.

Jessberger, E. K., Kissel, J., Fechtig, H., and Krueger, F. R.: 1986, 'On the average chemical composition of cometary dust', In *Proc. Comet nucleus sample return*, ESA-SP **249**, 27-30.

Jessberger, E. K., Christoforidis, A., and Kissel, J.: 1988, 'Aspects of the major element composition of Halley's dust', *Nature* **332**, 691-695.

Jessberger, E.K., Kissel, J., and Rahe, J.: 1989, 'The composition of comets', In *Origin and Evolution of Planetary and Satellite Atmospheres* (eds. S. K. Atreya, J. B. Pollack and M. S. Mathews), pp. 167-191. University Arizona Press, Tucson.

Jessberger, E. K. and Kissel, J.: 1991, 'Chemical properties of cometary dust and a note on carbon isotopes', In *Comets in the post-Halley era* (eds. R. Newburn *et al.*), 1075-1092. Springer Verlag, Heidelberg.

Jessberger, E.K., Bohsung, J., Chakaveh, S., and Traxel, K.: 1992, 'The volatile element enrichment of chondritic interplanetary dust particles', *Earth Planet. Sci. Lett.* **112**, 91-99.

Jessberger, E.K., Stephan, T., Rost, D., Arndt, P., Maetz, M., Stadermann, F.J., Brownlee, D.E., Bradley, J., and Kurat, G.: 1998, 'Properties of interplanetary dust: Information from collected samples', in *Interplanetary Dust* (eds. E. Grün, H. Fechtig, B. Gustafson) U. Arizona Press, in press.

Kissel, J.: 1986a, 'The GIOTTO particulate impact analyzer', *ESA* SP-1077, 67-83.

Kissel, J.: 1986b, 'Mass spectrometric studies of Halley comet', *Adv. Mass Spectr.* **1985**, 175-184.

Kissel, J. *et al.*: 1986c, 'Composition of comet Halley dust particles from Giotto observations', *Nature* **321**, 336-338.

Kissel, J. *et al.*: 1986d, 'Composition of comet Halley dust particles from Vega observations', *Nature* **321**, 280-282.

Kissel, J. and Krueger, F. R.: 1987a, 'Ion formation by impact of fast dust particles and comparison with related techniques', *Appl. Phys.* **A 42**, 69-85.

Kissel, J. and Krueger, F. R.: 1987b, 'The organic component in dust from comet Halley as measured by the PUMA mass spectrometer on board Vega 1', *Nature* **326**, 755-760.

Kortenkamp, S. and Dermott, S.F.: 1998, 'Accretion of Interplanetary Dust Particles by the Earth', *Icarus* **135**, 469-495.

Langevin, Y., Kissel, J., Bertaux, J.-L., and Chassefiere, E.: 1987, 'First statistical analysis of 5000 mass spectra of cometary grains obtained by PUMA 1 (Vega 1) and PIA (Giotto) impact ionization mass spectrometers in the compressed modes', *Astron. Astrophys.* **187**, 779-784.

Lawler, M., Brownlee, D., Temple, S., and Wheelock, M.: 1989, 'Iron, magnesium and silicon in dust from comet Halley', *Icarus* **80**, 225-242.

Lawler, M. and Brownlee, D.: 1992, 'CHON as a component of dust from comet Halley', *Nature* **359**, 810-812.

Maas, D., Krueger, F. R., and Kissel, J.: 1989, 'Mass and density of SILICATE-and CHON-type dust particles released by comet P/Halley', *Asteroids Comets Meteors* **III**, 389-392.

Mukhin, L., Dolnikov, G., Evlanov, E., Fomekova, M., Prilutsky, O., and Sagdeev, R.: 1991, 'Re-evaluation of the chemistry of dust grains in the coma of comet Halley', *Nature* **350**, 480-481.

Sagdeev, R., Evlanov, E., Fomenkova, M., Mukhin, L., Prilutsky, O., and Zubkov, B.: 1987, 'Composition of comet Halley dust particles based on PUMA instruments measurements in zero mode', *Space Research* **25**, 849-855.

Sagdeev, R., Evlanov, E., Fomenkova, M., Prilutsky, O., and Zubkov, B.: 1989, 'Small size dust particles near Halley's comet', *Adv. Space Res.* **9**, 263-267.

Schulze, H., J. Kissel, and E. K. Jessberger: 1997, 'Chemistry and mineralogy of comet Halley's dust', in *From Stardust to Planetesimals: Review Papers* (eds. Y. J. Pendleton and A. G. G. M. Tielens) Astronomical Society of the Pacific: San Francisco, Vol. **122**, 937-414.

Sekanina, Z., Hanner, M.S., Jessberger, E.K., and Fomenkova, M.: 1999, 'Composition of Halley's dust', In *Interplanetary Dust* (eds. E. Grün, H. Fechtig, B. Gustafson), U. Arizona Press, in press.

Solc, M., Jessberger, E. K., Hsiung, P., and Kissel, J.: 1987, 'Halley dust composition', *Proc. 10th Europ. Reg. Astron. Meeting* IAU **2**, 4750.

Utterback, N. and Kissel, J.: 1990, 'Attogram dust cloud a million kilometers from comet Halley', *Astron. J.* **100**, 1315-1322.

Zinner, E., Wopenka, B., Amari, S., and Anders, E.: 1990, 'Interstellar graphite and other carbonaceous grains from the Murchison meteorite: structure, composition and isotopes of C, N, and Ne', *Lunar Planet. Sci. Conf.* **21**, 1379-1380.

Address for correspondence: Elmar K. Jessberger, ekj@uni-muenster.de

THE SILICATE MATERIAL IN COMETS

MARTHA S. HANNER
Jet Propulsion Laboratory, California Institute of Technology, Pasadena CA 91109, USA

Abstract. Silicates in comets appear to be a mix of high–temperature crystalline enstatite and forsterite plus glassy or amorphous grains that formed at lower temperatures. The mineral identifications from the 10 and 20 μm cometary spectra are consistent with the composition of anhydrous chondritic aggregate IDPs. The origin of the cometary silicates remains puzzling. While the evidence from the IDPs points to a pre-solar origin of both crystalline and glassy components, the signatures of crystalline silicates appear in the spectra of young stellar objects only at a late evolutionary stage, when comets are the likely source of the dust.

Keywords: comets, Hale-Bopp, silicates, infrared spectra

1. Introduction

Silicates comprise a major part of the non-volatile material in comets. Silicate grains are also abundant in interstellar clouds and in the envelopes of oxygen-rich evolved stars. At present, it is not clear where the cometary silicates originated. Comets formed in regions of the solar nebula that were cold enough for interstellar silicate grains to have been incorporated without thermal alteration. However, their prior history in the solar nebula is not known. Were they shock heated during infall to the circumsolar disk? Were they altered by collisions? Was there extensive mixing between warm and cold regions of the solar nebula? Did small grains experience heating events similar to those that produced chondrules?

Up to the present, our knowledge of cometary silicates comes primarily from two sources: remote observations of infrared spectral features and in situ sampling of the dust composition during the Halley flybys. Yet, there is another very important means for learning about cometary dust. Particles released from comets are among the interplanetary dust particles (IDPs) swept up by the Earth and retrieved for laboratory analysis. If we can identify which IDPs are from comets, then we have a powerful means to study the structure, mineralogy, and history of the cometary silicates.

Comet Hale-Bopp has brought us a big step closer to deducing the mineralogy of cometary silicates and their relation to IDPs. In this paper, I summarize what is known about cometary silicates and discuss the link between cometary silicates, the chondritic aggregate interplanetary dust particles, and interstellar grains.

2. Silicate Spectral Features in Comets

Silicate particles produce a spectral feature near 10 μm due to stretching vibrations in Si–O bonds. Additional bending mode vibrations occur between 16 and 35 μm. The wavelengths and shapes of these features are diagnostic of the mineral composition. The 10 μm feature lies within the 8–13 μm atmospheric "window" allowing ground–based observations. Although 20 μm observations can also be made from the ground, the full 16–35 μm region is best observed from above the atmosphere.

A broad 10 μm emission feature is visible in filter photometry of many new and long-period comets, beginning with the first infrared observations of Comet Ikeya–Seki in 1965 (Becklin and Westphal, 1966, Ney 1982, Gehrz and Ney, 1992). The feature is strongest in active comets which have a strong scattered light continuum at visible wavelengths.

Low–resolution 8–13 μm spectra with good signal/noise have been acquired for about a dozen comets. (See Hanner et al., 1994a, for a review.) Six of the comets display a strong structured emission feature. These are long-period Comets Bradfield (1987 XXIX = C/1987 P1) (Hanner et al., 1994a), Levy (1990 XX = C/1990 K1) (Lynch et al., 1992a), Hyakutake, Hale–Bopp (C/1995 O1), (Hayward and Hanner, 1997; Hanner et al., 1998b; Wooden et al., 1999), new Comet Mueller (1994 I = C/1993 A1) (Hanner et al., 1994b) and P/Halley (Bregman et al., 1987; Campins and Ryan, 1989).

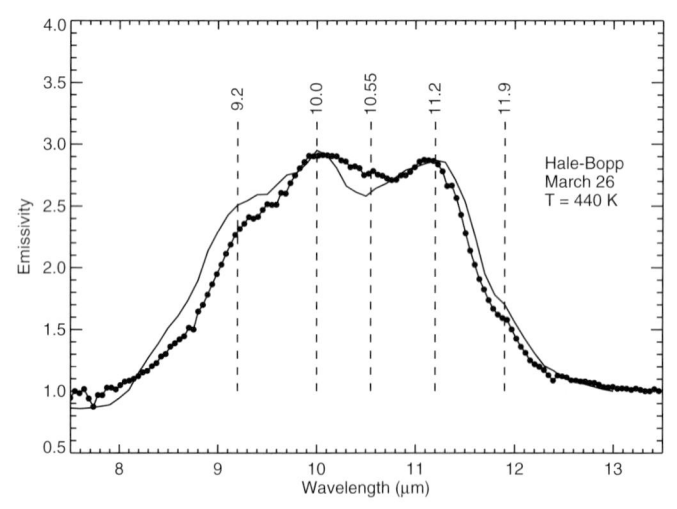

Figure 1. Silicate feature in Hale–Bopp (points) at $r = 0.92$ AU (Hanner et al., 1998b) and Halley (line) at $r = 0.79$ AU (Campins and Ryan, 1989). In each case, the total flux has been divided by a blackbody for the indicated temperature fit at 7.8-8.0 μm and 12.8-13.5 μm. The spectral shapes are remarkably similar, with a peak at 11.2 μm, broad maximum near 10 μm, shoulder near 9.2 μm, and possible structure near 10.5 μm. The inflection at 11.9 μm is also real.

By far the strongest emission feature is seen in Comet Hale–Bopp, (Fig. 1). One sees 3 peaks, at 9.2, 10.0, and 11.2 μm and minor features at 11.9 and 10.5 μm.

These peaks appear consistently in spectra taken with three different instruments under a wide range of observing conditions (Hanner *et al.*, 1998b). The spectral shape is similar in Comet P/Halley (Fig. 1) and the other comets cited above. However, the lower signal/noise in previous spectra meant that only the 11.2 μm peak was confidently identified (Bregman *et al.*, 1987; Campins and Ryan, 1989; Hanner *et al.*, 1994a).

Transmission spectra of various amorphous and crystalline silicates have been measured in the laboratory (e.g. Koike *et al.*, 1993; Dorschner *et al.*, 1995; Colangeli *et al.*, 1995; Day, 1979). Transmission spectra of the silicates in chondritic aggregate IDPs were reported by Sandford and Walker (1985). Direct emission spectra of both amorphous and crystalline pyroxene (enstatite) and olivine (forsterite) were obtained by Stephens and Russell (1979). Except for scattering in the wings, Stephens and Russell did not find any systematic differences between the transmission and emission spectra of their samples. Both transmission and emission spectra will be used here to interpret the comet spectra.

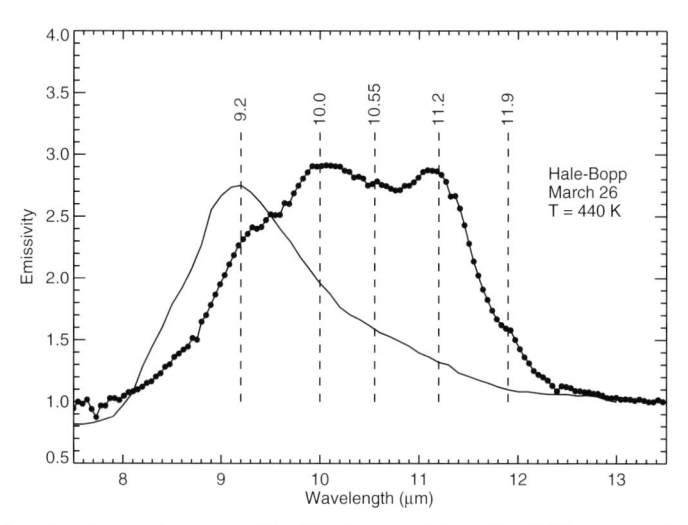

Figure 2. Emissivity of amorphous enstatite (Stephens and Russell, 1979) compared with the spectrum of Hale–Bopp (Hanner *et al.*, 1998b). The peak position at 9.2 μm matches the shoulder in the comet spectrum.

The 11.2 μm peak and the 11.9 μm shoulder are attributed to crystalline olivine, based on the good spectral match with the measured spectral emissivity of Mg–rich crystalline olivine (Stephens and Russell, 1979; Koike *et al.*, 1993). The 9.2 μm feature (Fig. 2) corresponds to amorphous, Mg–rich pyroxene (Stephens and Russell, 1979; Dorschner *et al.*, 1995). The broad maximum at 9.8–10.0 μm is similar to that seen in many interstellar sources and is most likely produced by amorphous or glassy olivine particles. The overall width of the cometary spectral feature, the relatively flat top from 10–11 μm, and the structure near 10.5 μm can best be fit with crystalline pyroxenes. Crystalline pyroxenes have considerable variety in their spectral shapes. The orthorhombic orthoenstatite measured by Stephens and Russell has a narrow feature sharply peaked at 9.90 μm. While orthoenstatite and clinoenstatite are found together, clinoenstatite is the more common form in chondritic aggregate IDPs (Bradley *et al.*, 1992). These pyroxene-rich IDPs have a broader spectral feature with major peaks in the 10-11 μm spectral region (Sandford and Walker, 1985).

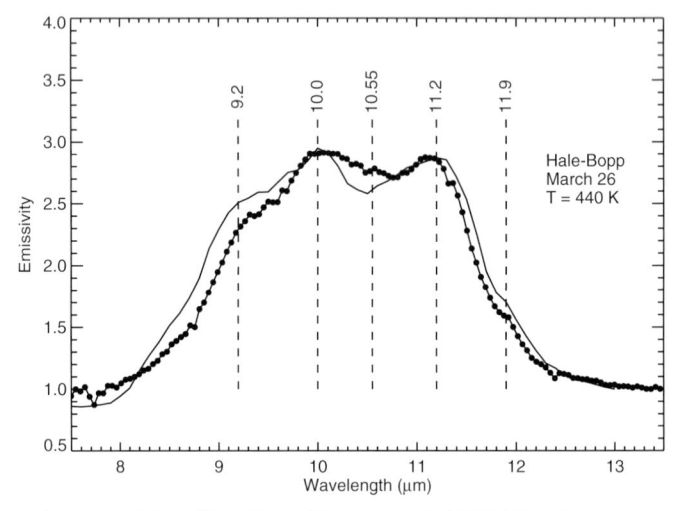

Figure 3. Composite spectral fit to Hale–Bopp (Hanner *et al.*, 1998b) based on measured emissivities of amorphous enstatite, amorphous olivine, and forsterite (Stephens and Russell, 1979) and "Key" pyroxene IDP (Sandford and Walker, 1985).

An example of a composite spectrum composed of these 4 components is compared with the Hale–Bopp spectrum in Fig. 3 (Hanner *et al.*, 1998b). The transmission spectrum of "Key" IDP (Sandford and Walker, 1985) was adopted for the crystalline pyroxene; the other emissivities are from Stephens and Russell (1979). Other fits using similar components are presented in Wooden *et al.* (1999).

It is not possible to quantify the relative abundances of these different silicate components for several reasons. 1) The laboratory measurements of spectral transmission and emissivity are only relative and do not give the quantitative mass absorption coefficients. 2) The strength of the feature depends on particle size. 3) Mie theory, in wide use for modeling the scattering and emission from small spherical particles, is not applicable for the crystalline silicates even when measured optical constants are available because of the strong dependence on particle shape near a resonance in the optical constants (Bohren and Huffman, 1983). 4) Thermal emission from a particle depends upon its temperature. The temperature is controlled by the balance between absorption of solar radiation and infrared emission and will vary with grain size, Mg/Fe ratio, and any admixture of dark material (Hanner *et al.*, 1998b; Wooden *et al.*, 1999). If the temperatures of the various silicate components are comparable, then the stronger mass absorption coefficient of forsterite implies it constitutes not more than 10% –20% of the silicates.

Silicates also have bending mode vibrations at 16–24 μm; the central wavelength depends on the mineral structure. A remarkable 16–45 μm spectrum of Comet Hale–Bopp at r = 2.9 AU was acquired with the SWS spectrometer on the Infrared Space Observatory (Crovisier *et al.*, 1997). Five peaks are clearly visible, corresponding in every case to laboratory spectra of Mg–rich olivine (Koike *et al.*, 1993). Airborne spectra of Comet P/Halley at r = 1.3 AU (the only other 16–30 μm spectra of a comet) show only weak olivine features at 28.4 and 23.8 μm (Herter *et al.*, 1987).

The inference from the spectra that the cometary silicates are Mg-rich is consistent with in situ results from Comet Halley. The Vega and Giotto space probes carried a time-of-flight mass spectrometer to measure the elemental composition of impacting dust particles (Kissel *et al.*, 1986; Jessberger *et al.*, 1988 and this volume). Two main grain types were recognized, silicate, or rock, particles dominated by ions of the major rock-forming elements Mg, Si, Ca, Fe, and CHON particles, enriched in the lighter elements, H, C, N, O. The rock particles displayed a wide range of Mg/Fe; in general, the silicates appeared to be depleted in iron, with some of the iron occurring as FeO.

The spectra of four other new comets discussed in Hanner *et al.* (1994a) are puzzling; each has a unique, and not understood, spectrum. For example, the feature is extremely broad in Wilson (1987 VII), suggesting a very amorphous silicate material. We may be witnessing the effect of cosmic ray damage to the outermost layer of the nucleus over the lifetime of the Oort cloud.

Short-period comets usually exhibit either a broad, weak 10 μm emission feature above a black-body continuum, such as 4P/Faye and 19P/Borrelly (Hanner *et al.*, 1996) or no feature at all, such as 23P/Brorsen–Metcalf (Lynch *et al.*, 1992b). These comets may lack a silicate feature because there is a deficiency of small grains rather than a lack of silicate material. However, a weak 11.2 μm peak was detected in the ISOCAM CVF spectrum of 103P/Hartley 2, the first time crystalline silicates have been observed in a short-period comet (Crovisier *et al.*, 1999).

3. Interplanetary Dust Particles from Comets

The submicron grain size, high Mg/Fe ratio, and mix of crystalline and glassy (or amorphous) olivine and pyroxenes have no counterpart in any meteoritic material, with the exception of the anhydrous chondritic aggregate interplanetary dust particles (IDPs). These are fine-grained heterogeneous aggregates having chondritic abundances of the major rock–forming elements; they comprise a major fraction of the IDPs captured in the stratosphere. Typical grain sizes within the aggregates are 0.1–0.5 μm. These aggregate IDPs are thought to originate from comets, based on their porous structure, high carbon content, and relatively high atmospheric entry velocities. The match between the mineral identifications in the Hale–Bopp spectra and the silicates seen in the IDPs strengthens the link between comets and anhydrous chondritic aggregate IDPs.

The chondritic aggregate IDPs divide into 3 classes based on their infrared spectra, dominated respectively by pyroxenes, olivine, and layer lattice silicates (Sandford and Walker, 1985), although a given IDP may contain some grains of all 3 types. The IDPs dominated by layer lattice silicates have lower atmospheric entry velocities and less porous structure, consistent with an asteroidal origin. The pyroxene- and olivine-rich IDPs of probable cometary origin are unequilibrated mixtures of high and low temperature constituents, containing small glassy or microcrystalline silicate grains and larger single crystals of enstatite or olivine (0.5–5 μm in size) along with carbonaceous material and FeS grains (Bradley et al., 1992).

Some of the crystalline silicate grains are enstatite whiskers, ribbons, and platelets with growth patterns that indicate direct vapor phase condensation from a hot gas (Bradley et al., 1983). Such growth patterns would not have formed if the crystals were annealed from glassy silicate grains. The pure Mg crystalline silicates forsterite and enstatite are predicted from thermodynamic models to be the first to condense in a hot gas at 1200–1400 K. A likely origin for the enstatite and olivine crystals is in the circumstellar outflows of oxygen–rich evolved stars.

The most intriguing component of the anhydrous chondritic aggregate IDPs are the GEMS, 0.1–0.5 μm glassy Mg silicate grains with embedded nanometer sized FeNi and Fe sulfide crystals (Bradley, 1994). The GEMS bear evidence of exposure to large doses of ionizing radiation, such as sputtered rims, Mg/Si/O gradients, and formation of reduced FeNi metal. This radiation exposure must have pre-dated their incorporation into comets, since the inferred dosage greatly exceeds their recent exposure in interplanetary space after release from the parent body. Some GEMS contain relict mineral grains with deeply ion-etched surfaces. GEMS constitute the major component of non-crystalline silicate grains in these IDPs.

Bradley (1994) argues for an interstellar origin of the GEMS, citing their high irradiation dosage, the relict microcrystals, and similarity to the inferred properties of interstellar silicates. The FeNi inclusions are sufficient for the superparamagnetic grain alignment of interstellar grains (Jones and Spitzer, 1967; Goodman and Whittet, 1995). Interstellar silicates, such as the Trapezium, show a smooth asymmetric silicate feature with a single 9.8 μm maximum consistent with glassy or amorphous olivine. Bradley et al. (1999) report that the infrared spectra of individual GEMS closely resemble the Trapezium spectrum.

4. Origin of the Cometary Silicates

As we have discussed, comets contain a mixture of silicates, both amorphous and crystalline, that do not necessarily share a common origin. Comets formed in regions of the solar nebula where the temperature was too low to sublimate refractory interstellar grains. Boss (1994, 1998) predicts that the temperature at $r \geq 5$ AU was ≤ 160 K. The high CO:H$_2$O ratio in comets implies that even interstellar ices may have survived. Thus, it is possible that *all* cometary silicates are of pre-solar origin. In the following paragraphs, evidence regarding the origin of amorphous and crystalline silicates is briefly discussed.

The GEMS described in Section 3 appear to constitute the major fraction of the non-crystalline silicates in cometary IDPs, and the evidence is quite strong that these are interstellar grains. Absorption spectra of GEMS show a 9.8 μm maximum typical of glassy olivine, although a 9.3 μm shoulder indicative of pyroxene is present in some samples (Bradley *et al.*, 1999), and this is consistent with the cometary spectra. It is likely, then, that the bulk of the amorphous cometary silicates are of interstellar origin.

The crystalline silicates are more controversial. Crystalline grains can form by direct condensation from the vapor phase at T = 1200–1400 K with very slow cooling or by annealing (heating) of amorphous silicate particles. While the enstatite whiskers and ribbons must have condensed from a hot gas, other crystalline grains in IDPs have no distinctive structure to distinguish between annealing or direct condensation. Direct condensation provides a natural explanation for Mg-rich silicates, because forsterite and enstatite are the first to condense in a hot gas and only react with Fe at lower temperatures. It is not so easy to explain why annealed grains should convert to pure forsterite and enstatite unless the glassy precursors had the correct stochiometry. Koike and Tsuchiyama (1992) created crystalline olivine by heating glassy Mg-silicate particles for 105 hours at 875 K. The annealing rate drops by orders of magnitude at lower temperatures (Hallenbeck *et al.*, 1998).

Heating is insufficient to anneal the silicates in the coma or on the nucleus. The 11.2 μm peak is observed at the same strength relative to the 10 μm maximum in both new and evolved comets over a wide range in heliocentric distance, from 4.6 AU (Hale-Bopp; Crovisier *et al.*, 1996), where the blackbody temperature is only 130 K, to 0.79 AU (P/Halley; Campins and Ryan, 1989). Nor does the 11.2 μm peak increase with time as the grains move outward through the coma. Thus, the crystalline grains must have been present in the solar nebula where the comets accreted.

The crystalline grains could not have been created by the short-term heating events that produced the chondrules. These millimeter sized components of chondritic meteorites required heating to about 1800 K followed by rapid cooling of order 1000 K/hr (Hewins, 1988), too rapid to allow crystal growth.

Direct grain condensation or annealing could have occurred in the hot inner solar nebula. Disk midplane temperatures ≥ 1000 K were reached inside about 1 AU, depending on mass infall rate (Chick and Cassen, 1997; Boss, 1998). Accretion shock heating of silicates was significant only at small $r \leq 1$ AU (Chick and Cassen, 1997). Transport of micron-sized grains from 1 AU to 10–30 AU is problematical, however. Cuzzi *et al.* (1993) have computed that small grains entrained in the outflowing gas near the midplane of the solar nebula could drift radially outward roughly 2–5 AU before accretion onto larger particles. Grain growth by accretion was a relatively rapid process, limiting the time available for radial diffusion. If sufficient mixing did take place to transport crystalline silicates to the comet formation zone, then other thermally processed components, such as refractory organics, could also have been transported.

Could the crystalline silicates have a pre-solar origin? Thanks to ISO, we now have a better picture of the distribution of crystalline silicates in astronomical sources. Forsterite is detected in oxygen-rich outflows of some evolved stars (Waters *et al.*, 1996). It is also seen in debris disks around young main-sequence objects such as β Pictoris (Knacke *et al.*, 1993) and in late-stage Herbig Ae/Be stars that are the precursors of β Pictoris systems (Waelkens *et al.*, 1996). The ISO spectrum of HD100546 is very similar to that of Comet Hale-Bopp (Malfait *et al.*, 1998). These systems are thought to have a population of sun-grazing comets, in order to explain transient gaseous emission features seen in their spectra (e.g. Grady *et al.*, 1997). However, crystalline silicates are not seen in the spectra of most young stellar objects (e.g. Hanner *et al.*, 1998a). Moreover, they are not present in spectra of the diffuse ISM or molecular clouds, such as the Trapezium. Grain destruction in the ISM is an efficient process. Thus, if the comet grains formed in circumstellar outflows, one must explain how they survived destruction in the ISM and why they are not apparent in spectra of the ISM or YSOs.

Thus, the origin of the crystalline cometary silicates remains puzzling. If they formed in the inner solar nebula, then their presence in comets requires significant mixing in the solar nebula. If they are circumstellar in origin, one has to explain why their spectral features are not visible in interstellar dust. The signatures of crystalline silicates appear in the spectra of young stellar objects only at a late evolutionary stage, when comets are the likely source of the dust.

Acknowledgements

This research was conducted at the Jet Propulsion Laboratory, California Institute of Technology, under contract with the National Aeronautics and Space Administration.

References

Becklin, E.E. and Westphal, J.A.: 1966, 'Infrared Observations of Comet 1965f', *ApJ* **145**, 445–452.
Bohren, C. E. and Huffman, D.R.: 1983, *Absorption and Scattering of Light by Small Particles*, John Wiley and Sons, New York.
Boss, A.P.: 1994, 'Midplane Temperatures and Solar Nebula Evolution', *Proc. Lunar Plan. Sci. Conf. 25th*, 149.
Boss, A.P.: 1998, 'Temperatures in Protoplanetary Disks', *Ann. Rev. Earth Planet. Sci.* **26**, 53–80.

Bradley, J.P.: 1994, 'Chemically Anomalous, Preaccretionally Irradiated Grains in Interplanetary Dust from Comets', *Science* **265**, 925.
Bradley, J.P., Brownlee, D.E., and Veblen, D.R.: 1983, 'Pyroxene Whiskers and Platelets in Interplanetary Dust: Evidence of Vapour Phase Growth', *Nature* **301**, 473–477.
Bradley, J.P., Humecki, H.J., and Germani, M.S.: 1992, 'Combined Infrared and Analytical Electron Microscope Studies of Interplanetary Dust Particles', *ApJ* **394**, 643–651.
Bradley, J.P., Keller, L.P., Snow, T., Flynn, G.J., Gezo, J.,Brownlee, D.E., Hanner, M.S., and Bowie, J.: 1999, 'An Infrared Spectral Match at 10 μm between GEMS and Interstellar Amorphous Silicates', *Science*, in press.
Bregman, J.D., Campins, H., Witteborn, F.C., Wooden, D.H., Rank, D.M., Allamandola, L.J., Cohen, M., and Tielens, A.G.G.M.: 1987, 'Airborne and Groundbased Spectrophotometry of Comet P/Halley from 5–13 Micrometers', *Astron. Astrophys.* **187**, 616–620.
Campins, H. and Ryan, E.: 1989, 'The Identification of Crystalline Olivine in Cometary Silicates', *ApJ* **341**, 1059–1066.
Chick, K.M. and Cassen, P.: 1997, 'Thermal Processing of Interstellar Dust Grains in the Primitive Solar Environment', *ApJ* **477**, 398–409.
Colangeli, L., Mennella, V., Rotundi, A., Palumbo, P., and Bussoletti, E.: 1995, 'Simulation of the Cometary 10 micron Band by Means of Laboratory Results on Silicatic Grains', *Astron. Astrophys.* **293**, 927–934.
Crovisier, J., Brooke, T.Y., Hanner, M.S., *et al.*: 1996, 'The Infrared Spectrum of Comet C/1995 O1 (Hale-Bopp) at 4.6 AU from the Sun', *Astron. Astrophys.* **315**, L385–L388.
Crovisier, J. Leech, K., Bockelée-Morvan, D., Brooke, T.Y., Hanner, M.S., Altieri, B. Keller, H.U., and Lellouch, E.: 1997, 'The Spectrum of Comet Hale–Bopp (C/1995 O1) Observed with the Infrared Space Observatory at 2.9 Astronomical Units from the Sun', *Science* **275**, 1904–1907.
Crovisier, J., Encrenaz, T., Lellouch, E., *et al.*: 1999, 'ISO Spectroscopic Observations of Short-Period Comets', in *The Universe as Seen by ISO* ESA SP-427.
Cuzzi, J.N., Dobrovolskis, A.R., and Champney, J.M.: 1993, 'Particle–Gas Dynamics in the Midplane of a Protoplanetary Nebula', *Icarus* **106**, 102–134.
Day, K.L.: 1979, 'Mid-Infrared Optical Properties of Vapor-Condensed Magnesium Silicates', *ApJ* **234**, 158–161.
Dorschner, J., Begemann, B., Henning, Th., Jager, C., and Mutschke, H.: 1995, 'Steps toward Interstellar Silicate Mineralogy II. Study of Mg-Fe-Silicate Glasses of Variable Composition', *Astron. Astrophys.* **300**, 503–520.
Gehrz, R.D. and Ney, E.P.: 1992, '0.7–23 micron Photometric Observations of P/Halley 1986 III and Six Recent Bright Comets', *Icarus* **100**, 162–186.
Goodman, A.A. and Whittet, D.C.B.: 1995, 'A Point in Favor of the Superparamagnetic Grain Hypothesis', *ApJ* **455**, L181–L184.
Grady, C.A., Sitko, M.L., Bjorkman, K.S., Perez, M.R., Lynch, D.K, Russell, R.W., and Hanner, M.S.: 1997, 'The Star-Grazing Extrasolar Comets in the HD100546 System', *ApJ* **483**, 449–456.
Hallenbeck, S.L., Nuth, J.A., and Daukantas, P.L.: 1998, 'Mid-Infrared Spectral Evolution of Amorphous Magnesium Silicate Smokes Annealed in Vacuum: Comparison to Cometary Spectra', *Icarus* **131**, 198–209.
Hanner, M.S., Lynch, D.K., and Russell, R.W.: 1994a, 'The 8–13 micron Spectra of Comets and the Composition of Silicate Grains', *ApJ* **425**, 274–285.
Hanner, M.S., Hackwell, J.A., Russell, R.W., and Lynch, D.K.: 1994b, 'Silicate Emission Feature in the Spectrum of Comet Mueller 1993a', *Icarus* **112**, 490–495.
Hanner, M.S., Lynch, D.K., Russell, R.W., Hackwell, J.A., and Kellogg, R.: 1996, 'Mid-Infrared Spectra of Comets P/Borrelly, P/Faye, and P/Schaumasse', *Icarus* **124**, 344–351.
Hanner, M.S., Brooke, T.Y., and Tokunaga, A.T.: 1998a, '8–13 Micron Spectroscopy of Young Stars', *ApJ* **502**, 871–882.

Hanner, M.S., Gehrz, R.D., Harker, D.E., Hayward, T.L., Lynch, D.K., Mason, C.G., Russell, R.W., Williams, Wooden, D.H., and Woodward, C.E.: 1998b, 'Thermal Emission from the Dust coma of Comet Hale–Bopp and the Composition of the Silicate Grains',*Earth, Moon, and Planets*, in press.

Hayward, T.L. and Hanner, M.S.: 1997, 'Ground-Based Thermal Infrared Observations of Comet Hale–Bopp (C/1995 O1) During 1996', *Science* **275**, 1907–1909.

Herter, T., Campins, H., and Gull, G.E.: 1987, 'Airborne Spectrophotometry of P/Halley from 16 to 30 Microns', *Astron. Astrophys.* **187**, 629–631.

Hewins, R.H.: 1988, 'Experimental Studies of Chondrules' in *Meteorites and the Early Solar System*, ed. J.F. Kerridge and M.S. Matthews, Univ. Arizona Press, Tucson, 660–679.

Jessberger, E.K., Christoforidis, A., and Kissel, J.: 1988, 'Aspects of the Major Element Composition of Halley's Dust', *Nature* **332**, 691–695.

Jones, R.V. and Spitzer, L.: 1967, 'Magnetic Alignment of Interstellar Grains', *ApJ* **147**, 943.

Kissel, J. *et al.*,: 1986, 'Composition of Comet Halley Dust Particles from Vega Observations', *Nature* **321**, 280–282.

Knacke, R.F., Fajardo-Acosta, S.B., Telesco, C.M., Hackwell, J.A., Lynch, D.K., and Russell, R.W.: 1993, 'The Silicates in the Disk of β Pictoris',*ApJ* **418**, 440–450.

Koike, C. and Tsuchiyama, A.: 1992, ' Simulation and Alteration for Amorphous Silicates with Very Broad Bands in Infrared Spectra', *MNRAS* **255**, 248–254.

Koike, C., Shibai, H., and Tuchiyama, A.: 1993, 'Extinction of Olivine and Pyroxene in the Mid– and Far–Infrared', *MNRAS* **264**, 654–658.

Lynch, D.K., Russell, R.W., Hackwell, J.A., Hanner, M.S., and Hammel, H.B.: 1992a, '8–13 μm Spectroscopy of Comet Levy 1990 XX', *Icarus* **100**, 197–202.

Lynch, D.K., Hanner, M.S. and Russell, R.W.: 1992b, '8–13 μm Spectroscopy and IR Photometry of Comet P/Brorsen–Metcalf (1989o) near Perihelion', *Icarus* **97**, 269–275.

Malfait, K., Waelkens, C., Waters, L.B.F.M., Vandenbussche, B., Huygen, E., and de Graauw, M.S.: 1998, 'The Spectrum of the Young Star HD100546 Observed with the Infrared Space Observatory', *Astron. Astrophys.* **332**, L25–L28.

Ney, E.P.: 1982, 'Optical and Infrared Observations of Bright Comets in the Range 0.5 to 20 microns', in *Comets*, ed. L.L. Wilkening, University of Arizona Press, Tucson AZ, 323–340.

Sandford, S.A. and Walker, R.M.: 1985, 'Laboratory Infrared Transmission Spectra of Individual Interplanetary Dust Particles from 2.5 to 25 microns', *ApJ* **291**, 838–851.

Stephens, J.R., and Russell, R.W.: 1979, 'Emission and Extinction of Ground and Vapor–Condensed Silicates from 4 to 14 microns and the 10 micron Silicate Feature', *ApJ* **228**, 780–786.

Waelkens, C., Waters, L.B.F.M., de Graauw, M.S., *et al.*: 1996, 'SWS Observations of Young Main-Sequence Stars with Dusty Circumstellar Disks', *Astron. Astrophys.* **315**, L245–248.

Waters, L.B.F.M., Molster, F.J., de Jong, T., *et al.*: 1996, 'Mineralogy of Oxygen-Rich Dust Shells' *Astron. Astrophys.* **315**, L361–L364.

Wooden, D.H., Harker, D.E., Woodward, C.E., Koike, C., Witteborn, F.C., McMurtry, M.C., and Butner, H.M.: 1999, 'Silicate Mineralogy of the Dust in the Inner Coma of Comet C/1995 O1 (Hale–Bopp) Pre– and Post–Perihelion', *ApJ* **517**, 1034–1058.

Address for correspondence: Martha S. Hanner, Mail Stop 183-501, Jet Propulsion Laboratory, Pasadena, CA 91109, USA, msh@scn1.jpl.nasa.gov

ON THE ORGANIC REFRACTORY COMPONENT OF COMETARY DUST

MARINA N. FOMENKOVA
*Center for Astrophysics and Space Sciences, University of California San Diego,
La Jolla, CA 92093-0424, USA*

Abstract. Cometary nuclei consist of ices intermixed with dust grains and are thought to be the least modified solar system bodies remaining from the time of planetary formation. Flyby missions to Comet P/Halley in 1986 showed that cometary dust is extremely rich in organics (∼50% by mass). However, this proportion appears to be variable among different comets. In comparison with the CI-chondritic abundances, the volatile elements H, C, and N are enriched in cometary dust indicating that cometary solid material is more primitive than CI-chondrites. Relative to dust in dense molecular clouds, bulk cometary dust preserves the abundances of C and N, but exhibits depletions in O and H. In most cases, the carbonaceous component of cometary particles can be characterized as a multi-component mixture of carbon phases and organic compounds. Cluster analysis identified a few basic types of compounds, such as elemental carbon, hydrocarbons, polymers of carbon suboxide and of cyanopolyynes. In smaller amounts, polymers of formaldehyde, of hydrogen cyanide and various unsaturated nitriles also are present. These compositionally simple types, probably, are essential "building blocks", which in various combinations give rise to the variety of involatile cometary organics.

1. Introduction

Convincing evidence has been accumulated (Mumma, 1997) that comets are the most primitive objects in the solar system, preserving pristine materials from the presolar molecular cloud and from the early stages of the proto-solar nebula. During the planetary formation phase, some volatile-rich planetesimals became comets while some accreted into planets. Later, the impacts of comets as well as the constant flux of interplanetary dust (a fraction of which undoubtedly has a cometary origin) contributed to the prebiotic organic inventory on Earth and other planets. Therefore, the chemical and physical properties of cometary organic materials shed light on:
- the nature of primordial interstellar organics present during the solar system formation;
- the processes of incorporation of volatiles and organics into the solar system bodies;
- the post-accretion flux of exogenous organic material in the solar system.

The major goal of comet research is to determine the chemical composition of cometary nuclei, and this is the major thrust of future spacecraft missions to comets. In the meantime, attempts to reconstruct the true composition of the nuclei deal with remote observations of cometary gas and dust released into the coma during a perihelion passage.

While the composition of an inorganic component of cometary dust has been successfully inferred from 10 micron spectra of comets (cf. Hanner, this volume), the nature of an organic component is more difficult to learn. The 3.4 μm emission feature observed in a number of comets (Brooke et al., 1991) was initially ascribed to hot organic grains (Chyba et al., 1989), but later was explained by the presence of methanol and another, as yet unidentified, gaseous carrier (Di Santi et al., 1995). The ability to find signatures of organic grains in comets by remote sensing remains an open question. Thus, in our quest for studying the composition of cometary organics, we have to rely upon the only available set of data: mass spectra of Comet Halley dust particles obtained by spacecraft missions to this comet more than 10 years ago (Kissel et al., 1986).

Three international spacecraft sent to Comet Halley made a close flyby of its nucleus in March 1996 (and three additional spacecraft studied the comet from much greater distances). The closest approach distance was 8800 km for VEGA-1, 8000 km for VEGA-2, and 600 km for GIOTTO, and the relative velocities were 76 km/sec, 72 km/sec and 68 km/sec correspondingly. Almost identical instruments — dust-impact time-of-flight mass spectrometers PUMA-1, PUMA-2 and PIA — were flown onboard in order to determine the elemental, isotopic, and, if possible, even chemical/mineralogical composition of cometary dust. In all three instruments, high-velocity collisions of dust particles with targets (made of silver or Ag-doped platinum) provided a source of ions for mass-spectrometric analysis. The general outline of the instruments' functioning and related problems are presented by Jessberger (this volume) in sufficient detail.

More than 5000 individual dust particles were measured by all three instruments. Here we present the main findings about carbonaceous components of Comet Halley dust based on the measurements of the PUMA-1 and PUMA-2 instruments. These results have been reported in the literature, and the reader is referred to an extensive list provided by Jessberger (this volume) for original references. Note that masses of the analyzed grains are very small (range from 5×10^{-17}g to 5×10^{-12}g - Fomenkova et al., 1991) and the total mass of the sample is just a few nanograms. However, these particles, found in the coma at 8 - 180 thousand km from the nucleus, originated independently from different parts of the nucleus and thus yield a representative sample of the whole comet.

2. Results

1. Grains measured during the flyby by PUMA-1 and PUMA-2 instruments spent at least a few hours in the coma. It is possible, therefore, that during that time their original composition was altered to some extent by heating and/or UV irradiation.

2. Most reliable data were obtained on the abundances of C and N, while not quite understood processes of hyper-velocity impact ionization may have influenced the measured abundances of H and O in some cases (cf. discussions by Jessberger *et al.*, 1988; Fomenkova *et al.*, 1994).
3. The following quantitative classification of grains into three major groups was proposed by Fomenkova *et al.* (1992). Particles with the ratio of C to any rock-forming element (Mg, Si, Fe, Ca *et al.*) less than 0.1 are included in the group Rock. These particles contain no organic/carbonaceous materials. Particles where this ratio is larger than 10 are categorized as CHON, the remainder are Mixed. The number of particles in each group in PUMA-1 and PUMA-2 data is, correspondingly: Rock, 430 and 161, CHON, 464 and 51, Mixed, 974 and 288.
4. The overall mass ratio of silicates to organics in Comet Halley dust is between 2 and 1 (Fomenkova and Chang, 1993).
5. Cometary solids represent the most carbon-rich material in the solar system. A few IDPs have the abundance of carbon comparable to that in cometary dust (Thomas *et al.*, 1993) and it is possible that these IDPs are of cometary origin.
6. On average, CHON and Mixed grains containing organic materials are heavier than Rock particles (Fomenkova *et al.*, 1992). It is possible that the structure of larger grains is similar to that of some IDPs: carbonaceous matrix with embedded small mineral particulates.
7. Particles containing some amounts of organics (CHON and Mixed) are relatively more abundant closer to the nucleus, in the central part of the coma, while Rock particles dominate outer regions of the coma. These observations suggest that more volatile components of the grains gradually evaporate releasing small silicate grains and providing an extended source for some gaseous species observed (A'Hearn *et al.*, 1986; Eberhardt *et al.*, 1987).

The next two results rely implicitly on the dust-to-gas weight ratio of 2:1 as has been determined for Comet Halley (Krankowsky and Eberhardt, 1990; Jessberger and Kissel, 1991).

8. Overall (dust+gas) abundance of C and O is solar, while H and N are depleted by factors of 600 and 2, correspondingly. Geiss (1987) estimated that N was depleted by a factor of 3, but his result was based on an estimated dust-to-gas weight ratio of 0.5 which was later revised upward.
9. Carbon and nitrogen occur predominantly in the dust, the partition between a solid phase and a gas phase being 2:1. On the other hand, H and O are two times more abundant in the gas than in the dust - as expected for a water-rich object.
10. The bulk abundances of rock-forming elements in the dust are solar (chondritic) within a factor of 2 (cf. Jessberger, this volume).
11. The bulk abundance of O in the cometary dust is about the same as its chondritic abundance, while the abundances of H, C, N in cometary dust are higher than

in carbonaceous chondrites. The higher abundance of volatiles in cometary dust points toward its more pristine nature.

The next items discuss the compositional make-up and properties of the specific carbonaceous compounds found among the dust grains (Fomenkova et al., 1994; Fomenkova and Chang, 1997).

12. The following specific carbonaceous compounds have been identified in the dust: pure carbon (C grains), polycyclic aromatic and highly branched aliphatic hydrocarbons ([H, C] grains), polymers of carbon suboxide and of cyanopolyynes ([C, O] and [C, N] grains).

13. The majority of grains ([H, C, N], [H, C, O], and [H, C, N, O] groups) contain heteropolymers and/or variable mixtures of carbon phases and complex organic compounds - various compounds which are consistent with structures of alcohols, aldehydes, ketones, acids and amino acids, and their salts. Exact make-up of these mixtures cannot be unambiguously identified from the available data. Note that the simplest members of those homologous series occur in the interstellar medium (Irvine, 1998), and members of a higher molecular weight were found in the Murchison meteorite (Cronin and Chang, 1993) and observed in laboratory experiments simulating the origin to comets ("cometary ice tholins" - McDonald et al., 1996).

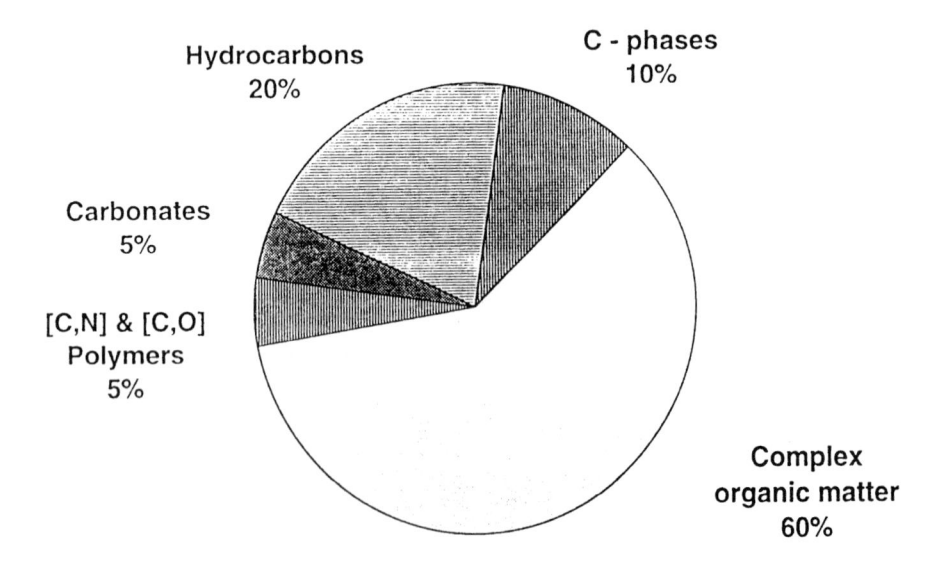

Figure 1. The distribution of carbon among different carbonaceous compounds of cometary dust.

14. Figure 1 illustrates the distribution of the total carbon content between different compounds.

15. Grains composed almost entirely out of carbon or containing pure C inclusions typically are in the mass range of $< 10^{-15}$ g, that is among the smallest observed. Based on unusually high $^{12}C/^{13}C$ ratios identified in such grains (Jessberger and Kissel, 1991), it was suggested that the carbon cometary grains may have a circumstellar rather that an interstellar origin (Fomenkova and Chang, 1995).
16. The proportion of pure carbon phases decreases when interstellar dust is being incorporated and processed into planetary materials: in the ISM up to 50% of the solid carbon is locked up in graphite or amorphous carbon, in Comet P/Halley carbon grains contain $\sim$10% of the total solid carbon, and in carbonaceous chondrites carbon phases account for only $\sim$2 % of the total carbon content.
17. The proportion of carbon bound in complex organic material increases with the degree of processing: $\sim$50% in interstellar grains and their mantles, 70-80% in Comet P/Halley, and 90-95% in carbonaceous meteorites.
18. The observed diversity of types of cometary organic compounds is consistent with the interstellar dust model of comets (Greenberg, 1982) and probably reflects the differences in the history of precursor dust.

Acknowledgements

The author thanks the International Space Science Institute for the travel grant to attend the workshop "The Origin and Composition of Cometary Material". The support from NASA Exobiology Program NAG5-4659 is gratefully acknowledged.

References

A'Hearn, M. F., et al.: 1986, 'Cyanogen jets in comet Halley', *Nature* **324**, 649-651.

Brooke, T. Y., Tokunaga, A., and Knacke, R.: 1991, 'Detection of the 3.4 μm emission feature in comets P/Brorsen-Metcalf and Okazaki-Levy-Rudenko and an observational summary', *Astron. J.* **101**, 268-278.

Chyba, C., Sagan, C., and Mumma, M. : 1989, 'The heliocentric evolution of cometary infrared spectra: results from an organic grain model' *Icarus* **79**, 362-381.

Cronin, J., and Chang S.: 1993, 'Organic matter in meteorites: molecular and isotopic analysis of the Murchison meteorite' in *The Chemistry of Life's Origin* (ed. J. M. Greenberg et al.) pp. 209-258, Kluwer.

Di Santi, M. et al.: 1995, 'Systematic observations of methanol and other organics in comet P/Swift-Tuttle: discovery of new spectral structure at 3.42 μm' *Icarus* **116**, 1-17.

Eberhardt, P. et al.: 1987, 'The CO and N_2 abundance in comet Halley',*Astron. Astrophys.* **187**, 481-484.

Fomenkova, M., Evlanov, E., Mukhin, L. and Prilutsky, O.: 1991, 'Determination of mass of comet Halley dust particles' *LPSC* **22**, 397-398.

Fomenkova, M., Kerridge, J., Marti, K., and McFadden, L.: 1992, 'Compositional trends in rock-forming elements of comet Halley dust' *Science* **258**, 266-269.

Fomenkova, M. and Chang, S.: 1993, 'Mass and spatial distribution of the carbonaceous components in comet Halley' *Lun. Plan. Sci. Conf.* **24**, 501-502.

Fomenkova, M., Chang, S. and Mukhin, L.: 1994, 'Carbonaceous components in the comet Halley dust' *Geochim. Cosmochim. Acta* **58**, 4503-4512.

Fomenkova, M., and Chang, S.: 1995, 'Chemical evolution of interstellar dust into planetary materials', *Lunar Plan. Sci. Conf.* **26**, 413-414.

Fomenkova, M. and Chang, S.: 1997, 'Carbonaceous components of organic/inorganic assemblages in comet Halley dust', *Meteoritics & Planetary Science* **32**, A44.

Geiss, J.: 1987, 'Composition measurements and the history of cometary matter', *Astron. & Astrophys.* **187**, 859-866.

Greenberg, J. M.: 1982, 'What are comets made of? A model based on interstellar grains' in *Comets* (ed. L. L. Wilkening), pp. 131-163, Univ. of Arizona Press.

Hanner, M.: 1999, 'The Silicate Material in Comets', *Space Sci. Rev.*, this volume.

Irvine, W.: 1998, 'Extraterrestrial organic matter: a review', in *Origins of Life and Evolution of the Biosphere* **28**, Kluwer Academic Publishers, pp. 365-383.

Jessberger, E.K., Cristoforidis, A., and Kissel, J.: 1988, 'Aspects of the major element composition of Halley's dust' *Nature* **332**, 691-695.

Jessberger, E.K. and Kissel, J.: 1991, 'Chemical properties of cometary dust and a note on carbon isotopes', in '*Comets in the Post-Halley Era*' (ed. R. Newburn *et al.*), pp. 1075-1092, Kluwer.

Jessberger, E.K.: 1999, 'Rocky Cometary Particulates: Their Elemental, Isotopic and mineralogical ingredients', *Space Sci. Rev.*, this volume.

Kissel, J. *et al.*: 1986, 'Composition of comet Halley dust particles from Vega observations', *Nature* **321**, 280-282, 336- 338.

Krankowsky, D. and Eberhardt, P.: 1990, 'Evidence for the composition of ices in the nucleus of comet Halley', in *Comet Halley: Worldwide Investigations, Results and Interpretations*, pp. 273-289. Ellis Horwood Ltd.

McDonald, G. D. *et al.*: 1996, 'Production and chemical analysis of cometary ice tholins', *Icarus* **122**, 107-117.

Mumma, M.: 1997, 'Organics in comets' in *Astronomical and Biochemical Origins and the Search for Life in the Universe* (eds. C. Cosmovici *et al.*), pp. 121-142, Editrice Compositori.

Thomas, K., Blanford, G., Keller, L., Klock, W. and McKay, D.: 1993, 'Carbon abundance and silicate mineralogy of anhydrous interplanetary dust particles' *Geochim. Cosmochim. Acta* **57**, 1551-1566.

Address for correspondence: Marina N. Fomenkova, mfomenkova@ucsd.edu

II: FROM COMA ABUNDANCES TO NUCLEUS COMPOSITION

FROM COMA ABUNDANCES TO NUCLEUS COMPOSITION

W. F. HUEBNER
Southwest Research Institute, San Antonio, TX 78228-0510, USA

J. BENKHOFF
DLR, Institut für Planetenerkundung, D-12484 Berlin, Germany

Abstract. A major goal of comet research is to determine conditions in the outer solar nebula based on the chemical composition and structure of comet nuclei. The old view was to use coma abundances directly for the chemical composition of the nucleus. However, since the composition of the coma changes with heliocentric distance, r, the new view is that the nucleus composition must be determined from analysis of coma mixing ratios as a function of r. Taking advantage of new observing technology and the early detection of the very active Comet Hale-Bopp (C/1995 O1) allows us to determine the coma mixing ratios over a large range of heliocentric distances.

In our analysis we assume three sources for the coma gas: (1) the surface of the nucleus (releasing water vapor), (2) the interior of the porous nucleus (releasing many species more volatile than water), and (3) the distributed source (releasing gases from ices and hydrocarbon polycondensates trapped and contained in coma dust).[*] Molecules diffusing inside the nucleus are sublimated by heat transported into the interior. The mixing ratios in the coma are modeled assuming various chemical compositions and structural parameters of the spinning nucleus as it moves in its orbit from large heliocentric distance through perihelion.

We have combined several sets of observational data of Comet Hale-Bopp for H_2O (from OH) and CO, covering the spectrum range from radio to UV. Many inconsistencies in the data were uncovered and reported to the observers for a reanalysis. Since post-perihelion data are still sparse, we have combined pre- and post-perihelion data. The resulting mixing ratio of CO relative to H_2O as a function of r is presented with a preliminary analysis that still needs to be expanded further. Our fit to the data indicates that the total CO release rate (from the nucleus and distributed sources) relative to that of H_2O is 30% near perihelion.

Keywords: Coma, Nucleus, Heat Diffusion, Gas Diffusion, Mother Molecules

1. Introduction

There are two major goals for comet research: One goal is to determine the chemical composition and physical structure of comet *nuclei* to reflect on chemical and thermodynamic conditions in the nebula in which comets formed, and a second goal is to determine physical properties of comet nuclei needed for mitigation of potentially hazardous objects (PHOs) of which comets are one component. Mitigation procedures critically depend on physical properties, such as material strengths of the nucleus (see, e.g., Huebner, 1999). We could mention additional

---

[*] Historically, coma gas source 3 was identified before source 2.

research goals such as understanding the nature of materials imported to young planetary surfaces to contribute to oceans, atmospheres, and even the origins of life. However, in all of these cases the materials have undergone chemical changes to such an extent that it is difficult to trace their metamorphosis.

Two constraints must be applied in the analysis of comet nuclei: (1) Observations of comae of evolved comets (as they exist now) and (2) models for formation of comet nuclei. Coma observations provide a strict, although not readily interpretable, constraint. We will discuss it in detail in Section 3, below. Comet formation models provide a more open, much less certain constraint.

2. Scenarios for Comet Nucleus Formation

The origin of the solar system is an active area of research. It is based on observations of solar system bodies including the Sun, planets, moons, asteroids, comets, and interplanetary dust, and on observations of dark interstellar clouds, star-forming regions, circumstellar disks, and indicators of extrasolar planets.

The origin of comet nuclei is usually studied within the framework of a cosmogonic model and is closely linked to the origin of the solar system. The physical details and the sequence for formation of comet nuclei and of the solar system are still vigorous research topics. However, there are a number of hypotheses that can be compared with results based on observations of disks surrounding young main-sequence stars (see, e.g., Waelkens, 1999) as analogs of our solar nebula and of interstellar clouds (see, e.g., Dartois *et al.*, 1998) as precursors of a presolar nebula.

The agreement of isotopic composition of basic refractory elements in the Sun, the Earth, and meteorites, and the similarity of element abundances in the Sun and in Jupiter, are evidence that all members of the solar system originated from the same mixture of material in the solar nebula. The elemental composition of *condensable* materials of at least one comet (Comet Halley) is also very close to solar composition (Grün and Jessberger, 1990; Jessberger and Kissel, 1991). Obviously this excludes hydrogen, the noble gases, and nitrogen, which is not very reactive and forms mostly noncondensable N_2 (cf. Geiss *et al.*, this volume). In addition, N_2 is an apolar molecule and therefore is very mobile and can easily escape if trapped in comet nuclei. Some molecular clouds (e.g., the Taurus molecular cloud), recognized as regions of star formation, contain stars that are similar to the Sun (a spectral type G2 star) in terms of mass and chemical composition. In terms of age they are much younger (10^5 to 10^7 years) compared to the Sun which is 4.6×10^9 years old. Theoretical models indicate that the evolution of a star is determined by its mass and composition of chemical elements. The similarity of all the parameters justifies viewing T-Tauri stars as an analog of the young Sun and to use the observational data of these stars to construct a theory for formation of the solar nebula and the solar system.

In molecular clouds, the refractory elements are tied up in dust grains. The nebula from which our solar system formed began with the collapse of a dense interstellar cloud starting at the interior and working its way outward (see left side of Fig. 1). Grains fell inward together with gas. During the collapse the cloud was spinning. Smaller companion fragments may have spun off. This spinning caused a flattening of the cloud to give it a disk-like appearance. As more molecules and dust from the cloud fell into the solar nebula they encountered increasingly higher densities and were slowed down from supersonic flow to subsonic flow, creating a shock – the nebular accretion shock. Material falling through the shock was heated. Temperatures in the shock may have been sufficiently high even in the outer solar nebula (Ruzmaikina, 1998) to transform amorphous water ice into crystalline ice (transition temperature $T \approx 135$ K). The shock may have vaporized ices (Engel *et al.*, 1990; Lunine *et al.*, 1991; Chick and Cassen, 1997; Cassen and Chick, 1997) and possibly some organic components in loosely aggregated dust particles. Even dissociation of some molecules cannot be ruled out. Radicals may have undergone chemical reactions. Grains may have been transported radially in the disk and consequently enhanced heterogeneous thermal chemistry. Long-term drag heating of particles that ensued the shock heating and x-rays and extreme UV radiation from the young Sun may have caused additional destruction of the interstellar composition but enhanced solar nebula chemistry. Other processes that are important for nebula chemistry, but only peripherally understood, include energetic chemistry associated with lightning and nebular flares.

During the time that the solar nebula was turbulent, mixing of different parts of the nebula may have occurred, and grains entrained by gas collided and aggregated. Small, fluffy particles formed and were held together by van der Waals forces. As turbulence faded, the particles precipitated toward the central plane of the disk. During this process the fluffy particles suffered further collisions and grew to centimeter size and larger.

Based on our present knowledge, formation of comet nuclei occurs in a cold (about 30 K to 50 K) medium with sufficient density where molecules condense on particles and particle aggregation leads to the formation of subnuclei. Thus, a major hypothesis for the origin of comet nuclei is that they formed icy planetesimals in the zone of the giant planets and beyond. Nuclei forming in the giant planet zone were subsequently scattered by gravitational interactions with the outer planets into the Oort cloud. Comet nuclei forming in the trans-Neptunian region make up what is now called the Edgewood-Kuiper (EK) belt. About 60 EK-belt objects have been discovered to date. Observational evidence exists for three dynamical classes of these Kuiper belt objects: (1) The classical Kuiper belt objects, (2) Plutinos in resonance orbits, and (3) an extended swarm of scattered objects such as 1996 TL66 (Jewitt *et al.*, 1998). While we do not expect to see differences in *elemental* abundance between these groups, differences in *molecular* abundances may be expected.

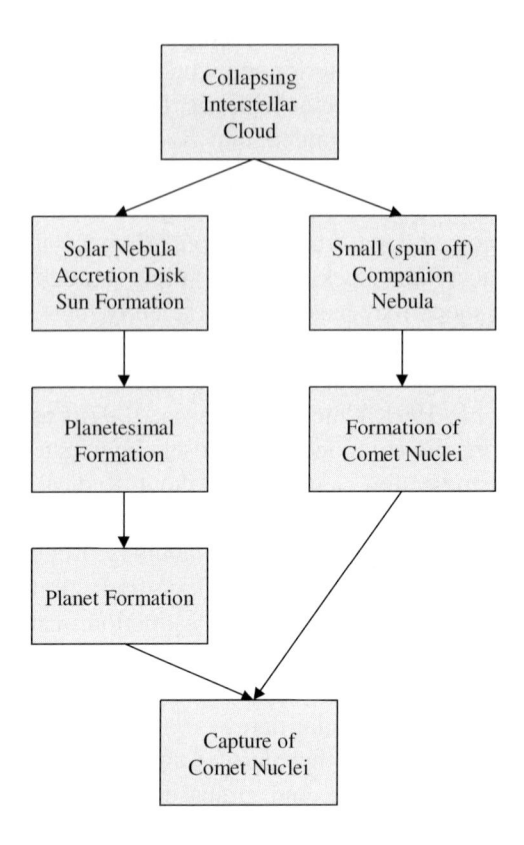

```
        Collapsing
        Interstellar
           Cloud
         /        \
Solar Nebula    Small (spun off)
Accretion Disk   Companion
Sun Formation     Nebula
     |               |
Planetesimal    Formation of
 Formation      Comet Nuclei
     |               |
Planet Formation     |
         \          /
          Capture of
          Comet Nuclei
```

Figure 1. Schematic presentation of likely processes in the solar nebula.

In the scenario just described, water ice would have to be in a crystalline phase since it condensed in the solar nebula (Kouchi *et al.*, 1994). More volatile ices would be in separate phases, and the molecular composition would be determined by chemical reactions in the solar nebula. These reactions could lead to many species similar to interstellar molecules, but they would not be molecules that survived accretion from the interstellar cloud. Yet, there are powerful arguments that the composition of comets is closely related to the composition of interstellar clouds (see, e.g., Greenberg, 1982, 1987; Greenberg and Grim, 1986; Mumma *et al.*, 1996; Irvine *et al.*, 1999; Eberhardt, 1999; Meier *et al.*, 1998). For a comparison of molecules found in comets versus those detected in the interstellar medium and in dark interstellar clouds, see Huebner and Benkhoff (1999a) and Irvine (this volume).

More in line with an interstellar origin of comet nuclei is their formation in a companion fragment of the solar or presolar nebula. Such a companion fragment may be a smaller cloud that has separated from the original cloud. It is not massive enough to form a star, but it is large enough to form comet nuclei. Physical processes for formation of comet nuclei in a companion nebula (Cameron, 1973) have not been investigated in more detail since the early work by Biermann and Michel (1978). However, we may expect that the molecular composition of comet nuclei forming in it may be interstellar since its accretion shock must have been much weaker. It also has the appealing feature that stellar perturbations can easily extract comet nuclei from the companion fragment. Comet nuclei that can be captured by the main cloud will have a random (isotropic) distribution similar to that of the Oort cloud. In this scenario, Oort cloud and Edgewood-Kuiper belt comets would have a distinctly different composition.

Conceptually, the origin of comets is equivalent to the formation of kilometer-sized bodies consisting of ice–dust mixtures. Figure 1 describes some of these processes.

3. Nucleus Composition from Coma Abundances

The old view was that coma abundance directly reflects the composition of the nucleus. This required chemical modeling of the coma coupled to gas-dynamic flow. In parallel with the chemical modeling was the modeling of entrainment, fragmentation, and vaporization of the dust.

The new view is that since the composition of the coma depends on heliocentric distance, distance-dependent analyses of the coma will provide a better evaluation of the composition of the nucleus. The mixing ratio of volatile mother molecules relative to H_2O changes with heliocentric distance, as amply illustrated by observations of Comet Hale-Bopp (C/1995 O1) by Biver et al. (1997), Bockelée-Morvan and Rickman (1999), and Crovisier and Bockelée-Morvan (this volume). The mixing ratios can change by factors as large as several hundred going from heliocentric distances of $r \approx 1$ AU to $r \approx 7$ AU. This requires all of the above modeling plus modeling of the heat and gas-flow in the nucleus (Huebner and Benkhoff, 1999a; Huebner and Benkhoff, 1999b). Figure 2 illustrates these requirements.

The left-most column in Fig. 2 indicates the type of process occurring in the parts of the comet as indicated in boxes to the right. The arrows show the more detailed processes. The processes most important for the discussions here are in the top half of the figure.

Energy input on the nucleus is solar radiation. The solar wind may reach the nucleus only at large heliocentric distances, but cannot reach the nucleus if the comet has a well-developed coma. Part of the solar radiation is scattered or absorbed by the coma gas and dust. The absorbed energy is reradiated in the infrared (IR) region of the spectrum. Thus, attenuated solar radiation (mostly visible light) reaches the nucleus. Except for very hard UV photons, the coma of a bright comet at 1 AU heliocentric distance is optically thick in the UV. Optical thickness reduces photodissociation and photoionization and thus inhibits chemical reactions in the

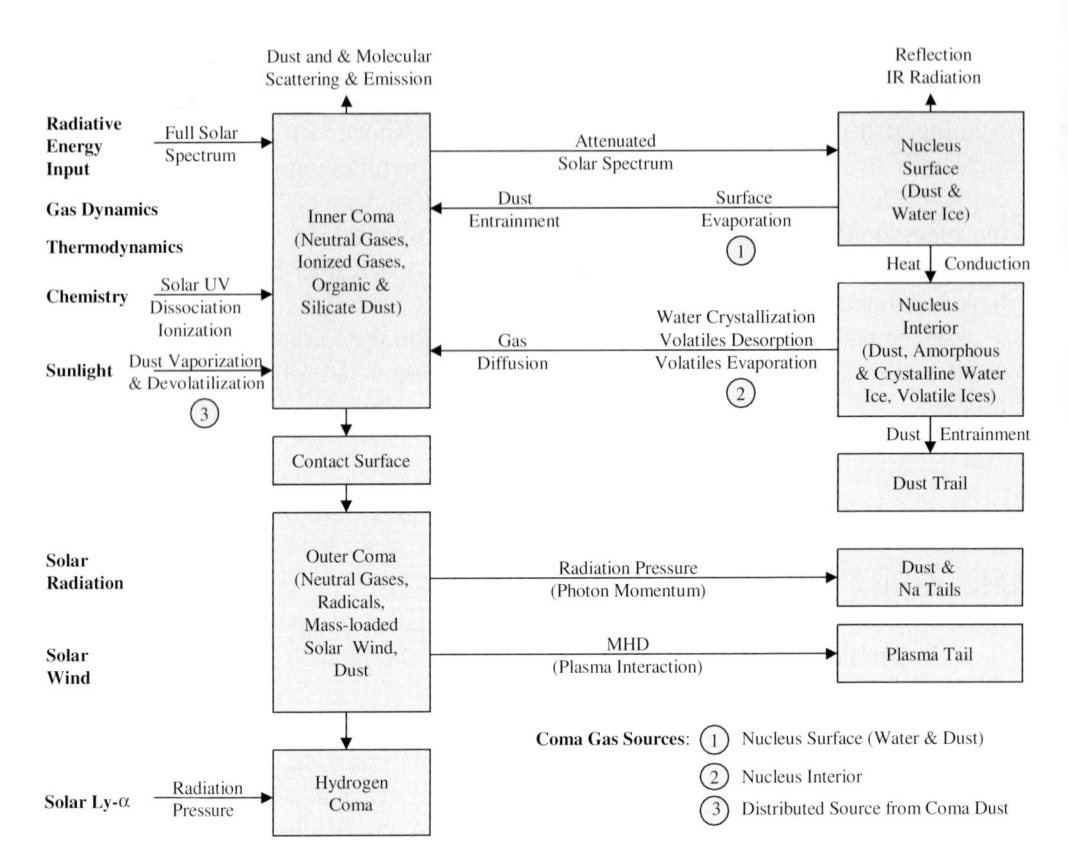

Figure 2. Radiative, physical, chemical, and thermodynamic processes in comets as they relate to the nucleus, coma, dust and plasma tails, and dust trails.

inner coma. The solar energy reaching the nucleus is to some small part reflected (low albedo of the nucleus), in part reradiated in the IR (high emissivity), and the rest is absorbed on the surface. Most of the energy absorbed on the surface of the nucleus is used to evaporate (sublimate) water ice. This is indicated by coma gas source 1 in Fig. 2. A smaller fraction is transported into the interior of the nucleus by solid state conduction and to a very small degree by radiation in pores. The amount of water vapor released depends on the dust cover on the surface. A few millimeters of dust cause most of the incident energy to be reradiated, leaving very little energy for sublimation of water ice. Since the solar energy input decreases with the square of the distance from the Sun, the reradiated energy will dominate the power balance at large heliocentric distances even for exposed icy surfaces. Figure 3 represents the maximum sublimation rate of water ice from the surface of a comet nucleus at the subsolar point, $\zeta = 0°$, with a Bond albedo $A = 4\%$, and is corrected to zero pressure. The curve can be scaled to other values for the angle of insolation (ζ) and albedo (A) by the appropriate *horizontal* shift as indicated on the abscissa. Vapor pressure and enthalpy for vaporization data, valid for the range $100 \leq T \leq 273.16$, are from Gibbons (1990). The release rates are maximum values because it is assumed that there is no dust on the surface, no heat conduction into the interior (and therefore no heat stored in the nucleus), and the return flux of gas to the surface is negligible.

Heat that is conducted into the interior of the porous nucleus may encounter ices more volatile than water ice. Figure 4 illustrates schematically the heat and gas diffusion in a comet nucleus, assuming only a simple mixture of water ice and CO ice for the purpose of illustration. Many, more different species are expected to be in a comet nucleus. As heat diffuses into the nucleus, each volatile constituent forms its own sublimation front depending on its enthalpy of vaporization. If amorphous ice is present, it will change to crystalline ice in an irreversible and exothermic process, forming a crystallization front for the phase transition. Gases trapped by the amorphous ice are released during the phase transition. If the ice is crystalline, then volatile ices are frozen as separate phases. As an ice species evaporates the gas pressure at the sublimation front increases toward its maximum (equilibrium) value at that temperature. The pressure forms a gradient that is negative above the sublimation front and positive below it (as seen from the center of the nucleus). This pressure gradient drives the gas flow. The gas flowing outward will diffuse through the comet nucleus and escape through its surface into the coma. This is labeled as coma gas source 2 in Fig 2. The gas flowing inward will recondense a short distance below the sublimation front. This recondensation occurs within a thermal skin depth and liberates energy. It also constricts the pores.

Because of heat and gas diffusion, the nucleus will be chemically differentiated in layers. The least volatile material (dust) will be at the top of the nucleus. It will be followed by a layer of dust and water ice. The next deeper layer will contain dust, water ice, and species such as CO_2. In the deepest layers we would find dust and all ices including the most volatile species (such as CO and CH_4).

At the surface of the nucleus, water and other, more volatile ices evaporate, leaving a layer of dust behind. This dust may be entrained by escaping gas or, if the gas flux cannot overcome gravity and adhesive forces, it can remain on the surface creating an inactive mantle. Dust particles entrained by the gas into the coma will heat up in sunlight, and the organic component (hydrocarbon polycondensates) will be vaporized. Hydrocarbon polycondensates and possibly some ice trapped in the fluffy dust particles constitute the distributed source of coma gas labeled 3 in Fig. 2. Polymerized formaldehyde (POM) is an important part in the dust to produce short-lived formaldehyde gas, which quickly dissociates into CO (Boice *et al.*, 1990; Eberhardt, this volume).

To obtain the CO gas release rates from the nucleus, the CO release rates from the distributed coma source must be subtracted from the total CO release rates. This is a difficult problem because of the scatter of the data and there may be more than one distributed source for CO.

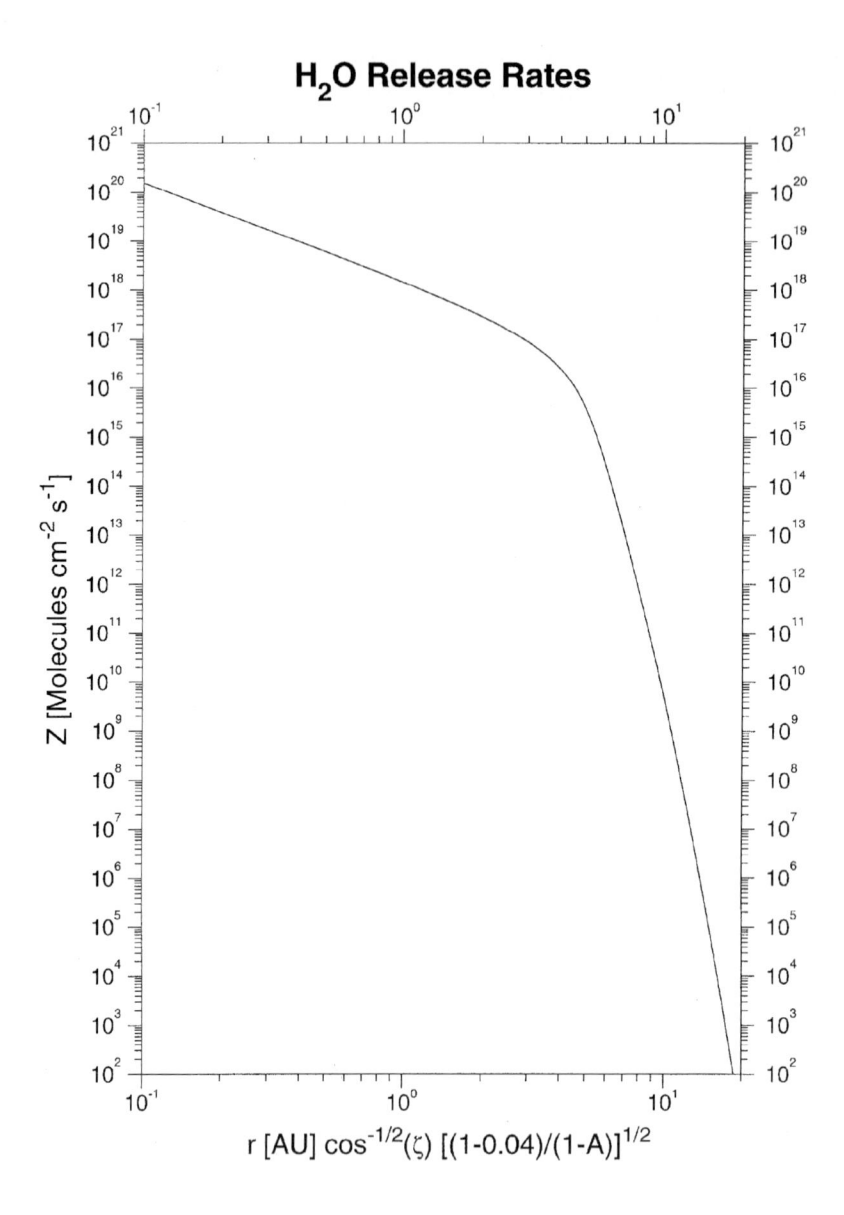

Figure 3. Maximum water release rates from the surface of a comet nucleus at the subsolar point $\zeta = 0°$, with Bond albedo $A = 4\%$, and corrected for zero pressure. The curve may be shifted horizontally for different albedo and different angle of insolation, as suggested by the labeling of the abscissa.

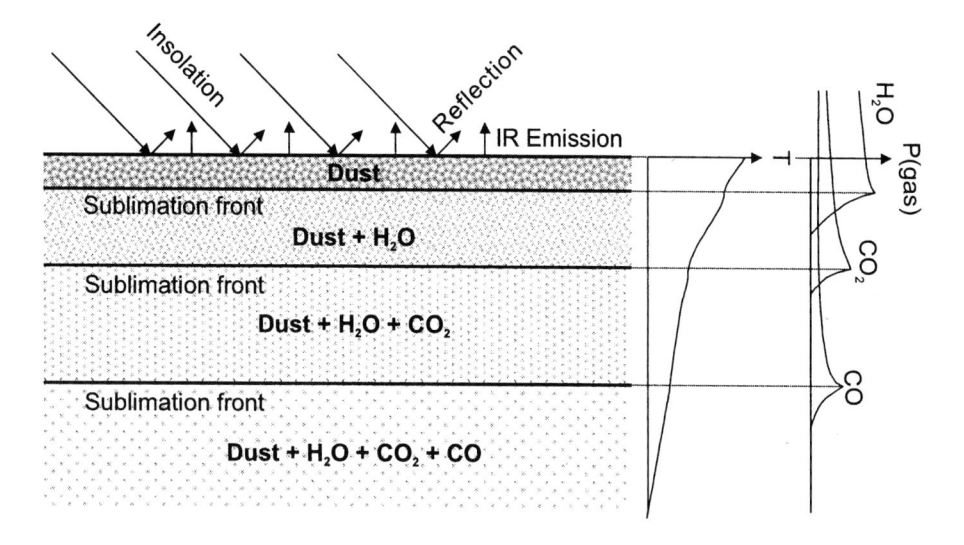

Figure 4. Schematic presentation of heat and gas diffusion in a porous comet nucleus.

There appears to be a dip in the gas release rates for many species (including H_2O) as observed in Comet Hale-Bopp inbound between approximately 1 and 2 AU. The cause for this dip may have several reasons: The orientation of the spin axis may cause a seasonal effect (Kührt, this volume), or dust may accumulate on the surface of the nucleus, reducing the efficiency of vaporization, at about 2 AU inbound but then get blown away at about 1 AU inbound as a result of increased insolation and the accompanying gas pressure increase. Alternatively, it has been suggested that icy particles contributed significantly to the gas release rates at large heliocentric distances (see, e.g., Bockelée-Morvan and Rickman, 1999). In Fig. 5 we have compiled the H_2O observations as obtained from OH by many (but by no means all) observers. The data include radio observations of Biver *et al.* (1997), Bockelée-Morvan and Rickman (1999), and Womack *et al.* (1997) with modifications as given by Biver *et al.* (1998) and Womack (1998). IR data were obtained from DiSanti *et al.* (1999). Data from the visual range of the spectrum were supplied by Schleicher *et al.* (1997). They had many data with large error bars at large heliocentric distances. Because of the large error bars, we averaged the data corresponding to the same heliocentric distance. The effect is that it reduces the weighting of the many points with large error bars when fitting the data to curves. UV data were taken from Weaver *et al.* (1997) with an update supplied by Weaver (1998). A cubic fit to the logarithm of the H_2O release rates fits the data best, but if extrapolated to larger heliocentric distances, it would exceed the maximum water release rates for $\zeta = 0°$. Forcing the cubic fit to approach the $\zeta = 0°$ curve at large r (see Fig. 5) tends to exaggerate the dip between $r = 1$ and 2 AU.

Hale - Bopp
H$_2$O Release Rates vs. Heliocentric Distance

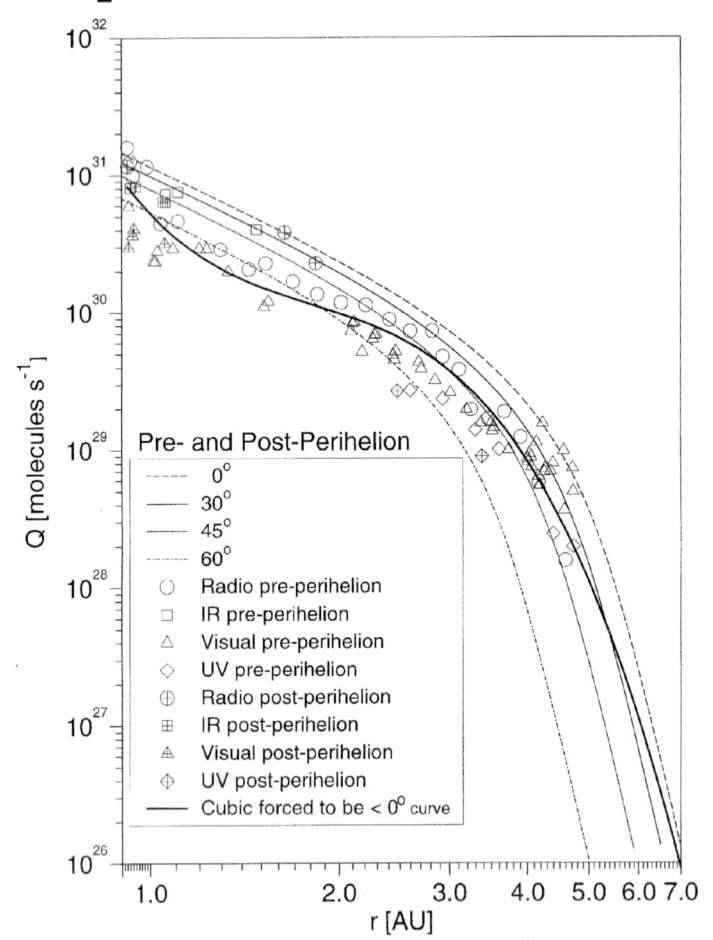

Figure 5. H$_2$O observational data from the radio to the UV range of the spectrum. Thin lines from Fig. 3 for angles of incidence of solar radiation at $\zeta = 0°$, $30°$, $45°$, and $60°$ as a guide. Heavy solid line is a fit to the logarithm of the release rates forcing them to approach the $\zeta = 0°$ curve at large heliocentric distances.

In Fig. 6 we compiled the CO observations from the radio range [Biver *et al.* (1997), Bockelée-Morvan and Rickman (1999), Womack *et al.* (1997), and Jewitt *et al.* (1996)] with modifications as given by Womack (1998), and the IR range from DiSanti *et al.* (1999). A cubic curve gives a satisfactory fit to the observational data.

Hale - Bopp
CO Release Rates vs. Heliocentric Distance

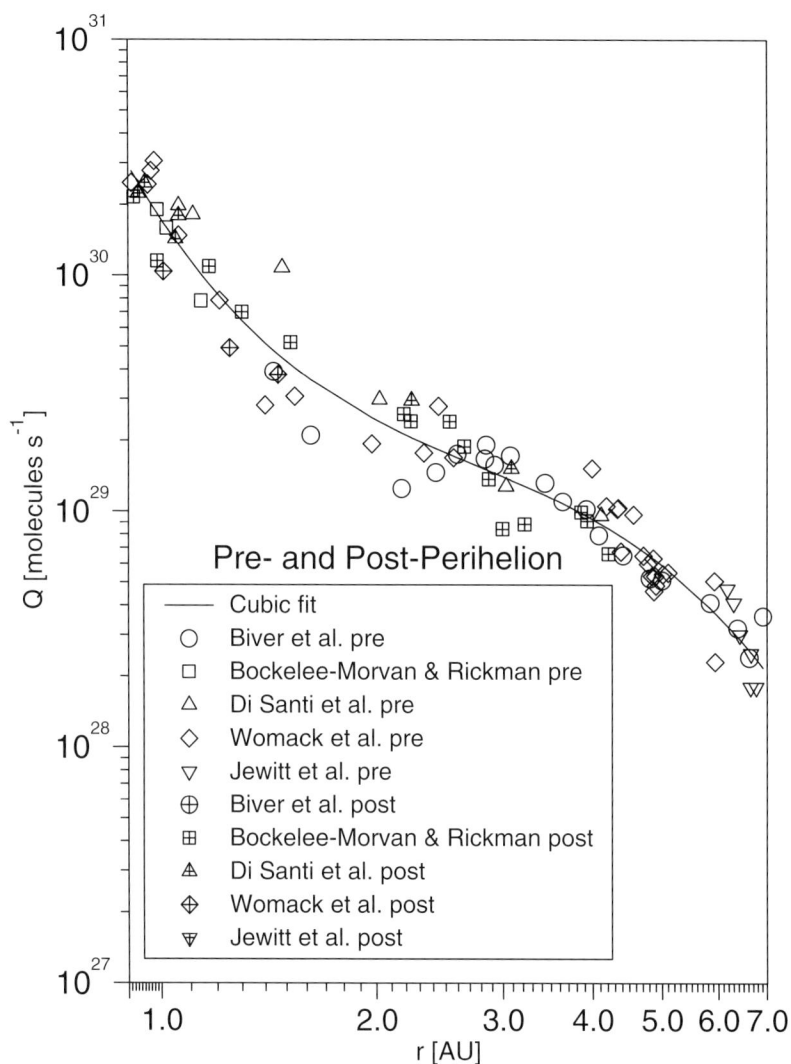

Figure 6. CO observational data from the radio and IR ranges of the spectrum. Solid line is a cubic fit to the logarithm of the release rate.

4. Conclusions

Figure 7 is an improved version with more observational data than earlier attempts permitted to produce a mixing ratio curve for CO relative to H_2O (Huebner and Benkhoff, 1999a, 1999b). Our fit to the data indicates that the total CO release rate (from the nucleus and the distributed sources combined) relative to that of H_2O is 30% (by number of molecules) near perihelion. The mixing ratio of release rates

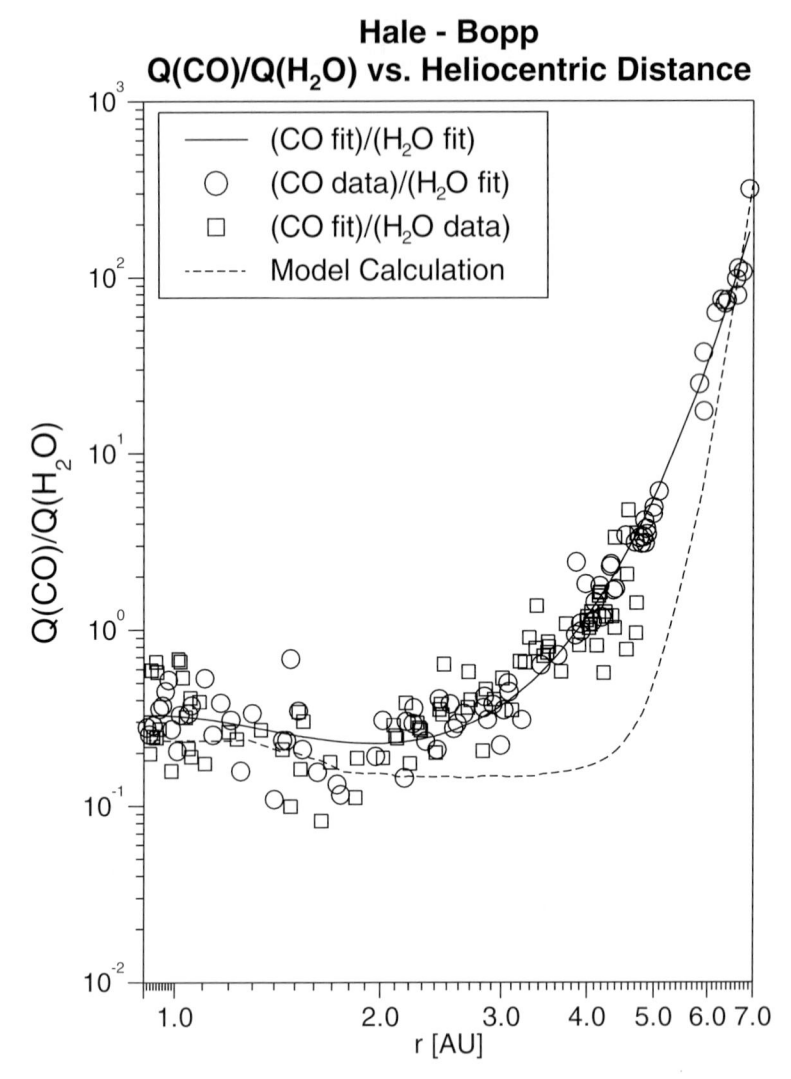

Figure 7. The mixing ratio (by number abundances) of CO relative to H_2O using the cubic fits to the observational release rates (solid curve) shown in Figs. 5 and 6. The heavy dashed curve is the result from model calculations for a mixture (by mass) of 35% amorphous H_2O, 7% CO_2, 13% CO (50% trapped in the amorphous ice), and 45% dust. The model result was deliberately chosen to be low because the CO from the distributed source has not yet been subtracted.

of CO/H_2O is much more sensitive then a comparison of CO and H_2O separately. The best results from our model calculations corresponds to a mixture (by mass) of 35% amorphous H_2O, 7% CO_2, 13% CO (50% trapped in the amorphous ice), and 45% dust. The CO_2 has very little influence on the results. This mixture is close to what one would expect from the *condensable* component of molecules forming at low temperatures from a mixture of elements with solar abundances. Although a better fit could have been achieved with a higher value of CO relative to H_2O in the nucleus, we opted for this preliminary value because the CO from the distributed source has not yet been subtracted. We want to emphasize that the results are still very preliminary.

The model results are too low in the region from $r \approx 2$ AU to 5 AU. After more post-perihelion data are analyzed, the pre-perihelion and post-perihelion data must be fitted separately and the distributed source for CO must be subtracted. Then model calculations as described above will be used in attempts to match the observational mixing ratios for more species using the appropriate spin axis orientation (Farnham *et al.*, 1998) to account for seasonal effects, which we believe will be important (cf. Kührt, this volume). We also point out that there are still significant discrepancies in thermal and gas diffusion models (Huebner *et al.*, 1999).

Acknowledgements

We gratefully acknowledge support from NASA grant NAG5-6785.

References

Biver, N., Bockelée-Morvan, D., Colom, P., Crovisier, J., Davies, J. K., Dent, W. R. F., Despois, D., Gérard, E., Lellouch, E., Rauer, H., Moreno, R., Paubert, G.: 1997, *Science* **275**, 1915.

Biver, N., Bockelée-Morvan, D., Colom, P., Crovisier, J., Germain, B., Lellouch, E., Davies, J. K., Dent, W. R. F., Moreno, R., Paubert, G., Wink, J., Despois, D., Lis, D. C., Mehringer, D., Benford, D., Gardner, M., Phillips, T.G., Gunnarsson, M., Rickman, H., Winnberger, A., Bergman, P., Johansson, L. E. B., Rauer, H.: 1998, *Earth, Moon, Planets*, in press. See also URL <http://www.ifa.hawaii.edu/~biver/prep/hbemp.dat>.

Biermann, L., Michel, K. W.:1978, *Moon Planets* **18**, 447.

Bockelée-Morvan, D., Rickman, H.: 1999, *Earth, Moon, Planets*, in press.

Boice, D.C., Huebner, W. F., Sablik, M. J., Konno, I.: 1990, *Geophys. Res. Lett.* **17**, 1813.

Cameron, A. G. W.: 1973, *Icarus* **18**, 407.

Cassen, P., Chick, K. M.: 1997, in *Astrophysical Implications of the Laboratory Study of Presolar Materials*, eds. T. Bernatowicz, E. Zinner, American Institute of Physics, New York, p.697.

Chick, K. M., Cassen, P.: 1997, *Astrophys. J.* **477**, 398.

Dartois, E. *et al.*: 1998, *Astron. Astrophys. Lett.*, in press.

DiSanti, M. A., Mumma, M. J., Dello Russo, N., Magee-Sauer, K., Novak, R., Rettig, T. W.: 1999, preprint.

Eberhardt, P.: 1999, in *Proc. IAU Colloq. 168, Cometary Nuclei in Space and Time*, ed. M. A'Hearn, Astron. Soc. Pacific Conf. Series, in press.

Engel, S., Lunine, J. I., Lewis, J. S.: 1990, *Icarus* **85**, 380.

Farnham, T. L., Schleicher, D. G., Blount, E. A., Ford, E.: 1998, *Earth, Moon, Planets*, in press.

Gibbons, C. J.: 1990, *Ann. Geophys.* **8**, 859.

Greenberg, J. M.: 1982, in *Comets*, ed. L. L. Wilkening, University of Arizona Press, Tucson, p. 131.

Greenberg, J. M.: 1987, *Adv. Space Res.* **7**, 33.

Greenberg, J. M., Grim, R.: 1986, in *20th ESLAB Symposium on the Exploration of Halley's Comet*, ESA SP-250, Vol. 2, p. 255.

Grün, E., Jessberger, E. K.: 1990, in *Physics and Chemistry of Comets*, ed. W. F. Huebner, Springer-Verlag, Heidelberg, p. 113.
Huebner, W. F.: 1999, in *International Seminars on Planetary Emergencies*, in press.
Huebner, W. F., Benkhoff, J.: 1999a, in *Proc. IAU Colloq. 168, Cometary Nuclei in Space and Time*, ed. M. A'Hearn, Astron. Soc. Pacific Conf. Series, in press.
Huebner, W. F., Benkhoff, J.: 1999b, *Earth, Moon, Planets*, in press.
Huebner, W. F., Benkhoff, J., Capria, M. T., Coradini, A., De Sanctis, M. C., Enzian, A., Orosei, R., Prialnik, D.: 1999, *Adv. Space Sci.* **23**, 1823.
Irvine, W. M., Schloerb, F. P., Crovisier, J., Fegley Jr., B., Mumma, M. J.: 1999, *Protostars and Planets* **IV**, in press.
Jessberger, E. K., Kissel, J.: 1991, in *Comets in the Post-Haley Era*, eds. R. L. Newburn, M. Neugebauer, J. Rahe, Kluwer Academic Publishers, Dordrecht, p. 1075.
Jewitt, D., Senay, M., Matthews, H.: 1996, *Science* **271**, 1110.
Jewitt, D., Luu, J., Trujillo, C.: 1998, *Astron. J.* **115**, 2125.
Kouchi, A., Yamamoto, T., Kozasa, T., Kuroda, T., Greenberg, J. M.: 1994, *Astron. Astrophys.* **290**, 1009.
Lunine, J. I., Engel, S., Rizk, R., Horanyi, M.: 1991, *Icarus* **94**, 333.
Meier, R., Owen, T. C., Jewitt, D. C., Matthews, H. E., Senay, M., Biver, N., Bockelée-Morvan, D., Crovisier, J., Gautier, D.: 1998, *Science* **279**, 1707.
Mumma, M. J., DiSanti, M. A., Dello Russo, N., Fomenkova, M., Magee-Sauer, K., Kaminski, C. D., Xie, D. X.: 1996, *Science* **272**, 1310.
Ruzmaikina, T.: 1998, private communication.
Schleicher, D. G., Lederer, S. M., Millis, R. L., Farnham, T. L.: 1997, *Science* **275**, 1913.
Weaver, H. A.: 1998, private communication.
Weaver, H. A., Feldman, P. D., A'Hearn, M. F., Arpigny, C., Brandt, J. C., Festou, M. C., Haken, M., McPhate, J. B., Stern, S. A., Tozzi, G. P.: 1997, *Science* **275**, 1900.
Womack, M., Festou, M. C., Stern, S. A.: 1997, *Astron. J.* **114**, 2789.
Womack, M.: 1998, private communication.
Waelkens, C: 1999, *Earth, Moon, Planets*, in press.

Address for correspondence: Walter F. Huebner, whuebner@swri.edu

ON THE PREDICTION OF CO OUTGASSING FROM COMETS HALE-BOPP AND WIRTANEN

ACHIM ENZIAN*
Jet Propulsion Laboratory, California Institute of Technology

Abstract. The gas flux from a volatile icy-dust mixture is computed using a comet nucleus thermal model in order to study the evolution of CO outgassing during several apparitions from long-period Comet Hale-Bopp and short-period Comet Wirtanen. The comet model assumes a spherical, porous body containing a dust component, one major ice component (H_2O), and one minor ice component of higher volatility (CO). The initial chemical composition is assumed to be homogeneous. The following processes are taken into account: heat and gas diffusion inside the rotating nucleus; release of outward diffusing gas from the comet nucleus; chemical differentiation by sublimation of volatile ices in the surface layers and recondensation of gas in deeper, cooler layers. A 2-D time dependent solution is obtained through the dependence of the boundary conditions on the local solar illumination as the nucleus rotates. The model for Comet Hale-Bopp was compared with observational measurements (Biver *et al.*, 1999). The best agreement was obtained for a model with amorphous water ice and CO, assuming that a part of the latter is trapped by the water ice, another part is condensed as an independent ice phase. The model confirms that sublimation of CO ice at large heliocentric distance produces a gradual increase in the comet's activity as it approaches the Sun. Crystallization of amorphous water ice begins at 7 AU from the Sun, but no outbursts were found. Seasonal effects and thermal inertia of the nucleus material lead to larger CO outgassing rates as the comet recedes from the Sun. In the second part of this work the model was run with the orbital parameters of Comet Wirtanen. Unlike Comet Hale-Bopp, the predicted CO outgassing from Comet Wirtanen is almost constant throughout its orbit. Such behavior can be explained by a thermally evolved and chemically differentiated comet nucleus.

Keywords: thermal modeling, comet nucleus, Hale-Bopp, Wirtanen

1. Introduction

The study of CO outgassing from comets provides clues about the chemical differentiation of the nucleus and of the origin of comets. Carbon monoxide is the second most abundant molecule, after water, in the ice mantles of interstellar grains (see Ehrenfreund, 1999). Hence, if cometary nuclei are formed by accretion of interstellar matter, they should have a high CO abundance ratio. However, during the early radiogenic heating of cometary nuclei in the Oort cloud a fraction of CO could have been lost due to the low sublimation temperature of CO (20-40 K). Another fraction of CO is lost during the thermal evolution that comets undergo in the planetary region of the solar system.

* National Research Council Award – NASA/JPL Research Associate

In this paper, the CO outgassing from Comets Hale-Bopp and Wirtanen is modeled and compared with measurements. Comet Hale-Bopp is ideal to test a comet nucleus thermal model due to its exceptional brightness and its early discovery at a heliocentric distance, $r_h = 7$ AU. For the first time, outgassing rates of molecular species including CO, CO_2, H_2CO, CH_3OH as well as the OH radical, a tracer of the water molecule (Roettger et al., 1990), could be intensively monitored during the entire cometary apparition. Comparison with observations of Comet Hale-Bopp allows one to calibrate free parameters in the thermal model, i.e. the thermal properties of the comet nucleus material. The calibrated comet model can then be used to estimate upper limits of the CO outgassing from short-period Comet Wirtanen. This comet is the current target of the International Rosetta Mission, which will study the onset and evolution of cometary activity from a heliocentric distance of 4 AU to perihelion at 1.08 AU.

2. Comet Nucleus Thermal Model

The comet nucleus thermal model assumes a spherical, porous body containing a dust phase, one major ice component (amorphous water ice), and one minor ice component of higher volatility (CO). The initial chemical composition within the nucleus is assumed to be homogeneous. When the solar flux reaches the comet nucleus material, a small part of it is reflected by the optically dark surface whereas a larger part is absorbed and heats the surface layer. One part of the absorbed flux is reradiated in the infrared, another part drives the sublimation of water ice and the remainder of the flux is diffused as a heat wave into the interior of the rotating nucleus. The relative magnitude of the different dissipation terms depends on the local illumination. The model includes chemical differentiation of volatile ices in the subsurface layers, release of trapped CO molecules during crystallization of amorphous water ice (Bar-Nun et al., 1985; Schmitt and Klinger, 1987; Jenniskens and Blake, 1994), gas diffusion through tiny pores and recondensation of gas in deeper, cooler layers. A 2-D time dependent solution is obtained through the dependence of the boundary conditions on the local solar illumination as the nucleus rotates.

Coma effects such as attenuation of the solar flux (see Weissman and Kieffer, 1984) and a return-flux of gas from the coma to the comet nucleus surface (see Crifo and Rodionov, 1997) are not taken into account. The estimates of the outgassing rates do not include the gas produced from sublimation of icy-dust grains in the coma (see Enzian et al., 1998). For a detailed mathematical description of our model the reader is referred to the papers by Enzian et al. (1997, 1999).

TABLE I
Model parameters. See also Huebner et al. (1999)

Semimajor axis (Hale-Bopp)	193.6 AU
Semimajor axis (Wirtanen)	3.115 AU
Orbit eccentricity (Hale-Bopp)	0.9953
Orbit eccentricity (Wirtanen)	0.655
Nucleus radius (Hale-Bopp)	35 km
Nucleus radius (Wirtanen)	0.6 km
Spin period	10 hours
Albedo	0.04
Infrared emissivity	0.96
Initial nucleus temperature	20 K
Hertz factor	0.01
Nucleus bulk density	500 kg m^3
Bulk thermal conductivity of dust	10 W m^{-1} K^{-1}
Pore radius	10^{-4} m
Chemical composition	
H$_2$O (Hale-Bopp)	0.40 (mass ratio)
H$_2$O (Wirtanen)	0.45
CO (Hale-Bopp)	0.05 (ice) + 0.05 (trapped)
CO (Wirtanen)	0.05 (ice)
Dust	0.50

3. Results and Discussion

3.1. COMET HALE-BOPP

The model for Comet Hale-Bopp considers the known orbital parameters. These parameters correspond to a perihelion of 0.91 AU, an aphelion of 386 AU and an orbital period of 2700 years. However, during most of this time the comet is too far from the Sun to experience significant thermal alteration. In this work the comet model is run for the part of the orbit which is closer than 50 AU to the Sun. Dynamical computations by Bailey et al. (1996) showed that Comet Hale-Bopp is not a new comet but that it approached the Sun already several times having perihelion distances of about 1 - 2 AU. In this work two apparitions were modeled. During the first apparition the nucleus spin axis is perpendicular to the orbital plane (zero obliquity, $\delta = 0°$). During the second orbit the nucleus spin axis is in the orbital plane ($\delta = 90°$). At perihelion and aphelion one pole faces the Sun. The

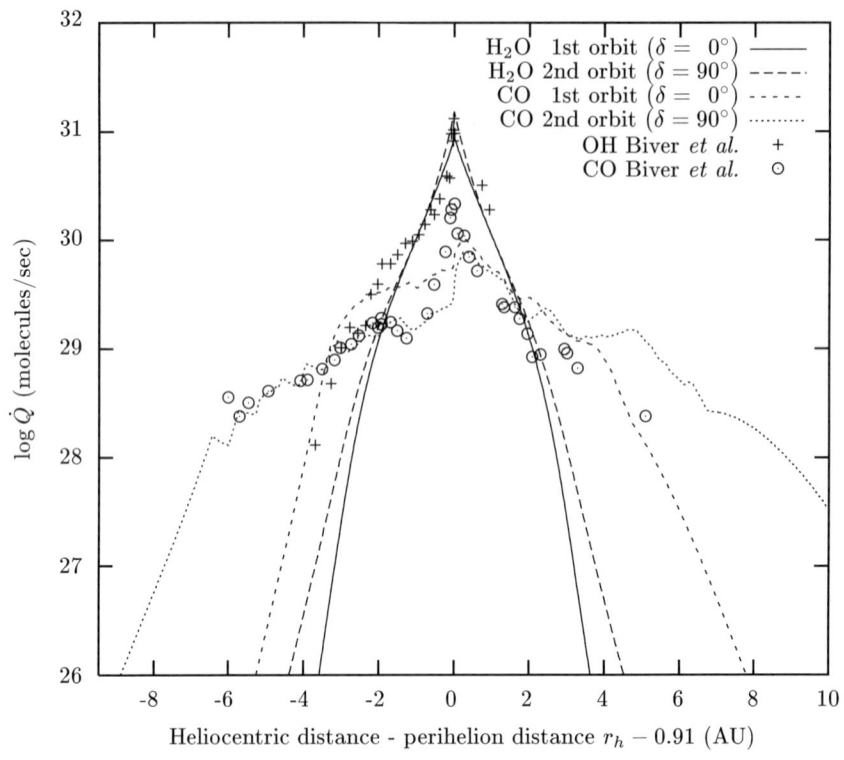

Figure 1. Evolution of the modeled CO and water outgassing rates compared with measured outgassing rates provided by Biver *et al.* (1999). The modeled comet nucleus has an albedo of 0.04 and a radius of 35 km, the radius suggested by CCD observations (Weaver and Lamy, 1999). The initial nucleus composition contains dust (50% by mass), amorphous water ice (40%), CO trapped in amorphous water ice (5%) and an additional small content of CO ice (5%) in an independent phase. The thermal evolution is computed for two apparitions from 50 AU pre-perihelion to 50 AU post-perihelion. During the first orbit the spin axis of the model comet nucleus is perpendicular to the orbital plane, whereas in the second orbit the spin axis lies in the orbital plane and points towards the Sun at perihelion.

latter spin axis corresponds roughly to the current estimated spin state (Sekanina and Boehnhardt, 1999). The spin state during previous apparitions is unknown as the spin evolves due to gravitational and non-gravitational forces. The first orbit is used to compute more realistic initial conditions and will not be discussed here in detail. Model parameters are given in Table I.

The evolution of the CO outgassing calculated by the comet nucleus thermal model for Comet Hale-Bopp is shown in Figure 1. The model confirms that at large heliocentric distance the activity is driven by sublimation of CO. With the release of CO the comet nucleus becomes chemically differentiated and the CO sublimation front moves to deeper layers. During the first apparition a layer of one meter thickness is depleted of CO. The thickness of this layer also depends on the latitude. The sublimation of CO ice produces a gradual increase in the comet's activity. This behavior is in agreement with a dust tail analysis of C/Hale-Bopp by Fulle *et al.* (1998). Dust tail modeling is the only tool that provides estimates of the dust grain ejection rate during the years preceding the comet's discovery. Outbursts such as computed by Prialnik (1997), using her comet nucleus thermal model, were found neither by Fulle *et al.* nor with this comet model. As the comet approaches the Sun the CO outgassing rate increases roughly as r_h^{-2}. Beyond 7 AU from the Sun the CO outgassing is driven by sublimation of CO ice. Closer to the Sun, crystallization of amorphous water ice begins and release of trapped CO becomes the major CO source. The crystallization front moves faster to deeper layers as the surface is eroded. The depth of the CO source increases as the crystallization front moves to deeper, cooler layers. This results in decreased CO production rates. Radio observations by Biver *et al.* (1998) showed that between 1.6 and 3 AU, the CO production rate stalled or even decreased before exhibiting a steep increase closer to the Sun. At about 2.5 AU from the Sun, sublimation of water ice from the surface layer becomes more and more efficient and water outgassing becomes the major driver of the comet's activity. Increasing water sublimation and surface erosion decrease the thickness of the layer between the surface and the CO source, resulting in increased CO outgassing rates close to perihelion. Another possible explanation for the steep increase in CO outgassing close to perihelion is due to a more effective heat conduction. With the sublimation of water ice, vapor diffusion becomes an efficient heat transport mechanism. An alternative explanation for the fast increase in the CO production rate close to perihelion was the hypothesis of a distributed source of CO in Comet Hale-Bopp which turned on gradually between 1.5 and 2 AU from the Sun. It was suggested by Huebner and Benkhoff (1999) that such a distributed source can be explained by photolysis of polymerized formaldehyde which is released from hydrocarbon polycondensates in the dust particles in the coma.

The post-perihelion outgassing rates for CO are generally larger than before perihelion. This behavior is explained by the thermal inertia of the layers between the surface and the CO sublimation front. A tilted nucleus spin axis such as considered by the model for the second orbit enhances seasonal effects. A tilted spin axis leads also to higher surface temperatures, in particular at large heliocentric distance. Beyond 3 AU from the Sun the internal heat flux becomes a major term in the surface energy balance. The result of the tilted spin axis is a larger heat transport to deeper layers and larger CO sublimation rates at large heliocentric distance.

The apparent disagreement between modeled and observed water production rates beyond 4 AU can be explained by additional sublimation of micron-sized icy grains in the coma. Such a process was also suggested based on the analysis of the evolution of the observed OH radio-line shape (Bockelée-Morvan and Rickman, 1999). Note, it was argued by Kührt (1999) that a smaller heat conductivity leads to a higher water outgassing rate from the nucleus. This possibility was taken into account in a sensitivity analysis. A smaller heat conductivity than the one used in this work, leads to a significantly lower CO outgassing rate that is incompatible with the measurements.

3.2. COMET 46P/WIRTANEN

Carbon monoxide has not been detected in Comet Wirtanen so far. However, it should be pointed out that due to a small nucleus (with a radius of about 600 m, Lamy 1996) and an instrumental detection limit (on the order of 4×10^{27} molecules/s, D. Bockelée-Morvan, personal communication), CO would not be detected even if CO outgassing was as efficient as in Comet Hale-Bopp. A significant improvement on the current detection limit is predicted for the MIRO experiment onboard the Rosetta spacecraft, to be launched in 2003.

The evolution of the CO production rate calculated by the model for Comet P/Wirtanen is shown in Figure 2. The nucleus composition contains initially dust (50% by mass), crystalline water ice (45%) and CO ice (5%). The spin axis of the model nucleus is perpendicular to the orbital plane. In the model CO sublimates from a subsurface layer. The depth of the sublimation front increases with the time and the production rate decreases. In contrast to the water production rate, the CO production rate is larger after perihelion passage than before perihelion. This behavior is explained by the thermal inertia of the layers between the surface and the CO sublimation front.

As no direct observations of CO exist for Comet Wirtanen, an empirical formula given by Di Folco and Bockelée-Morvan (1997) was used which has been derived for Comet Hale-Bopp from the correlation between the measured CO production $\dot{Q}_{CO}$ rate and the brightness m_h

$$\log \dot{Q}_{CO} = 30(\pm 0.04) - 0.256(\pm 0.009) \times m_{r_h} \qquad (1)$$

This expression is valid only for heliocentric distances larger than 3 AU where CO was observed to be the most abundant molecule in the coma of Comet Hale-Bopp. CO production rates derived from observed magnitudes of P/Wirtanen at $r_h > 3$ AU are shown in Figure 2. However, the dashed line in Figure 2 shows the apparent production rates corresponding to the expected magnitude of a 600 m cometary nucleus with an albedo of 0.04. From the comparison with the available observational data, it cannot be concluded that there was any CO production from P/Wirtanen in 1996. The upper limit for the CO outgassing rate based on the model results is $< 10^{26}$ molecules/sec (after five orbits), whereas the analysis of the observed magnitudes lead to an upper limit of the order of $< 5 \times 10^{25}$ molecules/sec.

It is believed that comet nuclei such as P/Wirtanen are strongly differentiated bodies as a result of their thermal evolution in a short-period comet orbit. Therefore, care must be taken in the interpretation of the model results, in particular with respect to the current or future CO outgassing rates. The assumption of an initially homogeneous composition is likely not justified. In addition, the limited

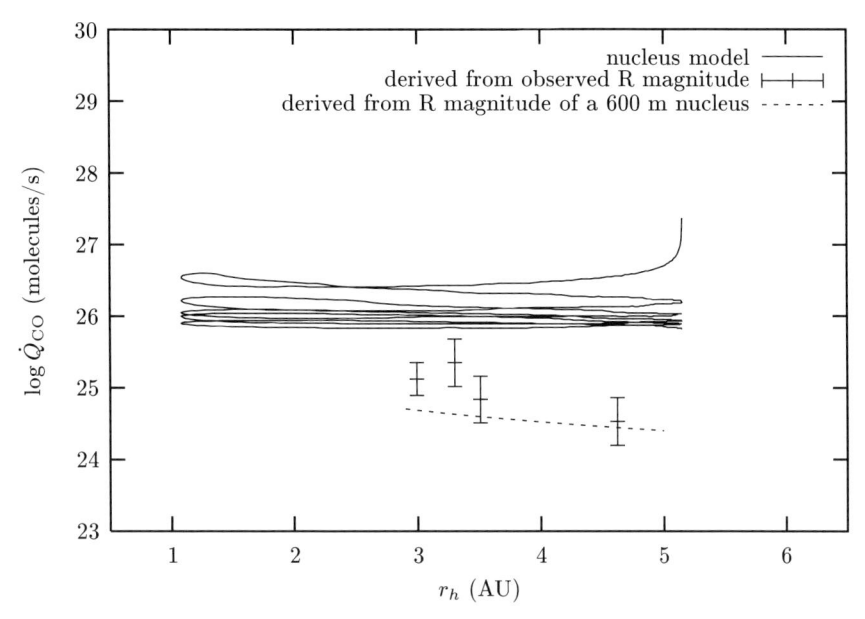

Figure 2. Comparison of modeled carbon monoxide outgassing with production rates derived from the visual brightness of Comet Wirtanen. Magnitude data are from a compilation of broad-band R magnitudes by Meech *et al.* (1997). The dashed line shows the apparent production rate corresponding to the expected magnitude of a 600 m cometary nucleus with an albedo of 0.04.

integration time over a few tens or hundreds of years, assuming the current orbits or even a trajectory from the Kuiper belt, might be too small compared to the age of the comet. We know that P/Wirtanen has made at least 9 orbits of the Sun since its discovery in 1947, and likely many more given the typical dynamical lifetime of Jupiter-family comets. Therefore the computed CO production rate should be considered as upper limit, with a lower limit of zero CO production. For a detailed analysis of the water production rate the reader is referred to the paper by Enzian *et al.* (1999).

4. Conclusions

Outgassing rates for CO were modeled for Comets Hale-Bopp and Wirtanen and are compared with observations. The model results show that Comet Hale-Bopp had been active prior to discovery at 7 AU from the Sun. This conclusion is in agreement with a detailed analysis of the comet's dust tail by Fulle *et al.* (1998). Until that time, the activity was driven by sublimation of CO ice. At about the time the comet was first observed, crystallization of amorphous water ice and release of trapped gases began. The increase in the comet's activity was found to be gradual and no outbursts were found. Further, possible CO outgassing from Comet Wirtanen was investigated. The calculated upper limits for carbon monoxide outgassing from a nucleus with a radius of 600 m is $< 10^{26}$ molecules/s (after five orbits). This rate depends on the thermal history and the possible chemical differentiation inside the nucleus of Comet Wirtanen.

Acknowledgements

This work was done at the Jet Propulsion Laboratory, California Institute of Technology, under contract with NASA. Data for the measured CO and OH outgassing rates were made available in electronic form by N. Biver. The author is very grateful to Dr. P. R. Weissman and three anonymous referees for useful discussions and comments. This work is supported in part by the NASA Planetary Geology and Geophysics Program.

References

Bailey, M., Emelyanenko, V.V., Hahn, G., Harris, N.W., Hughes, K.A., Muinonen, K., and Scotti, J.V.: 1996, 'Orbit evolution of comet 1995 O1 Hale-Bopp', *Mon. Notes Roy. Astron. Soc.* **281**, 916–924.
Bar-Nun, A., Herman, G. and Laufer, D.: 1985, 'Trapping and release of gases by water ice and implications for icy bodies', *Icarus* **63**, 317–332.
Biver, N., Bockelée-Morvan, D., Colom, P., Crovisier, J., Davies, J., Dent, W., Despois, D., Gérard, E., Rauer, H., Moreno, R., Paubert, G., Lis, D., Mehringer, D., Benford, D., Gardner, M. and Philips, T.: 1999, 'Long term evolution of the outgassing of comet Hale-Bopp from radio observations', *Earth, Moon, and Planets*, in press.
Bockelée-Morvan, D. and Rickman, H.: 1999, 'C/1995 O1 (Hale-Bopp): Gas production curves and their interpretation', *Earth, Moon, and Planets*, in press.
Crifo, J.-F., and Rodionov, A. V.: 1997, 'The dependence of the circumnuclear coma structure on the characteristics of the nucleus. I comparison between a homogeneous and an inhomogeneous spherical inucleus, with application to P/Wirtanen', *Icarus* **127**, 319–353.
Ehrenfreund, P.: 1999, 'An ISO view on interstellar and cometary ice chemistry', *Space Sci. Rev.*, this volume.
Enzian, A., Cabot, H. and Klinger, J.: 1997, 'A 2 1/2 D thermodynamic model of cometary nuclei: I Application to the activity of comet 29P/Schwassmann-Wachmann 1', *Astron. Astrophys.* **319**, 995–1006.
Enzian, A., Cabot, H. and Klinger, J.: 1998, 'Simulation of the water and carbon monoxide production rates of comet Hale-Bopp using a *quasi* 3D nucleus model', *Planet. Space Sci.* **46**, 851–858.
Enzian, A., Klinger, J., Schwehm, G. and Weissman, P.: 'Temperature and gas production distributions on the surface of a spherical model comet nucleus in the orbit of 46P/Wirtanen', *Icarus* **138**, 74–84.
Di Folco, E., and Bockelée-Morvan, D.: 1997, Thesis, Observatoire de Paris Meudon, Université Paris VII.
Fulle, M., Cremonese, G. and Bohm, C.: 1998, 'The pre-perihelion dust environment of C/1995 O1 Hale-Bopp from 14 to 4 AU', *Astron. J.* **116**, 1470–1477.
Huebner, W., and Benkhoff, J.: 1999, 'From coma abundances to nucleus composition', *Space Sci. Rev.*, this volume.

Huebner, W., Benkhoff, J., Capria, M., Coradini, A., Sanctis, M.D., Enzian, A., Orosei, R. and Prialnik, D.: 1998, Results from the Comet Nucleus Model Team at the International Space Science Institute, Bern, *COSPAR proceedings*, in press.
Jenniskens, P. and Blake, D.F.: 1994, 'Structural transitions in amorphous water ice and astrophysical implications', *Science* **265**, 753–756.
Kührt, E.: 1999, 'H_2O activity of comet Hale-Bopp', *Space Sci. Rev.*, this volume.
Lamy, P.L.: 1996, 'Comet 46P/Wirtanen', *IAU Circ.* **No 6478**.
Meech, K.J., Bauer, J. and Hainaut, O.: 1997, 'Rotation of comet Wirtanen', *Astron. Astrophys.* **326**, 1268–1276.
Roettger, E.E., Feldman, P.D. and A'Hearn, M.F.: 'Comparison of water production rates from UV spectroscopy and visual magnitudes for some recent comets', *Icarus* **86**, 100–114.
Schmitt, B. and Klinger, J.: 1987, 'Different trapping mechanisms of gases by water ice and their relevance for comet nuclei', In *Symposium on the Diversity and Similarity of Comets*, pages 613–619, Noordwijk, The Netherlands, ESA SP-278, 1987.
Sekanina, Z., and Boehnhardt, H.: 1999, 'Dust Morphology of Comet Hale-Bopp (C/1995 O1)', *Earth, Moon, and Planets*, in press, 1999.
Weissman, P.R. and Kieffer, H.H.: 1984, 'An improved thermal model for cometary nuclei', *Journal of Geophysical Research* **89**, C358–C364.
Weaver, H. and Lamy, P.: 1999, 'Estimating the Size of Hale-Bopp's Nucleus' *Earth, Moon, and Planets*, in press.

Address for correspondence: Achim Enzian, Jet Propulsion Laboratory, California Institute of Technology, MS 183-601, Pasadena, CA, 91109, USA, achim.enzian@jpl.nasa.gov

ON THE FLUX OF WATER AND MINOR VOLATILES FROM THE SURFACE OF COMET NUCLEI

J. BENKHOFF

DLR, Institut für Planetenerkundung, Rudower Chaussee 5, Bldg. 16.16, D-12484 Berlin, Germany

Abstract. Surface temperature and the available effective energy strongly influence the mass flux of H_2O and minor volatiles from the nucleus. We perform computer simulations to model the gas flux from volatile, icy components in porous ice-dust surfaces, in order to better understand results from observations of comets. Our model assumes a porous body containing dust, one major ice component (H_2O) and up to eight minor components of higher volatility (e.g. $CO, CH_4, CH_3OH, HCN, C_2H_2, H_2S$), The body's porous structure is modeled as a bundle of tubes with a given tortuosity and an initially constant pore diameter. Heat is conducted by the matrix and carried by the vapors. The model includes radially inward and outward flowing vapor within the body, escape of outward flowing gas from the body, complete depletion of less volatile ices in outer layers, and recondensation of vapor in deeper, cooler layers. From the calculations we obtain temperature profiles and changes in relative chemical abundances, porosity and pore size distribution as a function of depth, and the gas flux into the interior and into the atmosphere for each of the volatiles at various positions of the body in its orbit.

In this paper we relate the observed relative molecular abundances in the coma of Comet C/1995 O1 (Hale-Bopp) and of Comet 46P/Wirtanen to molecular fluxes at the surface calculated from our model.

Keywords: Comets, Ices, Energy Balance, Molecular Flux

1. Introduction

Comets spend most of their time at the edges of our solar system. Therefore, they very likely contain pristine material from the time the solar system was formed. The nucleus of the comet is the source of the gas and dust in the cometary coma. Surface areas of the nucleus covered with an insulating dust mantle will be hotter than more exposed icy areas, resulting in different amounts of heat transported into the interior. Due to this energy input the ices in the interior sublimate and diffuse through the pores of the low density nucleus. Some vapors recondense in the deeper interior and some escape into the coma. This leads to chemical differentiation of the nucleus and also to a variation of the mixing ratio of vapors from ices more volatile than water relative to water vapor in the coma as a function of heliocentric distance.

In this paper we will discuss the flux of water and the flux of vapors from ices more volatile than water that are essential to study the composition and chemical and physical evolution of a comet. The only in situ results available to study the chemical and physical properties of the nucleus of a comet are the investigations of Halley's Comet in 1986. More results can be expected in the near future from spacecraft missions designed to study comets. The European Rosetta mission will study Comet Wirtanen by escorting the comet on its pre-perihelion leg starting at about 3 AU. The mission will also carry a lander that will allow in situ investigations at the surface of the nucleus for the first time. Several future NASA missions, Stardust, Deep Space 1, Deep Space 4, and CONTOUR, will study the dust environment around the nucleus and the surface properties of the nucleus.

2. The Thermal Model

We assume a spherical (spin axis orientation perpendicular to the orbit plane) model comet nucleus on the orbit of either Hale-Bopp or Wirtanen. The results are calculated for a point located at the equator. The model calculations were carried out as follows: We started with a homogeneously mixed body at a constant temperature (10 K) and a constant mass density. Our model assumes a porous body, containing a dust component, one major ice component (amorphous H_2O), and several minor components of higher volatility ices (e.g. CO, CH_4, CO_2, CH_3OH, HCN, NH_3, H_2S, C_2H_2). The porosity of the body changes because of sublimation (vaporization) and recondensation of the ice. Heat is transferred into the interior of the body by solid-state heat conduction of the ice- and dust matrix, and by vapor flow through the porous matrix (the flow being driven by a vapor pressure gradient), as well as by recondensation. We solve the mass and energy equations for the different volatiles simultaneously (see Benkhoff and Huebner, 1995, 1996).

The energy conservation equation for the porous, icy, dusty layer is

$$\rho c \frac{\partial T}{\partial t} + \psi \rho_g c_g v \nabla T = \nabla \cdot (\kappa \nabla T) - \sum_{i=1}^{n} \Delta H_i q_i, \qquad (1)$$

where H_i and q_i are the enthalpies per unit mass of sublimation and the intrinsic mass release rates of vapor per unit volume of components i, respectively, κ is the thermal conductivity of the ice-dust matrix, T is the temperature, t the time, ψ the porosity, ρ the mean density, ρ_g the density of the gas, v the mean velocity of the gas evaporating from deeper layers and streaming through the crust, and c and c_g the average specific heats of the matrix and of the gas at constant volume, respectively. The surface temperature is calculated from the balance between the net incoming solar flux, losses from thermal reradiation, heat needed for sublimation or becoming free during condensation, and heat transport in and out of the shell

$$\frac{F_0(1-A)\cos\xi}{r^2} = \varepsilon \sigma T_s^4 + \kappa \nabla T_s + \sum_{i=1}^{n} \Delta H_i Q_i. \qquad (2)$$

In Eq.(2) A denotes the albedo, F_0 the solar constant, r the heliocentric distance in AU, ξ the local zenith angle of the Sun, ε the infrared emissivity, σ the Stefan-Boltzmann constant, and T_s the surface temperature, and Q the gas mass production rate at the surface. If there is no ice at the surface the last term at the right hand side of Eq.(2) becomes zero. The conservation of mass in an n-component system undergoing a phase change is given by the following equation:

$$\psi \frac{\partial \rho_{g_i}}{\partial t} + \nabla \cdot j_i = q_i \tag{3}$$

where, for each component i, ρ_g is the gas density, j the gas flux, and q_i is the internal gas production rate. For the boundary condition at the bottom of the shell we assume a constant heat flux and zero mass flux for the lower boundary. As a result of our calculations we obtain the temperature and abundance distribution in the interior, porosity and pore size distribution as a function of depth, the gas flux into the interior, and the gas flux through the surface of the nucleus into the coma for each of the volatiles at various positions of the comet in its orbit around the Sun.

3. Results

The energy input at the surface of the nucleus depends mainly on the sun-comet distance, the rotational state, spin period, scattering properties of the coma, and the reflectivity of the surface. The gas flux from the surface is strongly related to the effective energy. The effective energy is defined as the incoming energy minus the reradiated energy and is available for sublimation and internal heating. The reradiated energy is a strong function of the surface temperature (first term on the right hand side of Eq. 2). Rising surface temperatures indicate that the amount of reradiated energy increases at the expense of the effective energy.

Surface temperatures can be determined from the power balance. Figure 1a shows the surface temperature T versus heliocentric distance for the Jupiter-family model Comet Wirtanen where a dust covered surface is assumed. We are considering a point at the equator with a spin period of 6 hours. A spin starts when this point is subsolar (noon). Temperatures are shown at six specific times during one spin. At noon we obtain the highest surface temperatures of about 370 K at perihelion and of about 150 K at aphelion. The difference from the highest to the lowest temperature at one specific sun distance represents the daily temperature variations. At perihelion we obtain a difference of about 190 K between day and night. We assumed a thermal conductivity of the crust of about 0.1 $Wm^{-1}K^{-1}$ to obtain these. If we assume a hundred times smaller thermal conductivity of the crust, the gap between day and night temperatures increases significantly. At perihelion we observed day/night variations of about 300 K and at aphelion the difference is still 120 K (for more details see Benkhoff, 1998). The reason for this is that

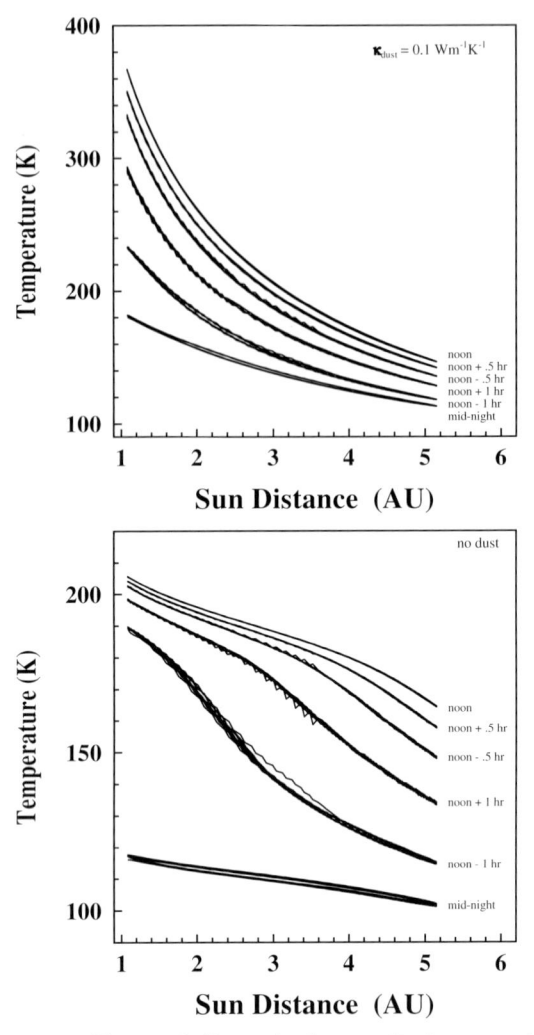

Figure 1. Surface temperature T versus heliocentric distance. During one spin (spin period 6 hrs) of the nucleus temperature values are shown at noon, half an hour before and after noon, one hour before and after noon, and at midnight. Results given are obtained from a model where a dust covered surface (crust) is assumed (a) and from a model where an active surface is assumed (b).

with a poorly heat conducting crust much more energy must be reradiated from the surface during the day. This increases the surface temperature. During the night, there is less energy available from internal cooling of the nucleus. As a result there is a rapid decline in night temperatures. With respect to active areas, areas with no crust on top of the surface, results will be significantly different. In Figure 1b the surface temperature T is plotted versus heliocentric distance for an unmantled area. At noon we obtain surface temperatures of 203 K at perihelion and of about 165 K at aphelion, respectively. Nighttime temperatures are lower, between 105

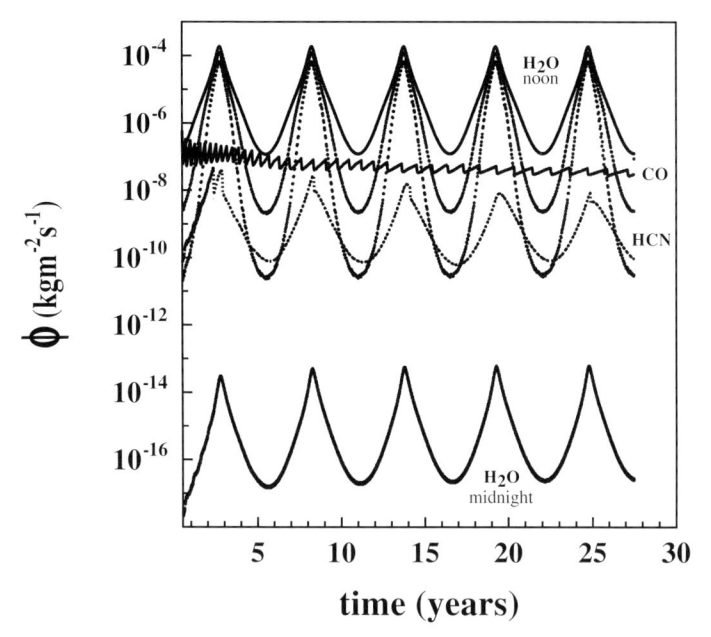

Figure 2. Mass fluxes of H_2O, HCN, and CO from the surface of a Wirtanen model comet as a function of time for a model assuming a dust-free surface. Time zero is the start of the evolution in our model. The time between about 16 and 27 years represent the fourth and the fifth orbit of the model comet around the sun. H_2O fluxes are shown at noon, midnight, and at two equidistantly spaced positions in between.

and 115 K. At perihelion the maximum temperature of about 203 K is almost identical to the free sublimation temperature of water. The lowest temperatures at night are about 115 K, which yields day/night variations of about 90 K at the smallest comet-sun distance.

For a model comet in the orbit of Hale-Bopp we would get perihelion noon temperatures of about 205 K assuming an active surface, and temperatures of about 415 K assuming a dust covered surface with a thermal conductivity of the crust of about 0.1 $Wm^{-1}K^{-1}$, respectively. Day to night variations at perihelion are about 300 K for the dust covered surface model and about 100 K for the active surface model.

The mass flux depends strongly on the amount of energy transported to the sublimation front of water. In the crust models this energy is too small to obtain flux rates which are able to fit measured results. From the active area calculations we obtain averaged (over one spin) perihelion H_2O fluxes on the order of about $10^{-4} kgm^{-2}s^{-1}$ (Figures 2 and 3). This value is determined for the equator. To obtain a total flux we assume a radius (40 km for Hale-Bopp and 0.6 km for Wirtanen), convert kilograms into molecules, and integrate from one pole to the other. For the integration we use the equator flux times cosine of the latitude.

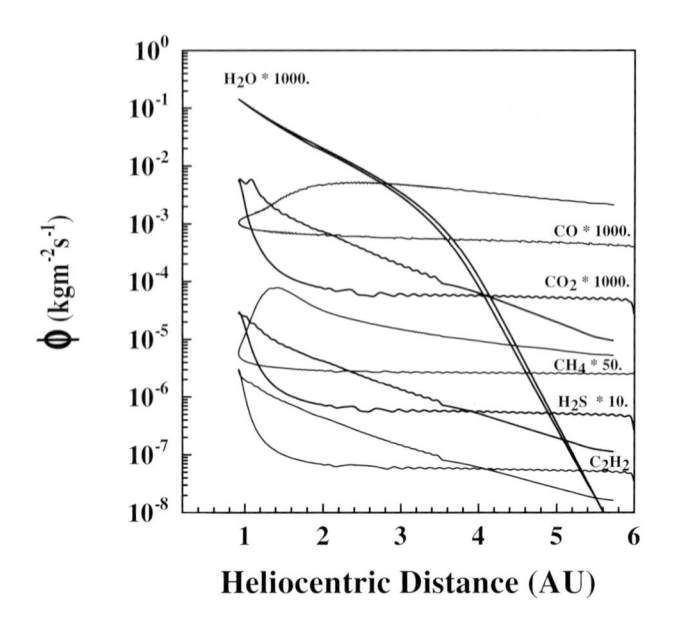

Figure 3. Mass fluxes of H_2O, H_2S, CO_2, C_2H_2, CH_4, and CO from the surface of the Hale-Bopp model comet as a function of time for a model assuming an active surface. Some lines are shifted by the factors given to make the plot better readable.

As a result we get an upper limit for the total flux. For Hale-Bopp we obtain a flux of $\approx 4 \times 10^{31}$ molecules s^{-1}, about the same order as Colom *et al.* (1998) derived from OH radio measurements and an order of magnitude higher than gas production rates derived from narrowband photometry (Schleicher *et al.*, 1998). For a Wirtanen model comet we obtain a total flux of 1.5×10^{28} molecules s^{-1}. This value is about the same as Fink *et al.* (1997) measured for Comet Wirtanen at perihelion.

In Fig. 2 the mass fluxes of H_2O, HCN, and CO from the surface of a model comet in the orbit of the Jupiter-family Comet Wirtanen are plotted as a function of time for models assuming an active surface. Time zero represents the start of the evolution in our model. The time between about 16 and 27 years represents the fourth and the fifth orbit of the model comet around the sun. Water sublimates at the surface and therefore it shows day/night variations. The lowest thin solid line shows the night flux of the water, which is extremely small close to perihelion. The uppermost line represents the fluxes at noon. At perihelion one obtains maximum values on the order of about 2×10^{-4} kg m^{-2} s^{-1}. The mass flux depends strongly on the amount of energy transported to the sublimation fronts, which in case of the minor volatiles lie from a few centimeters up to meters below the surface. The flux of HCN (Fig. 2) varies by 2-3 orders of magnitude during one orbit. Day/night temperature changes of the surface seem to have no influence on the flux, because the sublimation front is below the diurnal skin depth, but not below the orbital skin depth. Therefore we see the variations of the flux during the orbit. The maximum HCN flux at perihelion is about 5×10^{-8} kg m^{-2} s^{-1} or about 2×10^{24} molecules s^{-1}. This maximum value is about a factor of 2 lower than CN fluxes measured by Fink *et al.* (1997). Because of a sublimation front deeper than the orbital skin depth CO fluxes are almost constant and slightly decreasing with time. The reason is the inward migration of the sublimation front.

In Fig. 3 mass fluxes of H_2O, H_2S, CO_2, C_2H_2, CH_4, and CO from the surface of a model comet in the orbit of Hale-Bopp are plotted as a function of heliocentric distance. Lines are shifted to make the plot better readable. For water we obtain a flux of $\approx 2 \times 10^{-4}$ kg m^{-2} s^{-1}. Fluxes of H_2S, CO_2, C_2H_2, CH_4, and CO seem to be almost constant up to about 2 AU on the inbound leg. Day/night temperature changes of the surface and orbital energy input changes at the surface have no influence on the flux. Therefore, it is to assume that the sublimation fronts are below the diurnal and the orbital skin depth. Due to a higher mass release of water than of all the other volatiles starting at about $2 - 3$ AU the distance between the surface and the sublimation fronts of all the other ices within the nucleus decreases. This causes the observed flux of H_2S, CO_2, and C_2H_2 to increase by more than an order of magnitude. Gas release rates of CH_4 and CO start to increase around perihelion, because their sublimation fronts are much deeper than the front of the other minor volatiles. After perihelion gas release rates of all the ices decrease. The decrease of the fluxes of CH_4 and CO is delayed. Due to the position of the sublimation front within the nucleus and the thermal inertia of the nucleus material both ices reach their maximum flux after perihelion before declining at about 1.5 AU and 2.5 AU, respectively.

The results given for the mass release rates of all the ices other than water are different from what is observed. Biver *et al.* (1998) determined an increase of the gas production of almost all volatiles before perihelion and a decrease of the flux starting almost around perihelion. Therefore, the origin of all the ice fluxes must be close to or at the surface.

4. Conclusions

The mass flux depends strongly on the amount of energy transported to the sublimation fronts, which in case of the minor volatiles lie from a few centimeters up to meters below the surface. We need to assume almost 100% active surfaces to obtain flux rates which are able to fit measured results. For Hale-Bopp we obtain a water flux of $\approx 4 \times 10^{31}$ molecules s^{-1}, about the same order as Colom *et al.* (1998) derived from OH radio measurements and an order of magnitude higher than gas production rates derived from narrowband photometry (Schleicher et al., 1998). For a Wirtanen model comet we yield a total flux of 1.5×10^{28} molecules s^{-1}. This value is about the same as Fink et al. (1997) measured for Comet Wirtanen at perihelion. The flux of HCN of the Wirtanen models varies by 2-3 orders of magnitude during one orbit. The maximum HCN flux at perihelion is about 2×10^{24} molecules s^{-1}. This maximum value is within the range of CN fluxes measured by Fink et al. (1997).

From our Hale-Bopp models we found that due to higher mass release of water than of all the other volatiles starting at about 2 − 3 AU, sublimation fronts of H_2S, CO_2, and C_2H_2 get closer to the surface, causing a flux increase of more than an order of magnitude. Gas release rates of CH_4 and CO start to increase around perihelion, because their sublimation fronts are much deeper than the other minor volatiles. After perihelion gas release rates of all the ices decrease. The decrease of the fluxes of CH_4 and CO is delayed. The results given for the mass release rates of all the ices other than water are different from what is observed. Therefore, the origin of all the measured ice fluxes must be close to or at the surface. The source of the sublimation may come closer to the surface if the ice is trapped within amorphous ice. Then the mass release is not only controlled by the sublimation temperature but also from the temperature of the phase transition. It is also possible that the ice is not loosely mixed as assumed in our models. If the nucleus is made of complex aggregates containing several ice and maybe also dust components it is very likely that all the mass is released at the surface. From our calculations we found that the measurements of a Jupiter-family comet like Wirtanen can be explained with a layered nucleus model. In this model gases of minor volatiles come from sublimation fronts up to several meters below the surface. The story seems to be different for long-period comets like Hale-Bopp.

References

Benkhoff, J. and Huebner, W. F.: 1995, *Icarus* **114**, 348-354.
Benkhoff, J., and Huebner, W. F.: 1996, *Planet. Space Sci.* **44**, No. 9, pp. 1005-1013.
Benkhoff, J.: 1999, *Planet. Space Sci.* , in press.
Biver, N. et al.: 1999, *Earth Moon Planets*, in press.
Colom, P., Gérard, E., Crovisier, J., Bockelée-Morvan, D., Biver, N., and Rauer, H.: 1999, *Earth Moon Planets*, in press.
Fink, U., Hicks, M.D., Fevig, R. A., and Collins, J.: 1998, *Astron. Astrophys.*, **335**, L37 - L45.
Schleicher, D.G., Farnham, T. L., and Birch, P. V.: 1999, *Earth Moon Planets*, in press.

Address for correspondence: Johannes Benkhoff, DLR, Institut für Planetenerkundung, Rudower Chaussee 5, D-12484 Berlin, Germany, e-mail: Johannes.Benkhoff@dlr.de

MORPHOLOGICAL STRUCTURE AND CHEMICAL COMPOSITION OF COMETARY NUCLEI AND DUST

J. MAYO GREENBERG[1] and AIGEN LI[1,2]

1. *Raymond and Beverly Sackler Laboratory for Astrophysics at Leiden Observatory, Sterrewacht Leiden, Postbus 9504, 2300 RA Leiden, The Netherlands*
2. *Beijing Astronomical Observatory, Beijing 100012, P.R. China*

Abstract. The chemical composition of comet nuclei derived from current data on interstellar dust ingredients and comet dust and coma molecules are shown to be substantially consistent with each other in both refractory and volatile components. When limited by relative cosmic abundances the water in comet nuclei is constrained to be close to 30% by mass and the refractory to volatile ratio is close to 1:1. The morphological structure of comet nuclei, as deduced from comet dust infrared continuum and spectral emission properties, is described by a fluffy (porous) aggregate of tenth micron silicate core-organic refractory mantle particle on which outer mantles of predominantly H_2O ices contain embedded carbonaceous and polycyclic aromatic hydrocarbon (PAH) type particles of size in the of 1 - 10 nm range.

Keywords: ISM: dust - ISM: molecules - ISM: abundances - Comets: general - Comets: individual: Halley, Borrelly, Hale-Bopp - Stars: individual: β Pictoris - Stars: circumstellar matter - Stars: planetary systems

1. Introduction

Comets are unique objects in the solar system in that they most closely represent the matter out of which the solar system was born. This is now an accepted fact. This alone makes it an important reason for understanding their properties. However, the degree to which their properties are representative of pre-solar system matter remains a matter of debate. Establishing the connection between comets and pre-solar system matter is the ultimate goal of all our studies.

Unfortunately, comets have the annoying habit of hiding themselves in a blanket of gas and dust which makes it almost impossible to observe them directly. Although comet astronomers have devoted their major efforts towards observing and interpreting this blanket, it is really what is underneath that they are interested in — the nucleus of a comet. On the other hand, the dust and gas in the coma and tail do provide an important source of information on the nucleus from which they are ejected as a result of solar radiation and heating. So long as we are limited to remote observations there are many secondary properties of comets which we are obliged to work with in order to derive the primary properties. After all, what we are looking for are the material and morphological properties of the nucleus in order to work back to its beginnings.

The fact that comets evolve is a necessary evil because our aim is to discover the primordial comet material, and in terms of life's origins, at least back to 3.8×10^9 years ago. Thus the major effort in comet coma studies is to provide information on the nucleus:

$$coma\ gas + dust \Rightarrow comet\ nucleus.$$

However, we may also try to start with what we know about dust and gas in the collapsing protosolar cloud to infer how the comet nucleus might have formed initially:

$$dust\ and\ gas\ in\ collapsing\ protosolar\ cloud \Rightarrow comet\ nucleus.$$

Both procedures have their own problems. Classically, the composition of comet nuclei was derived primarily from the coma volatile molecules dominated by water (or OH). The dust was considered mostly in terms of its scattering properties from which empirical approximations were used to deduce a dust-to-gas ratio. The discovery of the silicate emission feature (Maas *et al.*, 1970) confirmed the existence of refractory material in comets along with volatiles (ices). The idea of organic refractories as a major comet nucleus constituent was first quantitatively introduced in the interstellar dust model of comets by Greenberg (1982). But it was the mass spectroscopic evidence of the Giotto/Vega space probes which provided the first proof that the refractory material in comet dust consisted of both the organic elements (O, C, N) as well as the rocky elements (Mg, Si, Fe) (Kissel *et al.*, 1986a, 1986b; Jessberger and Kissel, 1991). While the visual and ultraviolet emission of coma molecules excited, photolyzed and ionized by the solar radiation is used to deduce the volatile composition of the nucleus, it is the infrared radiation by the dust which is the remote observational data used to deduce the refractory components.

In this paper we first attempt to constrain the chemical composition of comet nuclei using both the interstellar dust model and the coma gas and dust observational data. Then we show how the infrared emission for several distinctly different types of comets bear a general resemblance to each other by reproducing the infrared emission of various comets (Halley – a periodic comet; Borrelly – a Jupiter-family short-period comet; Hale-Bopp – a long-period comet; and extra-solar comets in the β Pictoris disk) within the framework that all comets are made of aggregated interstellar dust.

2. Chemical Composition

The chemical composition of a comet nucleus has often been described as a dirty snow ball. This originated with the pioneering work of Whipple (1950, 1951) who suggested almost 50 years ago that a comet nucleus is a well defined "solid" object dominated by water ice. However, while this term is still being used today it does not provide an operationally adequate basis for understanding either the evolution of comet nuclei or the coma dust and molecules. For example, it could not anticipate the major "surprises" discovered by the space missions to Comet Halley. Nevertheless, Whipple anticipated two major features of our current perception. One was that he believed that "the relative abundance of the elements in comets should be typical of the universe at large, with the limitation that elements not freezing or forming compounds should be rare or absent".

It is indeed commonly assumed that comet nuclei, as the most primitive solar system objects, have an atomic composition which is well represented by solar system abundances of condensable elements (excluding hydrogen and helium, of course) and they have often been compared with carbonaceous chondrites, though the latter are certainly less primitive. Whipple also suggested that "... the meteoric materials should constitute about one-third or less the mass of cometary nuclei, the other two-thirds being made up largely of the hydrides of C, N and O." In fact, these latter resembled the "dirty ice" model of interstellar dust proposed by Van de Hulst (1949). We shall see that such a comet nucleus would contain more ice than can be accommodated within the solar system abundance constraints. In addition, until about 15 years ago there was a puzzling lack of carbon relative to oxygen in the coma (volatile) molecules (Delsemme, 1982). By coincidence, in that same volume, Greenberg (1982) published a nucleus chemical composition based on the silicate core-organic refractory mantle interstellar dust model which automatically accounted for the missing carbon as being contained in the complex organic molecules resulting from ultraviolet photoprocessing of the interstellar ices which, rather than contributing much to the coma gas molecules, were a major contributor to the dust. This component in fact constituted a major fraction of the nucleus mass, about equal to the silicates (Whipple's meteorics), based on the interstellar dust model.

In that first quantitative derivation of how the atoms are distributed in the molecules of the comet, no attempt was made, a priori, to fit comet data. Rather, the construction was a purely inductive one in that it was taken as given that, in the final stage of presolar cloud contraction, the interstellar dust, as it was then modeled and consistent with solar system abundances, would contain all the condensable atoms and would aggregate without further modification into comet nuclei.

This first purely interstellar dust model, primitive and naive as it was, actually worked surprisingly well in predicting the major surprises resulting from the space missions to Comet Halley – the large abundance of very small particles and the organics as a major fraction of the dust (Greenberg, 1986; McDonnell et al., 1986; Kissel et al., 1986a, b). Further analysis of the data showed that the organic (CHON) molecules had, on the average, a higher initial energy than the silicate ions which led Krueger and Kissel (1987) to infer a core-mantle structure of the dust particle (also see Lawler and Brownlee, 1992). The aggregated interstellar dust model has also served as a basis for further theoretical extensions in terms of morphological properties of porous nuclei (Greenberg, Mizutani, and Yamamoto, 1995).

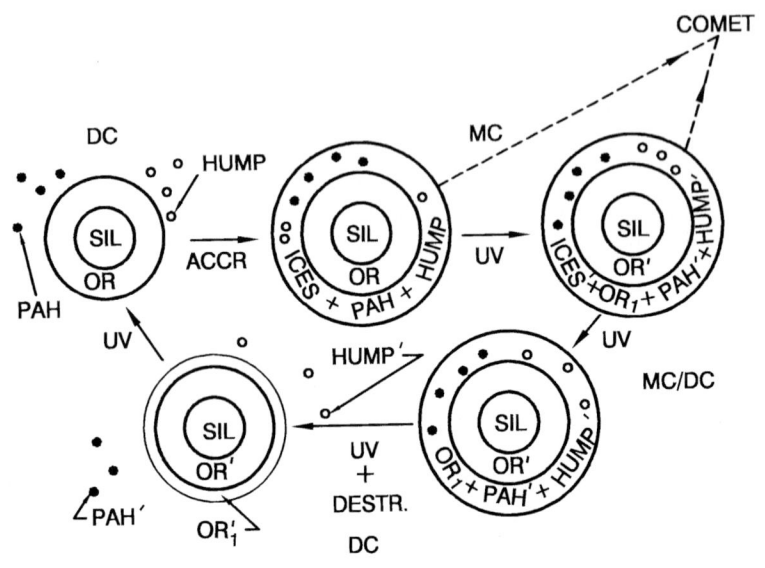

Figure 1. Schematic of evolution of the organic component of interstellar dust. DC = diffuse cloud phase; MC = molecular cloud phase; OR = average diffuse cloud organic refractory mantle material; OR_1 = "first generation" organics; components with a superscript $(')$ are modified forms. See text for additional descriptions.

The basic model of interstellar dust consists of three populations of particles (Li and Greenberg, 1997). The major mass is in tenth micron particles consisting of silicate cores with organic refractory (complex organic molecules) mantles. Additionally, there are very small carbonaceous particles/large molecules. In molecular clouds the large particles accrete additional mantles of frozen molecules, and in the dense clouds there is also accretion of the very small particle along with the "ices". This is schematically shown in Fig. 1. We note here that each full cycle taking about 10^8 yrs, as represented in Fig. 1, may occur up to ~ 50 times before leading to comet formation; i.e. a typical core-mantle particle survives as long as the ISM turnover time of 5×10^9 yrs (which is the mean time for consumption of the ISM by star formation) into and out of stars. The nature of the organic mantle material varies depending on whether the dust is in a low density diffuse cloud or a molecular cloud (Greenberg, 1999). There are significant variations in the relative proportions of C, N, O and H in the complex organics in different regions. In diffuse clouds the organic mantle is strongly depleted in oxygen and hydrogen, whereas in molecular clouds complex organic molecules are present with more abundant fractions of oxygen and hydrogen. Furthermore, the ratio of the mass of organic mantles to the silicate core is highly variable. We shall assume that the organic refractory mantles in the final stages of cloud contraction are most closely

TABLE I

Stoichiometric distribution of the elements in laboratory organics compared with the Comet Halley mass spectra of the organics alone normalized to carbon (Greenberg and Li, 1998a; Krueger and Kissel, 1987; and Krueger, private communication).

	Lab Organics			Halley		
	Volatile[†]	Refractory[†]	Total	PICCA(gas)	Dust	Total[‡]
C	1.0	1.0	1.0	1.0	1.0	1.0
O	1.2	0.6	0.9	0.8	0.5	0.6
N	0.05	> 0.01	>0.03	0.04	0.04	0.04
H	1.70	1.3	1.5	1.5	1.0	1.2

[†] Division between volatile and refractory is here taken at a sublimation temperature less than or greater than ~ 350 K respectively.
[‡] Assuming equal amounts of dust (refractory organics) and gas (relatively volatile organics).

represented by the properties obtained for Halley dust; i.e. $m_{or}/m_{sil} = 1$ and with an atomic distribution as given in Table 1 for comet dust organics.

It is often stated that the interstellar dust ice mantles contain molecular species which are a reasonable facsimile of comet coma species (see Crovisier, 1999). In fact, it is suggested that, if the dust molecules could be observed up to the time of comet nucleus formation, the construction of a nucleus model would be straightforward. The reason for this is that some of the coma molecules are difficult to associate with solar nebula chemistry using partially evaporated interstellar dust; e.g. HNC, CH_4 (Mumma et al., 1993; Irvine et al., 1996). Even though the dust mantle data is limited generally to prestellar molecular clouds or post (massive) star formation regions, we believe it is instructive to demonstrate the extent to which the comet nucleus molecular pattern is defined as compared with using coma molecules as a starting point. As in the case of coma molecules we note that there is quite a spread in abundances in dust ices. Using coma volatile data from Bockelée-Morvan (1997) and dust mantle ice data from Schutte (1996), the resulting molecular constituent abundances in the comet nucleus are derived and listed in Table II, rows (a) and (b) respectively. It is evident that a nominal water fraction of about 0.30 is impossible to escape whether we constrain the comet in terms of the core-mantle dust model or the comet coma volatile and dust composition both maintaining the limits of solar system abundances.

The refractory components are tightly constrained to consist of about 26% of the mass of a comet as silicates (a generic term for combinations of the elements

TABLE II

Distribution by mass fraction of the major chemical constituents of a comet nucleus: (**a**) as derived from comet volatiles, (**b**) as derived from dust ice mantles.

Mass Fraction	Materials								
	Sil.	Carb.	Organ.Refr	H_2O	CO	CO_2	CH_3OH	H_2CO	(other)
(a)	0.26	0.086	0.23	0.31	0.024	0.030	0.017	0.005	0.04
(b)	0.26	0.092	0.23	0.26	0.02	0.030	0.030	0.02	0.05

Si, Mg, Fe), 23% complex organic refractory material (dominated by carbon), and about 9% in the form of extremely small (attogram) carbonaceous/large molecule (PAH) particles. The remaining atoms are in an H_2O-dominated mixture containing of the order of 2 − 3% each of CO, CO_2, CH_3OH plus other simple molecules (see Greenberg, 1998, for details). The refractory to volatile (dust to gas in the nucleus) ratio is about 1:1, while the dust to H_2O ratio is $\approx$ 2 : 1. The maximum mean density of a fully packed nucleus would be $\approx 1.65 \, \text{g cm}^{-3}$ if one assumes nominal mass densities for the components: $\rho_{sil} = 2.8 \, \text{g cm}^{-3}$, $\rho_{or} = 1.4 \, \text{g cm}^{-3}$, $\rho_{ices} = 1.0 \, \text{g cm}^{-3}$. The morphological structure of the component materials, following the interstellar dust into the final stage of the presolar cloud contraction, is as tenth micron silicate cores with organic refractory inner mantles and outer mantles of "ices" with each grain containing many thousands of the attogram carbonaceous/large molecule particles embedded in the icy and outer organic fraction.

3. Halley: a Periodic Comet

The uniqueness of Comet Halley with regard to the dust was that for the first time three properties were simultaneously observed: chemical composition, size (mass) distribution, infrared emission. It was shown (Greenberg and Hage, 1990) that, in order to satisfy simultaneously such independent properties of Halley coma dust as: (1) 9.7 μm emission (amount and shape), (2) dust mass distribution, and (3) mass spectroscopic composition, one must represent the dust as very fluffy aggregates of submicron interstellar dust silicate core-organic refractory mantle particles.

The individual particle size and the organic refractory absorptivity in the visual provide the high temperatures required for the emission by the silicates which, alone (bare), are too cold to emit efficiently. An additional criterion required for the organic mantles is to account for the distributed CO (Eberhardt et al., 1987) by evaporating its C=O bearing molecules (Greenberg and Li, 1998a, 1998b; Li and Greenberg, 1999). The required degree of porosity (fluffiness of the aggregates) is defined by the fact that for a given mass the more porous the aggregate the more it acts like a *sum* of small particle rather than a *compact* particle of that mass, so that a larger fraction of the observed mass distribution provides silicate emission *as if* by submicron particles.

In summary, the result of the intertwining of the three basic observations is: (1) comet dust consists of aggregates of $\sim 0.1\,\mu$m silicate core-organic refractory mantle particles; (2) the average porosity of the comet dust is $0.93 < P < 0.975$. The inferred comet dust density is $0.08 \leq \rho_{cd} \leq 0.16\,\text{g cm}^{-3}$; i.e. $\rho_{cd} \approx 0.1\,\text{g cm}^{-3}$ is a reasonable canonical value. Note that ρ_{cd} [$= \rho_{solid} \times (1 - P)$, where ρ_{solid} is the mass density of compact particles] is only determined by the porosity of the aggregate, and is not size (mass) dependent (see Greenberg and Hage, 1990) except for masses $\leq 10^{-14}$ g which is the mean mass of a core-mantle interstellar dust particle. By considering the dust as comet nucleus material out of which all the volatiles, the very small (interstellar dust) particles (Utterback and Kissel, 1990), and about 1/2 of the original (relatively volatile) organic refractories were removed, the reconstituted comet nucleus density may be inferred to be $0.26 \leq \rho_c \leq 0.51\,\text{g cm}^{-3}$. Later works (Greenberg and Li, 1998a; Greenberg, 1998) modify these results slightly, but the bottom line is that Comet Halley dust has a density $\rho_{cd} \approx 0.1\,\text{g cm}^{-3}$ and its nucleus has a density $\rho_c \approx 0.3\,\text{g cm}^{-3}$. This is consistent with the low density suggestion proposed by Rickman (1986) based on the analysis of non-gravitational forces.

4. Borrelly (1994l): a Jupiter Family Short-Period Comet

The fluffy aggregate comet dust model is also applicable to short-period comets. As an example, we have calculated the dust thermal emission spectrum of Comet P/Borrelly (1994l), a Jupiter-family short-period comet (with an orbital period $P \simeq 7$ yrs), from $3\,\mu$m to $14\,\mu$m as well as the $10\,\mu$m silicate feature in terms of the comet modeled as a porous aggregate of interstellar dust (Li and Greenberg, 1998a). The fluffy aggregate model of silicate core-amorphous carbon mantle grains with a porosity $P = 0.85$ can match the observational data obtained by Hanner *et al.* (1996) quite well. It seems that, compared to the Halley dust, the dust grains of P/Borrelly appear to be relatively more processed (more carbonized), less fluffy, and richer in smaller particles.

Since P/Borrelly has passed through the inner solar system many more times than P/Halley and therefore been subjected much more to the solar irradiation, the dust grains *within the surface layer* of the nucleus could have been significantly modified. In particular, the organic refractory materials could have undergone further carbonization; namely, the organic materials, would partially lose their H, O, N atoms and thus become carbon-rich (Jenniskens *et al.*, 1993). This is supported by the results of the EURECA space experiments which have indicated the carbonization of the "first generation" organic refractory materials by solar irradiation (Greenberg *et al.*, 1995). Observations do show that some SP (Jupiter-family) comets are depleted in C_2 and C_3 (CN is approximately constant, but it is mostly produced from grains, see A'Hearn et al., 1995, for details).

The solar irradiation can also lead to a lower porosity than that of Halley dust due to the packing effect (Mukai and Fechtig, 1983; also see Smoluchowski et al., 1984). The dust size distribution could be weighted toward smaller grains; i.e., smaller grains are enhanced as a consequence of evaporation and subsequent fragmentation in the coma. There are both observational and theoretical indications of dust fragmentation in the coma of Comet P/Halley (see Greenberg and Li, 1998b, and references therein). If the fragmentation indeed results from the sublimation of volatile materials which act as "glue", one may expect relatively more drastic and more complete fragmentation in the coma of SP comets since volatiles are relatively depleted in SP comets (Weissman and Campins, 1993). A statistical analysis of fifteen comets indeed seems to suggest that the dust size distribution is somewhat steeper for short-period comets than for long-period comets (Fulle, 1999). However, at this point we are not able to generalize the dust properties of short-period comets. Systematic observations of the thermal emission spectra and the silicate features for a large set of samples are needed.

5. Hale-Bopp (C/1995 O1): a Very Large Long-Period Comet

Comet Hale-Bopp (C/1995 O1) is an exceptionally bright long-period comet (with an original period of $P \approx 4211$ yrs, after perturbation by Jupiter it was changed to $P \approx 2392$ yrs). It was visible even at a heliocentric distance of ~ 7 AU. Its strong activity and strong thermal emission features provide a rare opportunity to constrain the comet dust morphology, composition and size.

As a starting point, we have only modeled the 3–20 μm emission spectrum of Hale-Bopp of Feb. 20, 1997, obtained by Williams et al. (1997) when it was at a heliocentric distance $r_h = 1.15$ AU. Its spectrum (of Feb. 20, 1997) was characterized by the strongest 10 μm silicate feature ever observed for a comet. Its strong silicate emission feature has been generally interpreted as indicating the presence of an *unusually high abundance of small* ($< 1\,\mu$m, or equivalently $< 1.5 \times 10^{-11}$ g for *solid* silicate and $< 9.6 \times 10^{-12}$ g for *solid* carbon) grains (Hanner, 1997). Equally special was the highest ratio of the color temperature to the blackbody temperature (defined as *superheat* by Gehrz and Ney [1992]) derived by Williams et al. (1997). This has led to the suggestion that Hale-Bopp contains the smallest grains yet observed for any comet. Williams et al. (1997) estimated the mean Hale-Bopp dust size to be $\approx 0.4\,\mu$m (corresponding to 6.2×10^{-13} g for amorphous carbon, 9.4×10^{-13} g for silicate, respectively) in terms of solid separated silicate/carbon grains. This result did not take into account that the superheated thermal continuum and the strength of the silicate feature are not solely determined by grain sizes. The dust morphology, the optical properties of the dust components and the way in which the different dust components are mixed (Hanner *et al.*, 1996) are equally important.

We have calculated the dust thermal emission spectrum based on the model of comet dust consisting of porous aggregates of interstellar dust. As shown in Fig. 1 of Li and Greenberg (1998b), both the continuum emission and the 10 μm silicate feature are well matched (see Li and Greenberg, 1998b for details). The mean grain masses (derived by averaging over the size distributions) are $\sim 10^{-10} - 10^{-9}$ g, significantly different from the suggestion that Hale-Bopp is rich in submicron grains (Hanner, 1997; Williams *et al.*, 1997; see above). The presence of large numbers of very large particles in Hale-Bopp was confirmed by the submillimeter continuum emission observation (Jewitt and Matthews, 1999). It was argued that these large particles may dominate the total dust mass of the coma (Jewitt and Matthews, 1999). Assuming a spherically symmetric dust coma with uniform radial outflow, adopting the water production rate on Feb. 23, 1997 (Russo *et al.*, 1997) 4.3×10^{30} mols/s, and adopting an average dust outflow velocity of 0.12 km s^{-1} (calculated from $v_d \approx 0.05(r_h/6.82)^{-0.5}$ km/s, where r_h is the heliocentric distance in AU [Sekanina, 1996]). The dust-to-water mass production rate ratio was estimated to be as high as 41 or even higher (see Li and Greenberg, 1998b). However, one should keep in mind that the IR emission alone can not give a reliable dust production rate, since very large particles are too cold to contribute to the limited wavelength range of the infrared radiation considered here (as long as the size distribution for those cold particles is not too flat) and therefore the total mass of the large particles is not well constrained, as already noted by Crifo (1987).

The presence of a crystalline silicate component in Comet Hale-Bopp was explicitly demonstrated by both space (Crovisier *et al.*, 1997) and ground-based (Hayward and Hanner, 1997; Wooden *et al.*, 1999) observations. The fact that the crystalline silicate emission features are quite strong even when Hale-Bopp is at such large heliocentric distances as $r_h = 2.9$ AU (Crovisier *et al.*, 1997), $r_h = 4.2$ AU (Hayward and Hanner, 1997) is of particular interest because it may invoke serious questions on the role played by solar insolation. It seems likely to us that the crystallization did not occur before comet nucleus formation since, on the one hand, no evidence exists for the presence of crystalline silicates in the interstellar medium; and on the other hand, circumstellar dust which is partially crystalline (Waelkens *et al.*, 1996; Waters *et al.*, 1996) is unlikely to be directly incorporated into comets without first passing through the interstellar medium. We suggest as one possibility for the comet dust crystallized silicate component that the following 3 steps occur: (1) partially crystallized silicates are ejected into the ISM by stars; (2) as cores of the core-mantle particles, those silicates which are protected by the organic mantles are amorphized over the billions of years exposure to cosmic rays; (3) the state of amorphization of the silicates is metastable (microcrystalline ?) and recovery of crystallinity occurs at temperatures significantly less than < 1000 K, perhaps as low as 500 - 600 K which is easily achieved by the fluffy comet dust.

Another possibility, also occurring in the comet dust phase, has been described by Greenberg et al. (1996).

We have carried out calculations in which crystalline silicates are included in the model. Our calculations imply that the partially crystallized silicate model with the same dust parameters derived for the amorphous silicate model gives rise to prominent crystalline silicate features while the fit to the overall thermal emission spectrum is still maintained (see Li and Greenberg, 1998b).

6. The β Pictoris Disk: Comets in an External "Solar System"

One of the big surprises from the Infrared Astronomical Satellite (IRAS) was the so-called "Vega Phenomenon"; i.e., that some main sequence stars exhibit large infrared (IR) excesses over the blackbody emission of their photospheres (Aumann et al., 1984). It is generally believed that these IR excesses are attributed to the dust thermal emission of their circumstellar dust grains. Now it has become well established that the "Vega Phenomenon" is not exceptional but rather quite common among the main sequence stars (see Backman and Paresce, 1993). Beta Pictoris is an A5 star with the largest infrared excess among the "Vega-type stars" and an edge-on circumstellar dust disk (Smith and Terrile, 1984; Lagage and Pantin, 1994).

Various observational and theoretical evidences indicate that comets may exist in the disk (see Li and Greenberg, 1998c, and references therein). First of all, dynamical studies indicate that the grain destruction time scales due to grain-grain collisions, Poynting-Robertson drag and radiation pressure are shorter than the lifetime of β Pictoris (Backman and Paresce, 1993; Artymowicz, 1994). This indicates that there must be some dust source which continually replenishes the lost particles. Such a source was first suggested as due to comet-like bodies by Weissman (1984). Moreover, its silicate emission feature, showing a crystalline silicate feature at 11.2 μm superimposed on the broad amorphous silicate feature at 10 μm (Knacke et al., 1993), resembles that of some comets including Comet Halley and the most recent Comet Hale-Bopp (Campins and Ryan, 1989; Hanner et al., 1994; Hayward and Hanner, 1997; Crovisier et al., 1997; Wooden et al., 1999). This indicates the similarity of cometary dust with the β Pictoris dust. Furthermore, the systematic spectroscopic observations of gaseous elements in the visible or in the ultraviolet carried out by a French group showed that the spectral lines CaII, MgII, FeII, AlIII, AlII, CIV, CI and CO show strong time variations and are almost always redshifted relative to the stellar spectra (Vidal-Madjar et al., 1994, and references therein). These variations can be explained to result from the evaporation of the dust grains shed from comet-like bodies falling on the star as a consequence of planet perturbation (Beust et al., 1990).

We have developed a model for the β Pictoris dust disk which shows that the fluffy aggregate comet dust model is also applicable to extrasolar comets (see Li and Greenberg, 1998b). The basic idea is that the dust in the disk plane is continually replenished by comets orbiting the star where the dust may be quickly swept out by radiation pressure or spiral onto the star as a result of Poynting-Robertson drag. The initial dust shed by the comets is taken to be the fluffy aggregates of interstellar silicate core-organic refractory mantle dust grains (with an additional ice mantle in the outer region of the disk). The heating of the dust is primarily provided by the organic refractory mantle absorption of the stellar radiation. The temperature of some of the particles close to the star is sufficient to crystallize the initially amorphous silicates. The dust grains are then distributed throughout the disk by radiation pressure. The steady state dust distribution of the disk then consists of a mixture ranging from crystalline silicate aggregates to aggregates of amorphous silicate core-organic refractory mantle particles (without/with ice mantles) with variable ratios of organic refractory to silicate mass. The whole disk which extends inward to ~ 1 AU and outward to ~ 2200 AU is divided into three components which are primarily responsible respectively, for the silicate emission, the mid-infrared emission and the far infrared/millimeter emission. As a starting point, the grain size distribution is assumed to be like that observed for Comet Halley dust while in the inner regions the distribution of small particles is relatively enhanced which may be attributed to the evaporation and/or fragmentation of large fluffy particles. The dust grains which best reproduce the observations are highly porous, with a porosity around 0.95 or as high as 0.975. The temperature distribution of a radial distribution of such particles provides an excellent match to the $10 \, \mu$m amorphous and the $11.2 \, \mu$m crystalline silicate spectral emission as well as the excess continuum flux from the disk over a wide range of wavelengths. These models result in a total mass of dust in the whole disk $\sim 2 \times 10^{27}$ g of which only $10^{-5} - 10^{-4}$ is required to be heated enough to give the silicate excess emission.

7. Summary

Attempts have been made to generalize the chemical composition of comet nuclei based on the observation of cometary dust and volatiles and the interstellar dust model. The infrared emission of various comets can be matched within the framework that all comets are made of fluffy (porous) aggregates of interstellar dust. This is demonstrated by comparing results on Halley (a periodic comet), Borrelly (a Jupiter family short-period comet), Hale-Bopp (a long-period comet), and extra-solar comets in the β Pictoris disk.

Acknowledgements

We are grateful for the support by NASA grant NGR 33-018-148 and by a grant from the Netherlands Organization for Space Research (SRON). We thank

Drs. S.B. Fajardo-Acosta, M. Fulle, R.D. Gehrz, M.S. Hanner, W. Huebner, R.F. Knacke, C. Koike, M. Mueller for their kind help or useful suggestions. One of us (AL) would like to thank Leiden University for an AIO fellowship and the World Laboratory for a scholarship and the Nederlandse organisatie voor wetenschappelijk onderzoek (NWO) and the National Science Foundation of China for financial support.

References

A'Hearn, M.F., Millis, R.L., Schleicher, D.G., et al.: 1995, *Icarus* **118**, 223.
Artymowicz, P.: 1994, in *Circumstellar Dust Disks and Planet Formation* (edited by R. Ferlet and A. Vidal-Madjar), 47. Editions Frontieres: Gif sur Yvette.
Aumann, H.H., et al.: 1984, *Astrophys. J.* **278**, L23.
Backman, D.E. and Paresce, F.: 1993, in *Protostars and Planets* **III**, (edited by E.H. Levy, J.I. Lunine and M.S. Mathews), p. 208. University of Arizona Press, Tuscon.
Beust, H., Lagrange-henri, A.M., Vidal-Madjar, A. and Ferlet, R.: 1990, *Astron. Astrophys.* **236**, 202.
Bockelée-Morvan, D.: 1997, in *Molecules in Astrophysics: Probes and Processes* (edited by E.F. van Dishoeck), IAU Symp. **178**, p.222. Kluwer, Dordrecht.
Campins, H. and Ryan, E.: 1989, *Astrophys. J.* **341**, 1059.
Colangeli, L., Mennella, V., Di Marino, C., Rotundi, A. and Bussoletti, E.: 1995, *Astron. Astrophys.* **293**, 927.
Crifo, J.F.: 1987, in *Interplanetary Matter* (edited by Z. Ceplecha and P. Pecina), Czech. Academy of Sciences Report **G7**, 59.
Crovisier, J.: 1999, in *Formation and Evolution of Solids in Space* (edited by J.M. Greenberg and A. Li), p. 389. Kluwer, Dordrecht.
Crovisier, J., Leech, K., Bockelée-Morvan, D., et al.: 1997, *Science* **275**, 1904.
Delsemme, A.H.: 1982, in *Comets* (edited by L.L. Wilkening), P. 85. University of Arizona Press, Tucson.
Eberhardt, P., Krankowsky, D., Schulte, W., et al.: 1987, *Astron. Astrophys.* **187**, 481.
Fulle, M.: 1999, *Planet. Space Sci.*, in press.
Gehrz, R.D. and Ney, E.P.: 1992, *Icarus* **100**, 162.
Greenberg, J.M.: 1982, in *Comets* (edited by L.L. Wilkening), p. 131. University of Arizona Press, Tuscon.
Greenberg, J.M.: 1986, *Nature* **321**, 385.
Greenberg, J.M.: 1998, *Astron. Astrophys.* **330**, 375.
Greenberg, J.M.: 1999, in *Formation and Evolution of Solids in Space* (edited by J.M. Greenberg and A. Li), p. 53. Kluwer, Dordrecht.
Greenberg, J.M. and Hage, J.I.: 1990, *Astrophys. J.* **361**, 260.
Greenberg, J.M. and Li, A.: 1998a, in *Solar System Ices* (edited by B. Schmitt, C. de Bergh and M. Festou), Chapter 13, p. 337. Kluwer, Dordrecht.
Greenberg, J.M. and Li, A.: 1998b, *Astron. Astrophys.* **332**, 374.
Greenberg, J.M., Mizutani, H., and Yamamoto, T.: 1995, *Astron. Astrophys.* **295**, L35.
Greenberg, J.M., Li, A., Kozasa, T., and Yamamoto, T.: 1996, in *Physics, Chemistry, and Dynamics of Interplanetary Dust*, (edited by B.A.S. Gustafson, M.S. Hanner), ASP Conference Series, Vol. **104**, p. 497.
Greenberg, J.M., Li, A., Mendoza-Gómez, C.X., Schutte, W.A., Gerakines, P.A. and de Groot, M.: 1995, *Astrophys. J.* **455**, L177.
Hanner, M.S., Lynch, D.K. and Russell, R.W.: 1994, *Astrophys. J.* **425**, 274.

Hanner, M.S., Lynch, D.K., Russell, R.W., Hackwell, J.A. and Kellogg, R.: 1996, *Icarus* **124**, 344.
Hanner, M.S.: 1997, *BAAS* **29**, 1042.
Hayward, T.L. and Hanner, M.S.: 1997, *Science* **275**, 1907.
Irvine, W.M., Bockelée-Morvan, D., *et al.*: 1996, *Nature* **383**, 418.
Jenniskens, P., Baratta, G.A., Kouchi, A., de Groot, M.S., Greenberg, J.M. and Strazzulla, G.: 1993, *Astron. Astrophys.* **273**, 583.
Jessberger, E.K. and Kissel, J.: 1991, in *Comets in the Post-Halley Era* (edited by R.L. Newburn, M. Neugebauer and J. Rahe), p. 1075, Kluwer, Dordrecht.
Jewitt, D.C. and Matthews, H.E.: 1999, *Astron. J.* **117**, 1156.
Kissel, J., Sagdeev, R.Z., Bertaux, J.L., *et al.*: 1986a, *Nature* **321**, 280.
Kissel, J., Brownlee, D.E., Büchler, K., *et al.*: 1986b, *Nature* **321**, 336.
Knacke, R.F., Fajardo-Acosta, S.B., Telesco, C.M., *et al.*: 1993, *Astrophys. J.* **418**, 440.
Krueger, F.R. and Kissel, J.: 1987, *Naturwissenschaften* **74**, 312.
Lagage, P.O. and Pantin, E.: 1994, *Nature* **369**, 628.
Lawler, M.E. and Brownlee, D.E.: 1992, *Nature* **359**, 810.
Li, A. and Greenberg, J.M.: 1997, *Astron. Astrophys.* **323**, 566.
Li, A. and Greenberg, J.M.: 1998a, *Astron. Astrophys.* **338**, 364.
Li, A. and Greenberg, J.M.: 1998b, *Astrophys. J.* **498**, L83.
Li, A. and Greenberg, J.M.: 1998c, *Astron. Astrophys.* **331**, 293.
Li, A. and Greenberg, J.M.: 1999, *Space Sci. Rev.*, in press.
Maas, R.W., Ney, E.P. and Woolf, N.F.: 1970, *Astrophys. J.* **160**, L101.
McDonnell, J.A.M., Alexander, W.M., Burton, W.M., *et al.*: 1986, *Nature* **321**, 338.
Mukai, T. and Fechtig, H.: 1983, *Planet. Space Sci.* **31**, 655.
Mumma, M., Weissman, P. and Stern, S.A.: 1993, in *Protostars and Planets* III (edited by E.H. Levy, J.I. Lunine and M.S. Mathews), p. 1177. University of Arizona Press, Tucson.
Rickman, H.: 1986, in *Comet Nucleus Sample Return* (edited by O. Melitta), ESA SP-**249**, p. 195.
Russo, N.D.: 1997, *IAU Circ.* **6604**.
Schutte, W.A.: 1996, in *The Cosmic Dust Connection*, (edited by J.M. Greenberg), Kluwer, p. 1.
Sekanina, Z.: 1996, *Astron. Astrophys.* **314**, 957.
Smith, B.A. and Terrile: 1984, *Science* **226**, 1421.
Smoluchowski, R., Marie, M. and McWilliam, A.: 1984, *Earth, Moon, and Planets* **30**, 281.
Utterback, N.G. and Kissel, J.: 1990, *Astron. J* **100**, 1315.
Van de Hulst, H.C.: 1949, *Rech. Astr. Obs. Utrecht* **11**, part 2.
Vidal-Madjar, A., Lagrange-Henri, A-M., Feldman, P.D., *et al.*: 1994, *Astron. Astrophys.* **290**, 245.
Waelkens, C., Waters, L.B.F.M., de Graauw, M.S., *et al.*: 1996, *Astron. Astrophys.* **315**, L245.
Waters, L.B.F.M., Molster, F.J., de Jong, T., *et al.*: 1996, *Astron. Astrophys.* **315**, L36.
Weissman, P.R.: 1984, *Science* **224**, 987.
Weissman, P.R. and Campins, H.: 1993, in *Resources of Near-Earth Space* (edited by J. Lewis, M.S. Mathews and M.L. Guerrieri), p. 569. University of Arizona Press, Tucson.
Whipple, F.L.: 1950, *Astrophys. J.* **111**, 375.
Whipple, F.L.: 1951, *Astrophys. J.* **113**, 464.
Williams, D.M., Mason, C.G., Gehrz, R.D., *et al.*: 1997, *Astrophys. J.* **489**, L91.
Wooden, D.H., Harker, D.E., Woodward, C.E., *et al.*: 1999, *Astrophys. J.*, in press.

Address for correspondence: J. Mayo Greenberg, Raymond and Beverly Sackler Laboratory for Astrophysics at Leiden Observatory, Sterrewacht Leiden, Postbus 9504, 2300 RA Leiden, The Netherlands, greenber@strw.leidenuniv.nl

POLARIZATION OF LIGHT SCATTERED BY COMETARY DUST PARTICLES: OBSERVATIONS AND TENTATIVE INTERPRETATIONS

A.-CHANTAL LEVASSEUR-REGOURD

Université Paris VI and Service d'Aéronomie CNRS / IPSL, BP 3, 91371 Verrières, France

Abstract. Analysis of the polarization of light scattered by cometary particles reveals similarities amongst the phase curves, together with some clear differences: i) comets with a strong silicate emission feature present a high maximum in polarization, ii) the polarization is always slightly lower than the average in inner comae and stronger in jet-like structures. These results are in excellent agreement with the Greenberg model of dust particles built up of fluffy aggregates of much smaller grains. Also, they suggest the existence of different regions of formation, and of different stages of evolution for the scattering particles inside a given cometary coma.

Keywords: Comets, Dust particles, Scattering, Polarization

1. Introduction

The partial linear polarization of light emergent from cometary comae and tails has been noticed since Arago (1858). He precisely inferred the existence of scattering dust particles from his observations of a faint polarization in the light of the great comet of 1819. Polarization phase curves, which present the polarization as a function of the phase angle (comet-Sun, comet-observer angle), are of special interest since they do not require any normalization to a constant distance to the Sun and/or to the observer to allow comparisons between the results obtained for one single or even for many objects.

It was already noticed by Hanner (1980) that the polarization increases smoothly in the 40° to 90° phase angle range, and that the maximum is not the same for all comets. It was pointed out by Levasseur-Regourd (1992) and Chernova *et al.* (1993) that the polarization is always negative for small phase angles (intensity scattered perpendicularly to the scattering plane smaller than intensity scattered perpendicular to it), with an inversion near $(20 \pm 5)°$.

Some characteristics of the phase curves might provide clues to the physical properties of the scattering dust, although inferring them from the data is not an easy task. The slope at inversion has been suggested by many authors to decrease when the albedo increases, while an anticorrelation has been pointed out between the maximum in polarization (also very sensitive to the size of the particles) and the albedo.

2. Whole Coma Observations

Remote observations of relatively bright comets have provided a wide data base for polarization observations. The measurements need to be performed on the whole coma, with narrow filters to avoid the contamination by less polarized molecular bands. Once the data are separated into different wavelength ranges, significant polarization phase curves are obtained, which can be compared to one another.

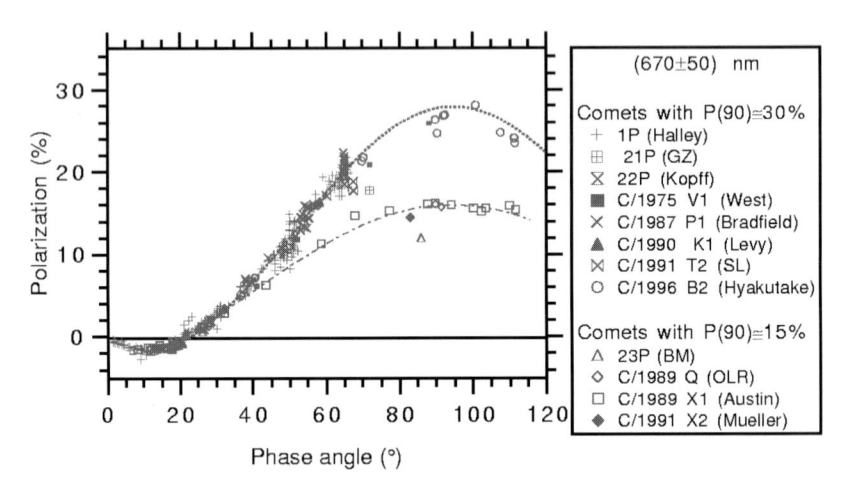

Figure 1. Cometary polarization in the red domain as a function of the phase angle. All the published observational data agree fairly well below 40°, while two classes at least are pointed out for larger phase angles. The dotted and dashed-dotted curves fit the data points for comets with a high maximum in polarization and a lower maximum in polarization, respectively.

Similar trends can be pointed out, with a polarization of about 0 percent near 0°, a slight negative branch and a wide positive branch with a near 90° maximum. Figure 1 shows that all the data agree fairly well within a given wavelength range below 40° phase angle, while, for larger phase angles, a dichotomy is pointed out between comets which have a maximum in polarization in the 25 to 30 percent range, and comets whose polarization at maximum is in the 10 to 15 percent range (Levasseur-Regourd *et al.*, 1996). This behavior is a clue to significant differences in the physical properties of the dust, comets presenting a high polarization being those that have been found by Hanner *et al.* (1994) to display a structured silicate feature, typical of small grains or very fluffy aggregates of such grains.

As observed for Comet 1P/Halley (Dollfus *et al.*, 1988) and later confirmed for other comets (Hadamcik *et al.*, 1995), the polarization slightly increases with increasing wavelength in the 500 to 1200 nm range. This increase, which grows for greater values of the polarization, seems to be too systematic to be only due to the contamination from molecular cometary emissions.

Bright Comet C/1995 O1 Hale-Bopp presented extremely high polarization levels, with a spectral dependence of the order of 1.1 percent per 100 nm at 45°

(Kiselev and Velichko, 1998). This high polarization, also noticed by other observers (e.g. Hadamcik *et al.*, 1998; Jockers *et al.*, 1998), was correlated with a strong silicate emission feature (Hanner, 1999). It could have come from some excesses of polarization in the jet-like features. Significant changes have indeed been observed, as detailed below, in the inner coma polarization of various comets.

3. Small Scale Observations

Drastic fluctuations in the inner coma polarization were detected by the Giotto Optical Probe Experiment during the flyby of Comets 1P/Halley and 24/P Grigg-Skjellerup (see e.g. Levasseur-Regourd *et al.*, 1993). These changes led us to carry on a programme of ground based polarization imaging.

High resolution polarization maps of comets such as C/1990 K1 Levy, 47P/Ashbrook-Jackson, or C/1995 O1 Hale-Bopp (Renard *et al.*, 1992, 1996; Hadamcik *et al.*, 1998) reveal halos corresponding to a lower polarization around the nucleus. This depolarization, noticed for distances to the nucleus approximately below 2000 km, could be attributed to some illumination by the nucleus, to some multiple scattering, or to the presence of dust particles with different physical properties hovering over the nucleus.

Polarization maps also enhance, as compared to brightness maps, jet-like structures corresponding to polarization levels higher than the average. This effect, already noticed by Eaton *et al.* (1988) for Comet 1P/Halley, could be produced by the alignment of elongated particles, or by the existence of freshly ejected dust particles with different physical properties.

4. Discussion and Interpretation

The shape of the polarization phase curves indicates that the scattering particles are irregular, and possibly made up of aggregates of smaller grains, in agreement with the Greenberg model (see e.g. Greenberg, 1996). The existence of at least two classes of comets is in good agreement with the existence of different regions of agglomeration of the cometary nuclei.

The two sets of values noticed in the near maximum polarization could correspond either to different compositions and albedos, or to different size distributions. Dark absorbing particles would be expected to lead to a higher polarization than silicates with similar sizes, as noticed from both observational and computational approaches (e.g. Goidet-Devel *et al.*, 1995; Xing and Hanner, 1997). However, since a clear silicate emission feature has been observed for such comets, it is most likely that the key difference is in the size distribution of the particles and of their aggregates.

The changes in polarization noticed inside a cometary coma may be interpreted in terms of temporal evolution, the particles observed in the jet-like structures being aggregates recently ejected from the nucleus. It may be noticed that, while the average "background" cometary particles, together with the particles in the jets, could present an increase in polarization with increasing wavelength, the particles in the inner comae present a decrease in polarization with increasing wavelength (Jockers et al., 1998). This behavior is reminiscent of the interplanetary dust particles, whose polarization decreases with increasing wavelength, and also with time as they spiral towards the Sun while losing some fluffiness (Levasseur-Regourd, 1995). The scattering properties observed in the inner comae indeed suggest the existence of relatively bigger and less porous aggregates, in agreement with the in situ observations of the Dust Impact Detector on board Giotto (McDonnell et al., 1991).

5. Conclusions and Perspectives

Although all the cometary polarization phase curves are typical of irregular aggregates, significant differences are found between comets, and even inside a cometary coma. These differences suggest both the existence of different accretion regions where the comets were formed, and different evolution stages of the particles present in a cometary coma. Some puzzling problems, such as the wavelength dependence of the polarization by cometary dust, remain unsolved.

It is unlikely that a large set of cometary dust samples from various comets and from various regions inside one cometary coma will be available in the next decade(s) through the sample return missions. Some of the interplanetary dust particles collected in the Earth's environment (IDP) are of cometary origin, but they are not likely to have preserved the cometary dust size distribution. It is thus of major interest to quantify the above-mentioned differences in terms of size distribution of the particles and of complex refractive index of their constituents, through models and laboratory measurements.

Computations of the scattering by irregular particles are now developed by various teams. We have computed the scattering properties of fractal aggregates, taking into account the fact that the growing process in the early solar system was likely to be triggered by inelastic collisions and to lead to fractal aggregates. For particles of fractal dimension about 2 or slightly below, built up of a very large number of dark absorbing spherical grains, the polarization curves are smooth and reproduce relatively well the observed behavior (Levasseur-Regourd et al., 1997).

However, many constraints appear in such computations, and it is of interest to perform direct measurements on characteristic particles, which permit the variation of structural parameters. To build in the laboratory realistic fractal particles, it is necessary to simulate the physical conditions prevailing in space. The purpose of future measurements by polar nephelometers in low pressure chambers, on board sounding rockets or space stations, is indeed to monitor aggregation processes representative of comet formation, and to document the temporal evolution of the scattering properties of such particles (see e.g. Levasseur-Regourd *et al.*, 1998).

Acknowledgements

It is a pleasure to thank M. Cabane, E. Hadamcik, V. Haudebourg, J.B. Renard and J.C. Worms for fruitful discussions on their observations, light scattering models, and laboratory measurements. Figure 1 was updated by E. Hadamcik. I am grateful to ISSI and to the convenors for the efficient organization of this stimulating workshop.

References

Arago, F.: 1858, *Les comètes*, Gide Editeur.
Chernova, G., Kiselev, N., and Jockers, K.: 1993, 'Polarimetric characteristics of dust particles as observed in 13 comets: comparisons with asteroids', *Icarus* **103**, 144-158.
Dollfus, A., Bastien, P., Le Borgne, J.M., Levasseur-Regourd, A.C., Mukai, T.: 1988, 'Optical polarimetry of P/Halley', *Astron. Astrophys.* **206**, 348-356.
Eaton, N., Scarrott, S.M., and Warren-Smith, R.F.: 1988, 'Polarization images of the inner regions of comet Halley', *Icarus* **76**, 270-278.
Goidet-Devel, B., Renard, J.B. and Levasseur Regourd, A.C.: 1995, 'Polarization of asteroids: Synthetic curves and characteristic parameters', *Planet. Space Sci.* **43**, 6, 779- 786.
Greenberg, M.: 1996, *The Cosmic dust connection*, NATO ASI Series **487**.
Hadamcik, E., Levasseur-Regourd, A.C., and Worms, J.C.: 1995, 'Polarimetric measurements of scattered light by dust grains in Earth and microgravity conditions', *Astron. Soc. Pac.* **104**, 391-394.
Hadamcik, E., Levasseur-Regourd A.C., and Renard J.B.: 1998, 'CCD imaging polarimetry of comet Hale- Bopp', *Earth, Moon, Plan.* (in press)
Hanner, M.S.: 1980, 'Physical characteristics of cometary dust from optical studies', *Solid particles in the solar system*, D. Reidel, 223-235.
Hanner, M.S., Lynch, D.K., Russel, R.W.: 1994, 'The 8-13 mm spectra of comets and the composition of the silicate grains', *Astrophys. J.* **425**, 275-285.
Hanner, M.S.: 1999, 'The silicate material in comets', *Space Sci. Rev.*, this volume.
Jockers, K., Rosenbush, V.K., Bonev, T., and Credner, T.: 1998, 'Images of polarization and color in the inner coma of comet Hale-Bopp', *Earth, Moon, Plan.* (in press)
Kiselev, N.N., and Velichko, F.P.: 1998, 'Polarimetry and photometry of comet Hale- Bopp', *Earth, Moon, Plan.* (in press)
Levasseur-Regourd, A.C.: 1992, 'Polarimetry of cometary dust', *Liège Intern. Astrophys. Coll.* **30**, 141-148.
Levasseur-Regourd, A.C.: 1995, 'Optical and thermal properties of zodiacal dust', *Astron. Soc. Pac.* **104**, 301-308.
Levasseur-Regourd, A.C., Goidet, B., Le Duin, T., Malique, C., Renard, J.B., Bertaux, J.L.: 1993, 'Optical probing of comet Grigg-Skjellerup dust from the Giotto spacecraft', *Planet. Space Sci.* **41**, 167-169.
Levasseur-Regourd, A.C., Hadamcik, E., Renard, J.B.: 1996, 'Evidence for two classes of comets from their polarimetric properties at large phase angles', *Astron. Astrophys.* **313**, 327-333.

Levasseur-Regourd, A.C., Cabane, M., Worms, J.C., Haudebourg, V.: 1997, 'Physical properties of dust in the solar system: relevance of a computational approach and of measurements under microgravity conditions', *Adv. Space Res.* **20**, 8, 1585-1594.

Levasseur-Regourd, A.C., Cabane, M., Haudebourg, V., and Worms, J.C.: 1998, 'Light scattering experiments under microgravity conditions', *Laboratory Astrophysics and Space Research*, Kluwer, 459-485.

McDonnell, J.A.M., Lamy, P.L., and Pankiewicz, G.S.: 1991, 'Physical properties of cometary dust', *Comets in the post-Halley era*, Kluwer, 1043-1073.

Renard, J.B., Levasseur-Regourd, A.C., Dollfus, A.: 1992, 'Polarimetric CCD imaging of comet Levy', *Ann. Geophys.* **10**, 288-292.

Renard, J.B., Hadamcik, E., and Levasseur-Regourd, A.C.: 1996, 'Polarimetric imaging of comet 47P/Ashbrook-Jackson', *Astron. Astrophys.* **316**, 263-269.

Xing, Z., and Hanner, M.S.: 1997, 'Light scattering by aggregate particles', *Astron. Astrophys.* **324**, 805-820.

Address for correspondence: A.-C. Levasseur-Regourd, aclr@aerov.jussieu.fr

CHANGES IN THE STRUCTURE OF COMET NUCLEI DUE TO RADIOACTIVE HEATING

D. PRIALNIK and M. PODOLAK
Department of Geophysics and Planetary Sciences
Tel Aviv University, Ramat Aviv 69978, Israel.

Abstract. The initial structure of a comet nucleus is most probably a homogeneous, porous, fine-grained mixture of dust and ices, predominantly water. The water ice is presumably amorphous and includes considerable fractions of occluded gases. This structure undergoes significant changes during the early evolution of the nucleus at large heliocentric distances, due to internal radiogenic heating. Structural changes occur mainly as a result of gas flow through the porous medium: the gas pressure that builds up in the interior is capable of breaking the fragile structure and altering the pore sizes and porosity. These effects are modeled and followed numerically, testing a large number of parameters.

Keywords: Comets: structure, composition, evolution

1. Introduction

Comets are usually believed to have preserved their primordial structure and composition, at least in their deep interior. Indeed much of the interest in the structure and composition of comet nuclei is due to the belief that, as pristine objects, they may hold important clues to the origin of the solar system. But comets are known to contain dust and if this dust includes radioactive isotopes, then these will provide a source of heat even while the comets are far from the sun (during early stages of evolution). The effect of radiogenic heating on the thermal structure and composition of comets was studied by Whipple and Stefanic (1966), Wallis (1980), Prialnik *et al.* (1987), Haruyama *et al.* (1993), Yabushita (1993), Prialnik and Podolak (1995, hereafter Paper I) and Podolak and Prialnik (1997). Heating due to the radioactive decay of ^{40}K, ^{235}U, ^{238}U, and ^{232}Th is found to cause a temperature rise of some tens of degrees. For very low thermal conductivities, a runaway increase in the internal temperature occurs and most of the ice in the nucleus crystallizes (if it is initially amorphous), while for a sufficiently large conductivity, the amorphous ice may be almost completely preserved. Melting in the interior of comet nuclei as a result of radiogenic heating may occur for a sufficiently low thermal conductivity and large cometary radii (≥ 200 km). Beside the long-lived radionuclides, an additional source of heating is that due to the decay of ^{26}Al, whose abundance in the Galaxy is estimated to be significantly high, ^{26}Al/^{27}Al $\approx 5 \times 10^{-5}$, (see e.g., Clayton, 1984), despite its very short half–life (7.2×10^5 yr). The resulting initial mass fraction of ^{26}Al in the solid material of the solar nebula is $\sim 7 \times 10^{-7}$ (in agreement with that derived from meteorite measurements); in fully grown comets it is uncertain, however, as it depends on their time of formation. Since for the giant planets the estimated formation time is between 1 and 16 million years (Pollak *et al.*, 1996) and since cometary size planetesimals took much less to form (possibly as little as a few hundred years), an initial ^{26}Al mass fraction of a few 10^{-8} should be a reasonable working assumption.

In Paper I we showed that the effect of radiogenic heating depends critically on the porosity of the nucleus. Thus a layered composition as a function of depth may be obtained with ices of higher volatility closer to the surface. Here we focus on the effect of radiogenic heating on the porous *structure* of the nucleus. We describe the numerical model and discuss the potential effect of radiogenic heating in §2; results of numerical computations are presented in §3 and the main conclusions are summarized in §4.

2. Brief Description of the Model

2.1. THE SET OF EVOLUTION EQUATIONS

Comet nucleus models assume some form of symmetry, for simplicity. Since we are concerned mainly with internal processes, we may adopt the simplest among such models: a one-dimensional, spherically symmetric nucleus.

The general composition of a comet nucleus includes water ice — amorphous and crystalline, water vapor, dust, and other volatiles, which may be frozen, free, or trapped in the amorphous ice. Let ρ denote the bulk mass density, and let the density of the various components be ρ_a (amorphous ice), ρ_c (crystalline ice), ρ_v (water vapor), ρ_d (dust), $\rho_{s,n}$ and $\rho_{g,n}$, where n runs over the different volatile species other than H_2O (CO, CO_2, *etc.*), in solid (s) or gaseous (g) form. The amorphous ice includes mass fractions f_n of trapped gases. The implicit assumption of the model is that all the components of the nucleus are in local thermal equilibrium and hence a unique local temperature $T(r)$ may be defined. The energy per unit mass u is given by $\rho u = \sum_\alpha \rho_\alpha u_\alpha$, where $u_\alpha(T)$ are the specific energies. We denote by $\mathbf{J}_\alpha$ mass fluxes, by $\mathbf{F}$ the heat flux, by q_α the rates of sublimation (condensation) and by $\lambda(T)$ the rate of crystallization of amorphous ice, given by Schmitt *et al.*(1989), by S the surface to volume ratio of the porous medium, by $\mathcal{P}$ the saturated vapor pressure, and by P the actual pressure. The set of equations that describes the evolution of a comet nucleus follows. The mass balance equations for H_2O are:

$$\frac{\partial \rho_a}{\partial t} = -\lambda(T)\rho_a, \qquad (1)$$

$$\frac{\partial \rho_c}{\partial t} = (1 - \sum_n f_n)\lambda(T)\rho_a - q_v, \qquad q_v = S(\mathcal{P}(T) - P)\left(\frac{m_{H_2O}}{2\pi kT}\right)^{1/2}, \qquad (2)$$

TABLE I
Basic data

Isotope	τ (years)	X_0	H (erg g^{-1})	$X_0 H$ (erg g^{-1})	$\tau^{-1} X_0 H$ (erg g^{-1} s^{-1})
^{40}K	1.82(9)	1.1(-6)	1.72(16)	1.89(10)	3.3(-7)
^{232}Th	2.00(10)	5.5(-8)	1.65(17)	9.08(9)	1.4(-8)
^{238}U	6.50(9)	2.2(-8)	1.92(17)	4.22(9)	2.1(-8)
^{235}U	1.03(9)	6.3(-9)	1.86(17)	1.17(9)	3.6(-8)
^{26}Al	1.06(6)	~5(-8)	1.48(17)	~7(9)	2.1(-4)

$$\frac{\partial \rho_v}{\partial t} + \nabla \cdot \mathbf{J}_v = q_v, \tag{3}$$

and similar equations for other volatiles. The energy conservation law is:

$$\frac{\partial}{\partial t}(\rho u) + \nabla \cdot (\mathbf{F} + \sum_\alpha u_\alpha \mathbf{J}_\alpha) = -\sum_\alpha q_\alpha \mathcal{H}_\alpha + \lambda(T)\rho_a(1 - \sum_n f_n)\mathcal{H}_{ac} + Q, \tag{4}$$

where $\mathcal{H}$ denotes latent heat of sublimation and $\mathcal{H}_{ac}$ is the energy released in crystallization of the amorphous ice and Q is the rate of radiogenic energy release,

$$Q = \rho_d \sum_j \tau_j^{-1} X_{0,j} \exp^{-t/\tau_j} H_j. \tag{5}$$

Here τ_j is the characteristic decay time of the $j'th$ radioactive isotope, $X_{0,j}$ is its initial mass fraction within the dust, and H_j is the energy released per unit mass upon decay, as given in Table I. The heat flux $\mathbf{F}$ is given by: $\mathbf{F} = -\psi(p)K(T)\nabla T$, where the thermal conductivity $K(T)$ represents the conductivity of the mixture of ices and dust and is corrected by a factor $\psi(p) < 1$ due to the reduced contact surface between grains. In simplest form, $\psi = 1 - p^{2/3}$, but it may be orders of magnitude smaller than unity (e.g., Steiner and Kömle, 1991). However, as heat conduction is found to be dominated by the gas flow (advection and transfer of latent heat), these corrections are not crucial. The gas flux in a porous medium depends both on the porosity and on the pore size (cf. Mekler *et al.*, 1990). When the pore size is small (i.e., the corresponding Knudsen number exceeds unity), the flow of gas is essentially a free molecular (Knudsen) flow. We use the algorithm described by Prialnik *et al.* (1993) to simulate the effect of pore growth due to internal gas pressure. The evolution of the porosity is obtained, locally, from the changes in density of the solids, which affect the pore sizes as well. Finally, the boundary conditions at the center of the nucleus are vanishing fluxes (mass and heat). At the surface R we have vanishing pressures and, since at large heliocentric distances d_H there is no surface sublimation,

$$\mathbf{F}(R,t) = -4\pi R^2[(1-A)L_\odot/(16\pi d_H(t)^2) - \epsilon\sigma T(R,t)^4], \tag{6}$$

TABLE II
Energy required for different processes

Process	T (K)	Δu (erg g^{-1})	H (erg g^{-1})	X_0 (^{26}Al)
H_2O sublimation	180	1.4(9)	2.8(10)	2.0(-7)
CO_2 sublimation	100	4.5(8)	5.9(9)	4.3(-8)
CO sublimation	30	5.0(7)	2.3(9)	1.6(-8)
H_2O melting	273	3.0(9)	3.3(9)	4.3(-8)
H_2O crystallization	130	7.4(8)	- 9(8)	5.0(-9)

where A is the albedo, $L_\odot$ is the solar luminosity, ϵ is the emissivity (of order 1) and σ is the Stefan-Boltzmann constant. The system of non-linear time-dependent second order partial differential equations is turned into an implicit difference scheme and solved iteratively (see Prialnik, 1992).

2.2. ANALYTICAL CONSIDERATIONS

In order to get an idea of the potential effect of radiogenic heating during the early evolution of comets, we list in Table II the energy per unit mass required by different processes (to be compared with the available energy listed in Table I). Since the initial ^{26}Al abundance is uncertain, we also list the mass fraction of ^{26}Al that would be required in order to supply this energy by itself. We conclude that none of the radioactive isotopes would be capable of evaporating the comet nucleus, but most of them supply sufficient energy for melting the ice or evaporating the other volatile species. However, if the radioactive energy is supplied at a low rate, its effect is bound to be far less significant, since the heat might be lost by radiation at the surface. A more realistic estimate of the effect of radiogenic heating is obtained from global energy considerations. If we neglect crystallization, as well as internal sublimation and gas flow, and integrate the energy equation (4) over the entire nucleus, we have:

$$\int \frac{\partial u}{\partial t} dm = M \sum_j \tau_j^{-1} X_{0,j} H_j e^{-t/\tau_j} - 4\pi R^2 \left[\sigma T(R)^4 - \frac{(1-A)L_\odot}{16\pi d_H^2} \right]. \quad (7)$$

Defining an equilibrium surface temperature

$$T_{eq}(d_H) = \left[\frac{(1-A)L_\odot}{16\pi d_H^2 \sigma} \right]^{1/4}, \quad (8)$$

we obtain

$$\langle \frac{du}{dt} \rangle = \sum_j \tau_j^{-1} X_{0,j} H_j e^{-t/\tau_j} - \frac{3\sigma}{R\rho}(T_s^4 - T_{eq}^4). \quad (9)$$

Now, the difference between T_{eq} and the actual surface temperature $T(R)$ is due to heat conduction into or out of the nucleus. Roughly,

$$\sigma T_s^4 - \sigma T_{eq}^4 = -K\frac{dT}{dr} \sim K\frac{T}{R}. \tag{10}$$

Hence

$$\langle\frac{du}{dt}\rangle = \sum_j \tau_j^{-1} X_{0,j} H_j e^{-t/\tau_j} - \frac{3KT}{R^2\rho}. \tag{11}$$

The maximal heating rate, obtained at $t = 0$, is $\tau_j^{-1} X_{0,j} H_j$; the numerical values of this factor for each radiogenic species are listed in Table I. The cooling rate,

$$\frac{3KT}{R^2\rho} \approx \frac{0.0016}{R_{km}^2} \text{ erg g}^{-1}\text{ s}^{-1} \tag{12}$$

for amorphous ice, is lower than the rate of heating for $R \gtrsim 3$ km in the presence of ^{26}Al, and for $R \gtrsim 70$ km in its absence. For crystalline ice the cooling rate is about 20 times higher and hence higher cometary radii are required for internal heating to take place. As the heating rate declines with time, it eventually becomes lower than the rate of cooling in all cases. In conclusion, for suitably large comets the internal temperature rises at the beginning up to a maximal value T_{max} and then falls, tending to T_{eq}. Obviously, a larger radius or a higher ^{26}Al content would lead to a higher T_{max}. The factor K/ρ in eq. (12) decreases with increasing porosity ($\psi(p) < 1 - p$) and hence T_{max} varies as the porosity. However, the presence of volatiles in the interior may reverse this trend, since gas flow, which increases with porosity, may contribute significantly to heat transfer. In addition, the pressure that gases exert could, in turn, affect the porous structure.

The thermal structure of a comet is thus determined by the competition between the rates of several different processes (cf. Paper I). At a depth of 1 km the rate of decay ^{26}Al becomes comparable to the conduction rate of amorphous ice, meaning that the ice might be barely heated; it would certainly be heated at larger depths, a few km and beyond. Eventually, the internal temperature would become sufficiently high for crystallization to set in, providing an additional internal heat source. At the same time, however, heat transfer would be more efficient, crystalline ice being a much better heat conductor than amorphous ice. Hence, only in still larger comet nuclei (beyond 10 km) would the internal temperature continue to rise. Eventually, most of the released energy will be absorbed in sublimation of ice from the pore walls. A steady state will develop, without further heating of the ice.

TABLE III
Initial parameter values

Parameter	Basic model	Range
Radius(km)	100	1 – 100
$X(^{26}Al)$	5×10^{-8}	$(0.1 – 5) \times 10^{-8}$
Porosity	0.1	0.05 – 0.9
Pore size (μm)	100	$1 – 10^4$
$f_{CO} = f_{CO_2}$	0.01	0.001 – 0.01

3. Results of Numerical Computations

Analytical considerations alone cannot lead to an evaluation of T_{max} and its consequences. We thus resort to numerical computations, using the code described in §2.1, with the conclusions of §2.2 guiding us in the choice of parameters. As a basic model we adopt a large comet, 100 km in radius, made of water ice and dust in equal mass fractions, with CO and CO_2 initially occluded in the amorphous ice. We assume the comet to be moving in a circular orbit at a heliocentric distance of 100 AU. Other parameters are listed in the second column of Table III. We follow the evolution of this model through the rise to T_{max} and decline from it. We then vary the parameters within the ranges listed in the third columns of Table III, and follow the evolution up to T_{max}, in order to test the effect of each parameter. [Full evolutionary computations for such a large number of models would have taken a prohibitive amount of time.]

3.1. Evolution of the Basic Model

At the beginning, as the radiogenic heat release is far more efficient than conduction, the temperature rises steadily. With it the rate of crystallization rises, and since this is an exoergic process, it escalates into a runaway. Eventually, the temperature becomes sufficiently high for sublimation to set in and the rise in temperature comes to a stop when the rate of energy release roughly equals the rate of absorption due to sublimation (part of the energy is transferred to the surface). As the rate of sublimation is hindered by the build-up of internal vapor pressure, the peak temperature attained is quite high, $\sim$ 260 K. The high (total) internal pressure, whose profile is shown in Figure 1, has the additional effect of enlarging the pores. The originally uniform pore size of 100 μm increases gradually to a few mm, starting at a depth of $\sim$ 70 km outwards. Due to the excellent (effective) thermal conductivity of vapor filled porous ice (see e.g., Prialnik, 1992) a flat temperature profile is obtained throughout a large inner part of the nucleus. Further out the temperature declines; the surface temperature is higher than T_{eq}.

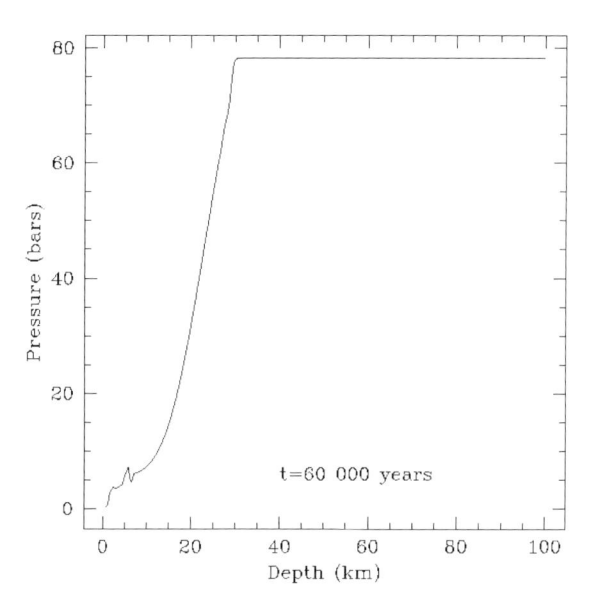

Figure 1. Pressure profile throughout the nucleus

The CO_2 released in the interior freezes when it encounters a temperature of about 100 K. Another temperature plateau, at the $\sim$ 100 K level, is maintained by the CO_2 thermostat. This process is transient, however, and gradually the newly formed CO_2 ice evaporates. At the adopted heliocentric distance of 100 AU, the temperature is not sufficiently low for CO to freeze as well. In the end, both the CO and the CO_2 are completely lost. Thereafter, the temperature profile is uniquely determined by the H_2O and the $\sim$ 260 K plateau extends up to about 15 km from the surface. At this depth evaporation reaches a peak, both because the temperature is high, and because the vapor is efficiently removed down the steep pressure gradient towards the surface. Thus, although in the interior the rate of sublimation (initially $\dot{X} < \tau^{-1} X_0 H/\mathcal{H} \sim 8 \times 10^{-7}$ yr^{-1}) is too low to alter the porosity significantly, at a depth of $\sim$ 15 km, due to vigorous ice sublimation, a layer of relatively high porosity and low water ice content is formed. Above this layer, which is a few kilometers thick, the temperature is too low for sublimation to be of importance and the initial porosity and ice mass fraction are maintained. The formation of such a 'weak' zone at some depth below the surface may have implications for the future history of the comet. Although the temperature and pressure will decline when the radiogenic energy source is exhausted, the altered porous structure will be preserved.

3.2. EFFECT OF INITIAL PARAMETERS

The evolutionary course described above is not necessarily typical of comets in general. In order to obtain a broader picture, we proceed to analyze the effect of

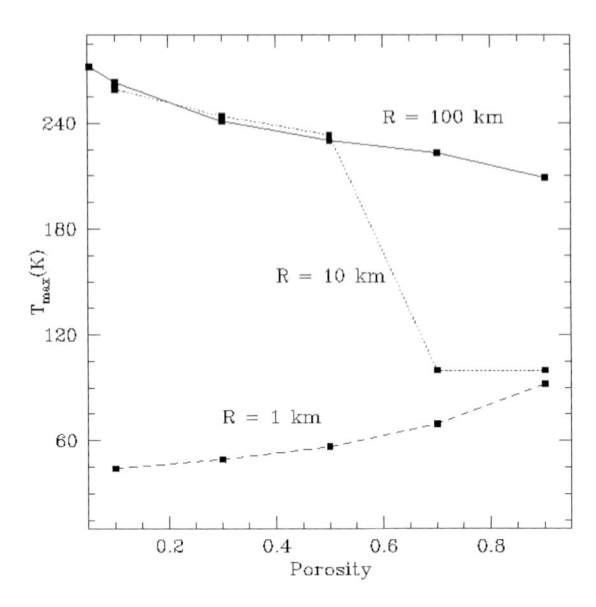

Figure 2. Maximal internal temperature attained as a function of porosity for comet nuclei of different radii, as marked.

the initial parameters. In Figure 2 we plot the maximal temperature attained as a function of porosity for three different cometary radii. Generally, the temperature rises at a rate determined by the competition between heat release and conduction, until steady state is achieved between heat release by radioactive decay on the one hand and conduction and heat absorption by sublimation on the other hand. For the small nucleus, the peak temperature attained increases with the porosity, reflecting the effect of the porosity on the thermal conductivity of the solid matrix. For the large nucleus an opposite effect is encountered: the peak temperature decreases as the porosity increases. This is due to the fact that at the high temperatures attained the ice crystallizes and releases the trapped gas. Heat transfer is now due predominantly to advection by the flowing gases and the advection rate obviously increases with porosity. For the intermediate size of 10 km the behavior changes trend at a porosity of about half. The fractional mass throughout which the temperature is close to T_{max}, as well as the time required to reach a steady configuration, depend on the initial ^{26}Al abundance, as shown in Figure 3. We note that a *threshold* initial abundance exists, of $\sim 10^{-8}$, below which the effect of ^{26}Al is almost negligible, and above which it is considerable, but does not increase appreciably with a further increase of $X_0(^{26}Al)$. As a function of pore size, the maximal temperature peaks at 10 μm and decreases for larger pores due to the increase in gas flow (and advection). If the average pore size is smaller than 10 μm, however, the temperature decreases again. This is due to the large surface to volume ratio that is obtained for small pores, which enhances the rate of sublimation. Finally, the amount of occluded gas affects the effective conductivity considerably: between $f \approx 0.02$

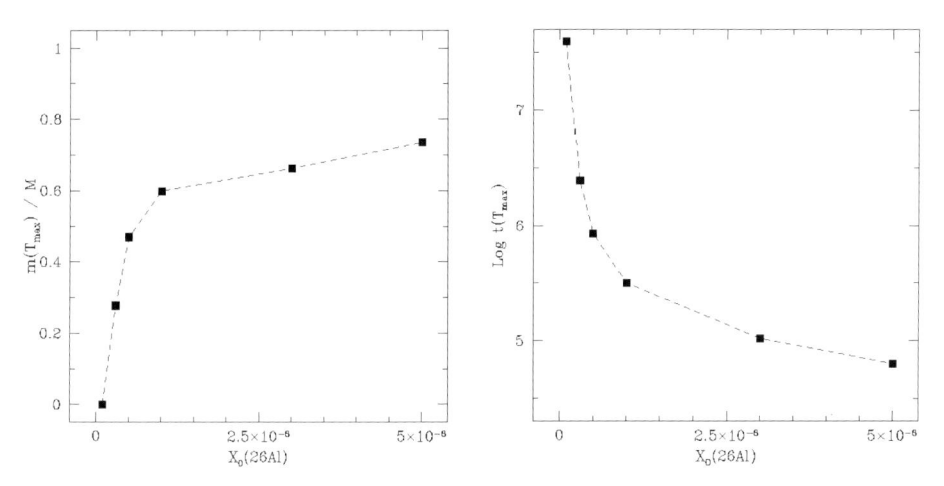

Figure 3. Mass fraction of nucleus throughout which the maximal temperature is attained (*left*) and time required to reach maximal temperature (*right*) as a function of the initial ^{26}Al abundance.

and $f \approx 0.1$ the peak temperature drops from over 260 K to less than 100 K. In the cases when the temperature remains low, so does the pressure, and pore breaking is less severe. The outflowing gases refreeze in the outer cold layers of the nucleus, closing the pores to some extent and leading to a layered composition (cf. Paper I).

4. Conclusions

Small comet nuclei, of order 1 km, do not undergo any long-lasting changes due to radiogenic heating. Large nuclei, a few tens of km or more in size, are significantly altered, both in composition and in structure. In large and very porous comets, made of gas-laden amorphous ice, the ice crystallizes and releases most of the trapped gas. Some of the gas that flows towards the cold outer region refreezes, producing a layered structure, with ices of more volatile species overlying ices of less volatile ones. If comet nuclei are very large, or their porosity is moderate, they reach sufficiently high temperatures for becoming completely crystallized, losing all gaseous species more volatile than water and, eventually, sublimating some of the water as well. In these cases, the porosity and the pore sizes are no longer homogeneous. As a rule, the average pore size increases towards the surface and regions of relatively high porosity may be formed. But although the volatiles initially frozen or trapped in the amorphous ice may be lost as the result of radiogenic heating, laboratory experiments (Bar-Nun *et al.*, 1988) show that some of the gas will remain trapped even as the ice crystallizes and will be released only when it sublimates. In very large nuclei, of very low porosity, the water ice may melt in a central core (possibly prompting the formation of organic molecules). The effect of radiogenic heating on comets of intermediate size is uncertain and depends critically on the properties and initial structure of cometary material. In principle, comets of very different sizes should exhibit different activity patterns when exposed to insolation (unless, of course, small comets are fragments of larger ones). However, we have seen that a major role in the early heating of comets is played by the short-lived isotope ^{26}Al. Thus, if large comets took a longer time to form than small ones, the distinction between them would be less conspicuous. In any case, with the possible exception of very large comets, the effect of radiogenic heating should be significant only if comets were formed on time scales of several million years.

References

Bar-Nun, A., Kochavi, E., and Laufer, D.: 1988, 'Trapping of gaseous mixtures by amorphous water ice', *Phys. Rev. B.* **38**, 7749–7754.

Clayton, D. D.: 1984, '^{26}Al in the interstellar medium', *Astrophys. J.* **280**, 144–149.

Haruyama, J., Yamamoto, T., Mitzutani H., and Greenberg, J. M.: 1993, 'Thermal history of comets during residence in the Oort cloud: effect of radiogenic heating in combination with the very low thermal conductivity of amorphous ice', *J. Geophys. Res. (Planets)* **98**, 15079–15088.

Mekler, Y., D. Prialnik, and Podolak, M.: 1990, 'Evaporation from a porous comet nucleus', *Astrophys. J.* **356**, 682–686.

Podolak, M., and Prialnik, D.: 1997, '^{26}Al and liquid water environments in comets', *Comets and the Origin of Life* (P. Thomas, C. Chyba, and C. McKay, eds.) Springer-Verlag, New York, pp. 259–272.

Pollack, J. B., Hubickyj, O., Bodenheimer, P., Lissauer, J. J., Podolak, M., and Greenzweig, Y.: 1996, 'Formation of the giant planets by concurrent accretion of solids and gas', *Icarus* **124**, 62–85.

Prialnik, D., A. Bar–Nun, and Podolak, M.: 1987, 'radiogenic heating of comets by ^{26}Al and implications for their time of formation', *Astrophys. J.* **319**, 992–1002.

Prialnik, D.: 1992, 'Crystallization, sublimation, and gas release in the interior of a porous comet nucleus', *Astrophys. J.* **388**, 196–202.

Prialnik, D., Egozi, U., Bar–Nun, A., Podolak, M., and Greenzweig, Y.: 1993, 'On pore size and fracture in gas–laden comet nuclei', *Icarus* **106**, 499–507.

Prialnik, D., and Podolak, M.: 1995, 'Radioactive heating of porous comet nuclei', *Icarus* **117**, 420–430.

Schmitt, B., S. Espinasse, R. J. A. Grin, J. M. Greenberg, and Klinger, J.: 1989, 'Laboratory studies of cometary ice analogues', *ESA SP* **302**, 65–69.

Steiner, G., and Kömle, N. I.: 1991, 'Thermal budget of multicomponent porous ices', *J. Geophys. Res. (Planets)* **96**, 18897–18902.

Wallis, M. K.: 1980, 'radiogenic heating of primordial comet interiors', *Nature* **284**, 431–433.

Whipple, F. L., and Stefanic, R. P.: 1966, 'On the physics and splitting of cometary nuclei', *Mem. Roy. Soc. Liege (Ser. 5)* **12**, 33–52.

Yabushita, S.: 1993, 'Thermal evolution of cometary nuclei by radioactive heating and possible formation of organic chemicals', *Mon. Not. R. Astron. Soc.* **260**, 819–825.

Address for correspondence: Dina Prialnik, dina@planet.tau.ac.il

III: ORIGIN OF COMETARY MATERIALS

SPECTROSCOPY BETWEEN THE STARS

G. WINNEWISSER and C. KRAMER
I. Physikalisches Institut
Universität zu Köln, D-50937 Köln, Germany

Abstract. The emission and absorption spectra of interstellar molecules are reviewed with special consideration of recent observational and technical advances in the shorter submillimeter wave region of the electromagnetic spectrum. Single-dish observations have contributed in the past probably most of the information about the structure of interstellar molecular clouds.

At present about 120 interstellar molecules have been identified in interstellar clouds and circumstellar envelopes, evidence of a rich and diversified chemistry. CO, the most abundant interstellar molecule and other diatomic molecules and radicals are found throughout molecular clouds, whereas the more complex molecules are found in high-density cores, which are often the sites of active star formation. These locations represent prime targets for the search for larger molecules, such as glycine. The ignition of young stars is accompanied by strong heating of the surrounding material by radiation and/or shocks, leading to photoevaporation of molecules depleted on dust grains driving a "hot core" chemistry, traceable by its rich organic chemistry and its prevailing high excitation conditions (up to about 2000 cm^{-1}).

However, in the list of detected interstellar molecules many simple hydrides are still missing, e.g. SH, PH, PH_2, etc., which constitute the building blocks for larger molecules. With the technological opening of the terahertz region ($\nu \sim 1$ THz corresponds to $\lambda \sim 0.3$ mm) to both laboratory and interstellar spectroscopy, great scientific advances are to be expected. Amongst these will be the direct detection of the lowest rotational transitions of the light hydrides, the low energy bending vibrations of larger (linear) molecules, and possibly the ring-puckering motion of larger ring molecules such as the polycyclic (multiring) aromatic hydrocarbons.

1. Introduction — Scope of Interstellar Molecules

The space between the stars is not void, but filled with interstellar matter, mainly composed of dust and gas, which gather in large interstellar clouds. In our Galaxy these interstellar clouds are distributed along a thin, but extended layer, which basically traces out the spiral structure of the Galaxy: the stars, the gas, the dust, and the radiation component. About 120 different molecules have been identified in interstellar molecular clouds, and in circumstellar sources. The majority of the interstellar molecules are carbon-containing organic substances. The close agreement with high resolution laboratory spectra furnishes the ultimate exacting test for definite identifications of them. Molecules have been detected throughout the electromagnetic spectrum, predominantly however, within the last 20 years in the microwave, submillimeter wave, and adjacent far infrared regions.

The list of interstellar molecules is not complete, in the sense that only polar molecules can be observed by microwave techniques, and some of the light hydrides are still missing due to a variety of problems, mainly technical. In addition, the lower rotational lines of light hydrides lie in the far-infrared region and thus cannot be observed from the ground, since radiation of these wavelengths — with a few exceptions — is completely blocked by the atmosphere. The molecules observed in the gas phase and on surfaces of dust particles are most likely produced from atoms, ions, and smaller molecules by local chemical processes, see e.g. Winnewisser and Herbst (1993), Herbst (1995). The results from the Infrared Space Observatory, ISO, obtained with the Short Wavelength Spectrometer, SWS, operated in the 2- to 20-μm region, are beautiful spectroscopic evidence for a variety of ice mantle and refractory grain core features, such as solid-state CO, CO_2, and silicates (van Dishoeck, 1998).

The chemically important or "biogenic" elements carbon, nitrogen, and oxygen have abundances which are fractions between 10^{-3} to 10^{-4} of the hydrogen abundance. Carbon, and in consequence the carbon-chain molecules, form the main architectural theme of the discovered interstellar molecules. Aside from the cyanopolyynes and their various derivatives, pure carbon cluster molecules have been observed in space as well: CCC (Hinkle *et al.*, 1988) and CCCCC (Bernath *et al.*, 1989) have been detected in the circumstellar shell of the late type giant carbon star IRC+10216. In addition, a tentative assignment of CCC towards the galactic center source Sagittarius B2, a star-forming region, has been reported (van Orden *et al.*, 1995). Very recently the laboratory gas-phase electronic spectrum of $CCCCCCC^-$ has been measured (Tuley *et al.*, 1998) and five lines show close coincidence with visible diffuse interstellar bands (DIBs), probably paving the way towards future identifications of DIBs.

In molecular clouds, two types of processes dominate by which molecules can be synthesized from precursor atoms and ions: gas phase chemical reactions and reactions that occur on the surfaces of dust particles. The latter mechanism is invoked for the formation of molecular hydrogen as well as for the synthesis of the larger interstellar molecules. Ion-molecule gas-phase reactions can explain the observed interstellar abundance of intermediate sized molecules such as H_2CO or CH_3OH. The size of detected gas-phase organic species stops at molecules with 13 atoms (see Table I). The question of abundances of still larger molecules remains an intriguing one. Evidence for larger molecules, yet nowhere the size of the cores of dust particles (perhaps 10^9 atoms in extent), will probably come from lines or bands in the terahertz or far infrared region. They are expected to arise from vibrational frequencies, e.g. low frequency bending vibrations. Although their unambiguous assignment is still lacking, large planar ring molecules, a special class of which are the **p**olycyclic **a**romatic **h**ydrocarbons — PAHs, have been proposed as carriers of some of the "unidentified infrared" features, occurring at 3050, 1610, 1300, and 885 cm^{-1} or, as positively charged PAHs, as carriers of the diffuse visible interstellar bands (DIBs). They are made up of fused benzine rings, e.g. naphthalene, $C_{10}H_8$, coronene, $C_{24}H_{12}$, and many others, thus resembling chunks of graphite.

TABLE I

Interstellar Molecules (Apr. 1999)

(number of atoms)

2	3	4	5	6	7	8	9	10	11	12	13
H_2	H_2O	NH_3	SiH_4								
NS	H_3^+	H_3O^+									
HD	H_2S										
HCl	N_2H^+										
OH	SO_2										
NaCl	HNO										
SO	SiH_2?										
KCl	H_2D^+										
NO	NH_2										
AlCl											
SiO											
AlF											
SiS											
PN											
SiN											
NH											
SO^+											
HF											
CH^+	HCN	H_2CO	HC_3N	CH_3OH	HC_5N	$HCOOCH_3$	HC_7N	CH_3C_5N	HC_9N		$HC_{11}N$
CH	HNC	HNCO	C_4H	CH_3CN	CH_3CCH	CH_3COOH	$(CH_3)_2O$	$(CH_3)_2CO$			
CC	C_2H	H_2CS	H_2CNH	CH_3NC	CH_3NH_2	CH_3C_3N	CH_3CH_2OH				
CN	C_2S	HNCS	H_2C_2O	CH_3SH	CH_3CHO	C_7H	CH_3CH_2CN				
CO	SiC_2^{ring}	C_3N	NH_2CN	NH_2CHO	H_2CCHCN	CH_3COOH	CH_3C_4H				
CSi	HCO	C_3H^{lin}	HCOOH	H_2CCH_2	C_6H	H_2C_6	C_8H				
CS	HCO^+	C_3H^{ring}	CH_4	C_5H	$C_2H_4O^{cyc}$						
CP	HOC$^+$?	C_3O	$H_2C_3^{ring}$	C_5O							
CO^+	OCS	C_3S	$H_2C_3^{lin}$	HC_2CHO							
	HCS^+	$HOCO^+$	CH_2CN	$H_2C_4^{lin}$							
	CO_2	HCCH	C_4Si	HC_3NH^+							
	C_2O	$HCNH^+$	HCCNC	C_5N							
	MgNC	HCCN	HNCCC								
	MgCN	H_2CN	CCCCC								
	CCC	CH_2D^+	H_2COH^+								
	CH_2	SiC_3									
	NaCN										

Possible detection (?)

NH_2CH_2COOH (glycine)
C_{60}^+ (ionized fullerene)

Evidence (by low res. IR) for:
- PAHs (Polycyclic aromatic hydrocarbons)
- Dust: Silicate CORE + MANTLE (CO, CH_4, H_2O)

It will be one of the tasks of future molecular detections to bridge the gap presently existing between 13 atomic molecules, PAHs and fullerenes of intermediate size (about 20 to 60 atoms), as well as interstellar dust. It seems highly unlikely that interstellar matter at large would have this "three-divided" size distribution. It remains one of the intricate questions to what length carbon chains can exist in space and at which number of carbon atoms the linear carbon chain does tend to change over into monocyclic or higher cyclic rings and fullerenes by spontaneous isomerization. For a recent overview of the chemistry of the interstellar medium, the reader is referred to e.g. Herbst (1995).

2. Interstellar Molecules — Cosmic Dimension

During the past years, overwhelming evidence has been gathered, mainly through high spectral and spatial resolution spectroscopic observations, that interstellar molecular clouds provide the birthplaces for stars and planetary systems. Since the recent discovery of a Jupiter-mass planet orbiting the star 51 Peg by Mayor and Queloz from the Observatoire de Genève (1995) there have been 13 new detections of massive Jupiter-like planets closely orbiting stars near the solar system (Marcy and Butler, 1998; Schneider, 1999) and the number is constantly growing. These discoveries suggest that planetary systems around sun-like stars might be a common occurrence. So far, all extraterrestrial planets have been identified via Keplerian Doppler shifts of optical spectral lines in their host stars. In addition, proto-planetary disks have been found and analyzed e.g. by millimeter wave spectra of organic molecules (Guilloteau et al., 1999). There is speculation that the prebiotic chemistry of the solar system's early planets was influenced to some degree via frequent collisions with comets, asteroids, and meteorites, which may have seeded the early Earth with interstellar molecules. Comets, for example, might act as "chemistry messengers" between the interstellar cloud out of which the solar system formed and the planets themselves, for it is highly unlikely that any complex interstellar molecule may have survived in the hot inner presolar nebula. Comets are thought to have formed during the early solar nebula, and thus their volatiles are considered pristine messengers from 4.5×10^9 years ago. Cometary spectra are, therefore, of direct interest for learning about possible conditions of the solar nebula itself and about the sublimation process within the comet during solar approach. Recently molecular spectra were recorded in Comet Hyakutake by various observatories including the **K**ölner **O**bservatorium für **S**ub-**M**illimeter **A**stronomie, KOSMA 3 m-submillimeter-telescope located on the Gornergrat near Zermatt in the central Swiss Alps. Amongst other molecules detected in Hyakutake with the KOSMA telescope, the emission of HCN was found to be variable by about 50% on a time scale of hours (Wouterloot et al., 1998), as shown in Fig. 1. This variability of the recorded line intensity suggests that material comes from the comet's surface in bursts rather than continuously, probably allowing considerable chemical processing at the comet's surface, directly prior to sublimation.

In cosmic dimension, two initial steps seem to be important for interstellar chemistry:

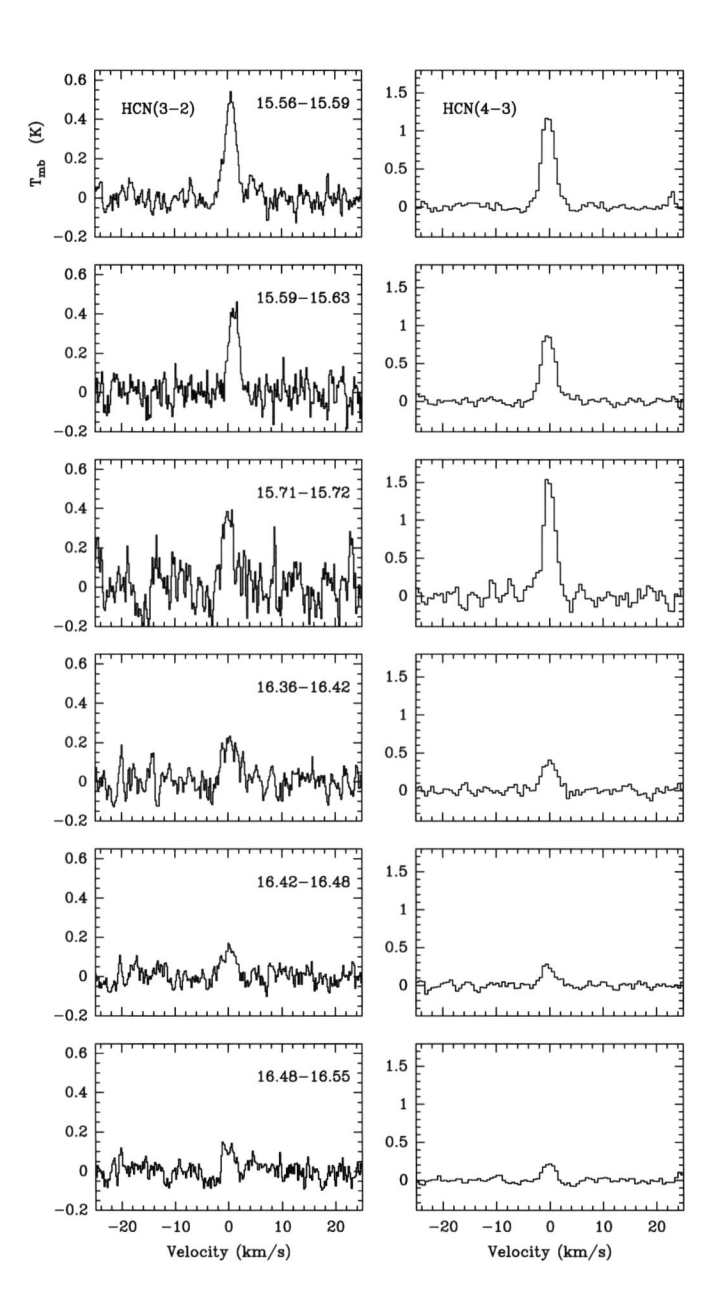

Figure 1. KOSMA (Kölner Observatorium für Submillimeter Astronomie) recording of the April 15, and 16, 1996, HCN time-dependent emission from Comet Hyakutake (1996 B2). The two HCN rotational transitions, $J = 3 - 2$ and $J = 4 - 3$, were observed simultaneously. The time difference between the two frames is a few hours and is shown in Universal Time in the upper right hand corners. The intensity variation of the line emission is apparent.

- *Cosmic evolution* which starts with the *Big Bang*, the creation of matter and in its wake the formation of the cosmic elements H, He, and D, followed in the early Universe by the formation of galaxies, stars, and the onset of stellar nucleosynthesis with the appearance of the chemical elements. Of particular importance are the emergence of the three "biogenic" atoms carbon, nitrogen and oxygen. Atomic carbon assumes a special role amongst the elements in the sense that its electronic configuration $1s^2 2s^2 2p^2$ bestows it with the unique property of forming long carbon chains and complex ring structures, so basic to the formation of life. These properties, well established on Earth, are not lost in the cold and dilute environment of interstellar space.
- *Chemical evolution* in a highly dilute, and cold interstellar medium has progressed surprisingly far and has produced a rich harvest of fairly complex interstellar and circumstellar molecules. Presently the largest one contains 13 atoms, however, still simple in comparison to biomolecules. Table I lists the known interstellar and circumstellar species. The list of presently known 120 molecules is subdivided into inorganic and organic species, i.e. all those which contain carbon. Although, amongst the presently known interstellar species, the carbon chain molecules dominate, it is to be noted that there are also radicals and molecules which contain functional groups, important for biomolecules: e.g. the NH radical, the amino (NH_2) group, and the carboxyl (COOH) group.

3. Line Detection Technique

The first two steps, the *cosmic* and the *chemical* evolution of the Universe, and their investigation by present techniques are intimately linked to the birth of modern spectroscopy. It was the fundamental discovery of the early 19th century that the Sun and the stars are made up of the same elements as the Earth, which led to the realization that spectroscopic techniques allow the observation of chemical processes in cosmic dimension.

The basic spectroscopic technique of matching specific wavelengths or frequencies from distant objects with those spectral features obtained in the laboratory dates back to the very beginning of spectroscopy and has remained in use up to the present day. Kirchhoff and Bunsen identified around 1860 the origin of the Fraunhofer absorption lines in the solar spectrum seen in 1814 by him for the first time: their cause were atoms in the chromosphere of the Sun. For example, the elements Cs, Rb, and He were detected this way. Kirchhoff and Bunsen note in one of their papers that the "almost inconceivable" sensitivity of spectroscopy would open "an entirely untrodden field, stretching far beyond the limits of the Earth or even the solar system". What a vision and foresight! The detection and identification of the Fraunhofer absorption lines remains one of the greatest spectroscopic achievements and marks the birth date of what we now call modern astrophysics.

The enormous technological and scientific advances which have occurred in the last 30 years have allowed two developments. First, they have made spectroscopic observations possible throughout virtually the entire electromagnetic spectrum — though with different spectral resolution and sensitivity — providing access to an overwhelming richness of spectral features. Second, they have pushed the accuracy of spectral line identification in some wavelength regions, e.g. at microwave, millimeter, and submillimeter wavelengths, to about 1 part in 10^9, thus virtually excluding erroneous identifications for all those cases where laboratory data exist. In Fig. 2 this matching technique of interstellar and laboratory recorded spectra is beautifully demonstrated for the band head of SO_2 near 663 GHz (Belov et al., 1998). In addition, the spatial resolution of millimeter wave observations has been pushed below the 1 arc sec limit by interferometric measurements which is important to resolve the hot cores of molecular clouds to disentangle the excitation conditions and the prevailing chemistry, particularly in view of the formation and existence of larger molecules.

Atoms and molecules, both ionized and neutral, exhibit spectral features throughout most of the electromagnetic spectrum. These spectral lines — in emission and/or absorption — carry detailed information on the physical and chemical conditions of the stellar and interstellar sources from which they emanate. The accurate determination of the ratio between atomic and molecular material remains one of the important parameters for any chemical model calculation, especially the ratio of neutral atomic carbon, C, to carbon monoxide, CO. In this connection, the abundance ratio between molecular hydrogen, H_2, and CO is the most important entity for the determination of molecular cloud masses. Since H_2 constitutes nearly all the mass of an interstellar cloud, it is also of importance for triggering star formation in cloud cores. Since H_2 does not carry a permanent electric dipole moment, its rotational spectra cannot easily be observed directly, except by very faint widely spaced $\Delta J = 2$ quadrupolar rotational transitions. In very cold molecular clouds, especially in cold cloud cores, H_2 transitions can not be detected in emission and CO often has to serve as a substitute by determining its abundance accurately. Recent work (Kramer et al., 1999) indicates that the abundance of CO stays rather constant in the outer layers of clouds, while its abundance drops in the innermost regions of cold pre-star forming cores, due to CO depletion and formation of CO ice surfaces on dust grains which permeate molecular clouds. Observations of CO ice features in the IR region strengthen this conclusion [see e.g. van Dishoeck, 1998].

Neutral atomic carbon has two outer electrons, which give rise to a singlet and triplet system. The triplet system is the lower of the two and is split into three fine-structure levels of differing total angular momentum J: ^3P_J, i.e. $^3P_0$, $^3P_1$, $^3P_2$. The separation of this fine-structure energy level diagram lies in the submillimeter wave region and leads to two observable, but forbidden magnetic dipole fine-structure transitions: $^3P_2 - ^2P_1$, at 809.34197 GHz and $^3P_1 - ^3P_0$ at 492.160651 GHz. These two transitions are of special astrophysical interest in the

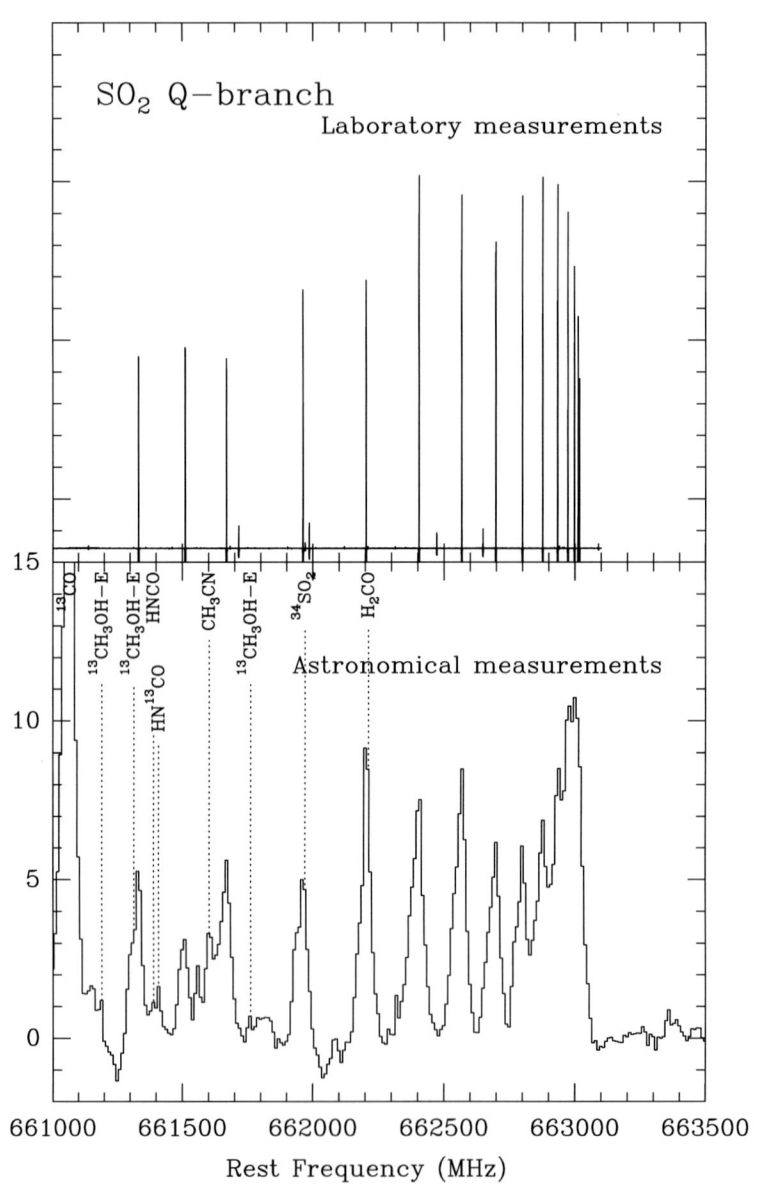

Figure 2. Comparison of the (K = 6) SO_2 Q branch between laboratory and interstellar recording towards Orion A. Additional spectral features have been assigned in the interstellar spectrum to different molecular species.

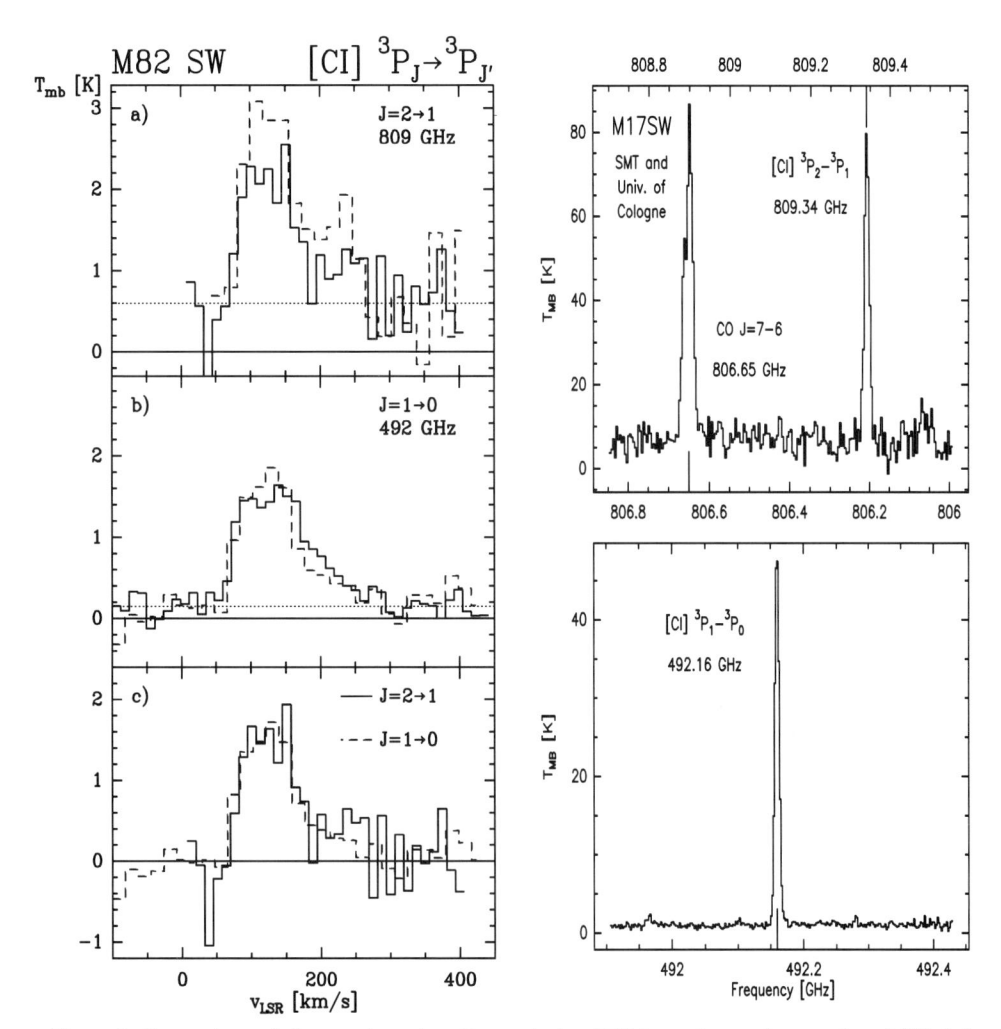

Figure 3. Comparison of the atomic carbon line emission [CI] from the star burst galaxy M82 (left hand side) and the galactic source M17SW (right hand side) obtained at the Heinrich Hertz submillimeter telescope (Stutzki *et al.*, 1997). Towards M82 the two fine-structure transitions were observed simultaneously with the Cologne-built dual channel SIS receiver. The dashed traces in a) and b) are the recorded line profiles of the center position of M82. The solid traces in a) and b) represent average spectra over various positions. No baseline correction has been applied. The dotted lines represent the continuum offset due to the known emission of interstellar dust. Part c) shows an overlay of the continuum subtracted $J = 2 - 1$ map average and the $J = 1 - 0$ peak spectrum. Towards M17SW a comparison with the CO $J = 7 - 6$ transition becomes possible through accidental recording in the other sideband.

low temperature regions of the interstellar medium, since they provide an important diagnostic tool, both for galactic and extragalactic atomic gas. The [CI] line intensity ratio is of particular astrophysical importance as it allows a derivation of the physical conditions of the emitting gas. In the optically thin regime this ratio is a direct measure of the excitation temperature and the density of the emitting gas. Stutzki *et al.* (1997) have for the first time detected these two lines simultaneously, both in galactic and extragalactic sources with our newly constructed dual channel SIS (superconducting-isolator-superconducting) waveguide mixer receiver. Fig. 3 displays the two carbon fine-structure lines towards the star burst galaxy M82, which is located at a distance of 3 Mpc. These measurements represent the first extragalactic detection of the $^3P_2 - ^3P_1$ fine-structure transition. The emitting atomic gas in M82 is relatively dense ($\sim 10^4$ cm^{-3}) and, with 50 K, relatively warm. Some fraction of the atomic gas may possibly reach temperatures of up to 150 K.

It might be noted in this context that recently the $^3P_2 - ^2P_1$ fine structure transition of ^{13}C was discovered (Keene *et al.*, 1998) towards Orion A with the aid of recently determined laboratory rest frequencies (Klein *et al.*, 1998).

Aside from well known, strong, and ubiquitous lines, as for example the transitions of CO, the assignment of spectral features to a certain species can be rather intricate. As an example of the complicated and detailed assignment process, we have presented in Fig. 2 a small portion of the Cal Tech Observatory 607 - 725 GHz high frequency spectral survey recorded at the CSO 10 m telescope (Schilke *et al.*, 1999) towards the Orion A nebula. Although most of the lines are assigned to known species (75%), these high sensitivity measurements reveal that deep integrations suffer from the confusion limit, i.e. the point where the identification of a line is hampered not by the system noise but rather by the number of weak lines overlapping each other. This is borne out in a high resolution scan of the SO$_2$ band head near 663 GHz. Fig. 2 displays both a laboratory spectrum of the $^rQ_6$ branch head and the appropriate Orion A spectral scan, where line positions of other additional molecular species have been marked.

In the optical region a spectral resolution of about one part in 10^5 is achieved and suffices since the density of the spectrum is not as high as in the millimeter wave region. Most of the 120 or more interstellar molecules (Table I) detected to date have been identified in the microwave, millimeter, and submillimeter wave part of the electromagnetic spectrum via their characteristic rotational transition frequencies. Molecular hydrogen, H$_2$, is the most abundant interstellar molecule followed by carbon monoxide, CO, with about 10^{-4} of the H$_2$ abundance. Other gas phase molecules have been detected in the infrared via characteristic vibrational frequencies or in the optical region by electronic transitions, i.e. diatomic molecules only. The possible identification of species larger than 13 atoms in size, e.g. PAHs and fullerenes, rests on IR features, assigned to localized fundamental vibrational frequencies of the stretching and bending modes. The most prominent interstellar emission feature at 3.29 μm ($\sim$ 3040 cm^{-1}) is assigned to the CH stretching mode (Allamandola, 1999). Similarly, the identification of interstellar dust features relies on the IR signatures, e.g. Sandford *et al.* (1993). Results from the ISO satellite have certainly added considerable new information (Whittet *et al.*, 1996; van Dishoeck *et al.*, 1998).

Traditionally, the primary instrument used in interstellar spectroscopy has been the single-dish telescope. A submillimeter telescope consists of a highly precise parabolic reflector, with a surface accuracy of about 20 μm, equipped with a highly sensitive radiometer positioned at the focal point of the antenna. Recently the panels and mount of the KOSMA 3 m-telescope were replaced by 16 highly precise aluminum panels with an individual surface accuracy of ~ 5 μm (Degiacomi *et al.*, 1995; Kramer *et al.*, 1998). The KOSMA telescope is dedicated to millimeter and submillimeter-wave astrophysics with observations mainly of interstellar atomic and molecular lines. Due to the excellent atmospheric conditions at the Gornergrat site, observations at the two highest atmospheric windows reachable from ground (> 660 GHz or < 450 μm and ~ 880 GHz or ~ 340 μm) have now been performed. Some selected examples of molecular line work and mapping results will be presented in the section on "Molecular Clouds".

In the millimeter and submillimeter wave region the radiometer (or detector) is usually composed of a specially designed frequency mixer element for heterodyne detection. The essential feature of the mixer element is its extremely non-linear current/voltage characteristic. In the past usually a GaAs Schottky barrier diode was used, working as a classical resistive mixer. The Superconductor-Insulator-Superconductor (SIS) tunnel junction makes use of the quasiparticle photon-assisted tunneling process and is used now up to about 1 THz. Above ~ 1 THz the hot electron bolometer comes into use as a mixer. A few groups worldwide, among them the Cologne laboratory, work on hot electron bolometers as mixers. They constitute the most promising heterodyne detection technique for frequencies above 1 THz. A NbN phonon cooled hot electron bolometer (HEB) working in the 350 μm atmospheric window (Kawamura *et al.*, 1999) was built by the Center for Astrophysics group and has recently been successfully used at the 10 m Heinrich-Hertz submillimeter telescope in Arizona to observe the interstellar [CI] $^3P_2$–$^2P_1$ transition and other important spectral lines. It is the first hot electron bolometer ever used at a telescope and its performance indicates that these devices hold great promise.

These devices, SIS- and HEB-mixers, are produced also at our Cologne laboratory and two SIS junctions have been put into the receiver employed for the detection of the extragalactic atomic carbon lines (Stutzki *et al.*, 1997). These SIS devices convert the frequency of the interstellar signal with the aid of a superimposed Local Oscillator (LO) frequency into an intermediate frequency (IF), which itself is amplified and fed into a spectrometer to display the received line radiation as a function of frequency. As frequency dispersive elements, the Cologne group developed acousto-optical spectrometers which are in use in the Cologne laboratories, at the KOSMA observatory, and several other observatories, as well as in the space mission SWAS (Submillimeter Wave Astronomy Satellite), which was launched in early 1999. A short overview of the scientific and technical developments is given e.g. in Winnewisser (1994). A recent summary of the field entitled "The Physics and Chemistry of Interstellar Molecular Clouds" is found in the *Proceedings of the 3rd Cologne-Zermatt Symposium* (Ossenkopf et al., 1999).

4. Molecular Clouds

A large fraction of the gas within our own Galaxy, the Milky Way, is found to be in the form of vast clouds of molecular hydrogen, H_2, hundreds of light years in extent and with masses up to $10^5 - 10^6$ solar masses, (1 $M_\odot$ equals about 2×10^{33} g). Within the denser molecular clouds ($> 10^3$ particles cm^{-3}), hydrogen appears entirely in molecular form as H_2, together with a variety of other less abundant elements (C, N, O, ...) which are also predominantly in molecular form. At the edges of the molecular clouds, in the interface region to interstellar space, the chemical content changes as one exits the molecular cloud: the molecular form gives way in a layered fashion to ionized molecular, atomic and ionized atomic material. These outer regions, often called photon dominated regions (PDRs), are affected by the UV light of nearby young stars, which provide the radiation field and thus the energy to break up the molecules and ionize the atoms. The chemical reactions are altered as well, in the sense that endothermic reactions can proceed and a shock driven chemistry takes place near star-forming regions. The molecular and density distributions are therefore far from homogeneous. In fact, all molecular clouds which have been investigated in great detail are found to break up into smaller fractal structures, giving rise to a "clumpy nature" of interstellar clouds. It is the molecular clumps which ultimately will form stars, given the right conditions, see e.g. Winnewisser *et al.* (1992).

Star formation sets in after parts of the molecular clouds have fragmented into gravitationally bound clumps which after a certain time are weaned from the parental molecular cloud. This process can take place as a purely random event, where the internal turbulent motion leads to an accidental clumping with subsequent isolated star formation. The Taurus region is a most typical example of accidental star formation. Other regions are more complex: the central part of the Cepheus Cloud, the shell-like HII region IC 1396, which is powered essentially by one O-type star, provides an area where the interaction between one individual luminous star and the surrounding matter can be studied. A second scenario, which may pertain more to giant molecular clouds, sees star formation in dense cloud cores in the wake of previous generations of newly born stars. The Orion Molecular Cloud furnishes probably the best studied example of this "triggered" star formation. Both cloud complexes will shortly be discussed in turn.

4.1. THE ORION REGION

The giant molecular cloud system in Orion lies at a distance of 450 pc and is one of the nearest and most prominent star formation regions in the northern sky, where low and high mass stars are forming. Due to its proximity, the Orion region has become the testing ground for observational and theoretical investigations of molecular cloud structures, star formation regions with associated outflows, and interfaces between ionized and neutral matter known as PDRs. For an overview see e.g. Genzel and Stutzki (1989).

Large scale maps of the Orion region in the $J = 1-0$ lowest rotational transition of CO (Maddalena *et al.*, 1986) reveal two clearly distinct, but very extended and massive molecular complexes: the southern cloud, Orion A, and the northern cloud, Orion B. Each of these two cloud complexes extends over more than 4 degrees in the sky, corresponding to 32 pc in linear dimension, and contains about 10^5 $M_\odot$. The peaks of the velocity integrated CO emission correspond to the cores of ongoing, massive star formation. The young massive stars are surrounded by HII regions, where the interstellar material is ionized due to the UV radiation of the stars. The PDRs are associated with warm CO, traceable via their mid-J rotational transitions. The molecular gas is heated indirectly, mainly through the photoelectric effect on dust grains and through IR-pumping of molecular hydrogen. The cooling of the inner parts of the clouds takes place primarily through the various CO rotational transitions, whereas the outer layers radiate through ionized carbon, C^+, and oxygen, O. It is interesting to note in this context that the line profiles of the fine structure line ($J = 1 - 0$) of atomic carbon, C, are very similar to those of e.g. ^{13}CO $J = 2-1$. This coincidence of the two line profiles suggests that both tracers are formed in about the same regions rather deep inside the clouds. These clouds in turn must exhibit significant sub-structure or clumpiness, allowing the far-UV to penetrate deeply inside and causing PDRs on the surfaces of these molecular clumps.

The most recent star forming activity includes the Orion A nebula itself (part of the southern cloud) and regions of the northern cloud Orion B, such as NGC 2024, the interface region with the optical nebula IC 434, the reflection nebula NGC 2023, and the Horsehead Nebula B33. With the KOSMA 3 m Submillimeter Telescope we have mapped large regions of the Orion complex in mid-J transitions of CO and some of its isotopomers. We are going to present here two selected areas: (i) the Orion A region (Figs.4, 5, 6) and (ii) the region NGC 2024 (Fig. 7) in the Orion B cloud (Fig. 4).

The most prominent region of the Orion A complex is the Orion Nebula, ionized by the four Trapezium stars. Close by, only about 0.1 pc behind the HII region, lies the densest part of the molecular cloud with the Orion-KL region at the core; the Kleinmann-Low region is a luminous cluster of compact infrared sources with two prominent protostars named after their discoverers, KL and BN, the Becklin-Neugebauer object. A pathfinder map of the region is presented in Fig. 5 with the

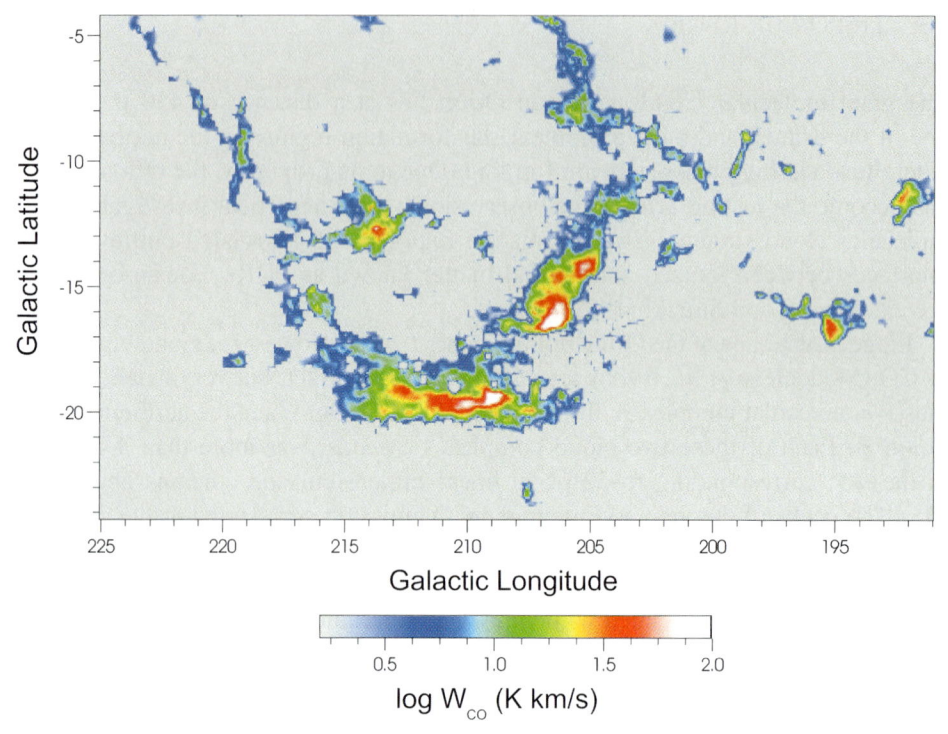

Figure 4. Orion region with the two molecular cloud complexes Orion A and Orion B seen in the CO $J = 1 - 0$ transition (Maddalena *et al.*, 1986).

Trapezium stars at the center. The Orion-Bar region represents an ionization front impinging on the molecular cloud (Fig. 6).

The CO line ratios of different rotational transitions and their appropriate line profiles found throughout the Orion molecular cloud region and in particular at the NGC 2024/IC 434 interface region suggest very high volume densities of more than 10^6 cm^{-3} (Fig.7). The gas apparently consists of small but very dense clumps suspended and "floating" in a fairly diluted interclump gas. In general, the density contrast between local densities and average densities is about 200. This high density ratio indicates that the clumps identified are not homogeneous entities but exhibit strong substructures amongst themselves (Bally *et al.*, 1987; Kramer *et al.*, 1996; Simon *et al.*, 1997; Köster *et al.*, 1994).

Observations of mid−J lines of the two major isotopomeric forms of CO, i.e. ^{12}CO and ^{13}CO, indicate additional constraints on the excitation conditions of the gas. The maps of Orion A and B obtained with the KOSMA telescope show that ^{13}CO($J = 6 - 5$) emission is found in the vicinity ($<$ 1 pc) of the HII regions associated with the locations of star formation. The line-widths are typically 5 kms^{-1} or less whereas the line temperatures reach 100K or more. These facts support the general picture that the molecular line emission arises in a moderately warm gas which is heated by the UV radiation field existing outside the HII region.

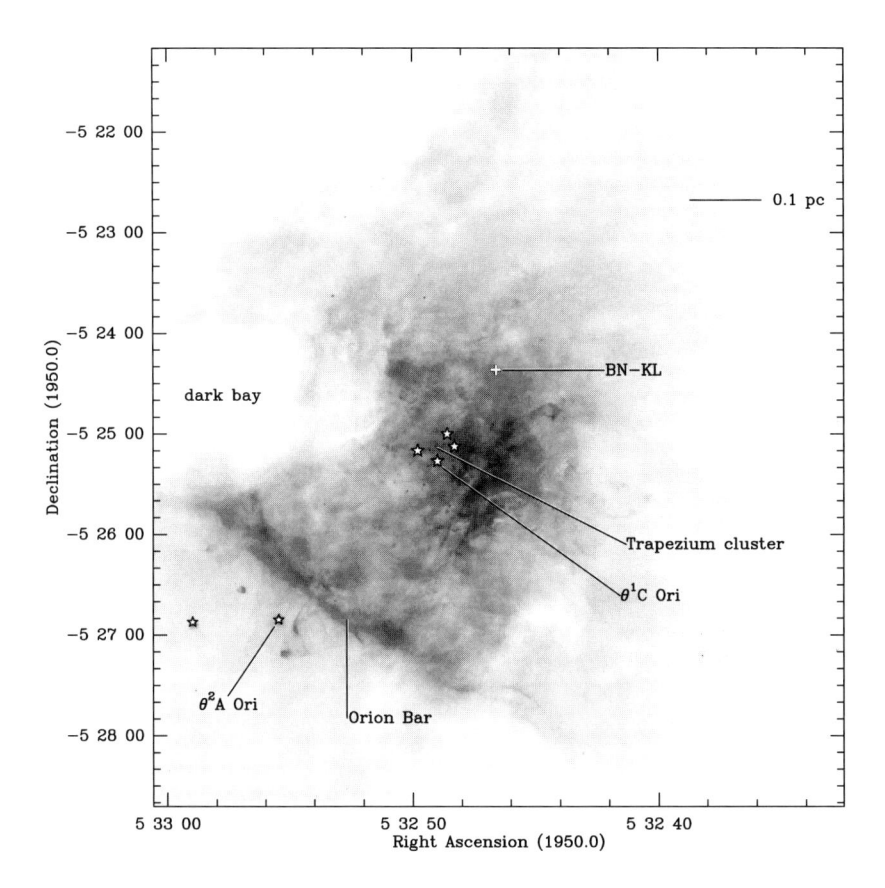

Figure 5. Path finder map of the central part of the Orion A nebula overlayed on an optical negative obtained with the Hubble space telescope (C.R. O'Dell and S.K. Wong, Rice University, STScI Press Release PR 95-45a).

4.2. THE CEPHEUS CORE — IC 1396

The Cepheus region located in the local (Orion) spiral arm is most remarkable in the sense that it consists of a giant molecular bubble where the individual, but relatively small molecular clouds, generally called globules, appear to be located on an expanding ring. All the globules exhibit bright rims, all facing the centrally located, exciting O-type star. This star also powers the associated HII region IC 1396. In comparison to the Orion A region, IC 1396 seems to be older because the central O star has freed itself from the parental molecular cloud which gave birth to it. The origin of the present ring-like arrangement of the globules is a consequence of the O-star formation which, due to stellar winds, is blowing out the remains of the once dense parental molecular cloud, causing the giant bubble. This area of the

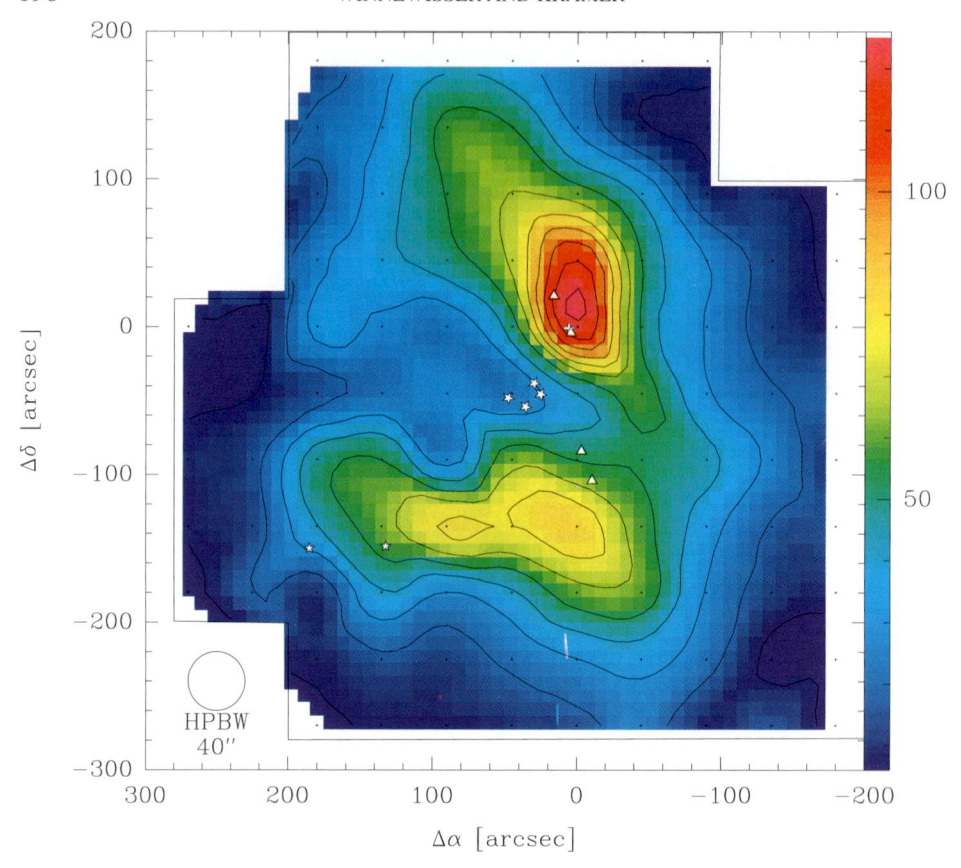

Figure 6. ^{13}CO $J = 6 - 5$ KOSMA map of the Orion A molecular cloud and the Orion Bar region. The position of the four Trapezium stars is marked in the central part of the map. The triangles mark the positions of the FIR dust continuum condensations observed by Mezger *et al.* (1988, 1992).

Cepheus region has been and is the subject of current investigations with different telescopes, including the KOSMA telescope. For a general background of the Cepheus region, the reader is referred to two rather recent publications (Weikard *et al.*, 1996; Patel *et al.*, 1998).

In Fig. 8 we present one of the very small and densest globules with the brightest rim, designated Rim A, in the light of the CO $J = 3 - 2$ transition observed with the KOSMA telescope. This globule is probably closest to the O-star (HD 206267) with a projected distance of 3.7 pc. The CO emission matches exactly the cloud as it appears on the optical photograph from the POSS atlas (Wouterloot and Winnewisser, 1999). The recent discovery of emission of CO $J = 6 - 5$ by the KOSMA telescope indicates that the Rim A globule must be heated externally from HD 206267. At present the detailed heating mechanisms are investigated.

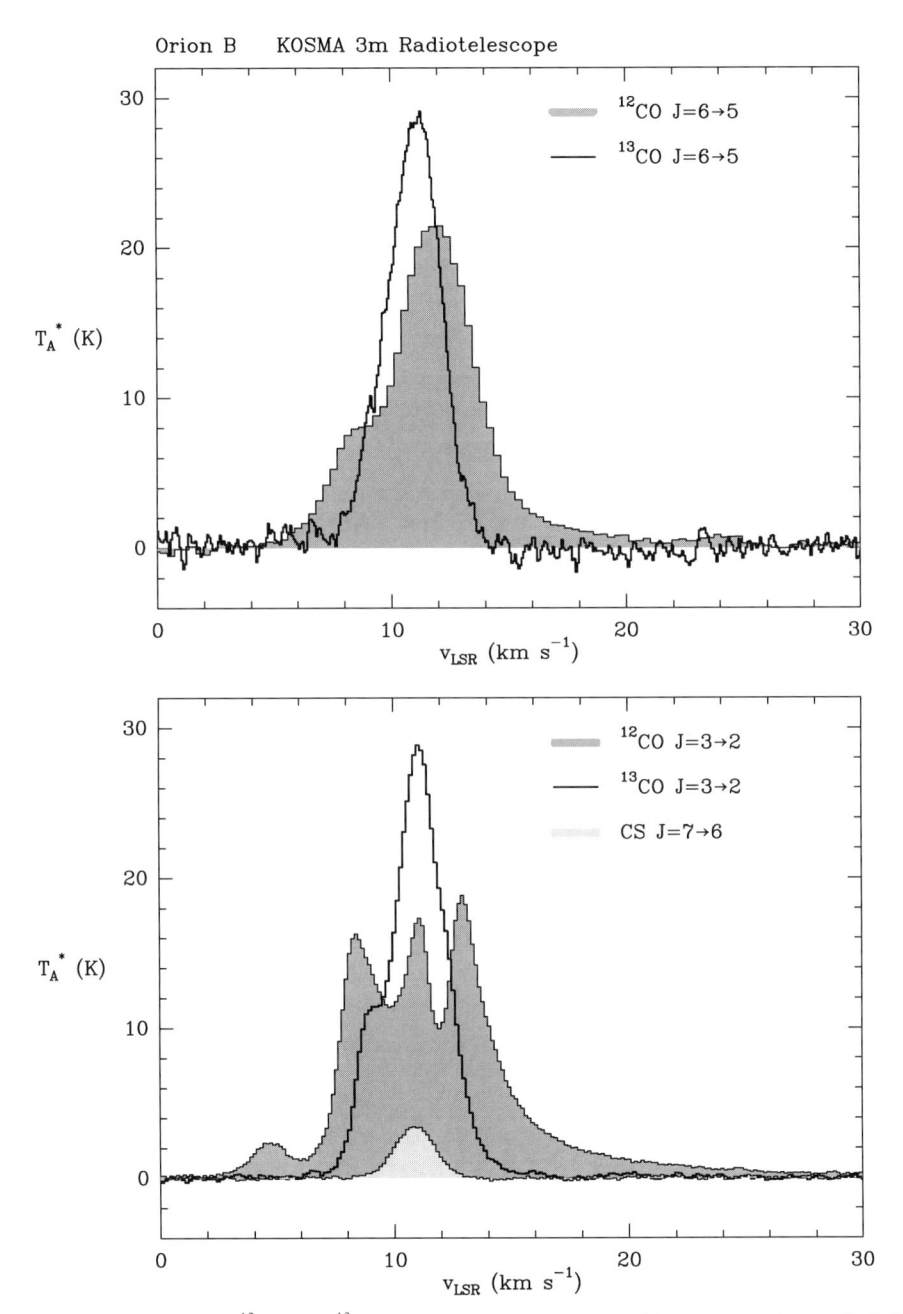

Figure 7. Line profiles of ^{12}CO and ^{13}CO $J = 6 - 5$ rotational transitions recorded towards Orion B with KOSMA. Plotted is the measured antenna temperature for each spectrometer channel against the velocity of the local standard of rest. The strong non-Gaussian line wings at positive velocities seen in ^{12}CO denote a molecular outflow. In addition, the ^{12}CO spectra show the signature of strong self absorption by cold foreground gas.

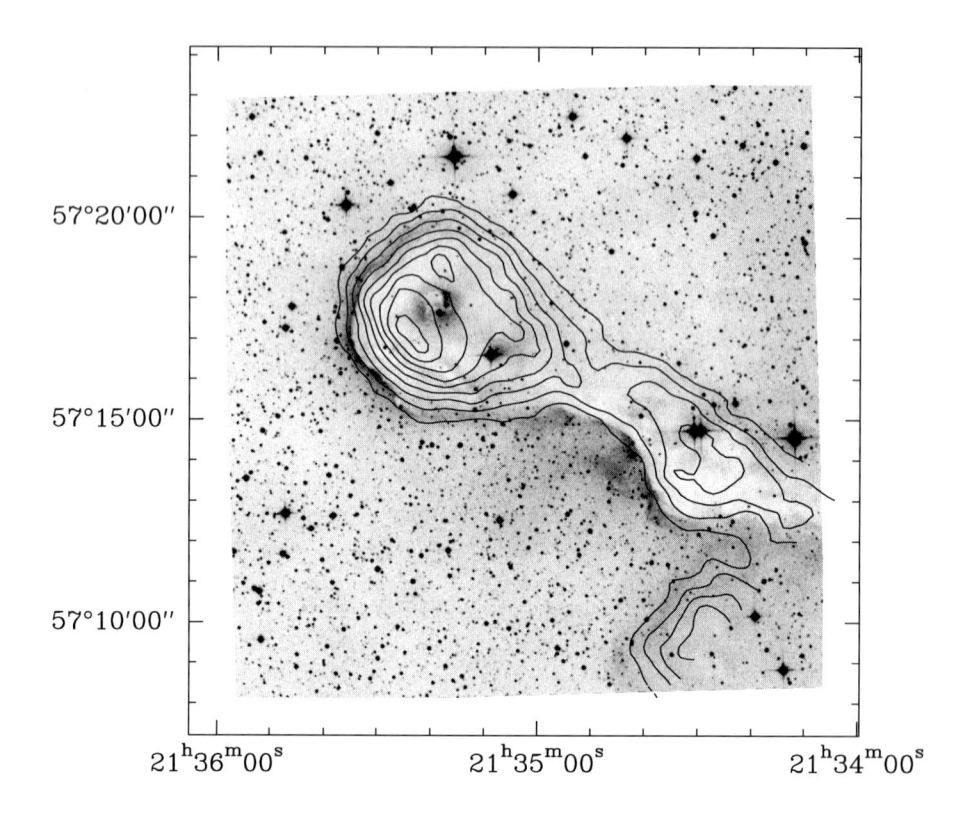

Figure 8. Small molecular cloud, named rim A, in the IC 1396 region is situated in Cepheus. In this overlay between radio map, observed with KOSMA, and optical photograph it is strikingly evident how closely the CO emission matches the optical contours. The total mass of the cloud is about 100 $M_\odot$.

5. Recent Laboratory Work — Terahertz Region

The early interstellar molecular detections greatly benefited from the vast amount of microwave and millimeter wave laboratory data accumulated in the literature around the world. However, interstellar detections have come to par with the published laboratory data and thus a real synergism occurs between space and laboratory spectroscopy. The major developments in high resolution gas phase laboratory spectroscopy are twofold: greatly improved sensitivity at frequencies up to about 150 GHz, and the quest for achieving higher frequencies, i.e. beyond 1 THz.

The terahertz region has recently been opened for broadband gas phase spectroscopy by the Cologne laboratory (see Winnewisser, 1995 and references therein). The goal hereby is to provide highly accurate spectral line frequencies in the terahertz region of higher rotational transitions of medium heavy molecules like

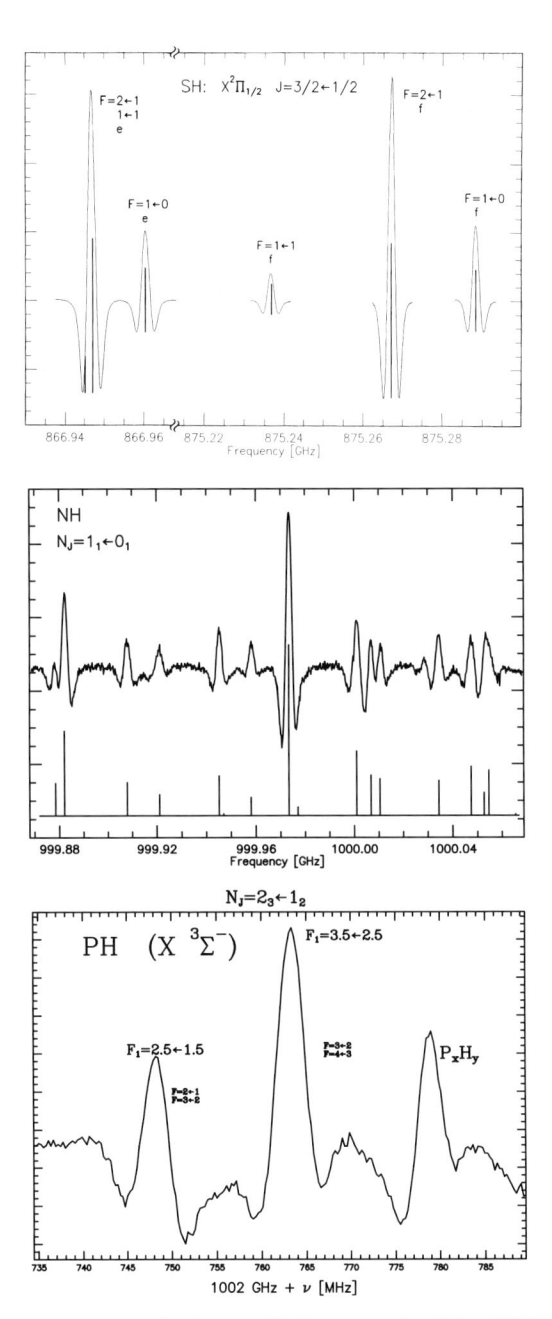

Figure 9. Terahertz spectra of three fundamentally important hydrides SH, NH, and PH recorded with the Cologne terahertz spectrometer.

CH$_3$OH, or low J transitions of lighter molecules, notably the light hydrides, as for example the diatomic hydrides SH, NH, PH, their isotopomers, and eventually OH, CH and others. The lowest rotational transitions of SH and SD have recently been detected in the laboratory (Klisch *et al.*, 1996). Another example is the spectrum of NH and ND (Klaus *et al.*, 1997; Takano *et al.*, 1998), the lowest rotational transitions of which occur near 1 and 0.5 THz. NH and ND possess a $^3\Sigma$ electronic ground state and consequently their $N = 1 - 0$ transition is split into 3 fine structure transitions, each of which displays electric quadrupole and magnetic hyperfine structure. Similarly, PH (Klisch *et al.*, 1998), the simplest phosphorous hydride, possesses a $^3\Sigma$ electronic ground state and displays a comparable rotational spectrum to NH. Parts of the rotational spectra of SH, NH, and PH around 1 THz are shown in Fig. 9a, 9b, and 9c, respectively. For a detailed explanation of the laboratory spectra, the reader is referred to the cited literature. Submillimeter laboratory spectroscopy has matured to the point where spectra up to 2 THz can be recorded in Doppler and sub-Doppler resolution with a frequency accuracy of about 1 kHz by employing saturation techniques (Winnewisser *et al.*, 1997).

The various hydrides are thought to be the basic building blocks of interstellar chemical networks and are therefore of prime astrophysical relevance. From the long list of basic hydrides, part of the rotational spectral pattern of three important radicals SH, NH, and PH have been displayed in Fig. 9. So far, they happen not to have been detected in interstellar clouds via their rotational spectra. One of the reasons for their non-detection by rotational spectroscopy is that most of their transitions fall into frequency regions with strongly reduced atmospheric transmission and are therefore difficult to observe from ground-based observatories. With the advent of SOFIA, the Stratospheric Observatory for Far Infrared Astronomy, in 2001, these atmospherically obscured frequency regions will become more accessible to interstellar observations and a new wave of important discoveries can be expected. These hydrides are potentially new interstellar species by which the higher temperature and density areas of molecular clouds can be tested, particularly near star formation and hot core regions. It should be noted that the interstellar detection of NH is in absorption in the optical region towards the star ς Per, and thus the detection pertains to diffuse, low density ($\leq 10^{-3}$ cm^{-3}) interstellar clouds.

6. Conclusion

Molecules are used via their rotational spectra to probe deep into the interior of interstellar molecular clouds to reveal their habitat and how they participate in star formation. In addition, differing physical and chemical conditions exist not only on the edges of clouds but also in areas of highly clumped material in the interior of the clouds, where external UV radiation of close-by sources, usually young stars, are important. Complex interstellar molecules are found in dense, warm cores located well inside the molecular clouds. For high resolution spectroscopy, the recently opened terahertz region ($\nu \sim 1000$ GHz; $\lambda \sim 300$ μm) opens the field to a set of new scientific inquiries, amongst which are the search for the interstellar distribution of important molecular building blocks, such as NH, NH_2, CH, CH_2, ... and many others, the detection of low lying bending vibrations of linear molecules, the direct detection of ring puckering motions, and the wide field of light hydride chemistry. All these effects are clearly traceable by high resolution submillimeter wave spectroscopy.

In particular, high spectral and spatial resolution measurements have greatly advanced our knowledge of molecular distribution and star formation. During star formation most of the infalling matter from the molecular cloud is redirected by the star and returned to the molecular cloud in a bipolar flow. However, it should be noticed that in the wake of star formation, planets form out of the very final material supplied from the parental molecular cloud to the newly forming star via an accretion disk. Thus scientists are attracted by its relevance to the formation of the solar system some 4.5×10^9 years ago and our very existence.

Acknowledgements

It is gratefully acknowledged that this work has been supported in part by the Deutsche Forschungsgemeinschaft via special research grant (Sonderforschungsbereich) SFB 301 and by the Ministry of Science and Technology of the State Nordrhein-Westfalen.

References

Allamandola, L.J., Bernstein, M.P., Sandford, S.A. and Walker R.L.: 1999, *Space Sci. Rev.*, this volume and references therein.
Bally, J., Stark, A.A., Wilson, R.W., Langer, W.D.: 1987, *Astrophys. J.* **312**, L45.
Belov, S.P., Tretyakov, M.Y., Kozin, I.N., Klisch, E., Winnewisser, G., Lafferty, W.J., Flaud, J.M.: 1998, *J. Mol. Spectrosc.* **191**, 17.
Bernath, P.F., Hinkle, K.H., Keady, J.J.: 1989, *Science* **244**, 562.
Degiacomi, C.G., Schieder, R., Stutzki, J., Winnewisser, G.: 1995, *Optical Engineering* **34** (9), 2701.
van Dishoeck, E.F., Blake, G.A.: 1998, *Ann. Rev. Astron. Astrophys.* **36**, 317.
van Dishoeck, E.F. et al.: 1998, In *Star Formation with the ISO Satellite*, ed. J. Yun and R. Liseau, **132**, 54.
Genzel, R., Stutzki, J.: 1989, *Annu. Rev. Astron. Astrophys.* **27**, 85.
Guilloteau, S., Dutrey, A., Simon, M.: 1999, *Astron. & Astrophys.* **348**, 570.
Herbst, E.: 1995, *Ann. Rev. Phys. Chem.* **46**, 27.
Hinkle, K.H., Keady, J.J., Bernath, P.F.: 1988, *Science* **241**, 1319.
Kawamura, C.Y., Tong, E., Blundell, R., Hunter, T.E., Goltsman, G., Cherednichenko, S., Voronov, B., Gershenzon E.: 1999, to appear in *IEEE Transactions on Applied Supercoductivity*.
Keene, J., Schilke, P., Koi, J., Lis, D.C., Mehringer, D.M., Phillips, T.: 1998, *Astrophys. J.* **494**, L107.
Klaus, Th., Takano, S., Winnewisser, G.: 1997, *Astron. & Astrophys.* **322**, L1.
Klein, H., Lewen, F., Schieder, R., Stutzki, J., Winnewisser, G.: 1998, *Astrophys. J.* **494**, L125.

Klisch, E., Klaus, Th., Belov, S. P., Dolgner, A., Schieder, R., Winnewisser, G., Herbst, E.: 1996, *Ap. J.* **473**, 1118.
Klisch, E., Klein, H., Winnewisser, G., Herbst, E.: 1998, *Z. Naturforsch.* **53a**, 1.
Köster, B., Störzer, H., Stutzki, J., Sternberg, A., Winnewisser, G.: 1994, *Astron. & Astrophys.* **284**, 545.
Kramer, C., Stutzki, J., Winnewisser, G.: 1996, *Astron. & Astrophys.* **307**, 915.
Kramer, C., Degiacomi, C.G., Graf, U.U., Hills, R.E., Miller, M., Schieder, R., Schneider, N., Stutzki, J., Winnewisser, G.: 1998, In: *Proc. of the Conference on "Advanced Technology MMW, Radio, and Terahertz Telescopes"*, Kona, Vol. **3350**.
Kramer, C., Alves, J., Lada, C., Lada, E., Sievers, A., Ungerechts, H., Walmsley, C.M.: 1999, *Astron. & Astrophys.* **342**, 257.
Maddalena, R.J., Morris, M., Moskowitz, J., Thaddeus, P.: 1986 *Astrophys. J.* **303**, 375.
Marcy, G.W., Butler, R.P.: 1998, *Ann. Rev. Astron. Astrophys.* **36**, 57.
Mayor, M., Queloz, D.: 1995, *Nature* **378**, 355.
Mezger, P.G., Chini, R., Kreysa, E., Wink, J.E., Salter, C.J.: 1988, *Astron. & Astrophys.* **191**, 44.
Mezger, P.G., Sievers, A., Zylka, R., Haslam, C.G.T., Kreysa, E., Lemke, R.: 1992, *Astron. & Astrophys.* **265**, 743.
Van Orden, A., Cruzan, J.D., Provencal, R.A., Giesen, T.F., Saykally, R.J., Boreiko, R.T., Beetz, A.L.: 1995, In *Proceedings of the Airborne Astronomy Symposium on the Galactic Ecosystemsystems; M.R. Haas, J.A. Davidson, and E.F. Erickson, Eds.; ASP Conference Series; The Astronomical Society of the Pacific: San Fransisco* **73**, 67.
Ossenkopf, V., Stutzki, J., Winnewisser, G.: 1999, eds., *Proceedings of the 3rd Cologne-Zermatt Symposium, "The Physics and Chemistry of Interstellar Molecular Clouds"*.
Patel, N.A., Goldsmith, P.F., Heyer, M.H., Snell, R.L., Pratap, P.: 1998, *Astophys. J.* **507**, 241.
Sandford, S.A., Allamandola, L.J., Geballe, T.R.: 1993, *Science* **262**, 400.
Schilke, P., Benford, D.J., Hunter, T.R., Lis, D.J., Phillips, T.G.: 1999, (submitted to *Astrophys. J. Supplement*).
Schneider, J.: 1999, WWW-page http://www.obspm.fr/planets/ *"The Extrasolar Planets Encyclopedia"*, Observatoire de Paris.
Simon, R., Stutzki, J., Sternberg, A., Winnewisser, G.: 1997, *Astron. & Astrophys.* **327**, L9.
Stutzki, J., Graf, U.U., Haas, S., Honingh, E.C., Hottgenroth, D., Jacobs, K., Schieder, R., Simon, R., Staguhn, J., Winnewisser, G., Martin, M., Peters, W., McMillan, J.: 1997, *Astrophys. J.* **477**, L33.
Takano, S., Klaus, Th., Winnewisser, G.: 1998, *J. Mol. Spectrosc.* **192**, 309.
Tuley, M., Kirkwood, D.A., Pachkov, M., Maier, J.P.: 1998, *Astrophys. J.* **506**, L69.
Weikard, H., Wouterloot, J.G.A., Castets, A., Winnewisser, G., Sugitani, K.: 1996, *Astron. & Astrophys.* **309**, 581.
Whittet, D.C.B., Schutte, W.A., Tielens, A.G.G.M., Boogert, A.C.A, de Graauw, T., et al.: 1996, *Astron. & Astrophys.* **315**, L357.
Winnewisser, G., Herbst, E., Ungerechts, H.: 1992, In: *Spectroscopy of the Earths Atmosphere and Interstellar Medium*, ed. K.N. Rao and A. Weber, Academic Press, **423**.
Winnewisser, G., Herbst, E.: 1993, *Rep. Prog. Phys.* **56**, 1209.
Winnewisser, G.: 1994, *Infrared Phys. Technol.* **35**, 551.
Winnewisser, G.: 1995, *Vibrational Spectroscopy* **8**, 241.
Winnewisser, G., Belov, S.P., Klaus, Th., Schieder, R.: 1997, *J. Mol. Spectrosc.* **184**, 468.
Wouterloot, J.G.A., Winnewisser, G.: 1999, unpublished results.
Wouterloot, J.K.G., Lingmann, A., Miller, M., Vowinkel, B., Winnewisser, G., Wyrowski, F.: 1998, *Planet. Space Sci.* **46**, 579.

Address for correspondence: Gisbert Winnewisser, winnewisser@ph1.uni-koeln.de; Carsten Kramer, kramer@ph1.uni-koeln.de

THE COMPOSITION OF INTERSTELLAR MOLECULAR CLOUDS

WILLIAM M. IRVINE
*Department of Physics and Astronomy, University of Massachusetts,
Amherst, MA 01003-4517, U.S.A.*

Abstract. We consider four aspects of interstellar chemistry for comparison with comets: molecular abundances in general, relative abundances of isomers (specifically, HCN and HNC), ortho/para ratios for molecules, and isotopic fractionation, particularly for the ratio hydrogen/deuterium. Since the environment in which the solar system formed is not well constrained, we consider both isolated dark clouds where low mass stars may form and the "hot cores" that are the sites of high mass star formation. Attention is concentrated on the gas phase, since the grains are considered elsewhere in this volume.

1. Introduction

In discussing the origin and composition of cometary material, a basic question is whether this material includes interstellar matter. More precisely, to what extent do cometary nuclei consist of relatively pristine interstellar molecular material, interstellar material processed to various degrees in the solar nebula, and/or compounds that were formed in solar nebula itself? Of course, all the matter in the solar system was interstellar in the elemental sense, but the degree to which molecular compounds survived the processes forming the nebular accretion disk should provide key information on the nature of the accretion process, as well as on the initial chemical conditions for the formation of the planets, asteroids and comets.

To answer these questions, it is clearly necessary to compare the composition of comets with that of the dense interstellar clouds in which stars form. But that is not a simple matter. Such clouds contain both gas and dust, each of which raises its own observational problems for astrochemists. Millimeter- and submillimeter-wavelength observations of the gas can be made with great sensitivity and extremely high spectral resolution, so that constituents with fractional abundances less than a part per million relative to CO can often be securely identified. On the other hand, there are selection effects, particularly the inability to observe non-polar molecules (Irvine *et al.*, 1985). In contrast, infrared observations can probe the composition of the grains, including volatiles present in icy mantles, but only for species with abundances $\gtrsim 0.5\%$ relative to the principal mantle constituent, water. Such observations can also detect polar and non-polar species in the gas phase, but with limited sensitivity and, for cold material, only if a suitable background source is present (e.g., Whittet, 1997; Van Dishoeck, 1998a).

Then there are issues about *to which* interstellar abundances comets should be compared, in order to answer questions about cometary origins. Did the Sun form as an isolated star, in which case the chemistry of cold, dense cores and regions containing low mass YSOs is relevant? Or did it form within a cluster in which high mass stars also formed, in which case the chemistry of so-called hot cores (see below) might be more directly related? It is clear from studies of Orion, for example, that abundant formation of low mass stars can accompany that of high mass stars (e.g., Adams et al., 1999; Walter et al., 1999). Moreover, data from extinct radionuclides in meteorites are consistent with the presence of relatively nearby massive stars at the time that the solar nebula was formed (Goswami and Vanhala, 1999). But the issue is far from resolved. Of course, the ideal situation would be to observe the composition of the disks around solar mass stars, but the angular resolution for such studies is not yet really available (a beginning has been made, e.g., by Dutrey et al. (1997).

In comparing the elemental abundances in comets with "cosmic abundances", it is important to note that recent studies indicate that the abundances of carbon, nitrogen and oxygen in the local interstellar medium are some 20-30% less than solar abundances (e.g., Mathis, 1997; Meyer, 1997). This has been interpreted in terms of a formation location for the Sun some 2 kpc closer to the Galactic Center than its current position (Wielen et al., 1996). This eases, but may not eliminate, the "missing oxygen" question; i.e., what are the reservoirs for oxygen in the interstellar medium in the gas phase and in the solid state (Van Dishoeck and Blake, 1998)?

In this review we shall concentrate on the gas phase of the interstellar medium, since the grains will be discussed by Allamandola (this conference; the polycyclic aromatic hydrocarbons (PAHs) will be included with the grains). We shall consider four aspects of interstellar chemistry for comparison with comets: molecular abundances, relative abundances of isomers (specifically, HCN and HNC), ortho/para ratios for molecules, and isotopic fractionation, particularly for hydrogen/deuterium. For completeness, a list of currently identified interstellar and circumstellar molecules is given in Table I.

2. The Evolution of Chemistry in Star-Forming Regions

2.1. Low Mass Star Formation

Isolated low mass stars form in the dense cores of dark clouds (e.g., Lada et al., 1993). The chemistry in these cold, UV-shielded regions is believed to be predominantly the result of gas phase ion-molecule reactions, although some neutral-neutral reactions are also known to be important. Models assume that the cores have evolved from more diffuse clouds, in which the composition is primarily atomic, apart from the dominant constituent, H_2. The chemistry evolves over time scales

TABLE I

Interstellar and circumstellar molecules

2 Atoms	3 Atoms	4 Atoms	5 Atoms	6 Atoms	7 Atoms	≥ 8 Atoms
H_2	C_2H	C_2H_2	C_4H	C_2H_4 *	C_6H	CH_3COOH
C_2	CH_2	l-C_3H	C_3H_2	H_2CCCC	HC_5N	$HCOOCH_3$
CH	HCN	c-C_3H	H_2CCC	CH_3OH	CH_2CHCN	CH_3C_3N
CH^+	HNC	NH_3	HCOOH	CH_3CN	CH_3C_2H	C_7H *
CN	HCO	HNCO	CH_2CO	CH_3NC	CH_3CHO	CH_3C_4H
CO	HCO^+	$HOCO^+$	HC_3N	CH_3SH	CH_3NH_2	CH_3OCH_3
CS	HOC^+	$HCNH^+$	CH_2CN	NH_2CHO	c-CH_2OCH_2	CH_3CH_2CN
OH	N_2H^+	HNCS	NH_2CN	HC_3HO		CH_3CH_2OH
NH	NH_2	C_3N	CH_2NH	C_5H		HC_7N
NO	H_2O	C_3O	CH_4	HC_3NH^+		CH_3C_4CN ?
NS	HCS^+	H_2CS	SiH_4 *	C_5N		CH_3COCH_3
SiC *	H_2S	C_3S	C_4Si *			HC_9N
SiO	OCS	HCCN	C_5 *			C_8H *
SiS	N_2O	H_3O^+	HCCNC			H_2C_6
SiN *	SO_2	H_2CN	HNCCC			$HC_{11}N$
SO	SiC_2 *	H_2CO	HCO_2H^+			
HCl	C_2S	c-SiC_3 *				
CP *	C_2O					
SO^+	C_3 *					
NaCl *	MgNC *					
AlCl *	MgCN *					
KCl *	NaCN *					
AlF *	HNO					
PN	H_3^+					
CO^+						
SiH ?						
HF ?						

Notes:

Recent detections include H_3^+ (Geballe and Oka, 1996),

c-CH_2OCH_2 (Dickens *et al.*, 1997),

H_2C_6 (Langer *et al.*, 1997),

C_5N (Guelin *et al.*, 1998),

C_7H (Guelin *et al.*, 1997),

CH_3COOH (Mehringer *et al.*, 1997),

$HC_{11}N$ (Bell *et al.*, 1997)

and HF (Neufeld *et al.*, 1997);

for previous identifications see Irvine (1998).

* Detected only in stellar envelopes

? Tentative detection

c- cyclic molecule

of order 10^5 years to produce a wide range of radicals, ions and neutral molecules, many of which are highly unsaturated. Isotopic fractionation, particularly for D/H, can be very large. An example is Taurus Molecular Cloud 1 (TMC-1), in which the largest inventory of species has been detected (e.g., Ohishi and Kaifu, 1998; Pratap et al., 1997; Irvine et al., 1991), including the heaviest known interstellar molecule ($HC_{11}N$; Bell et al., 1997). As the chemistry continues to evolve in models of such clouds, carbon is processed into CO and the abundance of organic molecules begins to decline. A more typical example of dense core may be L134N (L183), which contains lower abundances of the cyanopolyynes and has also been extensively studied (e.g., Dickens, 1998; Swade and Schloerb, 1992).

The time scale for accretion of gas phase molecules onto the grains in dense cores is of the same order as that at which the chemistry evolves, and infrared observations do indeed show that grains in these regions have acquired icy mantles (e.g., Whittet, 1997). This process is accelerated as cores collapse on their way to star formation. The chemical composition of the mantles evolves through surface reactions and processing by cosmic rays and UV irradiation (which can be generated by cosmic ray excitation of H_2). As long as the gas phase abundance of atomic hydrogen is not too low, its high mobility after sticking on the cold grains is thought to result in surface reactions which produce saturated species such as NH_3, H_2O, CH_4 and H_2S (Allmandola et al., 1999, this conference; Van Dishoeck and Blake, 1998; and references therein).

As a protostar forms in the core, it affects the surrounding gas and dust through both radiative heating and supersonic, typically bipolar outflows of gas. In consequence, the icy grain mantles may be sublimated and their constituents become accessible to observations of the gas phase. For low mass stars, however, the lines are weak and existing instrumental sensitivity limits the degree to which the composition can be determined. Ices that remain on the grains, particularly at larger distances from the forming star, become observable in absorption against the embedded IR source; H_2O, CO, CO_2, and "XCN" are detected (Van Dishoeck and Blake, 1998).

As the YSOs continue to evolve, the surrounding envelope is dispersed and the circumstellar accretion disk becomes observable. It is now established that a significant fraction of T Tauri (solar mass) stars are surrounded by such disks (e.g., Beckwith et al., 1996). The gas phase composition of this material is beginning to be probed by millimeter interferometers (see Van Dishoeck and Blake, 1998), but spatial resolution on solar nebula scales will mainly await completion of facilities such as the Millimeter Array.

Potentially important for comparison with comets are the abundances in the gas phase in the prestellar dense cores, since most of this material will accrete onto grains as the core collapses; the composition of the grains, in so far as this can be determined directly; and the composition of the post-sublimation gas, which will include constituents from the icy mantles. As indicated above, detailed information is available only on the first component, the quiescent gas, although the major constituents of the grain mantles can also be determined.

2.2. HIGH MASS STAR FORMATION

Since the Sun may have been born in a region of high mass star formation, we must also consider the chemistry of giant molecular clouds (GMCs). The quiescent gas in these regions is typically somewhat warmer than in isolated dark clouds ($\sim$25 K rather than $\sim$10 K). The composition of that gas appears not to differ greatly among GMCs (Ungerechts *et al.*, 1997; Bergin *et al.*, 1997). Likewise, there is a general similarity between these clouds and dark clouds (Ungerechts *et al.*, 1997), except for such particularly time-dependent or temperature-dependent effects as the abundance of highly unsaturated species, the HNC/HCN ratio, and possibly ortho/para ratios (e.g.,Minh and Irvine, 1991; Goldsmith *et al.*, 1986; Schilke *et al.*, 1992; Minh *et al.*, 1997). Of course, such differences are intriguing, since, if retained in comets, they might provide a way to distinguish among possible formation regions for the solar system.

Because of their high luminosity, massive protostars embedded in their natal molecular cloud provide sources against which the absorption by icy grain mantle constituents is best viewed (Chapter by Allamandola and Section III below).

The chemical differences observed between regions of isolated low mass star formation and regions of massive star formation in GMCs become most apparent as the massive protostars heat and shock their surrounding envelopes. So-called hot cores are produced, which exhibit a rich gas phase chemistry characterized by large abundances of fully saturated molecules such as H_2O, NH_3, and H_2S; complex oxygen and nitrogen containing organics such as CH_3OH, CH_3CH_2OH, $HCOOCH_3$, $(CH_3)_2O$, and CH_3CH_2CN; and (surprisingly) large deuterium fractionation for a variety of species including HDO, DCN, HDCO, CH_3OD and even D_2CO (Millar and Hatchell, 1998). Since this chemistry is widely believed to reflect and be driven by molecules sublimated from grain mantles, observations of hot cores provide a means of determining details of the composition of the mantles (albeit in a model dependent fashion). Unlike the case for isolated star formation, the physics and chemistry of a protostar in the cluster environment may be influenced by radiation, outflows, and shocks resulting from the formation of other stars.

3. Comparison of Chemical Abundances with Comets

Table II lists relative abundances for cometary volatiles, for a dark cloud, and for regions of both low mass and high mass star formation (note the normalization to *either* H_2O *or* CO). The comparison can be a little awkward, since gas phase water (the principal constituent of cometary ices) has not been detected in all these regions. Such detections are limited both by telluric absorption and by the low excitation of water in quiescent gas.

TABLE II
Comparative Abundances

Molecule	Comets	IS Ices	Hot Cores G34	Hot Cores Orion	Hot Cores W3	Low Mass: 16293	Low Mass: L134N
H_2O	100	100	100	> 100	–	–	–
CO	1-20	< 1-25	220	1000	1000	1000	1000
CO_2	3-20	12-24	–	2-10	–	–	–
H_2CO	0.1-4	≤ 4-7	0.2	0.1-1	0.02	–	0.03
CH_3OH	1-7	< 3-22	0.3	2	0.3	–	0.06
HCOOH	~ 0.05	≤ 3 ?	–	0.008	–	–	0.004
$HCOOCH_3$	~ 0.05	–	0.1	0.1	0.03	–	–
CH_3CHO	~ 0.5^a	–	0.003	–	–	–	0.008
HNCO	0.1	~ 2^b	–	0.06	0.02	–	–
NH_2CHO	~ 0.01	–	–	0.002	–	–	–
CH_4	~ 0.6	1-2	–	–	–	–	–
C_2H_2	0.1-0.3	< 10	–	3-10	–	–	–
C_2H_6	~ 0.3	< 0.4	–	–	–	–	–
C_3H_2	~ 0.1^a	–	–	–	–	–	0.02
NH_3	0.5-1	< 4-10?^{c}	20	8	–	–	0.7
HCN	0.05-0.2	< 6	0.009	4	0.05	–	0.09
HNC	0.01-0.04	–	< 0.002	0.02	0.003	–	0.3
CH_3CN	0.01-0.1	–	0.4	0.2	0.001	–	< 0.01
HC_3N	~ 0.03	–	0.0002	0.04	0.0006	–	0.005
CH_3CH_2CN	~ 0.03^a	–	–	0.3	–	–	–
H_2S	0.2-1.5	< 0.2	0.08	1	0.04	0.02	0.01
H_2CS	~ 0.02	–	0.4	0.01	0.006	0.002	0.008
OCS	0.2-0.5	0.04-0.3	0.2	0.5	0.009	0.07	0.025
SO_2	~ 0.1	< 0.2	0.1	0.6	0.04	0.02	0.03
SO	~ 0.5	–	0.02	0.5	0.05	0.04	0.07
CS	0.2	–	0.08	0.1	0.04	0.01	0.01
S_2	0.005	–	–	–	–	–	–
CH_3NH_2	< 0.15	–	–	–	–	–	–
CH_3CH_2OH	–	< 1	0.03	–	–	–	–

Abundances normalized to H_2O for comets, IS ices, and G34, and to CO for other sources. Cometary CO, H_2CO, HNC, and possibly OCS have extended as well as possible nuclear sources. Results not necessarily representative of all sources of given interstellar type; no claim is made to completeness for any specific source.

Data for IS Ices: Van Dishoeck and Blake (1998) and Langer et al. (1999);
Comets: Irvine et al. (1999), Crovisier and Bockelée-Morvan (1999), Altwegg et al. (1999);
G34.3+0.15: Millar et al. (1997) and Nummelin et al. (1998);
Orion: plateau, hot core and/or compact ridge from Van Dishoeck and Blake (1998), Sutton et al. (1995), Blake et al. (1987), Goldsmith et al. (1986), Johansson et al. (1984);
W3(H_2O): Helmich and van Dishoeck (1997);
IRAS 16293-2422: Blake et al. (1994);
L134N(C): Dickens (1998), Minh et al. (1995) and Ohishi et al. (1992).

^{a} Probable detection from mass spectrometry, so includes all isomers
^{b} From "XCN", assuming it arises from OCN-, which sublimes as HNCO
^{c} Detection in 1 source claimed by (Lacy et al., 1998).

However, the bulk of the water may well be frozen out onto the grains (Van Dishoeck, 1998a). Although there are variations in the abundances of other constituents both among comets and among the ices in various interstellar clouds, there is an interesting similarity between comets and the ISM. This similarity is even more pronounced if we include the gas phase abundances of a variety of minor constituents in hot cores, which may reflect the prior abundances in the icy grain mantles.

3.1. ISOMERIC ABUNDANCES

Are there molecular species which might act as chemical tracers of interstellar matter in comets? A prime characteristic of the chemistry in the quiescent interstellar medium is its extreme departure from thermochemical equilibrium. Classical markers of this low temperature ion-molecule chemistry are the presence of positive ions (not expected to survive upon incorporation into cometary nuclei), large isotopic fractionation (discussed below), and the high abundance of higher energy isomers relative to their lower energy counterparts (e.g., Herbst and Klemperer, 1976; Watson, 1980; Herbst, 1995). The most striking example of the latter is the pair HNC and HCN. The HNC/HCN ratio, which would be almost identically zero at equilibrium, can be greater than unity in cold clouds and exhibits a striking inverse temperature dependence (Goldsmith *et al.*, 1986; Schilke *et al.*, 1992).

The detection of HNC in Comet Hyakutake (Irvine *et al.*, 1996) thus appeared at first glance to indicate the survival of interstellar volatiles in the comet. However, subsequent observations of the HNC/HCN ratio in Comet Hale-Bopp showed a clear dependence on heliocentric distance, which has been matched by chemical models in which the HNC is not a nuclear constituent but is instead produced chemically in the coma (Rodgers and Charnley, 1998; Irvine *et al.*, 1998a; Irvine *et al.*, 1998b). Curiously, however, these models do not produce sufficient HNC to match the observations of Comet Hyakutake, in which the lower gas production produces a less favorable environment for chemical processing of the sublimating gas.

4. Ortho/Para Ratios

Another, somewhat puzzling, characteristic of cold interstellar clouds is the presence of ortho/para abundance ratios (OPR) for several molecular species which in some cases are, and in others are not, equilibrated at the local kinetic temperature (Minh *et al.*, 1997). For molecules like H_2CO, which contain equivalent hydrogen atoms, the ortho species is that for which the two proton spins are parallel, while the para species corresponds to the antiparallel case. The potentially diagnostic property of such molecules is that the ortho and para species are not interconverted by either radiative or non-destructive collisional processes. In so far as this is true, the two species are effectively two independent molecules. The same situation applies to the A and E symmetry states of symmetric top molecules like CH_3CN and CH_3CCH, which have 3 equivalent protons. For most cases of interest here (excluding H_2, which is not excited at the temperatures of quiescent clouds) the energy difference (expressed in K) between the lowest para and the lowest ortho state is comparable to the temperature of the gas and dust in the cloud, and vastly less than the energy released in a typical formation reaction for these species. It is then expected, apart from ill-understood quantum mechanical constraints, that the ortho/para ratio upon molecular formation in the gas phase will correspond to the ratio of statistical weights for the two species, which is the equilibrium value at high temperatures. Furthermore, this ratio should be preserved in the gas phase, apart from reactions with partners which exchange protons with the molecules of interest (typically slow relative to reactions which destroy (and reform) the ortho and para species). On the other hand, if the molecules of interest are formed on grains, or frozen onto and then sublimated from grains, the grain may act as a heat reservoir and bring the ortho/para ratio into equilibrium at the (low) temperature of the cloud.

OPRs have been reported for several species in cold clouds and GMCs (Table III). In the latter, the OPRs tend to reflect the expected ratio of statistical weights, corresponding to the warm gas and dust in these regions. An exception is H_2CS in the Orion Compact Ridge, whose lower value may reflect equilibration with and sublimation from grain mantles (Minh et al., 1991). Why the same equilibration is not seen for, e.g., water, is puzzling. In cold clouds such as TMC-1 the OPRs sometimes reflect equilibration at the cloud temperature, corresponding to a value less than the ratio of statistical weights, but sometimes seem to preserve the high temperature ratio which presumably reflects their origin in the gas phase. It should be borne in mind that measurement of the OPR is difficult, since it usually involves assumptions about the population of unobserved states and often involves the calibration uncertainties associated with observing at different frequencies and different times. Recently Dickens and Irvine (1999) reported that the formaldehyde (H_2CO) OPR appears to differ between cold clouds which lack embedded IR sources or outflows, and clouds which contain such evidence of the presence of low mass protostars. In the former case the OPR was at the high temperature limit expected for gas phase formation, while in the latter it was equilibrated at the temperature of the dust in the clouds. Moreover, in the sources containing protostars the formaldehyde abundance appeared to be systematically higher. Dickens and Irvine propose that in these sources formaldehyde has been synthesized on grains and sublimated as a result of the energy released by the protostar, while in the clouds without such embedded objects the observed H_2CO was produced in the gas phase.

To date the only molecule for which OPRs have been reported in comets is water, for which the only interstellar detections are in warm sources, where the OPR is apparently at the high temperature value (Van Dishoeck, 1998a).

TABLE III
Ortho/para ratios

Molecule	Stat. weight ratio	TMC-1	Quiesc. dark clouds	Dark clouds + Outflws	Orion CR	Other GMCs	Comets
H_2O	3				~ 3		2.5-3.2
H_2CO	3		2.6 ± 0.3	1.7 ± 0.3	3.2 ± 0.8		
H_2CCO	3	3.5 ± 0.6					
H_2CS	3	1.8 ± 0.3			1.8 ± 0.5	~ 3	
H_2C_3	3	5.9 ± 2.0					
H_2C_4	3	4.2 ± 1.5					
CH_3CCH	1	~ 1			~ 1	~ 1	
CH_3CN	1	0.75 ± 0.10					

Column 2 is ratio of statistical weights for ortho to para states; Orion CR is Compact Ridge; errors are 1σ. Data from Minh et al., 1991 (H_2CS); Ohishi et al., 1991 (H_2CCO); Kahane et al., 1984, Minh et al., 1995, and Dickens and Irvine, 1999 (H_2CO); Askne et al., 1984 (CH_3CCH); Kawaguchi et al., 1991 (H_2C_3, H_2C_4); Minh et al., 1993 (CH_3CN); Van Dishoeck, 1998a (H_2O). Madden (1990) finds values between 2.3 and 3.1 for C_3H_2, in 4 dark clouds but the small energy difference of 2.4 K between the lowest ortho and para states makes the results somewhat uncertain. For H_2O in three comets the OPRs seem to be distinctly less than the ratio of statistical weights, while in a fourth the OPR is close to 3 (typical uncertainties are $\sigma = 0.1$; see Irvine et al., 1999).

5. Isotopic Fractionation

One of the characteristics of low temperature chemistry in the ISM is large isotopic fractionation, particularly for hydrogen/deuterium. This is well understood in terms of the kinetics of relevant reactions, particularly (e.g., Van Dishoeck, 1998b)

$$H_3^+ + HD \rightarrow H_2D^+ + H_2 + 227 \text{ K} \tag{1}$$

$$CH_3^+ + HD \rightarrow CH_2D^+ + H_2 + 370 \text{ K} \tag{2}$$

which at low temperature transfer deuterium from its principal pool in HD into H_2D^+ and CH_2D^+, and hence into further trace constituents of the gas. The fractionation can be exceedingly large, up to 4 orders of magnitude relative to the cosmic D/H ratio, in cold clouds like TMC-1 (Table IV). As models predict, the observed fractionation differs for different molecular species. It is also predicted that the fractionation will evolve over time (Millar et al., 1989).

TABLE IV
Isotopic fractionation

Molecule	TMC-1	Hot Cores	Comets	Local ISM	Earth
HDO/H_2O	–	0.1-0.5	0.29-0.33		0.156
DCN/HCN	23	0.9-5	2.30 ± 0.40		
DNC/HNC	15				
DC_3N/HC_3N	15				
C_3HD/C_3H_2	100				
CH_2DCCH/CH_3CCH	60	< 40			
CH_2DCN/CH_3CN	–	∼ 10			
$HDCO/H_2CO$	15	∼ 100			
NH_2D/NH_3	< 20	∼ 60			
CH_3OD/CH_3OH		10-60	$< 10^a$		
D/H				0.016	
CH_2DOH/CH_3OD		$1.1 - 1.5^b$			
$D_2CO/HDCO$		∼ 0.02			
$^{12}C/^{13}C$				$59 - 68^c$	89
$H^{12}CN/H^{13}CN$			$\approx 100^d$		

Values above thick horizontal line multiplied by 1000. Data primarily from Van Dishoeck et al. (1993) and Irvine et al. (1999), and references therein.
Hot core ratios for CH_3CN, H_2CO, NH_3, CH_3OH, and $D_2CO/HDCO$ for Orion only; comet ratio for H_2O for 3 comets, for HCN only for Comet Hale-Bopp.
[a] limit includes CH_3OD and CH_2DOH (Eberhardt et al., 1994)
[b] Jacq et al. (1993)
[c] Lucas and Liszt (1998)
[d] for comets Halley and Hale-Bopp, thought to be consistent with the terrestrial value of 89; for Comet Hyakutake, the reported value of 34 may suffer from blending of the observed $H^{13}CN$ line with SO_2.

In view of the strong temperature dependence of reactions such as (1) and (2), it might be expected that D/H fractionation would be very small in hot cores. This is not observed to be the case (Table IV). The significant D/H ratios observed in the gas phase for several species in such regions are interpreted to be the result of sublimation of isotopically fractionated material off of icy grain mantles and the preservation of the fractionation for periods comparable to the age of the hot cores (Rodgers and Millar, 1996). The fractionation might have occurred in the cold gas which then was frozen out onto the grains in the initial collapse of these cores, or it may have partially occurred on the grain mantles themselves, where significant D/H fractionation can result from the predicted fractionation of gas phase atomic hydrogen (Tielens, 1983). In so far as the sublimed gas preserves the D/H ratios present in the grain mantles, the fractionation observed in hot cores can be compared to that found in comets, in particular in Hyakutake and Hale-Bopp (Table IV). A striking similarity is found for HDO/H_2O and for DCN/HCN.

The close approach of Comet Hyakutake and the large gas production from Comet Hale-Bopp allowed measurements to be made of the gas phase ratio for $^{13}C/^{12}C$, $^{15}N/^{14}N$, and $^{34}S/^{32}S$ in HCN and CS (Jewitt et al., 1997). Data for $^{18}O/^{16}O$ were obtained from mass spectrometer data for comet Halley (Balsiger et al., 1995; Eberhardt et al., 1995). The results in all cases were in agreement with the terrestrial values. This has been interpreted by some authors as showing that the corresponding volatiles have an origin in the solar nebula, rather than in the ISM, where the local $^{13}C/^{12}C$ ratio (for example) is somewhat lower than the terrestrial value. This interpretation is unwarranted. The ratios in the ISM will have evolved over the 4.5 billion years since solar system formation. Moreover, all of the observed ratios except that for the sulfur isotopes show gradients with galactocentric distance, and there is now evidence that the Sun formed some 2 kpc closer to the Galactic center than its present position (Wielen et al., 1996; Wielen and Wilson, 1997). It is thus expected that the bulk isotopic ratios in the solar system will differ somewhat from the corresponding values in the local ISM today.

Local chemical fractionation of the isotopic ratios, as opposed to large scale spatial or temporal variations in the Galaxy, can potentially distinguish between molecules that formed in the solar nebula rather than in the interstellar cloud from which the nebula condensed. Such chemical fractionation can clearly occur for D/H by reactions such as (1) and (2). Much smaller fractionation has been observed in the ISM for $^{13}C/^{12}C$, as a result of reactions such as

$$^{13}C^+ + {}^{12}CO \rightarrow {}^{13}CO + {}^{12}C^+ + 35K \qquad (3)$$

which will concentrate ^{13}C in CO at low temperatures (Langer et al., 1984; for sample observations, see, e.g., Taylor and Dickman, 1989). The opposite effect is produced by isotopic- dependent photodissociation at the boundaries of dense clouds, where ^{12}CO becomes optically self-shielding much more rapidly than the less abundant ^{13}CO. Unlike the case for hydrogen, where the principal reservoir of deuterium, HD, is much more abundant than other D-containing molecules, for carbon the overabundance of ^{13}CO can result in the depletion of ^{13}C in other carbon-bearing species such as H_2CO (Langer et al., 1984). Furthermore, *differential* $^{13}C/^{12}C$ fractionation for the nitrile carbon relative to the acetylenic carbons has been reported in HCCCN in TMC-1 (Takano et al., 1998), raising the possibility of tracing specific synthetic pathways for this and potentially for other species in dark clouds. Chemical fractionation for $^{18}O/^{16}O$, $^{15}N/^{14}N$, and $^{34}S/^{32}S$ is expected to be much smaller than for carbon in the ISM (Langer et al., 1984; Gusten and Ungerechts, 1985), although there are some suggestions that the sulfur ratio may differ between local dark clouds and GMCs (Pratap et al., 1997; Dickens, 1998).

Differences in the isotopic composition among different components in a comet would indeed provide important diagnostics for the origin of such material. Thus, some of the CHON grains measured by the GIOTTO mass spectrometers for comet Halley showed extremely large $^{12}C/^{13}C$ values (up to about 5000). These grains are almost certainly presolar (Eberhardt, 1998), although they are thought to originate from a specific nucleosynthetic (stellar) source, rather than to be the result of isotopic fractionation.

6. Discussion

There are some striking similarities between comets and dense interstellar clouds in chemical abundances, hydrogen isotopic fractionation, and other more subtle characteristics. Can we conclude that comets preserve interstellar volatiles?

6.1. ABUNDANCES

The abundances in interstellar icy grain mantles are remarkably similar to those reported in comets (Table II). The similarity is even closer if we assume that the gas phase abundance observed in hot cores for various trace constituents reflects the composition of sublimed grain mantles. An important question is the extent to which similar abundances could be produced by disequilibrium processes in the solar nebula itself (see Fegley, 1999, this conference).

6.2. HNC/HCN RATIO

The large values of the HNC/HCN abundance ratio in cold molecular clouds are a trademark of ion-molecule chemistry. It appears that the HNC in Comet Hale-Bopp, however, was produced in the coma, rather than being a nuclear constituent. In Comet Hyakutake the situation is unclear: it may be that the observed HNC is indeed present in the nucleus, in which case the intrinsic HNC abundance in comets must vary by at least a factor of 3, given the stringent upper limit on intrinsic HNC in Comet Hale-Bopp. Laboratory work is needed to determine whether HNC could survive in the nucleus over the age of the solar system, whether it would be interconverted to HCN during freeze-out onto grains and/or sublimation from grains, and whether it could be produced in the nuclear ices by solar irradiation.

6.3. ORTHO/PARA RATIO

The OPRs measured for H_2O in several periodic comets indicate equilibration at a low temperature (25-35 K), as do the OPRs observed in some dark clouds for species such as H_2CS and CH_3CN (Table III). However, other species in dark clouds (such as CH_3CCH and, for some clouds, H_2CO) exhibit OPRs corresponding to the high temperature limit. The situation in hot cores is likewise mixed, although most published results correspond to the high temperature limit. Clearly both further observations and laboratory studies of ortho-para interchange are needed before the cometary results can be interpreted in terms of surviving interstellar water.

6.4. ISOTOPIC FRACTIONATION

Isotopic anomalies clearly indicate that presolar material is preserved in meteorites (e.g., Zinner, 1997). It would hardly be surprising, therefore, to find such matter in comets, and this appears to be borne out by the large $^{12}C/^{13}C$ ratio in some Comet Halley grains. The question still to be answered is the degree to which interstellar/circumstellar molecular material is preserved in comets, particularly for volatiles. Relevant in this regard are the large values for HDO/H_2O (in comets Halley, Hyakutake, and Hale-Bopp) and DCN/HCN (in Comet Hale-Bopp.) Both these ratios indicate significant isotopic fractionation, and both are in the range observed for the same species in hot cores. However, there are some theoretical reasons to believe that the DCN/HCN ratio in the hot cores might have been diluted from value in the pre-existing icy grain mantles (Hatchell *et al.*, 1998); the observed value, while high, is significantly lower than values observed in cold clouds. Similar dilution is <u>not</u> predicted for the HDO/H_2O ratio in hot cores, which may therefore preserve the value from the grain mantles (Rodgers and Millar, 1996). This is difficult to verify, since the value of the ratio in cold clouds is not known.

7. Summary

Whether observed cometary abundances, ortho/para ratios, and isotopic fractionation might be produced in the solar nebula, rather than in the solar system's parent molecular cloud, is discussed elsewhere in this volume. Likewise, the relationship between observed cometary abundances in the coma, and true nuclear abundances, is often far from clear (e.g., Irvine *et al.*, 1999; Crovisier and Bockelée-Morvan, 1999). Although some combination of interstellar, partially processed, and nebular material may be present in comets, with the relative fractions depending on the heliocentric formation distance, further observational, laboratory and theoretical work on these issues is sorely needed.

Acknowledgements

The author is grateful to John Kerridge for helpful comments. This work was supported in part by NASA grant NAG5- 3653.

References

Adams, F., Meyer, M., Carpenter, J., Larson, R., and Hillenbrand, L.: 1999, in V. Mannings, A. Boss and S. Russell (eds.), *Protostars and Planets* **IV**, Tucson: Univ. Arizona Press, submitted.
Allamandola, L.J., Bernstein, M.P., Sandford, S.A. and Walker R.L.: 1999, *Space Sci. Rev.*, this volume.
Altwegg, K., Balsiger, H., and Geiss, J.: 1999, *Space Sci. Rev.*, this volume.
Askne, J., Höglund, B., Hjalmarson, Å., and Irvine, W.M.: 1984, *Astron. Astrophys.* **130**, 311-318.
Balsiger, H., Altwegg, K., and Geiss, J.: 1995, *J. Geophys. Res.* **100**, 5827-5834.
Beckwith, S.V.W. and Sargent, A.I.: 1996, *Nature* **383**, 139-144.
Bell, M.B., Feldman, P.A., Travers, M.J., McCarthy, M.C., Gottlieb, C.A., and Thaddeus, P.: 1997, *Astrophys. J.* **483**, L61-L64.
Bergin, E.A., Ungerechts, H., Goldsmith, P.F., Snell, R.L., Irvine, W.M., and Schloerb, F.P.: 1997, *Astrophys.J.* **482**, 267-284.
Blake, G.A., Sutton, E.C., Masson, C.R., and Phillips, T.G.: 1987, *Astrophys.J.* **315**, 621-645.
Blake, G.A., van Dishoeck, E.F., Jansen, D.J., Groesbeck, T.D., and Mundy, L.G.: 1994, *Astrophys. J.* **428**, 680-692.
Crovisier, J., and Bockelée-Morvan, D.: 1999, *Space Sci. Rev.*, this volume.
Dickens, J.E., Irvine, W.M., Ohishi, M., Ikeda, M., Ishikawa, S., Nummelin, A., and Hjalmarson, Å.: 1997, *Astrophys.J.* **489**, 753-757.
Dickens, J.E.: 1998, Ph.D. Dissertation, University of Massachusetts, Amherst.
Dickens, J.E., and Irvine, W.M.: 1999, *Astrophys.J.* **518**, 733-739.
Dutrey, A., Guilloteau, S., and Guelin, M.: 1997, *Astron. Astrophys.* **317**, L55-58.
Eberhardt, P., Meier, R., Krankowsky, D., and Hodges, R.: 1994, *Astron. Astrophys.* **288**, 315-329.
Eberhardt, P., Reber, M., Krankowsky, D., and Hodges R.: 1995, *Astron. Astrophys.* **302**, 301-316.
Eberhardt, P.: 1998, in *Asteroids, Comets, Meteors 1996*, COSPAR Colloquium 10, in press.
Fegley, B.: 1999, *Space Sci. Rev.*, this volume.
Geballe, T.R. and Oka, T.: 1996, *Nature* **384**, 334-335.
Goldsmith, P.F., Irvine, W.M., Hjalmarson, Å., and Ellder, J.: 1986, *Astrophys. J.* **310**, 383-391.
Goswami, J., and Vanhala, H.: 1999, in V. Mannings, A. Boss, and S. Russell (eds.), *Protostars and Planets* **IV**, Tucson: Univ. Arizona Press, in press.
Guelin, M., Cernicharo, J., Travers, M.J., McCarthy, M.C., Gottlieb, C.A., Thaddeus, P., Ohishi, M., Saito, S., and Yamamoto, S.: 1997, *Astron. Astrophys.* **317**, L1-L4.
Guelin, M., Neininger, N., and Cernicharo, J.: 1998, *Astron. Astrophys.* **335**, L1-L4.
Gusten, R., and Ungerechts, H.: 1985, *Astron. Astrophys.* **145**, 241-250.
Hatchell, J., Millar, T.J., and Rodgers, S.D.: 1998, *Astron. Astrophys.* **332**, 695-702.
Helmich,F.P., and van Dishoeck, E.F.: 1997, *Astron. Astrophys. Suppl.* **124**, 205-253.
Herbst, E.: 1995, *Ann. Rev. Phys. Chem.* **46**, 27-53.
Herbst, E., and Klemperer, W.: 1976, *Phys. Today* **29**, 32-39.
Irvine, W.M., Schloerb, F.P., Hjalmarson, Å., and Herbst, E.: 1985, in D.C. Black and M.S. Matthews (eds.), *Protostars and Planets* **II**, Tucson: Univ. Arizona Press, pp. 579-620.
Irvine, W.M., Ohishi, M., and Kaifu, N.: 1991, *Icarus* **91**, 2-6.
Irvine, W.M., Bockelée-Morvan, D., Lis, D., Matthews, H.E., Biver, N., Crovisier, J., Davies, J.K., Dent, W.R.F., Gautier, D., Godfrey, P.D., Keene, J., Lovell, A.J., Owen, T.C., Phillips, T.G., Rauer, H., Schloerb, F.P., Senay, M., and Young, K.: 1996, *Nature* **383**, 418-420.
Irvine, W.M.: 1998, *Origins Life Evol. Biosphere* **28**, 365-383.
Irvine, W.M., Bergin, E.A., Dickens, J.E., Jewell, D., Lovell, A.J., Matthews, H.E., Schloerb, F.P., and Senay, M.: 1998a, *Nature* **393**, 547-550.
Irvine, W.M., Dickens, J.E., Lovell, A.J., Schloerb, F.P., Senay, M., Bergin, E.A., Jewitt, D., and Matthews, H.E.: 1998b, *Faraday Disc.* **109**, 475-492 and 510-512.

Irvine, W.M., Schloerb, F.P., Crovisier, J., Fegley, B., Jr., and Mumma, M.J.: 1999, in V. Mannings, A. Boss, and S. Russell (eds.), *Protostars and Planets* **IV**, Tucson: Univ. Arizona Press, in press.
Jacq, T., Walmsley, C. M., Mauersberger, R., Anderson, T., Herbst, E., and De Lucia, F.C.: 1993, *Astron. Astrophys.* **271**, 276-281.
Jewitt, D.C., Matthews, H.E., Owen, T.C., and Meier, R.: 1997, *Science* **278**, 90-93.
Johansson, L.E.B., Andersson, C., Ellder, J., Friberg, P., Hjalmarson, Å., Höglund, B., Irvine, W.M., Olofsson, H., and Rydbeck, G.: 1984, *Astron. Astrophys.* **130**, 227-256.
Kahane, C., Frerking, M.A., Langer, W.D., Encrenaz, P., and Lucas, R.: 1984, *Astron. Astrophys.* **137**, 211-222.
Kawaguchi, K., Kaifu, N., Ohishi, M., Hirahara, Y., Ishikawa, S., Yamamoto, S., Saito, S., Takano, S., Vrtilek, J.M., Thaddeus, P., and Irvine, W.M.: 1991, *Publ. Ast. Soc. Japan* **43**, 607-619.
Lacy, J.H., Faraji, H., Sandford, S.A., and Allamandola, L.J.: 1998, *Astrophys. J.* **501**, L105-L109.
Lada, E., Strom, K.M., and Myers, P.C.: 1993, in E.H. Levy and J.I. Lunine (eds.), *Protostars and Planets* **III**, Tucson: Univ. Arizona Press, pp. 245-277.
Langer, W.D., Graedel, T.E., Frerking, M.A., and Armentrout, P.B.: 1984, *Astrophys. J.* **277**, 581-604.
Langer, W.D., Velusamy, T., Kuiper, T.B.H., Peng, R., McCarthy, M.C., Travers, M.J., Kovacs, A., Gottlieb, C.A., and Thaddeus, P.: 1997, *Astrophys. J.* **480**, L63-L66.
Langer, W.D., van Dishoeck, E.F., Bergin, E.A., Blake, G.A., Tielens, A.G.G.M., Velusamy, T., and Whittet, D.C.B.: 1999, in V. Mannings, A. Boss and S. Russell (eds.), *Protostars and Planets* **IV**, Tucson: Univ. Arizona Press, submitted.
Lucas, R., and Liszt, H.: 1998, *Astron. Astrophys.* **337**, 246-252.
Madden, S.C.: 1990, Ph.D. Dissertation, University of Massachusetts, Amherst.
Mathis, J.S.: 1997, in Y.J. Pendleton and A.G.G.M. Tielens (eds.), *From Stardust to Planetesimals*, Astron. Soc. Pacific Conf. Ser. **122**, 87-96.
Mehringer, D.M., Snyder, L.E., Miao, Y., and Lovas, F.J.: 1997, *Astrophys. J.* **480**, L71-L74.
Meyer, D.H.: 1997, in E.F. van Dishoeck (ed.), *Molecules in Astrophysics: Probes and Processes*, Dordrecht: Kluwer, pp. 407-420.
Millar, T.J., Bennett, A., and Herbst, E.: 1989, *Astrophys. J.* **340**, 906-920.
Millar, T.J., Macdonald, G.H., Gibb, A.G.: 1997, *Astron. Astrophys.* **325**, 1163-1173.
Millar, T.J. and Hatchell, J.: 1998, Faraday Disc. **109**, 15-30.
Minh, Y.C. and Irvine, W.M.: 1991, *Astrophys. Space Sci.* **175**, 165-169 .
Minh, Y.C., Irvine, W.M., and Brewer, M.K.: 1991, *Astron. Astrophys.* **244**, 181-189.
Minh, Y.C., Irvine, W.M., Ohishi, M., Ishikawa, S., Saito, S., and Kaifu, N.: 1993, *Astron. Astrophys.* **267**, 229-232.
Minh, Y.C., Dickens, J.E., Irvine, W.M., and McGonagle, D.: 1995, *Astron. Astrophys.* **298**, 213-218.
Minh, Y.C.: 1997, in E.F. van Dishoeck (ed.), *Molecules in Astrophysics: Probes and Processes*, Dordrecht: Kluwer, pp. 173-182.
Neufeld, D.A., Zmuidzinas, J., Schilke, P., and Phillips, T.G.: 1997, *Astrophys. J.* **488**, L141-L144.
Nummelin, A., Dickens, J.E., Bergman, P., Hjalmarson, Å., Irvine, W.M., Ikeda, M., and Ohishi, M.: 1998, *Astron. Astrophys.* **337**, 275-286.
Ohishi, M., Kawaguchi, K., Kaifu, N., Irvine, W.M., Minh, Y.C., Yamamoto, S., and Saito, S.: 1991, in A.D.Haschick and P.T.P. Ho (eds.), *Atoms, Ions, and Molecules: New Results in Spectral Line Astrophysics*, Astron. Soc. Pacific Conf. Ser. **16**, pp. 387-392.
Ohishi, M., Irvine, W.M., and Kaifu, N.: 1992, in P.D. Singh (ed.), *Astrochemistry of Cosmic Phenomena*, Dordrecht: Kluwer, pp. 171-177.
Ohishi, M. and Kaifu, N.: 1998, *Faraday Disc.* **109**, 205-216.
Pratap, P., Dickens, J.E., Snell, R.L., Miralles, M.P., Bergin, E.A., Irvine, W.M., and Schloerb, F.P.: 1997, *Astrophys. J.* **486**, 862-885.
Rodgers, S.D., and Millar, T.J.: 1996, *Monthly Notices Roy. Astron. Soc.* **280**, 1046-1054.
Rodgers, S.D., and Charnley, S.B.: 1998, *Astrophys. J.* **501**, L227-L230.

Schilke, P., Walmsley, C.M., Pineau des Forets, G., Roueff, E., Flower, D.R., and Guilloteau, S.: 1992, *Astron. Astrophys.* **256**, 595-612.
Sutton, E.C., Peng, R., Danchi, W.C., Jaminet, P.A., Sandell, G., and Russell, A.P.G.: 1995, *Astrophys. J. Suppl.* **97**, 455-496.
Swade, D.A. and Schloerb, F.P.: 1992, *Astrophys. J.* **392**, 543-550.
Takano, S., Masuda, A., Hirahara, Y., Suzuki, H., Ohishi, M., Ishikawa, S., Kaifu, N., Kasai, Y., Kawaguchi, K., and Wilson, T.L.: 1998, *Astron. Astrophys.* **329**, 1156-1169.
Taylor, D.K., and Dickman, R.L.: 1989, *Astrophys. J.* **341**, 293-298.
Tielens, A.G.G.M.: 1983, *Astron. Astrophys.* **119**, 177-184.
Ungerechts, H., Bergin, E.A., Goldsmith, P.F., Irvine, W.M., Schloerb, F.P., and Snell, R.L.: 1997, *Astrophys .J.* **482**, 245-266.
Van Dishoeck, E.F., Blake, G.A., Draine, B.T., and Lunine, J.I.: 1993, in E.H. Levy and J.I.Lunine (eds.), *Protostars and Planets* **III**, Tucson: Univ. Arizona Press, pp. 163-241.
Van Dishoeck, E.F., and Blake, G.A.: 1998, *Ann. Revs. Ast. Astrophys.* **36**, 317-368.
Van Dishoeck, E.F.: 1998a, *Faraday Disc.* **109**, 31-46 and 496-497.
Van Dishoeck, E.F.: 1998b, in T.W. Hartquist and D.A. Williams (eds.), *The Molecular Astrophysics of Stars and Galaxies — A Volume Honouring Alexander Dalgarno*, Oxford: Oxford University Press, in press.
Walter, F., Alcala, J., Neuhauser, R., Sterzik, M., and Wolk, S.: 1999, in V. Mannings, A. Boss and S. Russell (eds.), *Protostars and Planets* **IV**, Tucson: Univ. Arizona Press, in press.
Watson,W.D.: 1980, in B.H. Andrew (ed.), *Interstellar Molecules*, Dordrecht: Reidel, pp. 341-353.
Whittet, D.C.B.: 1997, *Origins Life Evol. Biosphere* **27**, 101-113.
Wielen, R., Fuchs, B., and Dettbarn, C.: 1996, *Astron. Astrophys.* **314**, 438-447.
Wielen, R. and Wilson, T.L.: 1997, *Astron. Astrophys.* **326**, 139-142.
Zinner, E.: 1997, in T. Bernatowicz and E. Zinner (eds.), *Astrophysical Implications of the Laboratory Study of Presolar Materials*, New York: American Inst. Physics, pp. 3-26.

Address for correspondence: W.M. Irvine, irvine@fcrao1.phast.umass.edu

EVOLUTION OF INTERSTELLAR ICES

LOUIS J. ALLAMANDOLA, MAX P. BERNSTEIN, SCOTT A. SANDFORD and
ROBERT L. WALKER

Astrochemistry Laboratory, NASA Ames Research Center, MS 245-6, Mountain View, CA 94035-1000, USA

Abstract. Infrared observations, combined with realistic laboratory simulations, have revolutionized our understanding of interstellar ice and dust, the building blocks of comets. Ices in molecular clouds are dominated by the very simple molecules H_2O, CH_3OH, NH_3, CO, CO_2, and probably H_2CO and H_2. More complex species including nitriles, ketones, and esters are also present, but at lower concentrations. The evidence for these, as well as the abundant, carbon-rich, interstellar, polycyclic aromatic hydrocarbons (PAHs) is reviewed. Other possible contributors to the interstellar/pre-cometary ice composition include accretion of gas-phase molecules and *in situ* photochemical processing. By virtue of their low abundance, accretion of simple gas-phase species is shown to be the least important of the processes considered in determining ice composition. On the other hand, photochemical processing does play an important role in driving dust evolution and the composition of minor species. Ultraviolet photolysis of realistic laboratory analogs readily produces H_2, H_2CO, CO_2, CO, CH_4, HCO, and the moderately complex organic molecules: CH_3CH_2OH (ethanol), HC(=O)NH$_2$ (formamide), CH_3C(=O)NH$_2$ (acetamide), R-CN (nitriles), and hexamethylenetetramine (HMT, $C_6H_{12}N_4$), as well as more complex species including amides, ketones, and polyoxymethylenes (POMs). Inclusion of PAHs in the ices produces many species similar to those found in meteorites including aromatic alcohols, quinones and ethers. Photon assisted PAH-ice deuterium exchange also occurs. All of these species are readily formed and are therefore likely cometary constituents.

1. Interstellar ice

Interstellar gas and dust, including ices and ice-and organic-mantled refractory dust grains, comprise the primary stuff from which the solar system formed. It is important to understand interstellar ice composition and photochemistry since these ices are considered important cometary building blocks. Furthermore, they contain more material and are more chemically complex than the interstellar gas-phase material.

Gaseous species readily condense onto cold, 10 K grains in dense molecular clouds forming ice mantles. Interestingly, mantle composition does not reflect gas phase composition or abundances. Differences in relative sublimation rates and chemical reactivities, complicated by photochemical processing, produce ice compositions strikingly different from the gas. New compounds are formed when reactive gaseous species condense on the grain surfaces and the ices are energetically processed by UV radiation and cosmic rays (cf. Greenberg and Li, 1999, elsewhere

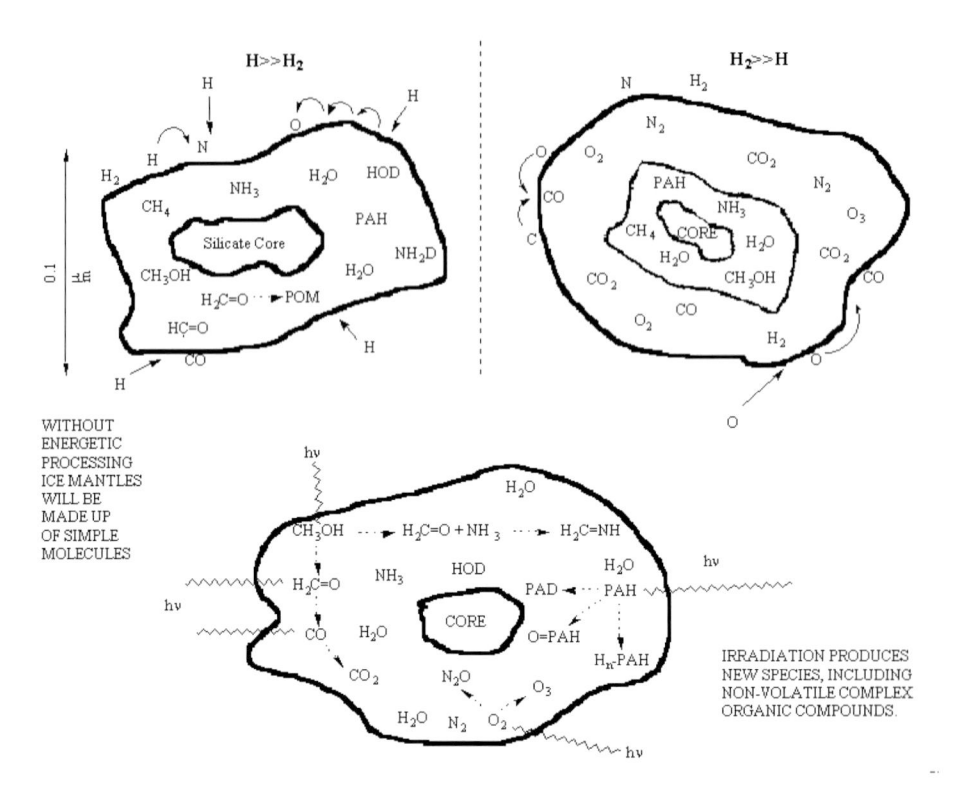

Figure 1. Schematic drawings of the types of ice mantles expected in dense molecular clouds. In regions where the H/H$_2$ ratio is much greater than one (top left), surface reactions tend to be reducing and favor the production of simple hydrides of the cosmicly abundant O, C, and N. In contrast, oxidized forms of these species are favored in regions where this ratio is much less than one (top right). Irradiation and thermal processing (bottom) of these ice mantles creates considerably more complex species and, ultimately, non-volatile residues.

in this book; Greenberg *et al.*, 1972; Tielens and Hagen, 1982; d'Hendecourt *et al.*, 1985; Brown and Charnley, 1990; Moore *et al.*, 1983; Bernstein *et al.*, 1995). Some of these processes are schematically shown in Figure 1.

Since hydrogen is 3 to 4 orders of magnitude more abundant than the next most abundant reactive heavier elements such as C, N, and O, overall grain surface chemistry is moderated by the H/H$_2$ ratio. In regions where this ratio is large, H atom addition (hydrogenation) dominates and species such as H$_2$O, NH$_3$, and CH$_4$ are expected to be prominent. If the H/H$_2$ ratio is substantially less than one, however, reactive species such as O and N are free to interact with one another forming molecules such as CO, CO$_2$, O$_2$, and N$_2$. Thus, two qualitatively different types of ice mantle are expected to be produced by grain surface reactions, one dominated by polar, H-bonded molecules and the other dominated by non-polar, or only slightly polar, highly unsaturated molecules. Figure 1 also shows the first generation of products one might expect upon photolysis (UV irradiation) of these mantles. This picture of interstellar grain mantle formation and evolution is supported by the observational evidence summarized below.

Interstellar ice compositions are revealed through their infrared (IR) spectra. A star situated in or behind a molecular cloud can generate a continuous IR emission spectrum. As this radiation passes through the cloud, the intervening molecules in the gas and dust absorb at their characteristic frequencies. Since ice features tend to dominate such spectra, interstellar ice composition can be readily analyzed by making spectral comparisons with ices prepared in a laboratory under realistic interstellar conditions. Such simulations are carried out at NASA Ames and elsewhere in high-vacuum chambers where thin layers of mixed- molecular ices, comparable in thickness to those of the mantles on interstellar grains ($\sim$0.05 μm), are frozen on cold substrates. In a typical experiment, the spectrum of the sample is measured before and after several periods of exposure to UV radiation and thermal cycling. These spectra are then compared directly with interstellar spectra in order to identify ice composition, determine molecular abundances, and probe ice evolution. Detailed descriptions of the approach can be found in Allamandola and Sandford (1988), and Bernstein *et al.* (1995).

Figure 2 shows a comparison between the spectrum of W33A, a protostar embedded in a molecular cloud (Willner, 1977; Capps *et al.*, 1978) with the laboratory spectra of interstellar ice analogs. The excellent matches between the interstellar absorption features with laboratory absorption spectra, as illustrated in Figure 2, represent the basis of our knowledge of interstellar ices. Interestingly, until quite recently, more was known of interstellar ice composition-grains hundreds of light years away-than of cometary ices in our own Solar System! A brief summary of the major components follows.

H_2O (water) - H_2O is the dominant ice component in dense clouds. At present, five interstellar features have been detected which fit reasonably well with laboratory H_2O ice spectra (c.f. Leger *et al.*, 1983; Tielens *et al.*, 1984; Smith *et al.*, 1989, 1993; Omont *et al.*, 1990). The 3280 cm^{-1}(3.07μm) band is typically one of the strongest in the interstellar spectra.

CH_3OH (methanol) - Molecular cloud spectra often contain a prominent absorption near 1460 cm^{-1}(6.85μm) as shown in Figure 2. It was suggested early on that much of this absorption might be due to the CH deformation mode vibration of methanol (Hagen *et al.*, 1980; Tielens *et al.*, 1984). Although supported by laboratory studies, the unequivocal identification of methanol required the detection of its other bands (Grim *et al.*, 1991; Allamandola *et al.*, 1992). CH_3OH abundances deduced from all of the features taken together suggest that other species contribute to the interstellar absorption around 1460 cm^{-1}. Aliphatic organic compounds and carbonates are reasonable candidates (cf. Allamandola and Sandford, 1988).Gas

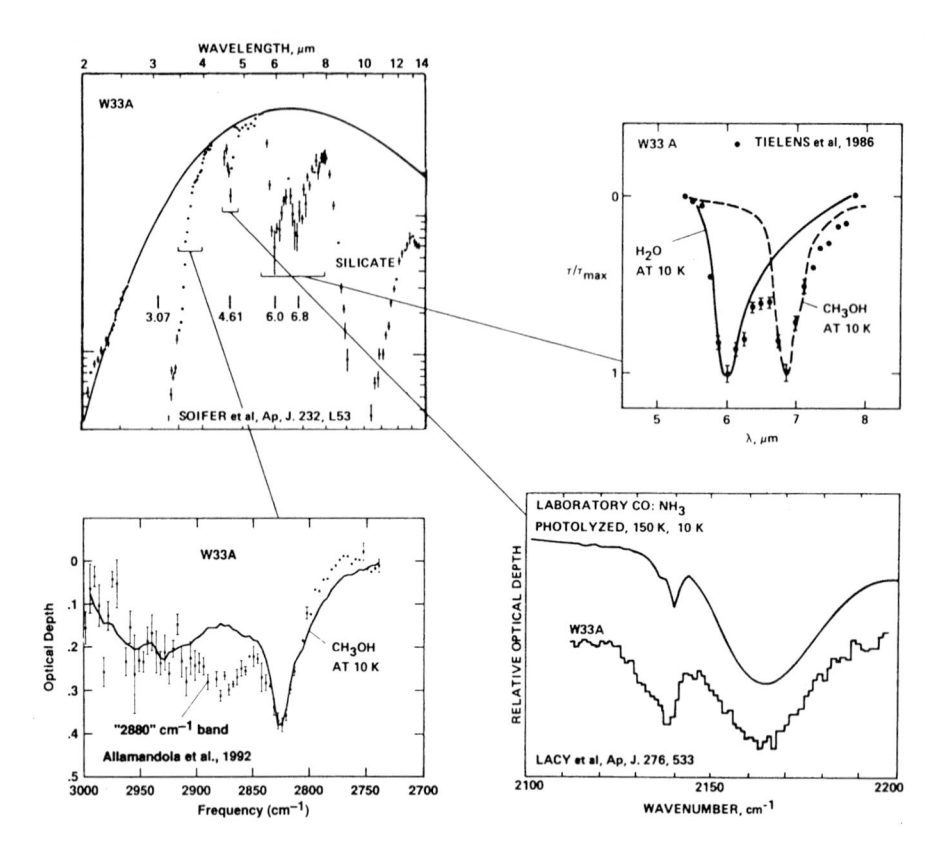

Figure 2. Comparisons of laboratory analog spectra with spectra from the object W33A, a protostar deeply embedded in a dense molecular cloud. Upper left- The dots trace out the interstellar spectrum and the solid line corresponds to the quasi-blackbody emission spectrum thought to be produced by the protostar. The strong absorption near 10 μm is due to the silicate grains in the cloud, and the excess absorption labeled "2880" is thought to arise from interstellar microdiamonds. The other absorptions are produced by interstellar ices. These bands are presented on expanded scales and compared to laboratory spectra in the surrounding frames. Lower left- The solid line is due to methanol in a laboratory ice. Lower right- The upper smooth line corresponds to a laboratory analog comprised of CO (sharp band) and XCN (broad band). Upper right- The solid and dashed lines correspond to spectra of H_2O and CH_3OH respectively.

phase methanol enhancements have been found in star and planet forming regions of dense clouds where the CH_3OH is thought to be liberated from warming ices.

CO (carbon monoxide) - After H_2O, the most studied interstellar ice component is probably carbon monoxide. CO has a characteristic absorption feature near 2140 cm^{-1} (4.67μm), as shown in Figure 2. Its position, width, and profile are a sensitive function of the ice matrix in which the CO is frozen (Sandford *et al.*, 1988; Elsila *et al.*, 1997). Many, but not all, of the lines-of-sight that contain H_2O

ice also contain CO ice and the relative strengths of the H_2O and CO bands indicate CO/H_2O ratios ranging from 0.0 to as much as 0.3 (Lacy *et al.*, 1984; Tielens *et al.*, 1991; Chiar *et al.*, 1994, 1995, 1998). Although a few of the CO bands have positions and profiles consistent with CO frozen in H_2O-rich matrices, most lines-of-sight exhibit profiles indicative of CO frozen in both non-polar matrices, i.e., ices thought to be dominated by molecules such as CO, CO_2, O_2, N_2, and CH_4; and polar, H_2O dominated matrices. These are the two sorts of mantle predicted on the basis of the H/H_2 ratio discussed earlier and sketched in Figure 1. The fact that H_2O, a highly polar molecule, is the most abundant molecule in the ices along *all* these lines-of-sight, but that the CO is generally in ices dominated by *non-polar* molecules provides clear evidence for the existence of multiple chemical environments within individual clouds. Elsila *et al.* (1997) and Ehrenfreund *et al.* (1996, 1997) have shown that these non-polar ices are likely to be dominated by N_2, O_2, and CO_2. In the more quiescent regions as much as 40% of the cosmic N and 25% of the cosmic O could be in the form of frozen N_2 and O_2. If comets indeed contain unmodified interstellar ices, these highly volatile species could drive cometary activity at large heliocentric distances.

"**XCN**" - The spectra of a limited number of lines-of-sight through dense clouds contain a broad, often weak, feature near 2165 cm^{-1}(4.62μm) (Figure 2; Lacy *et al.*, 1984; Tegler *et al.*, 1993, 1995). While the statistics are currently poor, there is an indication that this feature is present, or at least prominent, only in the spectra of objects embedded within clouds which are associated with protostellar and protoplanetary regions and not in the spectra of background stars (Tegler *et al.*, 1995). This suggests that, as with methanol, the band carrier is associated with the local environment of the embedded star. Laboratory experiments show that the interstellar band can be reproduced by photolyzing ice mixtures containing C and N (Lacy *et al.*, 1984; Tegler *et al.*, 1993; Bernstein *et al.*, 1995). This, plus the frequency, implicate the C≡N functional group in a larger molecule, and suggest that some form of energetic processing is needed to produce it in the interstellar medium. Isotope and ion trapping experiments make a good case for assigning this feature to OCN^- (Grim and Greenberg, 1987a; Schutte and Greenberg, 1997; Demyk *et al.*, 1998).

H_2 (hydrogen) - Molecular hydrogen (H_2) may have been detected along the line-of-sight to WL5 in the ρ Ophiucus cloud (Sandford *et al.*, 1993). This is a deeply embedded object which produces gas-phase CO bands which overlap the possible H_2 band. Confirmation awaits high resolution spectroscopy. In any event, laboratory studies of interstellar ice analogs show that ion and UV irradiation produce, and efficiently trap, frozen H_2 (Moore and Hudson, 1992; Sandford and Allamandola, 1993b). This process can be very efficient, producing H_2 abundances up to one third of the H_2O. Thus, hydrogen could have a high abundance in interstellar and cometary ices.

TABLE I

Composition and abundances of interstellar ice (relative to H_2O) compared to that deduced in Comets Halley, Hyakutake, and Hale-Bopp. The species listed above HCO in Column Two have been definitely detected, those below have been tentatively identified. The evidence is good for all of these species. Cometary abundances are from: Halley - Altwegg et al. (1999); Hyakutake and Hale-Bopp - Crovisier and Bockelée-Morvan (1999). See text for interstellar ice references.

MOLECULE	INTERSTELLAR ICE Abundance	COMET PARENT MOLECULES		
		Halley	Hyakutake	Hale-Bopp
H_2O	100	100	100	100
CO (polar ice)	1-10			
CO (non-polar ice)	10-40	17	6-30	20
CH_3OH	<4-10	1.25	2	2
CO_2	1-10	3.5	< 7	6
XCN	1-10	—	—	—
NH_3	5-10	1.5	0.5	0.7
H_2	~1	—	—	—
CH_4	~1	< 0.8	0.7	0.6
HCO	~1	—	—	—
H_2CO	1-4	3.8	0.2-1	1
N_2	10-40	—	—	—
O_2	10-40	—	—	—
OCS or CO_3	few	0.2	0.1	0.3

Other Species - Based on limited telescopic observations, laboratory studies of ice analogs, and theoretical chemistry models, a number of other molecular species are suspected of being present in interstellar ices in quantities on the order of a few percent relative to H_2O. Until recently, the following species have been tentatively identified in small numbers of objects: CH_4 (methane, Lacy et al., 1991), CO_2 (carbon dioxide, d'Hendecourt and Jordain de Muizon, 1989) HCO and H_2CO (formyl radical and formaldehyde, cf. Tielens and Allamandola, 1987; Schutte et al., 1996), OCS (carbonyl sulfide, Palumbo et al., 1995) or CO_3 (carbon trioxide, Elsila et al., 1997), N_2 and O_2 (Elsila et al., 1997; Ehrenfreund et al., 1996, 1997), and possibly ketones and/or aldehydes (Tielens and Allamandola, 1987). The diatomic species S_2, of relevance to comets as it has been detected in a cometary coma, is also readily made upon irradiation of an interstellar ice analog (Grim and Greenberg, 1987b).

Interstellar NH_3 ice has now been definitely detected, with a concentration between 5 and 10% that of the water (Lacy et al., 1998). Recent IR spectral observations by the ISO satellite have shown that frozen CO_2 is ubiquitous in dense clouds, having an abundance varying between roughly 10 and 15% of the H_2O, and that CH_4 is also a common interstellar ice component, typically at the few percent level (DeGrauw *et al.*, 1996; Boogert *et al.*, 1997). Over the next few years, ISO spectra will certainly deepen our understanding of interstellar ice. These interstellar ice constituents and their average abundances are compared to those of comets in Table I.

2. Gas Phase Accretion and Gas-grain Chemistry

Of course, any gas phase species observed in dense clouds by radio (or infrared) techniques should also be present in the grain mantles (e.g. Irvine, 1999; Winnewisser and Kramer, 1999). At the low temperatures characteristic of these environments, all molecules should be strongly depleted by condensation onto the ice (cf. Sandford and Allamandola, 1993b). This includes the polycyclic aromatic hydrocarbons (PAHs) which are known, through their infrared emission, to be ubiquitous and abundant gas phase species throughout the interstellar medium (cf. Allamandola *et al.*, 1989; Puget and Leger, 1989; Brook *et al.*, 1999). However, as shown in Table II, the much lower abundances of the specific gas-phase molecules known from radio observations require that direct accretion of most complex gas-phase species plays a very minor role in determining interstellar ice composition. This relationship also holds for the radio-quiet CH_4 as shown by recent ISO observations (Boogert *et al.*, 1997).

As mentioned earlier, the solid CO spectral characteristics along lines-of-sight which probe quiescent portions of molecular clouds can be quite distinct from those closely associated with embedded protostars. The narrow CO band that is characteristic of non-polar ices seems to be associated with lines-of-sight which sample the colder, quiescent regions of dense clouds.

This is consistent with the picture that the ices in the protostellar environment have probably undergone substantially more radiative processing and are somewhat warmer (i.e., above 30 K) than those ices along lines of sight which sample quiescent regions. Since the protostellar environment is also the formation site of new planetary systems, the interstellar/precometary ice composition in these regions is of particular relevance to comets. These are the regions in which the XCN and CH_3OH bands are prominent, suggesting that these materials are associated more with the radiation-rich and somewhat warmer environment of the protostar than with the molecular cloud itself. It is now clear that CH_3OH is often the second or third most abundant component in these and cometary ices (cf. Reuter, 1992; Allamandola *et al.*, 1992). The high abundance of methanol is of key importance since its presence drives a rich interstellar photochemistry (Allamandola *et al.*, 1988; Bernstein *et al.*, 1995) and gas phase chemistry (Charnley *et al.*, 1992). Furthermore, since methanol has profound effects on the physical behavior of H_2O-rich ices, this may have important implications for their structural (Blake

TABLE II

Comparison between the gas-phase and solid-state abundances for several molecular species normalized to hydrogen. This comparison shows that the interstellar ices contain the bulk of interstellar polyatomic molecules.

	MOLECULAR ABUNDANCE WITH RESPECT TO HYDROGEN			
Molecule	Gas phase TMC-1	Gas phase OMC-1	Ice NGC 7538 IR9	Ice/Gas Ratio
CO	8×10^{-5}	5×10^{-5}	6×10^{-6}	0.12
H_2O	—	—	6×10^{-5}	?
CH_3OH	2×10^{-9}	1×10^{-7}	6×10^{-6}	60-3000
NH_3	2×10^{-8}	—	6×10^{-6}	300
CO_2	5×10^{-8}	—	8×10^{-6}	160
CH_4	—	—	6×10^{-7}	?

Gas Phase Values: TMC-1 – Ohishi *et al.* (1992); OMC-1 – Blake *et al.* (1987); CO_2 – van Dishoek *et al.* (1996)
Ice mantles: Allamandola *et al.* (1992); CO_2 – DeGrauw *et al.* (1996); CH_4 – Boogert *et al.* (1997)

et al., 1991; Ehrenfreund *et al.*, 1998) and vaporization behavior (Sandford and Allamandola, 1993b).

Thorough recent reviews on interstellar ices can be found in Sandford *et al.* (1996) and Schutte (1996).

3. Interstellar Ice Evolution

As discussed earlier, the ices in dense molecular clouds are irradiated by UV photons and cosmic rays, breaking and rearranging chemical bonds within the ice. This is an important process since it can create complex molecular species that cannot be made via gas phase and gas-grain reactions at the low temperatures and pressures characteristic of dense clouds. The reason for this is that solid-phase reaction kinetics, stoichiometry, and energetics favor complexity and chemical diversity.

As an example, for the last three pages we consider the photochemical evolution of an interstellar ice analog comprised of $H_2O:CH_3OH:CO:NH_3$ (100:50:10:10). Exposure to UV destroys several species (particularly methanol) and creates others such as: HCO, H_2CO, CH_4, CO_2, XCN, etc. (Allamandola *et al.*, 1988; Bernstein *et al.*, 1995), all of which have been identified in interstellar ices (Table I). Presently, the strongest evidence that radiation processing is important, at least in some locations within dense clouds, is provided by the "XCN" feature (Figure 2), which

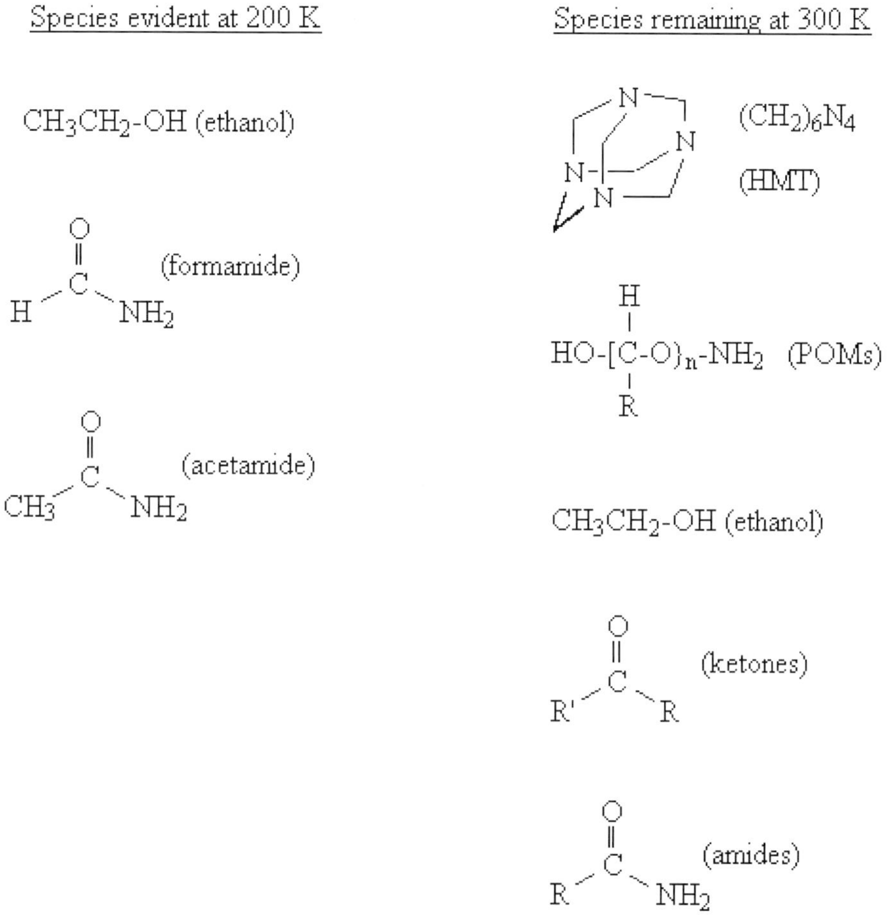

Species evident at 200 K

CH$_3$CH$_2$-OH (ethanol)

formamide: H-C(=O)-NH$_2$

acetamide: CH$_3$-C(=O)-NH$_2$

Species remaining at 300 K

(CH$_2$)$_6$N$_4$ (HMT)

HO-[C(H)(R)-O]$_n$-NH$_2$ (POMs)

CH$_3$CH$_2$-OH (ethanol)

R'-C(=O)-R (ketones)

R-C(=O)-NH$_2$ (amides)

Figure 3. Compounds produced by the 10 K UV photolysis of the realistic interstellar ice analogs H$_2$O:CH$_3$OH:NH$_3$:CO (100:50:10:10) and (100:50:1:1). Figure adapted Bernstein *et al.* (1995).

cannot be explained by any of the more abundant species predicted by gas and gas-grain chemical models, but which is readily made by the radiative processing of laboratory ices containing both C and N. For detailed descriptions of the UV driven chemical evolution of ice, see Briggs *et al.* (1992), Bernstein *et al.* (1995), Gerakines *et al.* (1996) and references therein. Ultraviolet processing of the non-polar ices of quiescent cloud regions produces CO$_2$, N$_2$O, O$_3$, CO$_3$, HCO, H$_2$CO, and possibly NO and NO$_2$ (Elsila *et al.*, 1997).

The same ice irradiation processes that produce the simpler molecules such as CO, CO_2, HCO, H_2CO, CH_4, etc., also produce far more complex species. The parent species and more volatile photoproducts sublime upon warm up to about 200 K under vacuum, leaving a mixture of less volatile species behind. The species identified in the material remaining on the window at 200 K and at room temperature are shown in Figure 3. These species, which result from the irradiation of laboratory ices which mimic the polar ices associated with protostellar environments differ from those produced by irradiating realistic analogs of the ices in the quiescent regions. Clear-cut evidence for these compounds in interstellar clouds is presently lacking, although some of the spectral structure detected in the 2000-1250 cm^{-1} region is consistent with their presence.

The residue which remains on the window at room temperature is of particular interest to the cometary community. This organic residue is rich in the cage molecule Hexamethylenetetramine (HMT, $C_6H_{12}N_4$), with lesser amounts of polyoxymethylene-related species (POMs), amides, and ketones (Bernstein et al., 1995). [As an aside, based on Bernstein et al.'s suggested HMT synthesis route which prominently invokes the reactive intermediate methyleneimine, Dickens et al. (1997) searched for this species and found it to be widespread throughout the ISM.] The dominance of HMT in photolyzed methanol-rich residues is in sharp contrast to the organic residues produced by irradiating ices which do not contain methanol (Agarwal et al., 1985), or the family of organic molecules produced in thermally promoted polymerization-type reactions in unirradiated realistic ice mixtures (Schutte et al., 1993). In non-CH_3OH containing ices, HMT is only a minor product in a residue dominated by a mixture of POM related species. POM-like species have been suggested as an important organic component detected in the coma of Comet Halley (Huebner et al., 1989).

Lastly, given that PAHs are known to be widespread and abundant throughout the interstellar gas, we have studied the photochemistry of PAH-containing ices. With an abundance of 10^{-7} with respect to H, PAHs are more abundant than all of the gas- phase polyatomic molecules combined (e.g. Table II). While the UV photolysis of PAHs in interstellar ices modifies only a small fraction of the interstellar PAH population, this change is significant. As shown below, the principle PAH reaction pathways induced by PAH photolysis in H_2O ices are: hydrogenation which produces H_n-PAHs; oxidation which produces ketones (quinones), alcohols, and ethers; and deuteration when D_2O is present in the ice (Bernstein et al., 1996, 1999; Sandford et al., 1999).

These alteration processes have dramatically different effects on the chemical nature of the parent. Hydrogen-atom addition transforms some of the edge rings into cyclic-aliphatic hydrocarbons, thereby creating molecules with both aromatic and aliphatic character and decreasing the overall degree of aromaticity. Ketone or aldehyde formation opens up an entire new range of possible chemical reactions that were impossible in the parent PAH.

These results are also important for the cometary community. First, PAH photochemistry in interstellar/pre-cometary ices is a potentially important contributor to the richness of Solar System chemistry. A strong connection of this material with the carbonaceous fractions of meteorites (and by implication comets) may well exist. For example, complex organic molecules similar to those produced here have been identified in meteorites (Hahn *et al.*, 1988; Cronin and Chang, 1993) and oxidized polycyclic aromatic hydrocarbons are present in organic extracts from the Murchison meteorite (Deamer, 1997). Furthermore, photon-assisted PAH deuteration may bear on the deuterium enrichments of PAHs seen in meteorites (Kerridge *et al.*, 1987) and in IDPs (Clemett *et al.*, 1993; McKeegan *et al.*, 1985). The formation histories of these extraterrestrial materials are not well understood, although the presence of deuterium enrichments in many of the classes of these compounds has been taken to implicate an interstellar origin (Kerridge *et al.*, 1987; Cronin *et al.*, 1993). Thus, the photolytic processes discussed here may well have contributed significantly to this meteoritic inventory.

4. Conclusion

Infrared observations, combined with realistic laboratory simulations, have revolutionized our understanding of interstellar ice and dust. These materials are of particular relevance here since these are the building blocks of comets and also because comets are thought to have brought most of the volatiles and complex organics to the early Earth. Interestingly, the similarity between the interstellar ice constituents and relative abundances with the list of known cometary constituents is remarkable. This strong similarity strengthens the case for applying interstellar ice studies directly to the study of comets.

Ice composition in molecular clouds depends on local conditions. In areas associated with star, planet, and comet formation, ices comprised of polar species and entrapped volatiles such as H_2O, CH_3OH, CO, CO_2, NH_3, XCN (OCN^-), H_2, and H_2CO are most important. In quiescent regions, the ices are dominated by non-polar, volatile species such as O_2, N_2, CO, and CO_2. Non-polar interstellar ices are far less important in the warm (by interstellar standards) star-formation regions due to the high volatility of the individual components. Photolysis of the polar, H_2O-rich ices produces H_2, H_2CO, CO_2, CO, CH_4, HCO, and the moderately complex organic molecules: CH_3CH_2OH (ethanol), $HC(=O)NH_2$ (formamide), $CH_3C(=O)NH_2$ (acetamide), R-CN and R-NC (nitriles and isonitriles), and hexamethylenetetramine (HMT, $C_6H_{12}N_4$), as well as more complex species including polyoxymethylene and related species (POMs), amides, and ketones. Inclusion of PAHs (also known to be ubiquitous throughout the interstellar medium) in the irradiated ices produces many species similar to those found in meteorites including aromatic alcohols, quinones and ethers. Photon assisted PAH-ice deuterium exchange also occurs. All of the above species are readily formed and thus likely cometary constituents at the 0.1 to 1 percent level.

References

Agarwal, V. K., Schutte, W., Greenberg, J. M., Ferris, J. P., Briggs, R., Connor, S., van de Bult, C. P. E. M., and Baas, F.: 1985, *Origins of Life* **16**, 21-40.
Allamandola, L. J., and Sandford, S. A.: 1988, In *Dust in the Universe* (eds. M. E. Bailey and D. A. Williams), pp. 229-263. Cambridge Univ. Press, Cambridge.
Allamandola, L. J., Sandford, S. A. and Valero, G.: 1988, *Icarus* **76**, 225-252.
Allamandola, L. J., Tielens, A. G. G. M., and Barker, J. R.: 1989, *Astrophys. J. Suppl.* **71**, 733-755.
Allamandola, L. J., Sandford, S. A., Tielens, A. G. G. M., and Herbst, T. M.: 1992, *Astrophys. J.* **399**, 134-146.
Altwegg, K., Balsiger, H. and Geiss, J.: 1999, *Space Sci. Rev.*, this volume.
Bernstein, M.P., Sandford, S.A., Allamandola, L.J., Chang, S. and Scharberg, M.A.: 1995, *Astrophys. J.* **454**, 327-344.
Bernstein, M.P., Sandford, S. A., and Allamandola, L. J.: 1996, *Astrophys. J.* **472**, L127
Bernstein, M.P., Sandford, S.A., Allamandola, L.J., Gillette, J.S., Clemett, S.J., and Zare, R.N.: 1999, *Science* **283**, 1135.
Blake, G.A., Sutton, E.C., Masson, C.R., and Phillips, T.G.: 1987, *Astrophys. J.* **315**, 621.
Blake, D., Allamandola, L., Sandford, S., Hudgins, D. and Freund, F.: 1991, *Science* **254**, 548-551.
Boogert, A.C.A., Schutte, W. A., Helmich F.P., Tielens A.G.G.M., and Wooden, D.H.: 1997, *Astron. Astrophys.* **317**, 929,
Briggs, R., Ertem G., Ferris, J.P., Greenberg, J.M., McCain, P.J., Mendoza-Gomez, C.X., and Schutte, W.: 1992, *Origins of Life and Evolution of the Biosphere* **22**, 287- 307
Brook, T. Y., Sellgren, K., and Geballe, T.R.: 1999, *Astrophys. J.* in press
Brown, P. D. and Charnley, S. B.: 1990, *Mon. Not. Roy. Astron. Soc.* **244**, 432-443.
Capps, R.W., Gillett, F.C. and Knacke, R.F.: 1978, *Astrophys. J.* **226**, 863-868.
Charnley, S.B., Tielens, A. G. G. M. and Millar, T. J.: 1992, *Astrophys. J.* **399**, L71-L74.

Chiar, J.E., Adamson, A.J., Kerr, T.H., and Whittet, D.C.B.: 1994, *Astrophys. J.* **426**, 240.
Chiar, J.E., Adamson, A.J., Kerr, T.H. and Whittet, D. C. B.: 1995, *Astrophys. J.* **455**, 234.
Chiar, J.E., Gerakines, P.A., Whittet, D. C. B., Pendleton, Y.J., Tielens, A.G.G.M., Adamson and Boogert, A.C.A.,: 1998, *Astrophys. J.* **498**, 716.
Clemett, S., Maechling, C., Zare, R., Swan, P., and Walker, R.: 1993, *Science* **262**, 721.
Cronin, J. R., Pizzarello, S., Epstein, S., and Krishnamurthy, R. V.: 1993, *Geochim. Cosmochim. Acta* **57**, 4745.
Cronin, J.R. and Chang, S.: 1993, in *The Chemistry of Life's Origins*, eds. J. M. Greenberg, C. X. Mendoza-Gmez, & V. Pirronello, (Kluwer Academic Publishers: Netherlands) p. 209
Crovisier, J. and Bockelée-Morvan, D.: 1999, *Space Sci. Rev.*, this volume.
Deamer, D. W.: 1997, *Microbiology and Molecular Biology Reviews* **61**, 239.
DeGrauw, *et al.*: 1996, *Astron. Astrophys.* **315**, L345.
Demyk, K., Dartois, E., d'Hendecourt, L., Jourdain de Muizon, M., Heras, A., Breitfellner, M.: 1998, *Astron.Astrophys.* **339**, 553.
d'Hendecourt, L. B., Allamandola, L. J. and Greenberg, J. M.: 1985, *Astron. Astrophys.* **152**, 130.
d'Hendecourt, L. B. and Jordain de Muizon, M.: 1989, *Astron. Astrophys.* **223**, L5-L8.
Dickens, J.E., Irvine, W.M., DeVries, C.H., and Ohishi, M.: 1997, *Astrophys. J.* **479**, 307.
Ehrenfreund, P., Boogert, A.C.A., Gerakines, P.A., Schutte, W.A., Tielens, A.G.G.M., and van Dishoek, E.F.: 1996, *Astron. Astrophys.* **315**, L341.
Ehrenfreund, P., Boogert, A.C.A., Gerakines, P.A., Tielens, A.G.G.M., and van Dishoek, E.F.: 1997, *Astron. Astrophys.* **328**, 649.
Ehrenfreund, P., Dartois, E., Demyk, K., and d'Hendecourt, L.B.: 1998, *Astron. & Astrophys.* **339**, L17.
Elsila, J., Allamandola, L.J., and Sandford, S.A.: 1997, *Astrophys. J.***479**, 818-838.
Gerakines, P. A., Schutte, W.A., and Ehrenfreund, P., 1996: *Astron. Astrophys.* **312**, 289.
Greenberg, J. M. *et al.*: 1972, *Mem Soc Roy Sci. de Liege 6e* serie 3.
Greenberg, J. M.: 1978, In *Cosmic Dust* (ed. J. A. M. McDonnell), pp. 187-294. J. Wiley, New York, NY.
Greenberg, J. M. and Li, A.: 1999, *Space Sci. Rev.*, this volume.
Grim, R.J.A., and Greenberg, J.M.: 1987a, *Astrophys. J. Lett.* **321**, L91-L96.
Grim, R.J.A., and Greenberg, J.M.: 1987b, *Astron Astrophys.* **181**, 155-168.
Grim, R.J.A., Baas, F., Geballe, T.R., Greenberg, J.M., and Schutte, W.A.: 1991, *Astron. Astrophys.* **243**, 473-477.
Hagen, W., Allamandola, L. J. and Greenberg, J. M.: 1980, *Astron. Astrophys.* **86**, L3-L6.
Hahn, J. H., Zenobi, R., Bada, J. L., and Zare, R. N.: 1988, *Science* **239**, 1523.
Huebner, W.F., Boice, D.C., and Korth, A.: 1989, *Adv. Space Res.* **9**, 29.
Irvine, W. M.: 1999, *Space Sci. Rev.*, this volume.
Kerridge, J. F., Chang, S., and Shipp, R., 1987: *Geochim. Cosmochim. Acta* **51**, 2527.
Lacy, J. H., Baas, F., Allamandola, L. J., Persson, S. E., McGregor, P. J., Lonsdale, C. J., Geballe, T. R. and van der Bult, C. E. P.: 1984, *Astrophys. J.* **276**, 533-543.
Lacy, J., Carr, J., Evans, N., Baas, F., Achtermann, J. and Arens, J.: 1991, *Astrophys. J.* **376**, 556-560.
Lacy, J., Faraji, H., Sandford, S.A., and Allamandola, L.J.: 1998, *Astophys. J.* **501**, L105.
Leger, A., Gauthier, S., Defourneau, D. and Rouan, D.: 1983, *Astron. Astrophys.* **117**, 164-169.
McKeegan, K. D., Walker, R. M., and Zinner, E.: 1985, *Geochim. Cosmochim. Acta* **49**, 1971.
Moore, M. H., Donn, B., Khanna, R. and A'Hearn, M. F.: 1983, *Icarus* **54**, 388-405.
Moore, M. H. and Hudson, R.: 1992, *Astrophys. J.* **401**, 353-360.
Mumma, M.J., Stern, S.A., and Weissman, P.R.: 1993, in *Protostars and Planets* **III**, eds E.H. Levy, J.I. Lunine, , and M.S. Matthews, (University of Arizona Press, Tucson), 1177.
Ohishi, M., Irvine, W. M., and Kaifu, N.: 1992, in *Astrochemistry of Cosmic Phenomena*, ed. P.D. Singh, (Kluwer, Dordrecht), 171.

Omont, A., Moseley, S. H., Forveille, T., Glaccum, W. J., Harvey, P. M., Likkel, L., Loewenstein, R. F., and Lisse, C. M.: 1990, *Astrophys. J.* **355**, L27-L30.
Palumbo, M. E., Tielens, A. G. G. M., and Tokunaga, A. T.: 1995, *Astrophys. J.* **449**, 674-680.
Puget, J. L. and Leger, A.: 1989, *Ann. Rev. Astron. Astrophys.* **27**, 161-198.
Reuter, D. C.: 1992, *Astrophys. J.* **386**, 330-335.
Sandford, S. A., Allamandola, L. J., Tielens, A. G. G. M. and Valero, G.: 1988, *Astrophys. J.* **329**, 498-510.
Sandford, S. A., Allamandola, L. J. and Geballe, T. R.: 1993, *Science* **262**, 400-402.
Sandford, S. and Allamandola, L. J.: 1993a, *Astrophys. J. (Letters)* **409**, L65-L68.
Sandford, S. A. and Allamandola, L. J.: 1993b, *Astrophys. J.* **417**, 815-825.
Sandford, S.A.: 1996, *Meteoritics & Planetary Science* **31**, 449-476.
Sandford, S. A., Bernstein, M. P., Allamandola, L. J., Gillette, J. S., and Zare, R. N.: 1999, *Astrophys. J.*, in press.
Schutte, W. A., Allamandola, L. J. and Sandford, S. A.: 1993, *Icarus* **104**, 118-137.
Schutte, W. A.: 1996, In *The Cosmic Dust Connection* (ed. J. M. Greenberg), D. Reidel, Dordrecht, TheNetherlands, p. 1.
Schutte, W.A., Gerakines, P.A., Geballe, T.R., van Dishoek, E.F., and Greenberg, J.M.: 1996, *Astron. Astrophys.* **309**, 633-647.
Schutte, W.A. and Greenberg, J.M.: 1997, *Astron. & Astrophys.* **317**, L43.
Smith, R. G., Sellgren, K. and Tokunaga, A. T.: 1989, *Astrophys. J.* **344**, 413-426.
Smith, R. G., Sellgren, K. and Brooke, T. Y.: 1993, *Mon. Not. Roy. Astron. Soc.* **263**, 749-766.
Tegler, S. C., Weintraub, D. A., Allamandola, L. J., Sandford, S. A., Rettig, T. W. and Campins, H.: 1993, *Astrophys. J.* **411**, 260-265.
Tegler, S. C., Weintraub, D. A., Rettig, T. W., Pendleton, Y. J., Whittet, D. C. B. and Kulesa, C. A.: 1995, *Astrophys. J.* **439**, 279- 287.
Tielens, A. G. G. M. and Hagen, W.: 1982, *Astron. Astrophys.* **114**, 245-260.
Tielens, A. G. G. M., Allamandola, L. J., Bregman, J., Goebel, J., d'Hendecourt, L. B. and Witteborn, F. C.: 1984, *Astrophys. J.* **287**, 697-706.
Tielens, A. G. G. M. and Allamandola, L. J.: 1987, In *Physical Processes in Interstellar Clouds* (eds. G. E. Morfill and M. Scholer), pp. 333-376. D. Reidel, Dordrecht, The Netherlands
Tielens, A. G. G. M., Tokunaga, A. T., Geballe, T. R. and Baas, F.: 1991, *Astrophys. J.* **381**, 181
van Dishoek *et al.*:1996, *Astron. Astrophys.* **315**, L349
Willner, S. P.: 1977, *Astrophys. J.* **214**, 706-711.
Winnewisser, G. and Kramer, C.: 1999, *Space Sci. Rev.*, this volume.

Address for correspondence: Louis J. Allamandola, lallamandola@mail.arc.nasa.gov

AN ISO VIEW ON INTERSTELLAR AND COMETARY ICE CHEMISTRY

PASCALE EHRENFREUND

Leiden Observatory, P O Box 9513, 2300 RA Leiden, The Netherlands

Abstract. The ISO-SWS instrument offering a large wavelength coverage and a resolution well adapted to the solid phase has changed our knowledge of the physical-chemical properties of ices in space. The discovery of many new ice features was reported and the comparison with dedicated laboratory experiments allowed the determination of more accurate abundances of major ice components. The presence of CO_2 ice has recently been confirmed with the SWS (Short Wavelength Spectrometer) as a dominant ice component of interstellar grain mantles. The bending mode of CO_2 ice shows a particular triple-peak structure which provides first evidence for extensive ice segregation in the line-of-sight toward massive protostars. A comparison of interstellar and cometary ices using recent ISO data and ground-based measurements has revealed important similarities but also indicated that comets contain, beside pristine interstellar material, admixtures of processed material. The investigation of molecules in interstellar clouds is essential to reveal the link between dust in the interstellar medium and in the Solar System.

Keywords: Interstellar dust, comets, ices, ISO

1. Introduction

Interstellar ices cover grains in cold dense clouds. Simple ice species, such as H_2O, CO_2, CO, CH_3OH, NH_3, CH_4, OCS and possibly HCOOH and H_2CO are currently detected by ground-based and satellite observations in the infrared (IR) in the interstellar medium (see e.g. Whittet *et al.*, 1996; Lacy *et al.*, 1998; Dartois *et al.*, 1999a). The cold dust around protostars is the primary source for infrared spectroscopy of ices in interstellar space. In dense clouds ices accrete on the grain surface in amorphous form due to the low temperature. The frozen molecules cause characteristic mid-infrared (2.5 - 20 μm) absorption bands in the spectra of sources obscured by dense cloud extinction. The profiles of infrared bands of frozen molecules depend on the structure and composition of the ice. Therefore, comparison with laboratory data, recorded under simulated interstellar conditions, allows us to constrain the grain structure and mantle composition. In the following we highlight the most recent results of interstellar and cometary ices from the Infrared Space Observatory ISO.

2. Recent Observations of the Infrared Space Observatory ISO

2.1. ABUNDANT CO_2 IN INTERSTELLAR ICES

A relatively high abundance of solid CO_2, namely 12-20% compared to H_2O ice, has been recently reported ubiquitously in the interstellar medium (de Graauw et al., 1996; d'Hendecourt et al., 1996; Gürtler et al., 1996; Whittet et al., 1998). The CO_2 bending mode, currently observed toward at least 50 targets, shows in many cases a very peculiar triple-peak structure.

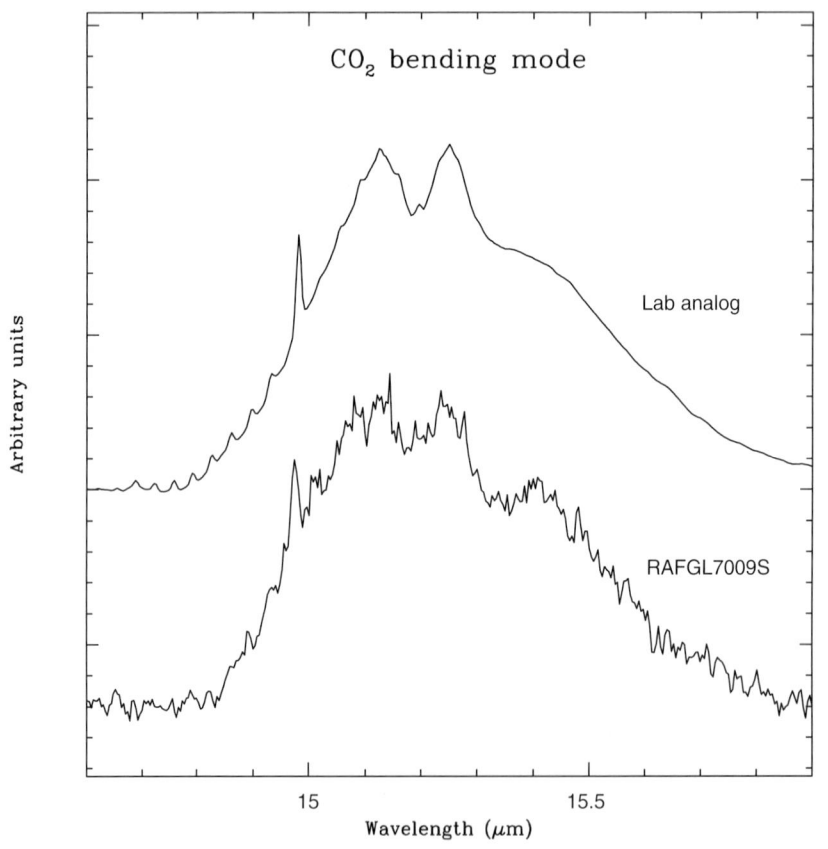

Figure 1. The ISO-SWS06 spectrum of the CO_2 bending mode (at 15.2 μm) toward the massive protostar RAFGL7009S is compared with laboratory data of an ice mixture containing equal amounts of H_2O, CH_3OH and CO_2, heated to 105 K. The sharp peak at 14.98 μm is due to gas phase CO_2. The symmetrical double-peak can be attributed to annealed CO_2 and the shoulder at 15.4 μm to CO_2-CH_3OH complexes (Ehrenfreund et al., 1998a; Dartois et al., 1999b).

This unique profile of the CO_2 bending mode could only be reproduced in the laboratory by thermal annealing and by adding significant amounts of CH_3OH to H_2O-CO_2 mixtures (Ehrenfreund et al., 1998a; 1999). Fig. 1 shows the remarkable comparison of an annealed ice mixture containing CO_2, CH_3OH and H_2O with the ISO-SWS spectrum toward the protostar RAFGL7009S. The shoulder at 15.4 μm appearing during the annealing process has been identified with CO_2-CH_3OH complexes (Ehrenfreund et al., 1998a; Dartois et al., 1999b). This CO_2 profile is observed toward many high mass protostellar objects and suggests extensive ice segregation and partly crystallized ices in such environments.

Comet Hale-Bopp showed CO_2 outgassing at notable heliocentric distance; it was first detected at r_h=4.59 AU. The inferred production rate of 1.3×10^{28} s^{-1} corresponds to 0.3 that of CO. CO_2 was further observed by ISO at 2.9 AU inbound and 3.9 AU post-perihelion with a similar mixing ratio relative to CO (Crovisier et al., 1996, 1997). Hale-Bopp has shown a CO-driven coma far from the Sun and an H_2O driven coma at 3 AU. Since CO_2 followed the trend of CO between 5-3 AU, its release seems not to be controlled by H_2O sublimation (Bockelée-Morvan and Rickman, 1999). This kind of cometary activity is consistent with the notion that the ice is, in fact, segregated.

2.2. MOLECULAR OXYGEN

The principal oxygen-bearing species in diffuse and dense clouds have been subject of considerable discussion (van Dishoeck and Blake, 1998). Recent measurements of the OI band in the UV indicate values in the diffuse interstellar medium of only two-thirds of the solar oxygen abundance (Meyer et al., 1998). At the same time ISO satellite data and ground-based observations allowed to determine accurate abundances of oxygen-bearing species, such as gaseous O, O_2 and H_2O, as well as solid-state species including H_2O, CO_2 and CO ice toward many targets (Whittet et al., 1996).

Using those recent measurements we tried to estimate how much O_2 ice could be found on icy grain mantles in dense clouds. The search for solid O_2 toward the most promising cold clouds with ISO-SWS (and ground-based observations) led to an upper limit of ~5% of the total oxygen abundance in the form of O_2 ice in those regions (Vandenbussche et al., 1999). The recent balloon experiment PIROG 8 searched for the 425 GHz gas-phase O_2 line in NGC 7538 and W51 (Olofsson et al., 1998). No emission could be detected and the inferred upper limits on the O_2/CO ratio are 0.04 and 0.05 (3 σ) for these regions. Very recent observations with the sub-mm satellite SWAS confirmed the low abundance of gaseous O_2 in cold clouds, such as NGC 7538 IRS9. Furthermore, the amount of oxygen found in "cold" H_2O toward such targets is negligible. The low abundances of solid and gaseous O_2 suggest that OI is the dominant form of oxygen in dense clouds. Those results are compatible with upper limits of O_2/H_2O < 0.5 % measured by the Giotto spacecraft for Comet Halley.

2.3. OTHER IMPORTANT ICE COMPONENTS

Other ice species which have important implications for comets have been revisited with ISO. Solid CH_4 was detected with an abundance of ~1-4 % and OCS with

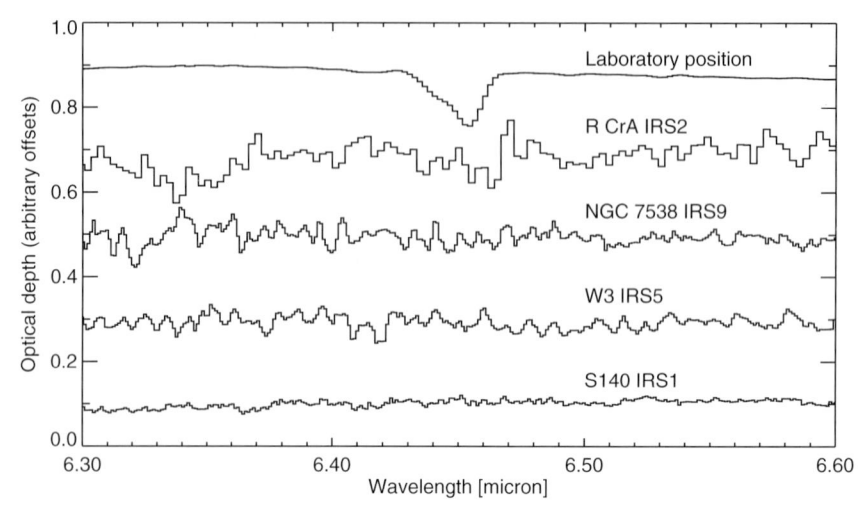

Figure 2. The upper panel indicates the laboratory position of the fundamental transition of molecular O_2 at 6.45 μm. This transition is infrared inactive, but becomes weakly infrared active in the solid state (Ehrenfreund *et al.*, 1992). The search for solid molecular O_2 toward cold clouds with ISO was not successful and led to an upper limit of $\sim 25\%$ solid O_2 relative to H_2O.

an abundance of 0.2 % relative to H_2O ice toward several protostars (Boogert *et al.*, 1996; Ehrenfreund *et al.*, 1997a). Observations of the 6 μm absorption band of NGC7538 IRS9 by ISO-SWS showed that this feature cannot be matched using only the OH bending mode of H_2O ice. A good match was obtained by Schutte *et al.* (1996) by adding the carbonyl C=O stretching mode absorption of organic acids, specifically formic acid. Recently, another, weaker band was found in the SWS spectrum of W 33A at 7.24 μm that corresponds well with the CH deformation mode of solid HCOOH (Schutte *et al.*, 1999). According to these measurements the HCOOH abundance in the interstellar medium equals $\sim 3\%$ of solid H_2O in a few objects. Upper limits for C_2H_6 in interstellar ices from ISO measurements were recently obtained by Boudin *et al.* (1998). ISO observations of gaseous counterparts indicate very low CO_2 and CH_4 abundances of $< 2 \times 10^{16}$ cm^{-2} and $\sim 5 \times 10^{16}$ cm^{-2}, respectively (Dartois *et al.*, 1998; van Dishoeck, 1998). The gas phase abundance of H_2O varies strongly in cold and warm clouds (van Dishoeck, 1998).

3. What Did We Learn From ISO About Interstellar Ice Chemistry?

ISO allowed to estimate accurate abundances for many interstellar ice species. High resolution profiles of the CO_2, H_2O and other bands enabled us to infer the grain mantle composition and temperature conditions in the line of sight toward protostars (Whittet *et al.*, 1996; d'Hendecourt *et al.*, 1996). Laboratory data of simple and annealed ices can provide excellent fits to astronomical data and indicate that UV radiation is efficiently attenuated in dense clouds and may play only a minor role in the processing of ices. According to the line-of-sight conditions, such as density and temperature, grains may have layered mantles.

There is strong observational evidence that apolar molecules such as CO cover grain mantles in cold, dense environments far from the protostar (Chiar *et al.*, 1998). Most of the missing nitrogen in the interstellar medium is expected to be present in the form of gaseous N_2, which may freeze out on grain mantles in cold regions. Solid N_2 is rather inert and shows no significant fingerprints in the profile of strong bands such as CO and CO_2 (Ehrenfreund *et al.*, 1997b). Its fundamental transition, which is weakly infrared active in the solid phase, falls adjacent to the strong CO_2 stretching mode at 4.28 μm. A substantial amount of solid N_2 embedded in apolar grain mantles can therefore not be excluded, but will be hard to estimate. ISO results in combination with ground-based observations showed that only small amounts of solid O_2 are present in apolar ices.

ISO has revealed the presence of extensive ice segregation toward massive protostars. This new ice type shows a morphology dominated by ice clusters of pure ice, complex and clathrate formation and seems to be ubiquitous in protostellar regions (Ehrenfreund *et al.*, 1999). The different grain populations currently identified in the interstellar medium have strong implications for cometary outgassing. Recently measured cometary volatiles, and their relative proportions, are not compatible with interstellar gas phase values but are similar to many solid state interstellar abundances (Mumma, 1997; Ehrenfreund *et al.*, 1997a). There are some notable exceptions, however, such as the large amounts of CH_3OH and NH_3 recently observed from the ground toward certain protostars (Dartois *et al.*, 1999a; Lacy *et al.*, 1998). In any event, if interstellar icy grains which contain segregated H_2O, CO_2, and CH_3OH are incorporated unaltered into comets, these species should be observed in comets at varying heliocentric distances because of their very different sublimation temperatures. The same applies for apolar ice mantles (containing CO, N_2 and some O_2) which are amorphous, but which suffer from high volatility (Ehrenfreund *et al.*, 1998b). As we have said, this is consistent with certain aspects of cometary activity. However, traces of even very volatile apolar molecules trapped in polar ices (H_2O-dominated) can be retained at much higher temperature and would evaporate only at much closer heliocentric distances. It is likely that comets also contain processed solar system materials (Fegley, this volume), this picture of interstellar grains being incorporated into comets is very powerful, and future models of cometary composition and activity should include the new ISO view on interstellar ice chemistry.

References

Bockelée-Morvan, D., Rickman, H.: 1999, *Earth, Moon and Planets*, in press.
Boogert, A.C.A. *et al.*: 1996, *A&A* **315**, L377.
Boudin, N., Schutte, W.A., Greenberg, J.M.: 1998, *A&A* **331**, 749.

Chiar, J., Gerakines, P.A., Whittet, D.C.B.: 1998, *ApJ* **498**, 716.
Crovisier, J. *et al.*: 1996, *A&A* **315**, L385.
Crovisier, J. *et al.*: 1997, *Science* **275**, 1904.
Dartois, E., d'Hendecourt, L., Boulanger, F. *et al.*: 1998, *A&A* **331**, 651.
Dartois, E., Schutte, W., Geballe, T. *et al.*: 1999a, *A&A* **342**, L32.
Dartois, E., Demyk, K., Ehrenfreund, P. *et al.*: 1999b, *A&A*, submitted.
de Graauw, Th. *et al.*: 1996, *A&A* **315**, L345.
d'Hendecourt, L. *et al.*: 1996, *A&A* **315**, L365.
Ehrenfreund, P., Breukers, R., d'Hendecourt, L., Greenberg, J.M.: 1992, *A&A* **260**, 431.
Ehrenfreund, P. *et al.*: 1997a, *Icarus* **130**, 1.
Ehrenfreund, P., Boogert, A., Gerakines *et al.*: 1997b, *A&A* **328**, 649.
Ehrenfreund, P., Dartois, E., Demyk, K., d'Hendecourt, L.: 1998a, *A&A* **339** L17.
Ehrenfreund, P., Boogert, A., Gerakines, P., Tielens, A.G.G.M.: 1998b, *Faraday Discussion* **No. 109**, The Royal Society of Chemistry, London, 463
Ehrenfreund, P., Kerkhof, O., Schutte, W. *et al.*: 1999, *A&A*, in press.
Gürtler, J., Henning, T., Kömpe, C., *et al.*: 1996, *A&A* **315**, L189.
Lacy, J., Faraji, H., Sandford, S., Allamandola, L.: 1998, *ApJ*, **501**, L105.
Meyer, D.M., Jura, M., Cardelli, J.A.: 1998, *ApJ* **493**, 222.
Mumma, M.J. 1997, in *From Stardust to Planetesimals*, Astronomical Society of the Pacific 122, eds. Pendleton Y. and Tielens, A.G.G.M, 369
Olofsson, G., Pagani, L., Tauber J.: 1998, *A&A* **339**, L81.
Schutte, W.A., Tielens, A.G.G.M., Whittet, D.C.B., *et al.*: 1996, *A&A* **315**, L333.
Schutte, W.A., Boogert, A.C.A., Tielens, A.G.G.M., *et al.*: 1999, *A&A* **343**, 966.
van Dishoeck, E.F.: 1998, *Faraday Discussion* **No. 109**, The Royal Society of Chemistry, London, 31.
van Dishoeck, E.F., Blake, G.A.: 1998, *ARA&A* **36**, 317.
Whittet, D.C.B. *et al.*: 1996, *A&A* **315**, L357.
Whittet D.C.B., Gerakines P.A., Tielens A.G.G.M., *et al.*: 1998, *ApJ* **498**, L159.
Vandenbussche, B., Ehrenfreund, P., Boogert, A., *et al.*: 1999, *A&A* **346**, L57.

Address for correspondence: Pascale Ehrenfreund, Leiden Observatory, P O Box 9513, 2300 RA Leiden, The Netherlands, pascale@strw.leidenuniv.nl

CHEMICAL AND PHYSICAL PROCESSING OF PRESOLAR MATERIALS IN THE SOLAR NEBULA AND THE IMPLICATIONS FOR PRESERVATION OF PRESOLAR MATERIALS IN COMETS

BRUCE FEGLEY JR.

Planetary Chemistry Laboratory, Department of Earth and Planetary Sciences, Washington University, St. Louis, MO 63130-4899 USA

Abstract. Chemical and physical processes in the outer solar nebula are reviewed. It is argued that the outer nebula was a chemically active environment with UV photochemistry and ion-molecule chemistry in its low density regions and grain-catalyzed chemistry in Jovian protoplanetary subnebulae. Presolar material was altered to greater or lesser extent by these spatially and temporally variable processes, which mimic many features of interstellar chemistry. Experiments, models, and observations are recommended to address the questions of presolar versus nebular dominance in the outer solar nebula and of how to distinguish interstellar and nebular sources of cometary volatiles.

Keywords: comets, solar nebula, isotopes, photochemistry, ion-molecule chemistry, grain catalyzed chemistry, Jovian protoplanetary subnebulae, thermochemistry

1. Introduction

Before being accreted into comets, presolar materials (e.g. grains and molecules formed in cool stellar outflows, in supernova ejecta, and molecules formed in interstellar molecular clouds) were altered to varying extents by chemical and physical processes in the solar nebula. These processes probably included, but were not necessarily limited to, heating and thermochemical reactions, UV photochemistry, ion-molecule reactions, and physical mixing with and dilution by other nebular materials. These processes occurred in the accretion shock, the solar nebula, and the Jovian protoplanetary subnebulae embedded in the outer nebula.

The extent of this reprocessing probably varied spatially and temporally resulting in different proportions of presolar and nebular materials accreted by comets formed at different radial distances in the nebula. The varying efficiencies of nebular processes combined with the different properties (e.g., vapor pressure, chemical reactivity) of presolar materials (gases, ices, organics, rocky grains) plausibly determined the extent of processing.

Three models can be considered for the origin of cometary materials: (1) the pristine interstellar grain model in which no chemical alteration, physical mixing, or dilution happens to accreted presolar material, (2) the complete chemical equilibrium model in which presolar material is altered and chemically equilibrated at the prevailing pressure and temperature conditions in the outer solar nebula, and (3) an intermediate model in which nebular reprocessing of presolar materials, and physical mixing with and dilution by nebular materials varies in efficiency. I adopt the third model as probably being the most reasonable one.

In this paper, I review observations of the chemical and isotopic composition of cometary volatiles and chemical and physical processes in the outer solar nebula. This leads to two broad questions: What was the balance between presolar and nebular chemistry in the outer solar nebula? How can we distinguish interstellar from nebular chemistry if the same processes occur in both environments? I conclude by recommending laboratory experiments, theoretical models, and astronomical observations that can help to address these questions.

2. Chemical and Isotopic Composition of Cometary Volatiles

Figure 1, based on Eberhardt (1998) and Irvine *et al.* (1999), shows abundances of cometary volatiles relative to water. This figure emphasizes volatiles for which the most data are available; more comprehensive discussion is given in the two papers cited and elsewhere in this volume.

Water. This is the most abundant volatile in all comets where it has been measured, e.g., 80 mol % in Halley (Eberhardt, 1998). The dominance of water is broadly consistent with solar abundances because oxygen is the third most abundant element (after H and He) and is about twice as abundant as carbon in solar material. Because water is the dominant volatile and perhaps also a host phase for less abundant volatiles which may be enclathrated in it and/or absorbed on it, the physical state (discussed in other chapters), isotopic composition, and ortho-para ratio of cometary water is very important for understanding the origin of cometary volatiles. The HDO/H_2O ratios in Hale-Bopp, Halley, and Hyakutake correspond to $D/H \sim 31 \times 10^{-5}$, similar to D/H values for water in hot cores of molecular clouds (Irvine *et al.*, 1999), and about twice the D/H of 15.6×10^{-5} in terrestrial standard mean ocean water (SMOW). The $^{18}O/^{16}O$ in Halley water is $(2.03 \pm 0.15) \times 10^{-3}$, the same within error as in SMOW (Eberhardt, 1998). The ortho-para ratios and deduced spin temperatures are 2.5 ± 0.1 (29 K) in Halley, 3.2 ± 0.2 (> 50 K) in Wilson, 2.45 ± 0.10 (25 K) in Hale-Bopp, and 2.7 ± 0.1 (35 K) in Hartley 2 (Irvine *et al.*, 1999). The interpretation of the ortho-para ratio in water is unclear. The D/H ratio has been used to argue for an interstellar origin, but as discussed later, comet-like HDO/H_2O ratios may also be produced by UV photochemistry in the outer nebula. Furthermore, high D/H ratios only indicate low temperature isotopic exchange. Ion-molecule reactions have low activation energies $\sim kT$ (where k is Boltzmann's constant) and yield high D/H ratios for hydrides in interstellar molecular clouds and hot cores. But, the same chemistry may also occur in cold, low density regions of protoplanetary disks (Aikawa *et al.*, 1998).

Carbon compounds. Carbon is the fourth most abundant element in solar material and oxidized carbon compounds (CO, CH_3OH, CO_2, H_2CO) are the next most abundant volatiles (after water) observed in comets. As Figure 1 shows the

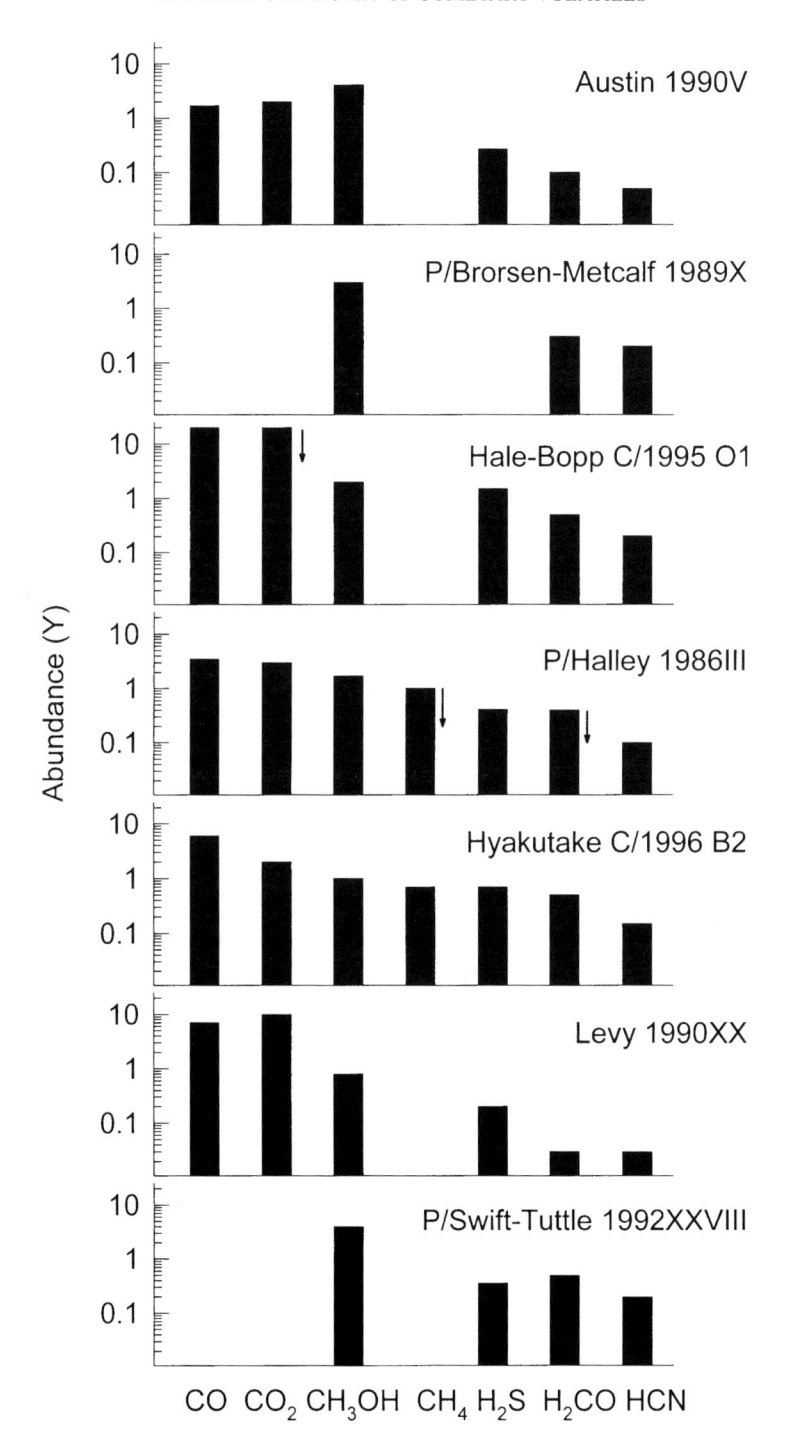

Figure 1. The abundances of different volatiles relative to water in several comets. The data are taken from Eberhardt (1998) and Irvine *et al.* (1999); see these two reviews for additional data on volatiles observed in only one or two comets. Arrows indicate that the value shown is an upper limit.

abundances of oxidized carbon compounds vary somewhat from comet to comet. Some species such as CO and H_2CO are partitioned between nuclear and extended sources (Eberhardt, 1998). Hydrocarbons (CH_4, C_2H_2, C_2H_6) were definitively observed in Hale-Bopp and Hyakutake (Irvine et al., 1999); carbonaceous grains were observed in Halley (Schulze et al., 1997) and presumably exist in other comets as well. The ratio of reduced/oxidized carbon volatiles is $<$ 1 in all comets where both families of compounds are observed, and is intermediate between the ratio of reduced/oxidized carbon compounds due to thermochemistry in the solar nebula and in Jovian protoplanetary subnebulae (Prinn and Fegley, 1989). The cometary CH_3OH/H_2O ratios are similar to CH_3OH/H_2O ratios in hot cores (typically a few to several percent). But CH_3OH/H_2O ratios of about 40% and 5-22% are observed in two sources (Dartois et al., 1999). The upper limit on D/H for CH_3OH in Halley is lower than D/H for CH_3OH in hot cores (Irvine et al., 1999).

Nitrogen compounds. The nitrogen volatiles observed in comets include NH_3, HCN, HNC, CH_3CN, HC_3N, HNCO, and NH_2CO. Molecular N_2 has not been detected directly but its abundance in Halley (N_2/H_2O $\sim$0.02%) was inferred from observations of N_2^+. The NH_3/H_2O ratio in Halley is $\sim$1% (Eberhardt, 1998). The derived NH_3/N_2 ratio is $\sim$50, intermediate between the values in the solar nebula ($\sim$0.01) and Jovian protoplanetary subnebula ($\sim$200) (Fegley, 1993). In the comets where nitrogen compounds are observed, the ratio of reduced/oxidized volatiles is $>$ 1. The $^{14}N/^{15}N$, $^{12}C/^{13}C$, and D/H ratios for HCN in Hale-Bopp have been measured. The D/H ratio is $(230 \pm 40) \times 10^{-5}$ and is consistent with DCN/HCN ratios in hot cores of interstellar clouds. The $^{12}C/^{13}C$ ratio (111 ± 12) and the $^{14}N/^{15}N$ ratio (323 ± 46) are terrestrial within the large uncertainties (Irvine et al., 1999). It is important to measure D/H ratios for NH_3 and HCN in other comets, and for other nitrogen volatiles. Some minor species such as HNCO, may be wholly or partially produced in the coma, as is the case for HNC in Hale-Bopp and Hyakutake (Irvine et al., 1999), and may not be primordial.

Sulfur compounds. The database for sulfur volatiles in comets is limited. Reduced (H_2S, OCS, H_2CS, CS, S_2, S) and oxidized (SO, SO_2) volatiles have been observed; CS_2 is the inferred (but as yet unseen) parent of CS (Eberhardt, 1998; Irvine et al., 1999). The solar S/O ratio is 2.1%. Depending on the oxidation state of carbon and the amount of oxygen tied up in rock, H_2O could have only one third of the solar oxygen abundance giving a (total) S/H_2O ratio slightly over 6%. The observed H_2S/H_2O ratios range from 0.2 − 1.6%, and H_2S seems to be a major, but perhaps not always the dominant, sulfur volatile in comets. Carbonyl sulfide is observed in Hale-Bopp and Hyakutake at $\sim$0.3-0.5% and most OCS in Hale-Bopp is from an extended source. Sulfur monoxide and SO_2 are observed in Hale-Bopp at $\sim$0.5% and $\sim$0.1%, respectively, and at least part of the SO is derived from SO_2 photolysis in the coma (Irvine et al., 1999). Thioformaldehyde (H_2CS) was observed in Hale-Bopp at $\sim$0.02%, and S_2 was observed in IRAS-Araki-Alcock and Hyakutake; the origin of these species is unclear (Irvine et al., 1999).

3. Solar Nebula Evolution and Thermal Structure

Before discussing the nebular chemistry of cometary volatiles, we need a model of the physical conditions in the solar nebula and of how these conditions changed with time. Any such model is necessarily imperfect, but astronomical observations of protoplanetary disks and young stars, geochemical analyses and geochronological dating of meteorites, and theoretical modeling of accretion disks can be combined to provide a consistent (and hopefully correct) picture. Here I adopt the recent model of Cameron (1995), which divides the evolution of our solar nebula into four stages, which can be summarized as follows.

Stage 1: Molecular cloud collapse. During this stage the nebular disk is assembled of infalling material from the collapsing molecular cloud core. This stage lasts for a few times 10^5 years. The disk mass is greater than the initial protosolar mass, but most (if not all) of the matter in the disk ultimately goes into the proto-Sun and is not preserved in meteorites or comets.

Stage 2: Disk dissipation. The Sun forms during this stage, which lasts for about 50,000 years (i.e., the Sun accretes at a rate of about 2×10^{-5} solar masses per year). The disk mass is less than the mass of the proto-Sun. Most matter falling onto the accretion disk is transported through the disk into the proto-Sun. Thus, there is a general mass transport inward. There is also an angular momentum transport outward in the (geometrically thin) disk. The major dissipation (i.e., transport) mechanisms include spiral density waves, disk-driven bipolar outflows, and the Balbus-Hawley magnetic instability. Cameron (1995) discusses why turbulent viscosity driven by thermal convection is less important than these other dissipative mechanisms, but other modelers (e.g., Willacy *et al.*, 1998) rely on turbulent viscosity as the main dissipative mechanism. The amount of outward mass transport, that would tend to "contaminate" the outer nebula with products from thermochemical processing in the innermost few AU of the nebula, is controversial and uncertain. Near the end of stage 2, some disk material survived and is preserved in meteorites (e.g., ^{26}Al-bearing minerals). Some materials in comets may date from this stage as well.

Stage 3: Terminal accumulation of the Sun. The final accumulation of the Sun occurs during this stage, which lasts for about $(1 - 2) \times 10^6$ years and the accumulation rate decreases from about 10^{-7} to 10^{-8} solar masses per year. The proto-Sun becomes a classical T Tauri star in this phase. Planetary accretion (almost complete for Jupiter and Saturn, and less advanced for the other planets) also occurs during this stage. Products of thermochemical reactions in the Jovian and Saturnian sub-nebulae may contaminate the outer nebula and be accreted into comets.

Stage 4: Loss of nebular gas. The Sun becomes a weak line T Tauri star in this stage, which lasts $(3 - 30) \times 10^6$ years, and is no longer accreting material from the disk. The T Tauri wind removes gas in the inner nebula and photoevaporation due mainly to UV radiation from the T Tauri wind removes gas in the outer nebula. UV photochemistry and ion-molecule chemistry may be important disequilibrating processes in the outer nebula, but depend on the nebular column density, which is poorly constrained and time dependent (Cameron, 1995). Gas-grain chemistry throughout the nebula ceases sometime (although not necessarily at the same time everywhere) during this stage.

In this model of nebular evolution, the accumulation of the Sun consumed essentially all of the material accreted by the nebular disk during the early stages of its history. Comets and all the other bodies in the solar system were then assembled from material accreted after the Sun had formed (i.e., during the time from the end of stage 2 into stage 4). I think the conclusion that most of the material in solar system bodies dates from the latter part of nebular history is robust, whether or not one accepts all details of Cameron's (1995) model or prefers a three stage model as he had proposed earlier. If this conclusion is adopted, the effects of UV radiation and other disequilibrating mechanisms (cosmic rays, X-winds) on nebular chemistry should be considered. The picture of the outer nebula as a cold, chemically inert region is almost certainly over simplified.

The temperature and pressure in the nebula during its latter history were important for chemical and physical reprocessing of presolar material. Three models of temperature and pressure as a function of radial distance in the solar nebula are contrasted in Figure 2. These three models are derived using different assumptions and differ in detail. However, they all predict low temperatures that agree within a factor of two and pressures that agree within an order of magnitude in the outer solar nebula. None of the models extend out as far as 1000 AU, the outer edge of the Kuiper Belt, and background values of temperature ($\sim$10 K) and pressure (10^3 cm^{-3} number density giving $\sim 10^{-18}$ bar at 10 K) are assumed there.

In fact, the radial extent of the solar nebula, the lowest temperatures attained in the outer nebula, and how the nebular disk blended into the surroundings are poorly constrained. The presence of N_2, CO, and CH_4 ices on Triton and Pluto, and of CH_4 ice on the Kuiper Belt Object 1993SC show that temperatures were low enough to condense these ices in the outer solar nebula (i.e., about 25 K for a wide range of plausible pressures). If these bodies formed at or near their present locations then temperatures were $\sim$25 K at 30-40 AU when Triton, Pluto, and 1993SC formed. However, temperatures could not drop below background temperature.

Finally, it is important to note that all three models are snapshots at one point in time when the snowline, where water ice condenses, was at Jupiter's orbit (5.2 AU). This is also the time when Jupiter started to form because water ice was the glue needed for its runaway accretion (Stevenson and Lunine, 1988). Observations of asteroids with hydrous spectral features (Jones et al., 1990), and time-dependent calculations of nebular thermal structure (e.g., Ruden and Lin, 1986) suggest that later in time the snowline moved inward to the asteroid belt.

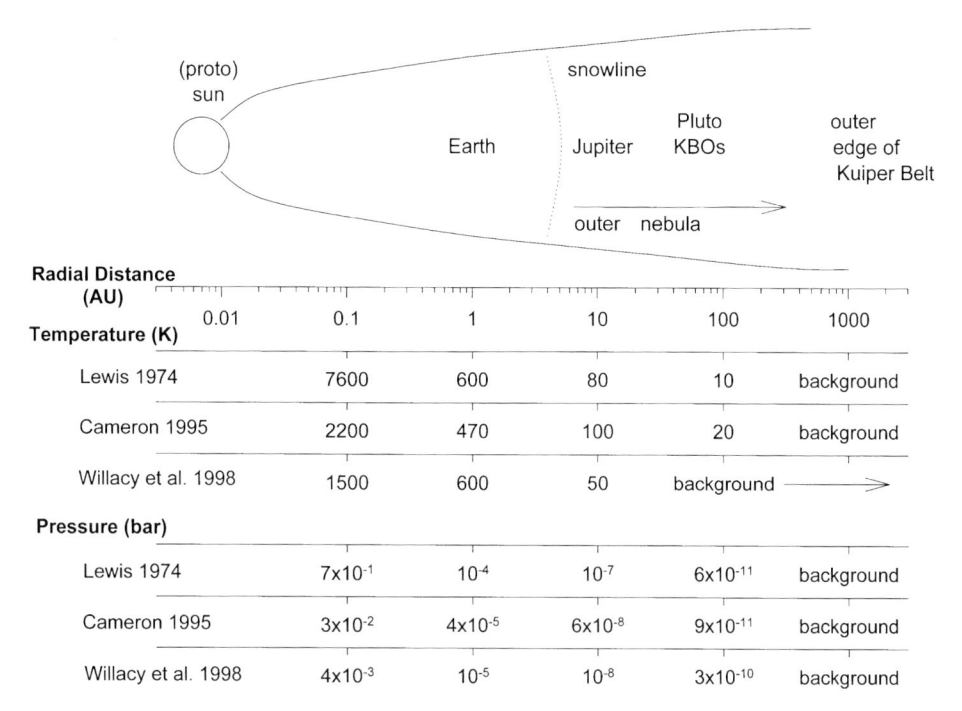

Radial Distance (AU)	0.01	0.1	1	10	100	1000
Temperature (K)						
Lewis 1974		7600	600	80	10	background
Cameron 1995		2200	470	100	20	background
Willacy et al. 1998		1500	600	50	background	→
Pressure (bar)						
Lewis 1974		7×10^{-1}	10^{-4}	10^{-7}	6×10^{-11}	background
Cameron 1995		3×10^{-2}	4×10^{-5}	6×10^{-8}	9×10^{-11}	background
Willacy et al. 1998		4×10^{-3}	10^{-5}	10^{-8}	3×10^{-10}	background

Figure 2. A schematic diagram comparing the radial variation of temperature and pressure for three solar nebula models. The distance scale is extended to 1000 AU to include the Kuiper Belt. The background temperature and pressure are about 10 K and $\sim 10^{-18}$ bar (a number density of 10^3 per cm^3). In the cartoon at the top the calculated snowline, where water vapor condenses to water ice, divides the solar nebula into an inner (< 5.2 AU) and an outer (≥ 5.2 AU) region. As discussed in the text, the position of the snowline moves inward as the nebula cools, but it was at or possibly beyond 5.2 AU when water ice provided the glue for the runaway accretion that led to the formation of Jupiter (at its present location or possibly at a greater radial distance).

4. Reprocessing of Presolar Materials in the Outer Solar Nebula

Presolar dust, gas, and ice accreted by the solar nebula were subjected to more or less chemical reprocessing and physical mixing before accretion into comets and other bodies. As schematically illustrated in Figure 3, with the exception of Jovian protoplanetary subnebulae, heating and thermochemical reactions were generally less important in the outer nebula than in the inner nebula. Chemical reaction rates between neutral atoms, radicals, and molecules vary exponentially with temperature, and the temperatures beyond the snowline were less than about 160 K (this is slightly higher or lower depending on the total pressure). Thus chemical reactions with activation energies $\gg kT$ would proceed at insignificant rates in the outer nebula.

However, it is unlikely that the outer nebula was simply an "icebox" that preserved interstellar (and other presolar) matter in a pristine state. Highly volatile

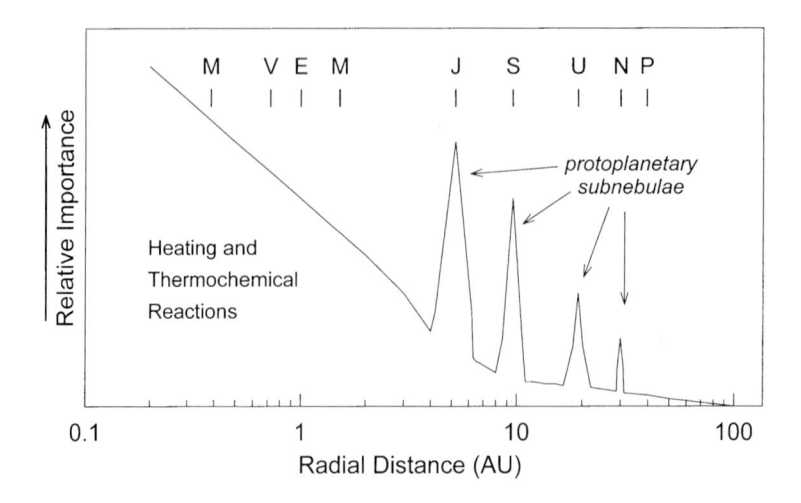

Figure 3. A cartoon illustrating the relative importance of heating and thermochemical reactions as a function of radial distance in the solar nebula. The symbols along the top show the present positions of the planets. The hypothesized Jovian protoplanetary subnebulae are hotter, higher density regions where thermochemical and grain-catalyzed reactions are favored relative to the cooler, lower density outer solar nebula.

ices (e.g., CO, N_2, CH_4, O_2) may have been partially or completely vaporized in the accretion shock or in the outer nebula itself. Other volatile ices may have been partially or completely vaporized depending on the distance at which grains were accreted. Atoms and radicals trapped in interstellar ices could diffuse and react once temperatures were a few tens of K above the background temperature. Volatiles that were vaporized may have reacted and undergone isotope exchange with nebular gas prior to recondensation.

Heating and thermochemical reactions. The outer solar nebula was possibly contaminated with thermochemical products from the Jovian protoplanetary subnebulae (e.g., by cometesimals ejected into the surrounding solar nebula or by gas exchange) and from the inner solar nebula as well. Prinn and Fegley (1989) attributed the small amounts of CH_4 and NH_3 in comets to sub-nebular ices because CO and N_2 are the dominant carbon and nitrogen ices expected in the solar nebula and in the interstellar medium. The importance of contamination of the outer solar nebula by thermochemical products from Jovian protoplanetary subnebulae and/or the inner solar nebula is twofold.

First, the isotopic compositions of presolar gas and grains may have been altered by mixing with nebular materials with normal isotopic compositions before their incorporation into comets. In general, these nebular materials could have been thermochemically processed in Jovian protoplanetary subnebulae or in the inner solar nebula. (In some cases, such as CH_4 and NH_3, species are only produced in the Jovian protoplanetary subnebulae because their production is kinetically inhibited in the solar nebula.) Hydrides from Jovian protoplanetary subnebulae and the inner nebula would have D/H ratios slightly higher than the D/H ratio of 2×10^{-5} for hydrogen in the solar nebula, because these gases were formed by chemical reactions in a warmer environment. The D/H ratio of 7.5×10^{-5} for CH_4 on Saturn's satellite Titan (Coustenis *et al.*, 1998) agrees with this model.

In contrast, hydrides formed at lower temperatures by ion-molecule chemistry would have significantly higher D/H ratios. A mixture of hydrides formed at higher and lower temperatures would have an intermediate D/H ratio. The D/H ratio of $\sim 31 \times 10^{-5}$ for H_2O in comets Hale-Bopp, Halley, and Hyakutake may indicate mixing of higher and lower D/H water. The upper limit on D/H for CH_3OH in Halley also may indicate mixing of CH_3OH from two sources.

Second, disequilibrium assemblages of oxidized and reduced molecules may have been produced by mixing presolar material with thermochemically processed nebular material. A mixture of CH_4 and NH_3 ices from Jovian protoplanetary subnebulae and CO and N_2 ices (presolar or nebular) has been proposed to explain the CO/CH_4 and N_2/NH_3 ratios in comets. These mixtures are disequilibrium assemblages and are probably also heterogeneous. However, the extent of mixing from the inner to outer nebula, and the mechanisms for this mixing are controversial (Prinn, 1990; Stevenson, 1990). The contamination of the outer nebula by thermochemical products from the Jovian protoplanetary subnebulae is also debated (see parts of the discussion following the paper by Fegley and Prinn, 1989).

Grain Catalyzed Chemistry. Industrial Fischer-Tropsch (FT) reactions use a Fe or Co based catalyst to convert a CO/H_2 gas mixture into hydrocarbons such as gasoline. Oxygenated organics such as methanol (CH_3OH) can also be formed in significant amounts depending on the reaction conditions and catalysts. In the late 1960s and early 1970s Anders and colleagues studied FT type reactions using Fe based catalysts and smixtures of CO, H_2, and NH_3 (Anders *et al.*, 1973; Hayatsu and Anders, 1981). They produced a variety of organic compounds and argued that the organic compounds in meteorites were products of nebular FT reactions. This conclusion is now controversial because Anders' results do not match the chemical and isotopic composition of organic compounds in the Murchison CM2 chondrite (e.g., see Cronin and Chang, 1993).

However, recent work shows that Fe and Fe-based catalysts (iron meteorites, ordinary chondrites, magnetite) convert CO to CH_4 in $CO-H_2$ gas mixtures with the solar C/H ratio at one bar total pressure and temperatures of a few hundred degrees (Fegley, 1998; Fegley and Hong, 1998). Iron-catalyzed NH_3 synthesis occurs at similar temperatures in N_2-H_2 mixtures with 10 times the solar N/H ratio at one bar pressure (Fegley, 1998). The experimental conditions match (P, T, C/H ratio) or come close (N/H ratio of 10 times solar) to those for the Jovian subnebula. The results support the subnebula chemistry models of Prinn and Fegley (1989). Grain-catalyzed chemistry is also thought to produce saturated compounds observed in circumstellar outflows from cool carbon stars (Glassgold, 1996).

Fegley (1993) suggested that grain-catalyzed chemistry may have produced some of the volatile carbon compounds and refractory organic matter in comets.

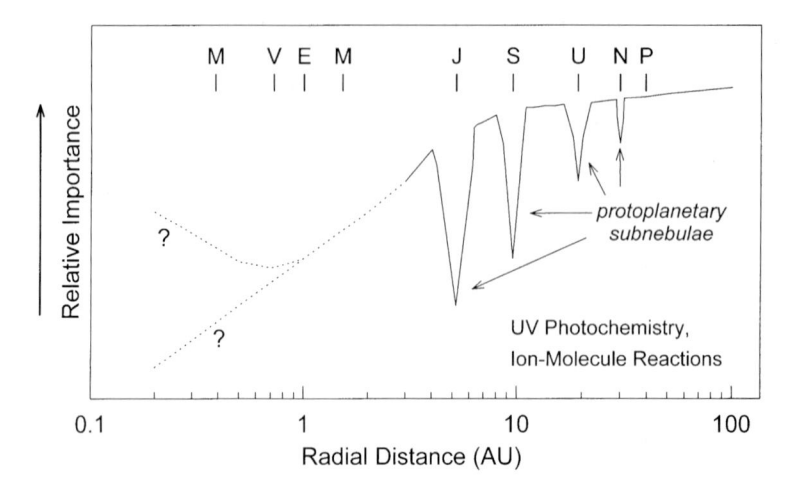

Figure 4. A cartoon showing the relative importance of disequilibrating processes (e.g. UV photochemistry and ion-molecule chemistry) as a function of radial distance in the solar nebula.

Conversion of CO to CH_4 is observed in a solar gas, but more work is needed to test this proposal. For example, no significant amounts of other hydrocarbons or oxygenated organics were produced in the initial experiments studying CO hydrogenation in a solar gas. However, other hydrocarbons, CO_2, methanol, and other oxygenated organics are produced in industrial FT reactions and may also be produced in FT reactions in a solar gas under some conditions. The lowest effective temperatures for the grain catalyzed CO to CH_4 and N_2 to NH_3 conversions are also unknown and need to be measured.

UV Photochemistry. Disequilibrating processes such as UV photochemistry and ion-molecule chemistry are potentially important during the later stages of nebular evolution. The cartoon in Figure 4 indicates that disequilibrating reactions were plausibly more important in the outer nebula than in the inner nebula. (The higher density Jovian protoplanetary subnebulae are the exception.) UV photochemistry can potentially drive isotopic exchange reactions (enrichment or depletion). The UV photon flux and amount of vertical mixing (midplane to nebular photosphere) needed to reach photochemical steady-state throughout the gas depend on the nebular column density. Yung *et al.* (1988) considered photochemistry driven by UV starlight and modeled the net photochemical reactions

$$HD + H_2O \rightarrow HDO + H_2 \tag{1}$$

$$HD + CH_4 \rightarrow CH_3D + H_2 \tag{2}$$

in optically thin regions of the solar nebula. They calculated D/H fractionation factors of 26 and 115 for reactions (1) and (2), respectively, at 150 K. Thus a D/H ratio of 52×10^{-5}, which is larger than the observed D/H ratio in cometary water, could be produced at photochemical steady-state.

Gladstone (1993) and Gladstone and Fegley (1997) proposed that solar HI Lyα (121.6 nm), He I (58.4 nm), and He II (30.4 nm) UV radiation backscattered by H atoms in the interplanetary medium was an important UV source in the outer solar nebula and did preliminary modeling showing that photodissociation rates of H_2O, CH_4, NH_3, and H_2S could be significant. Photolysis of H_2O is a source of OH radicals, which are important oxidizing agents for driving conversions of reduced carbon, nitrogen, and sulfur to the oxidized forms. Photolysis of H_2O, CH_4, NH_3, and H_2S is also a source of H atoms which may drive conversions of unsaturated to saturated compounds, carbonaceous dust to CH_4, and of oxidized carbon species (e.g., $CO_2 + H \rightarrow CO + OH$). Beyond the snowline, water vapor, the major UV opacity source, condenses out, and grains become the major source of UV opacity. This raises the potential for UV photoprocessing of ices, perhaps analogous to the current conversion of water ice to O_2 and ozone (O_3) observed on the three outermost Galilean satellites and on two Saturnian satellites (Rhea, Dione). Disequilibrium chemistry driven by UV light may proceed in the outer nebula and reprocess presolar gas and grains. Nebular photochemistry may also produce species thought to be diagnostic of interstellar molecular cloud chemistry.

Ion-molecule chemistry. Ion-molecule reactions driven by galactic cosmic rays and/or short-lived nuclides (^{26}Al) may occur in low density regions of the outer solar nebula. These reactions may lead to D/H fractionation in the nebula and produce the same molecules observed in interstellar molecular clouds. Depending on the nebular column density, ion-molecule chemistry may be limited to the nebular skin or occur throughout sufficiently low density regions in the outer nebula. ion-molecule chemistry would probably be less important in Jovian protoplanetary subnebulae than in the lower density solar nebula itself.

The key factor determining whether or not ion-molecule chemistry was important in the outer solar nebula is the fractional degree of ionization. When this is high ion-molecule chemistry is relatively important. In turn, the ionization fraction depends on the balance between the ionization rate and the neutralization rate. These rates are poorly constrained and the fractional degree of ionization in the outer solar nebula is model dependent and controversial.

In the nominal model of Aikawa *et al.* (1998) the ionization rate and the grain size are the same as in interstellar molecular clouds. Significant chemical reprocessing occurs in 10^6 years in this model. Initially, CO, N_2, H_2O, and atomic S are assumed to be the major gases of these elements and they are reprocessed into oxidized and reduced species. For example, CO is converted to CO_2, H_2CO, HCN, CH_4, C_2H_2, C_3H_4; N_2 is converted to NH_3, NO, NO_2, HCN, HC_3N; H_2O is converted to O_2; and S atoms are converted to OCS, SO, SO_2, CS, H_2CS, and H_2S. This mixture of reduced and oxidized species matches some features of the chemistry of cometary volatiles.

In contrast, Willacy et al. (1998) calculate that ion-molecule chemistry was relatively unimportant in the outer solar nebula because they compute a lower fractional degree of ionization due to effective neutralization of ions by grains. Their chemistry is dominated by neutral-neutral reactions, although several of the products are the same as those calculated by Aikawa et al. (1998).

5. Concluding Remarks and Recommendations

As described above, presolar materials were probably subjected to a variety of chemical and physical processes before being accreted into comets. Several of these processes such as UV photochemistry and ion-molecule chemistry in cold, low density regions of the outer nebula mimic interstellar chemistry and produce the same suite of molecules and isotopic fractionations. Products from thermochemical reactions and grain-catalyzed chemistry in Jovian protoplanetary subnebulae can mix with and alter the isotopic compositions of presolar materials. It is unlikely that nebular processes were efficient enough to completely reprocess all presolar material before accretion into comets. But as Figures 3 and 4 suggest, as thermochemistry becomes insignificant in the cold, low density outer nebula, UV photochemistry and ion-molecule chemistry become significant. What was the balance between presolar and nebular chemistry in the outer solar nebula?

Many of the volatiles in comets are attributed to interstellar chemistry because the same species are also observed in ices in interstellar molecular clouds. In some cases, a high D/H ratio supports this correspondence (e.g., HCN in Hale-Bopp) because similar D/H ratios are also observed in the interstellar molecular clouds. However, how can we distinguish interstellar from nebular chemistry if the same processes occur in both environments? Below I suggest some experiments, models, and observations that can help to address these two questions:

1. The D/H ratio in hydrous minerals on asteroids and in ices on satellites of the Jovian planets, Pluto, Charon, Centaur objects, and Kuiper Belt objects should be measured by remote sensing and/or in situ methods. These data can show if D/H values of hydrous and icy bodies in the solar system exhibit trends with radial distance and/or hypothesized formation location (solar nebula vs. subnebulae).
2. Similar data for the ortho-para ratio of water ice on cold outer solar system bodies (Triton, Charon, Kuiper belt objects) may show a relationship between the ortho-para spin temperature and D/H ratio.
3. Measurements of the D/H ratios in several volatiles from one comet and of the same volatile in a large number of comets, optimally from different dynamical families can assess intra- and inter-comet variability. This dataset would help to distinguish a high temperature origin (low D/H ratio from subnebula or inner nebula chemistry) from a low temperature origin (high D/H ratio from ion-molecule chemistry in an interstellar cloud or low density region of the outer

solar nebula) for H_2O, CH_4, NH_3, H_2S, hydrocarbons, and other H-bearing species.
4. High spatial resolution observations of protoplanetary disks are needed to measure abundances of neutrals, ions, and isotopic ratios (D/H) as a function of radial distance. Resolutions of a few percent of the disk radius are needed. The current angular resolution corresponds to ∼100 AU in the Taurus molecular cloud, the nearest star-forming region. This is insufficient to see if ion-molecule reactions and UV photochemistry occur in protoplanetary disks (Aikawa *et al.*, 1998).
5. Theoretical modeling of ion-molecule chemistry and UV photochemistry in the outer nebula is needed to predict steady state abundances of volatile O, C, N, and S species and isotopic ratios for comparison with cometary observations.
6. Laboratory studies of D/H exchange kinetics for cometary volatiles and ortho-para conversion kinetics for water, and of low temperature grain-catalyzed chemistry in a solar gas are needed for nebular chemistry models.

Acknowledgements

I thank A.G.W. Cameron, K. Lodders, J. Nuth, P. Ehrenfreund, and my co-authors on our PPIV book chapter (J. Crovisier, W. Irvine, M. Mumma, P. Schloerb) for comments, K. Lodders for drafting the figures, R. Osborne for editing, and the International Space Science Institute for the invitation to this workshop. This research is supported by Grant NAG5-6366 from the NASA Planetary Atmospheres Program.

References

Aikawa, Y., Umebayashi, T., Nakano, T., and Miyama, S.: 1998, 'Molecular Evolution in Planet Forming Circumstellar Disks', *Faraday Discuss.* **109**, 281-301.
Anders, E., Hayatsu, R., and Studier, M.H.: 1973, 'Organic Compounds in Meteorites', *Science* **182**, 781-790.
Cameron, A.G.W.: 1995, 'The First Ten Million Years in the Solar Nebula', *Meteoritics* **30**, 133-161.
Coustenis, A., Salama, A., Lellouch, E., Encrenaz, Th., de Graauw, Th., Bjoraker, G.L., Samuelson, R.E., Gautier, D., Feuchtgruber, H., Kessler, M.F., and Orton, G.S.: 1998, 'Titan's Atmosphere from ISO Observations: Temperature, Composition and Detection of Water Vapor', *Bull. Amer. Astron. Soc.* **30**, 1060.
Cronin, J.R., and Chang, S.: 1993, 'Organic Matter in Meteorites: Molecular and Isotopic Analyses of the Murchison Meteorite', in *The Chemistry of Life's Origins*, ed. J.M. Greenberg, C.X. Mendoza-Gomez, and V. Pirronello, Kluwer, Dordrecht, pp. 209-258.
Dartois, E., Schutte, W., Geballe, T.R., Demyk, K., Ehrenfreund, P., DHendecourt, L.: 1999, 'Methanol: The second most abundant ice species toward the high-mass protostars RAFGL7009S and W 33A', *Astron. Astrophys.* **342**, L32-35.

Eberhardt, P.: 1998, 'Volatiles, Isotopes and Origin of Comets', in *Asteroids, Comets, Meteors 1996*, *COSPAR Colloquium* **10**, in press.

Fegley, B., Jr. and Prinn, R.G.: 1989, 'Solar Nebula Chemistry: Implications for Volatiles in the Solar System', in *The Formation and Evolution of Planetary Systems*, eds. H. Weaver and L. Danly, Cambridge Univ. Press, Cambridge, UK, pp. 171-211.

Fegley, B., Jr.: 1993, 'Chemistry of the Solar Nebula', in *The Chemistry of Life's Origins*, ed. J.M. Greenberg, C.X. Mendoza-Gomez, and V. Pirronello, Kluwer, Dordrecht, pp. 75-147.

Fegley, B., Jr.: 1998, 'Iron Grain Catalyzed Methane Formation in the Jovian Protoplanetary Subnebulae and the Origin of Methane on Titan', *Bull. Amer. Astron. Soc.* **30**, 1092.

Fegley, B., Jr., and Hong, Y.: 1998, 'Experimental Studies of Grain Catalyzed Reduction of CO to Methane in the Solar Nebula', *EOS Trans. AGU* **79**, S361-362.

Gladstone, G.R.: 1993, 'Photochemistry in the Primitive Solar Nebula', *Science* **261**, 1058.

Gladstone, G.R., and Fegley, B., Jr.: 1997, 'Lya Photochemistry in the Primitive Solar Nebula', *EOS Trans. AGU* **78**, F538-539.

Glassgold, A.E.: 1996, 'Circumstellar Photochemistry', *Annu. Rev. Astron. Astrophys.* **34**, 241-277.

Hayatsu, R., and Anders, E.: 1981, 'Organic Compounds in Meteorites and their Origins', *Topics in Current Chemistry* **99**, 1-39.

Irvine, W.M., Schloerb, F.P., Crovisier, J., Fegley, Jr., B., and Mumma, M.J.: 1999, 'Comets: A Link Between Interstellar and Nebular Chemistry', in *Protostars and Planets* **IV**, ed. A. Boss, V. Mannings, and S. Russell, University of Arizona Press, Tucson, AZ, in press.

Jones, T.D., Lebofsky, L.A., Lewis, J.S., and Marley, M.S.: 1990, 'The Composition and Origin of the C, P, and D Asteroids: Water as a Tracer of Thermal Evolution in the Outer Belt', *Icarus* **88**, 172-192.

Lewis, J.S.: 1974, 'The Temperature Gradient in the Solar Nebula', *Science* **186**, 440-443.

Prinn, R.G.: 1990, 'On Neglect of Non-linear Momentum Terms in Solar Nebula Acretion Disk Models', *Astrophys. J.* **348**, 725-729.

Prinn, R.G., and Fegley, B., Jr.: 1989, 'Solar Nebula Chemistry: Origin of Planetary, Satellite, and Cometary Volatiles', in *Origin and Evolution of Planetary and Satellite Atmospheres*, ed. S.K. Atreya, J.B. Pollack, and M. Matthews, University of Arizona Press, Tucson, AZ, pp. 78-136.

Ruden, S.P., and Lin, D.N.C.: 1986, 'The Global Evolution of the Primordial Solar Nebula', *Astrophys. J.* **308**, 883-901.

Schulze, H., Kissel, J., and Jessberger, E.K.: 1997, 'Chemistry and Mineralogy of Comet Halley's Dust', in *From Stardust to Planetesimals*, ed. Y.J. Pendleton and A.G.G.M. Tielens, ASP Conf. Ser. 122, pp. 397-414.

Stevenson, D.J.: 1990, 'Chemical Heterogeneity and Imperfect Mixing in the Solar Nebula', *Astrophys. J.* **348**, 730-737.

Stevenson, D.J., and Lunine, J.I.: 1988, 'Rapid Formation of Jupier by Diffusive Redistribution of Water Vapor in the Solar Nebula', *Icarus* **75**, 146-155.

Willacy, K., Klahr, H.H., Millar, T.J., and Henning, Th.: 1998, 'Gas and Grain Chemistry in a protoplanetary Disk', *Astron. Astrophys.* **338**, 995-1005.

Yung, Y.L., Friedl, R.R., Pinto, J.P., Bayes, K.D., and Wen, J.S.: 1988, 'Kinetic Isotopic Fractionation and the Origin of HDO and CH_3D in the Solar System', *Icarus* **74**, 121-132.

Address for correspondence: Bruce Fegley, Planetary Chemistry Laboratory, Department of Earth and Planetary Sciences, Washington University, St. Louis, MO 63130-4899 USA, bfegley@levee.wustl.edu

RARE ATOMS, MOLECULES AND RADICALS IN THE COMA OF P/HALLEY

J. GEISS
International Space Science Institute, Hallerstr. 6, CH-3012 Bern, Switzerland
Max-Planck-Institut für Kernphysik, Saupfercheckweg 1, D-69029 Heidelberg, Germany

K. ALTWEGG and H. BALSIGER
Physikalisches Institut, Universität Bern, Sidlerstr. 5, CH-3012 Bern, Switzerland

S. GRAF
International Space Science Institute, Hallerstr. 6, CH-3012 Bern, Switzerland

Abstract. We have searched for rare molecules and radicals in the coma of P/Halley using the ion data obtained by IMS-Giotto. Whereas our established methods were used in the ionosphere, a new model was developed for the interpretation of the ion data in the outer coma. $Ne/H_2O < 1.5 \times 10^{-3}$ was determined in the coma of the comet. Upper limits for the production of Na were derived from the very low abundance of Na^+. Methyl cyanide and (probably) ethyl cyanide were identified with abundances of $CH_3CN/H_2O = (1.4 \pm .6) \times 10^{-3}$ and $C_2H_5CN/H_2O = (2.8 \pm 1.6) \times 10^{-4}$. These results and upper limits for other N-bearing species confirm that nitrogen is depleted in the Halley material. C_4H was identified and a point source strength of $C_4H/H_2O = (2.3 \pm .8) \times 10^{-3}$ was derived. Our upper limit for C_3H is lower than the abundance of C_4H. This is in agreement with the enhanced abundances of C_nH species with even numbers of C-atoms found in interstellar molecular clouds, suggesting that the C_4H in Halley was synthesized under molecular cloud conditions. Thus, C_4H and other organics with unpaired electrons may turn out to be indicators for a molecular cloud origin of cometary constituents.

1. Introduction

The composition of molecules, radicals and ions in cometary comae evolves primarily through photodissociation and -ionization, ion-molecule reactions, and dissociative recombination of ions and electrons. In order to obtain information on the composition and nature of cometary source materials, the chemical reaction network has to be combined with appropriate flow parameters and thermodynamic data in the coma. If we designate number densities of neutrals and ions by z_i and y_i, respectively, the steady-state continuity equation for the density of the ion 1 in a flux tube is given by

$$\frac{d(AVy_1)}{ds} = A(q_1 - d_1y_1) \tag{1}$$

where s is the length coordinate along the flux tube, $A(s)$ the flux tube cross section, and $V(s)$ the flow speed.

If diffusion across the flux tube boundary is neglected, the source density q_1 and the destruction rate d_1 are given by

$$q_1 = \sum_i \lambda_{1i} z_i + \sum_{i,j} k_{1ij} y_i z_j \qquad (2)$$

$$d_1 = k_{1e} n_e + \sum_{i,j} k_{i1j} z_j \qquad (3)$$

- λ_{1i} are the photoionization rate constants producing the ion 1;
- k_{1ij} are the ion-molecule reaction rate constants producing the ion 1;
- $k_{1e} n_e$ is the recombination rate for the ion 1;
- k_{i1j} are the ion-molecule reaction rate constants destroying the ion 1.

Electron impact ionization is not important inside the contact surface (Körösmezey et al., 1987; Haeberli et al., 1996); its possible importance in the outer coma is discussed in section 4.

V and A as a function of the nucleocentric coordinate **r** can be obtained by MHD or other theory (Schmidt et al., 1988; Ip and Axford, 1990; Huebner et al., 1991; Gombosi et al., 1996; Haeberli et al., 1997). Alternately, these flow parameters can be estimated from kinematic models that are based on observation.

The chemical models commonly in use include more than a hundred neutral and ionized species, which are interrelated by several hundred reaction equations. However, for the interpretation of ion measurements many of these reactions are of minor importance. In fact, for all the ion species discussed in this paper, $q_1(r)$ and $d_1(r)$ are dominated by one or two terms in the Equations 2 and 3. The rate constants for these dominating reactions are listed in Table I. The other rate constants are found in the references given therein.

For the cometary neutrals we assume radial flow with a constant speed $V_n = 0.9$ km/s, except for atomic Na (cf. Section 4.2). Destruction of neutrals is dominated by photolysis. In the source terms of the continuity equations of the neutrals, we include point sources and/or extended sources, and the products of photodissociation (cf. Geiss et al., 1991).

The ion flow pattern and the chemical conditions in the coma vary strongly as a function of the nucleocentric coordinate **r**. This is reflected in the absolute density and relative abundances of the cometary ions along the inbound trajectory of Giotto towards Halley (Figure 1). For the purpose of interpreting the ion abundance data measured along Giotto's trajectory, we distinguish three regions:

Region 1 is the ionosphere, i.e. the volume inside the contact surface that was crossed by Giotto at $R = 4600$ km (R is the distance between the spacecraft and the nucleus, cf. Neubauer et al., 1986 and Balsiger et al., 1986). We use in this region our well tested model and code (cf. Geiss et al., 1991; Altwegg et al., 1993). The ion speed is taken to be constant at $V = 0.9$ km/s, and the expansion is assumed to be radially symmetric, i.e. $A \sim 1/r^2$.

TABLE I

Rate Constants

Reaction	$\lambda, k^{1)}$	Ref.
$Ne + \gamma \to Ne^+ + e^-$	0.165	c,d
$Ne^+ + H_2O \to$ Products	0.74	a
$Na + \gamma \to Na^+ + e^-$	7.2	d
$Na + Ions \to Na^+ +$ Products	> 1	a, f
$Na^+ + e^- \to Na + \gamma$	0.003	b
$C_4 + H_3O^+ \to C_4H^+ + H_2O$	1.1	b
$C_4H^+ + e^- \to C_4 + H$	300.	b
$C_4H + H_3O^+ \to C_4H_2^+ + H_2O$	1.1	b
$C_4H_2^+ + e^- \to$ Products	300.	b
$C_4H_2 + H_3O^+ \to C_4H_3^+ + H_2O$	1.1	b
$C_4H_3^+ + e^- \to$ Products	300.	b
$CH_3CN + \gamma \to$ Products	8.3	g
$CH_3CN + H_3O^+ \to CH_4CN^+ + H_2O$	4.7	a
$CH_4CN^+ + e^- \to$ Products	300.	b
$C_2H_5CN + H_3O^+ \to C_2H_6CN^+ + H_2O$	4.7	e
$C_2H_6CN^+ + e^- \to$ Products	500.	e
$CH_3NH_2 + H_3O^+ \to CH_3NH_3^+ + H_2O$	3.	e
$CH_3NH_3^+ + e^- \to$ Products	500.	e
$C_3N + H_3O^+ \to C_3NH^+ + H_2O$	2.0	b
$C_3NH^+ + e^- \to$ Products	300.	b
$HC_3N + H_3O^+ \to H_2C_3N^+ + H_2O$	3.8	b
$H_2C_3N^+ + e^- \to$ Products	620.	b

[1] Photo reaction rates λ in 10^{-6} s^{-1} at 0.9 AU; ion-molecule reaction rates and recombination rates k in 10^{-9} cm^{3}s^{-1}

a: Anicich and Huntress (1993);

b: Millar et al. (1991);

c: Huebner and Lyon (1992);

d: Geiss (1998);

e: estimate;

f: effective rate constant for total ion density;

g: Bockelée-Morvan and Crovisier (1985).

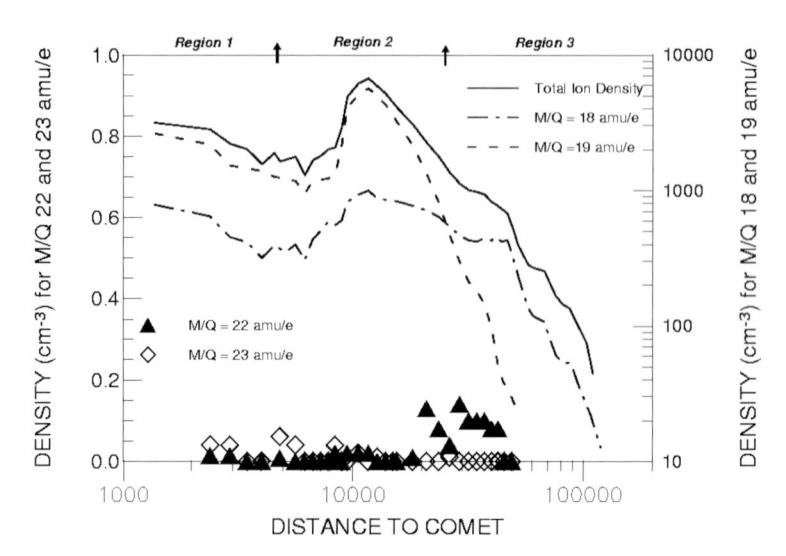

Figure 1. Densities of ions with highest and lowest abundances in the coma of Comet Halley showing the three principle regions of ion density and composition discussed in section 1. The figure demonstrates the low background and good mass resolution of the Giotto IMS/HIS Sensor which results in a dynamic range of $\sim 10^5$.

Region 2 is dominated by the "chemical pile-up" effect, which at the time of Giotto's encounter with P/Halley caused a maximum in ion density at $R = 10000$ to 15000 km. Haeberli *et al.* (1995) have developed a flow model for region 2 and showed unambiguously that the maximum ion density at 10000 to 15000 km is caused by an increased electron temperature which results in a dramatic decrease in the recombination rates of molecular ions. The increase in electron temperature in this region was not measured but indirectly derived from the H_3O^+/H_2O^+ ratio. According to Haeberli *et al.* (1996), the increase is caused by an in-flux of solar wind electrons into this region, but other possible causes have also been discussed.

Region 3 is the outer coma. We place the inner boundary of this region at $R = 25000$ km, where the H_3O^+ density falls below the H_2O^+ density (Figure 1), signaling the diminishing role of protonation reactions. A model for interpreting measured ion abundances anywhere in the outer coma in terms of source strengths and source distributions is presented in Section 3.

2. Instrument and Data

In this paper, we derive source strengths (or upper limits thereof) for several rare molecules and radicals using the ion density measurements obtained with the HIS sensor of the IMS Giotto experiment in the coma of P/Halley. HIS is a magnetic mass spectrometer (Balsiger *et al.*, 1987) that covers the range from $M12^+$ to $M57^+$

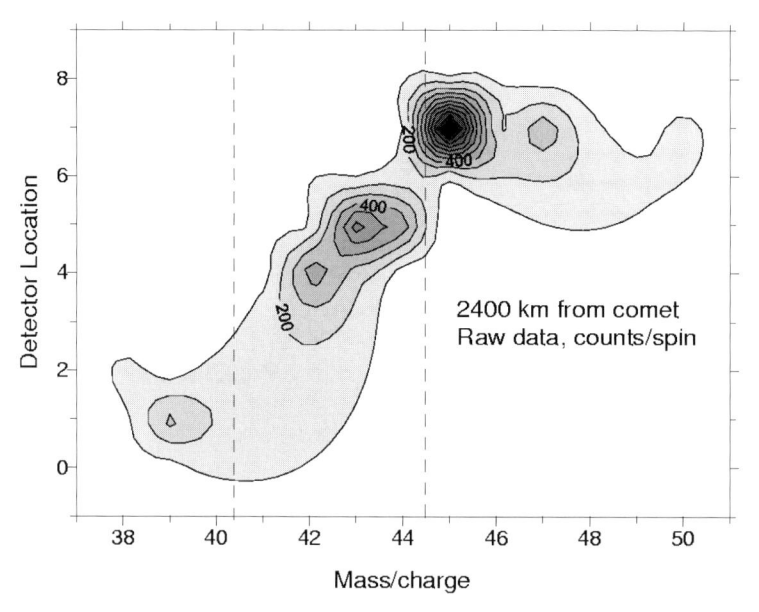

Figure 2. Contour plot of ion counts in the range M38$^+$ to M50$^+$. The mass/charge ratios given at the ordinate are determined from laboratory calibration. Changes in the direction of the ion track, marked by dashed lines, are due to scanning mode changes that were pre-programmed in order to avoid drifting out of the detector range (cf. Balsiger *et al.*, 1987). The recognizable maxima are principally due to protonated molecules or radicals (cf. Altwegg *et al.* (1999) and Table IV of this work), namely: C_3H_2 (at M39$^+$), CH_3CN (at M42$^+$), CHCHO + C_3H_6 (at M43$^+$), CH_3CHO + CS (at M45$^+$), C_4H (at M50$^+$) and probably HCOOH + C_2H_5OH (at M47$^+$).

(for unspecified ions, e.g. ions with M/Q = 12 amu/e, we use the notation M12$^+$). The mass resolution is remarkably good, even in the upper part of the mass range (Figure 2). The instrument function, i.e. the conversion of count rates to ion abundances for specific M/Q ratios were determined by laboratory calibration (Meier, 1988).

3. Model for the Outer Coma

In the outermost region of the coma the ion flow is approximately in the solar wind direction. If A, V, q_1 and d_1 as a function of s are given by theory and/or observation, Equation 1 can be solved.

In order to achieve a separation of variables, we write y_1 as a product

$$y_1 = f \cdot y_1^a \quad (4)$$

where y_1^a depends only on the physical parameters at the location of the spacecraft and f takes account of the relative changes of these parameters along the flux tube.

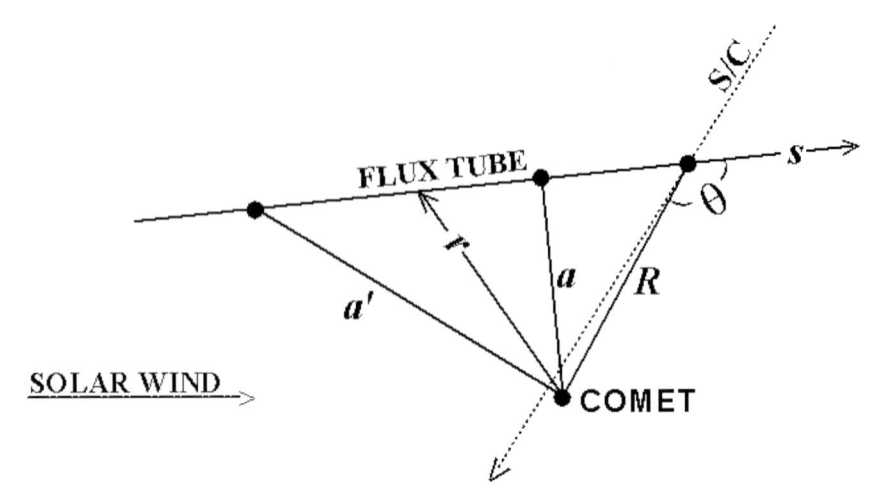

Figure 3. Geometry of the model for the outer coma that leads to an analytical expression connecting local ion density with source strength (eq. 7).

For obtaining an analytical expression for y_1^a, we start out from the conditions in the distant coma and make the following assumptions (cf. Geiss *et al.*, 1987):
1. The flux tube under investigation is straight (cf. Figure 3).
2. A and **V** are constant over the interval of the flux tube where production and destruction are significantly contributing to the density of the ion y_1 measured at the location of the spacecraft.
3. The source density q_1 and the destruction rate d_1 are proportional to $1/r^2$,

$$q_1, d_1 \propto r^{-2}. \tag{5}$$

For y_1^a Equation 1 has then the form of

$$\frac{dy_1^a}{c_0 + c_1 y_1^a} = \frac{ds}{s^2 + a^2} \tag{6}$$

a is the closest distance between the flux tube and the nucleus (Figure 3). Integration yields

$$y_1^a(R) = \frac{q_1(R)}{d_1(R)}\{1 - e^{-\frac{R}{l}\frac{\theta}{\sin\theta}}\} \tag{7}$$

θ is the angle between the local direction of the flux tube and the spacecraft-comet line in the nucleocentric frame of reference (cf. Figure 3). The local destruction length is given by

$$l = V/d_1(R) \tag{8}$$

TABLE II
Parameters for Outer Coma Model ; Giotto Encounter with P/Halley

Ion	R 10^3 km	$V(R)$ km/s	θ degrees	a 10^3 km	a' 10^3 km	f
Ne^+	30	3.5	130	23	30	0.9
Na^+	30	3.5	130	23	37	0.6
Ne^+	50	12	135	41	50	0.8
Na^+	50	12	135	41	52	0.6

There are two limiting cases for y_1^a:

1. For strong destruction, (7) becomes

$$y_1 = y_1^a = \frac{q_1(R)}{d_1(R)} \quad , \quad l \ll R\frac{\theta}{\sin\theta} \tag{9}$$

Equation (9) corresponds to local reaction equilibrium; i.e. we have $f = 1$.

2. If destruction can be neglected, (7) reduces to

$$y_1^a = q_1(R)\frac{R}{V}\frac{\theta}{\sin\theta} \quad , \quad l \gg R\frac{\theta}{\sin\theta} \tag{10}$$

In this case f depends only on variations in A. Changes in V have no influence on f, since mass spectrometers essentially measure fluxes: i.e. the product density × speed.

The "form factor" f takes account of deviations from the conditions 1 and 2 given above. Since the relevant data, e.g. the changes in flux tube cross section and the ion speed, exist only along the spacecraft trajectory, f has to be derived from theory.

We have calculated approximate f values from the relative changes in the flux tube cross section and ion speed given by Gombosi *et al.* (1996). In Table II we give the relevant flow parameters which we shall apply in Section 4 for the interpretation of the Ne^+ and Na^+ observations in Region 3. The angle θ and $V(R)$ are from observation (Altwegg et al., 1993). The θ-values are in fairly good agreement with the theoretical values obtained by Gombosi *et al.* (1996).

The distances from which ions reach the spacecraft are limited by destruction, by the $1/r^2$ dependency of the source term (eq. 5), and by the decreasing width of the flux tube in the upstream direction. Thus, the observed ions are produced in a limited r-interval. For $\theta \geq \pi/2$ and $l \geq R|\cos\theta|$ its lower limit is at $r = a$, where a is the closest distance between the flux tube and the nucleus (Figure 3). The a' values given in Table II (cf. Figure 3) are calculated such that 80 % of the measured y_1 ions are produced in the flux tube between $r = a$ and $r = a'$.

The flow lines resulting from the Gombosi *et al.* (1996) theory are approximately straight in the r-interval from a to a' (cf. Figure 3). The deviations of the actual flux tube from a straight line are $\Delta r/r < 0.03$ for all cases listed in Table II. This and the fact that the f values are not far from unity show that $y_1^a(R)$ obtained from equation 7 is a good first order approximation and that the factorisation used here (eq. 4) is a practical procedure. In the case of minor constituents, equation 7 allows us to calculate $y_1^a(R)$ and adopt f from an existing flow model.

If a' is the same for two species, their "form factors" f are identical, and therefore, the ratio of the two corresponding source strengths is independent of the f values and can be derived in good approximation from the analytical expressions for y_1^a (eq. 7).

The first term of q_1 (eq. 2) and the last term of d_1 (eq. 3) fulfill the condition 3 (eq. 5) in good approximation. For the neon and sodium sources discussed in section 4, these terms dominate the source density and the destruction rate, respectively. Since the condition given in eq. 5 needs to be fulfilled only in a limited r-range, one can also always approximate the last term in equation 2 or the first term in equation 3 by a $1/r^2$ function.

The model presented in this section enables us to derive information on the composition and distribution of sources from ion abundance data at any location in the outer coma.

The Giotto IMS experiment has provided us with valuable composition data in the outer coma of P/Halley which have not yet been fully exploited. We expect to address the following problems with the model introduced in this section: (a) Determine the origin of the C^+ and S^+ ions which are detectable even several hundred thousand km from the nucleus; (b) estimate the relative abundances of the elements C, N and S from the measured abundances of C^+, CO^+, N^+ and S^+ in the outer coma; (c) interpret the evolution of the water group ions (H_3O^+, H_2O^+, OH^+, O^+) to nucleocentric distances of 700000 km, taking into account H_2O and other parent molecules; (d) study gases released from grains (cf. Geiss *et al.*, 1987); (e) analyze whether conclusions regarding extended sources are influenced by jets or other non-spherical release of gas or dust (note that the travel time of gas or dust to these regions is of the order of the 2-day rotation period of P/Halley).

The gas production of P/Wirtanen is expected to be much weaker than the one of P/Halley, and consequently, outer coma models for interpreting ion abundance data will assume an even greater importance for Rosetta.

TABLE III

Result for Neon and Sodium from $Regions$ 1 and 3

Source [1]	R (km)	Ion	$y_1(R)$ (y_1)	$y_1(R)/z_1(R)$ (cm^{-3})	Q_1/Q_{H_2O} (10^{-2})
^{22}Ne (P)	2400	^{22}Ne$^+$	< 0.02	0.0024	< 1.4 × 10^{-3} [2]
^{22}Ne (P)	4500	^{22}Ne$^+$	< 0.02	0.083	< 1.4 × 10^{-3} [2]
^{20}Ne (P)	30000	^{20}Ne$^+$	< 0.3	0.22	< 5.0 × 10^{-3} [2]
^{22}Ne (P)	30000	^{22}Ne$^+$	< 0.15	0.22	< 3.6 × 10^{-2} [2]
^{22}Ne (P)	50000	^{22}Ne$^+$	< 0.02	0.44	< 3.3 × 10^{-2} [2]
Na (P)	2400	Na$^+$	< 0.05	1.96	< 3.0 × 10^{-7}
Na (P)	4500	Na$^+$	< 0.05	3.73	< 5.6 × 10^{-7}
Na$^+$ (E)	2400	Na$^+$	< 0.05	—	< 4.0 × 10^{-5}
Na$^+$ (E)	4500	Na$^+$	< 0.05	—	< 4.1 × 10^{-5}
Na$^+$ (E)	30000	Na$^+$	< 0.015	—	< 1.0 × 10^{-5}
Na$^+$ (E)	50000	Na$^+$	< 0.01	—	< 1.6 × 10^{-4}

[1] P = Point Source; E = Extended Source (equation 11)
[2] $Q_{Ne(total)}/Q_{H_2O}$

4. Results and Conclusions

4.1. NEON

Assuming a point source for highly volatile neon, we have evaluated the source strengths of ^{20}Ne and ^{22}Ne in four R-intervals around 2400 km and 4500 km inside the contact surface and around 30000 km and 50000 km in the outer coma. The count rates at M20$^+$ were corrected for H$_2^{18}$O$^+$ and H$_2$DO$^+$. At M22$^+$, contributions from ions other than ^{22}Ne$^+$ are neglected. Assuming that cometary neon would have the solar ^{20}Ne/^{22}Ne ratio of 13.7 (Geiss *et al.*, 1972; Kallenbach *et al.*, 1997), we find that ^{22}Ne gives better upper limits than ^{20}Ne, except for the measurement at 30000 km.

The results are given in Table III. As expected, the data closest to the nucleus yield the most sensitive upper limits. Thus, the value adopted in Table IV is based mainly on these data. Krasnopolsky *et al.* (1997) reported Ne$_{ice}$/Ne$_{PSC}$ ≤ 0.04 for Hale-Bopp (cf. Stern, 1999). Our upper limit for neon in Halley is about five times lower (Table IV).

TABLE IV
Summary of Results

Source [1]	Relative Source Strengths	Relative to the Abundance in the Protosolar Cloud (PSC) [2]
Ne (P)	$Ne/H_2O < 1.5 \times 10^{-3}$	$Ne/Ne_{PSC} < 8 \times 10^{-3}$ [3]
Na (P)	$Na/H_2O < 7 \times 10^{-7}$	$Na/Na_{PSC} < 2 \times 10^{-4}$
Na^+ (E)	$Na^+/H_2O < 3 \times 10^{-5}$	$Na^+/Na_{PSC} < 8 \times 10^{-3}$
C_3H (P)	$C_3H/H_2O < 1 \times 10^{-3}$	(C in C_3H)/$C_{PSC} < 4 \times 10^{-3}$
C_4 (P)	$C_4/H_2O < 5 \times 10^{-4}$	(C in C_4)/$C_{PSC} < 3 \times 10^{-3}$
C_4H (P)	$C_4H/H_2O = (2.3 \pm .8)10^{-3}$	(C in C_4H)/$C_{PSC} = 1.3 \times 10^{-2}$
C_4H_2 (P)	$C_4H_2/H_2O < 9 \times 10^{-4}$	(C in C_4H_2)/$C_{PSC} < 5 \times 10^{-3}$
CH_3NH_2 (P)	$CH_3NH_2/H_2O < 1.5 \times 10^{-3}$	(N in CH_3NH_2)/$N_{PSC} < 7 \times 10^{-3}$
CH_3CN (P)	$CH_3CN/H_2O = (1.4 \pm .6)10^{-3}$	(N in CH_3CN)/$N_{PSC} = 6.4 \times 10^{-3}$
C_2H_5CN (P)	$C_2H_5CN/H_2O = (2.8 \pm 1.6)10^{-4}$	(N in C_2H_5CN)/$N_{PSC} = 1.2 \times 10^{-3}$
C_3N (P)	$C_3N/H_2O < 1 \times 10^{-3}$	(N in C_3N)/$N_{PSC} < 5 \times 10^{-3}$
HC_3N (P)	$HC_3N/H_2O < 4 \times 10^{-4}$	(N in HC_3N)/$N_{PSC} < 1.8 \times 10^{-3}$

[1] P = Point Source; E = Extended Source (equation 11)
[2] 50 % cometary O assumed in H_2O
[3] relative to the ice phase (cf. Krasnopolsky et al., 1997)

4.2. ATOMIC SODIUM

We have evaluated upper limits for the number density of Na^+ (cf. Table III) and calculated source strengths for two hypothetical sources:

1. A point source of atomic sodium: The acceleration due to radiation pressure by sunlight is significant for Na. We have estimated its effect using $a = 2.0 \times 10^{-4}$ km/s^2 at 0.9 AU (Banaszkiewicz, 1999). Inside the contact surface the combined effects of radiation pressure and drag from the main gas leads to a minor increase in Na density along the Giotto trajectory. We neglect this increase because it would only somewhat reduce the upper limit for the Na point source strength. The drag decreases with $1/r^2$. Thus, the radiation pressure dominates beyond $r = 10^4$ km, and all Na coming from a point source should have crossed Giotto's trajectory closer to the nucleus than 30'000 km.

2. An extended source of sodium compounds that are released from grains producing Na^+ with a delay distance of r_1:

$$q_1 = \frac{Q_1}{4\pi r^2 (r_2 - r_1)} \{1 - e^{-\frac{r}{r_1}}\} \tag{11}$$

Equation 11 is normalized such that Q_1 is the total (spherically symmetric) production rate of Na^+ up to $r = r_2$. We chose $r_1 = 20000$ km and $r_2 = 10^6$ km.

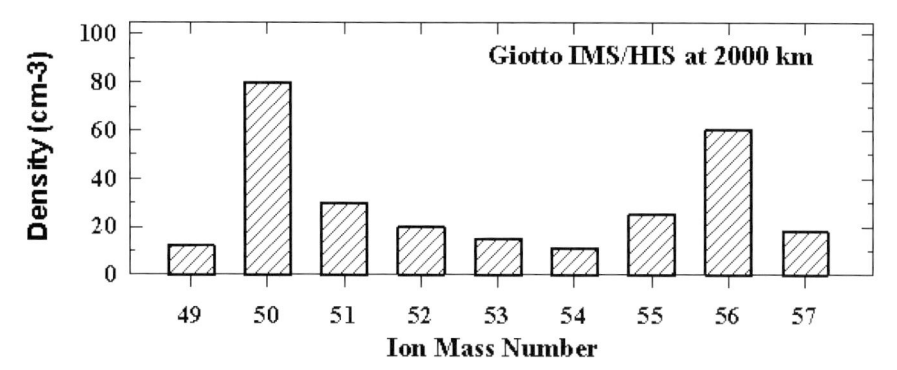

Figure 4. Ion densities in the range M49$^+$ to M57$^+$ in Halley's coma at a distance of 2000 km from the nucleus. The two highest peaks M50$^+$ and M56$^+$ are mainly due to protonated C$_4$H, and C$_2$H$_5$CN, respectively.

We note that outside the contact surface not only photons, but also electrons could play a significant role in the production of Na$^+$. The results are presented in Table III. As expected, the data inside the contact surface are more sensitive for the point source, while the data in the outer coma are more sensitive for the extended source. We have included the resulting upper limits for both types of sodium sources in Table IV. Although these sources are hypothetical in nature, the extremely low values obtained show that very little is released relative to the expected sodium abundance in cometary material.

Our upper limits for the two hypothetical sources of Halley (Table III and IV) can be compared with the data obtained from the sodium tails of Hale-Bopp. The sodium production rate in the narrow tail (Cremonese *et al.*, 1997) and in the diffuse tail (Wilson *et al.*, 1998) together was about 10^{-5} of the water production rate at 0.98 AU (cf. Cremonese, 1999).

4.3. C$_4$H (BUTADIYNYL)

M50$^+$ is the most abundant ion in the range M48$^+$ to M57$^+$ (cf. Figure 4). C$_3$N$^+$ and protonated NaCN and C$_4$H$_2^+$ are the candidates for the M50$^+$ ion.

There are three compelling arguments against NaCNH$^+$: (1) Inside the contact surface, the density of the M50$^+$ ions is at least 10^3 times higher than the density of Na$^+$. Although we do not quantitatively know the photodissociation and -ionization rate constants that lead from NaCN to Na$^+$, it seems very unlikely that the Na$^+$/NaCNH$^+$ ratio would remain below 10^{-3}; (2) an extended source of the type given in (11) would not be compatible with the M50$^+$ data; (3) if M50$^+$ would be mainly NaCNH$^+$, 20 to 100 percent of the total content of sodium (in ices plus grains) would have to be released as NaCN within $R = 1500$ km. We argue that these three points rule out any significant contribution of NaCNH$^+$ to M50$^+$.

Reaction of C$_3$N with H$_3$O$^+$ (Table I) leads to a high C$_3$NH$^+$/C$_3$N$^+$ ratio. Thus, the low abundance of M51$^+$ rules out a significant contribution of C$_3$N to M50$^+$.

This leaves $C_4H_2^+$. Biacethylen (C_4H_2) has a high proton affinity, i.e. reaction with H_3O^+ would lead to a high $C_4H_3^+/C_4H_2^+$ ratio, which is not observed (see Figure 4). Thus, we conclude that the main source of $C_4H_2^+$ is the radical C_4H. Its abundance is included in Table IV.

The well defined decrease of $M50^+$ with r shows that much of the C_4H is due to a point source (or a source close to the nucleus), and it implies a lifetime $> 10^4$ s for C_4H. Our preliminary investigation of the chemical properties of C_4H shows that its reaction with H_2O is endothermic ($\Delta H = 99$ kJ/mol) and that the lowest dissociation channel of C_4H is the reaction $C_4H \rightarrow C_4 + H$ at 5.3 eV. The first excited state of C_4H that could lead to this dissociation needs a photon energy of 5.9 eV. Our preliminary transition moment calculations yield a lifetime of 10^5 s to 10^6 s at 0.9 AU.

For C_3H we give an upper limit in Table IV. The dissociation energy of $C_3H \rightarrow C_3 + H$ is only 3.85 eV. Therefore the photodissociation rate of C_3H could be considerably higher than the C_4H rate. Assuming $\lambda < 10^{-3}$ s^{-1} for C_3H and taking photon-absorption in the near nucleus coma into account (cf. Giguere and Huebner, 1978), we arrive at the upper limit given in Table IV.

In Interstellar Molecular Clouds (IMCs) C_4H has a relatively high abundance (cf. Irvine, 1999; Winnewisser and Kramer, 1999). Also model calculations (Herbst and Leung, 1989; Millar *et al.*, 1991) predict C_4H to be among the most abundant organic species, under some IMC conditions even more abundant than H_2CO or any other C_nH_m with $n > 1$. The C_nH with an even number of C-atoms have systematically higher abundances than those with an odd number of C-atoms. This odd-even effect and a particularly high C_4H abundance is also predicted by theory, resulting from a carbon insertion process (Herbst and Leung, 1989).

Our data indicate that the odd-even effect exists also in the C_nH radicals of Halley, i.e. C_4H is much more abundant than C_3H (Table IV). Since the reaction of C_4H with H_2O is endothermic, it could have been incorporated into the ice phase at low temperature and, surviving the long storage time and sublimation process, emerges now as a point source in the coma of Halley.

The odd-even effect in the C_nH abundance is a very specific result of IMC chemistry. Thus, the C_nH group and other unsaturated radicals with unpaired electrons could turn out to be indicators for an IMC origin of cometary material. We intend to further investigate this question.

4.4. Nitrogen Compounds

Using the IMS/HIS data in the ionosphere of P/Halley, we have searched for nitrogen compounds in the M/Q range above 30 (cf. Figures 4 and 5) and were able to identify methyl cyanide (CH_3CN) and (probably) ethyl cyanide (C_2H_5CN). The source strengths of these molecules are given in Table IV, along with upper limits for some other nitrogen bearing species.

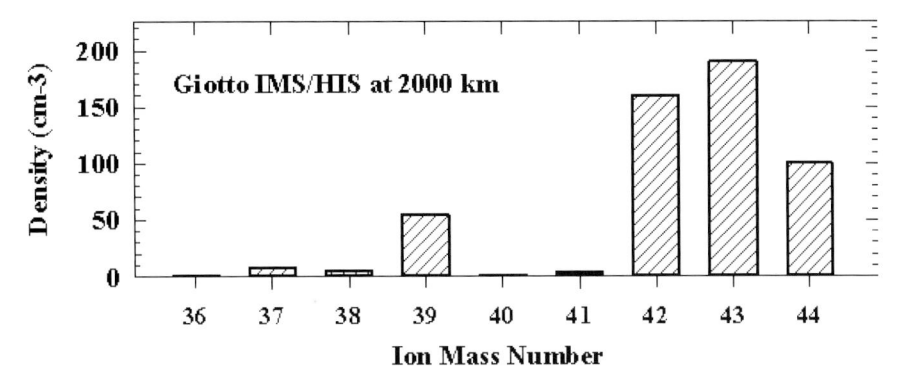

Figure 5. Ion densities in the range M36$^+$ to M44$^+$ in Halley's coma at a distance of 2000 km from the nucleus. The ions at M39$^+$, M42$^+$ and M43$^+$ are principally due to protonated C_3H_2 (Altwegg et al., 1999), CH_3CN (this work) and $CHCHO + C_3H_6$ (not discussed in this paper). M44$^+$ represents mainly CO_2^+.

Most organic molecules with M > 30 have a higher proton affinity than water and therefore, protonated molecules dominate the ion mass spectra. If organic molecules are made up of H, C and O, protonation leads to odd ion mass numbers, if such molecules contain one nitrogen atom, protonation gives ions with even mass numbers.

Next to the M18$^+$ ion (made up of ~50 percent of NH_4^+), M42$^+$ is the most abundant even numbered ion observed inside the contact surface. M42$^+$ is well separated from its neighbours (Figure 2). Ionized C_3H_6 and CH_2CO or protonated CH_3CN, CHCO or C_3H_5 could contribute to M42$^+$. A major contribution from C_3H_6 and CH_2CO is ruled out because these molecules react with H_3O^+, which would result in a much higher M43$^+$/M42$^+$ ratio than is observed (Figure 5). We also rule out protonated CHCO or C_3H_5 as the major contributors to M42$^+$, because this would mean exceptionally high abundances for these radicals (cf. the abundances of C_2H or C_4H_n given by Altwegg et al. (1999) and in Figure 4 of this paper). The mass spectra in Figures 4 and 5 show that the radicals C_3H_5 or CHCO would have to have higher abundances even than the unsaturated hydrocarbons C_3H_2, C_3H_4, C_4H_2 and C_4H_4, or propynal (HC_3HO). If the IMC chemistry is a guide to the systematics in comets of radicals with an unpaired electron, then low abundances of CHCO and C_3H_5 are expected. The model calculations of Millar et al. (1991) give much less protonated CHCO than protonated C_4H, C_3H or CH_3CN. As for C_3H_5, there is the general observation that highly unsaturated organics are much more abundant in IMCs than the more saturated ones like C_3H_5. Thus, we conclude that about 80 percent of M42$^+$ is CH_4CN^+ and obtain the source strength for methyl cyanide given in Table IV. The identification of CH_3CN is further supported by the presence of M42$^+$ in the chemical pile-up region, which is expected from the low destruction rate of this molecule (Table I).

The CH_3CN abundance in the comets Hyakutake and Hale-Bopp (cf. Crovisier and Bockelée-Morvan, 1999) is considerably lower than our value for Halley. We intend to further study this and the differences in the abundances of other rare species between these three comets.

Our arguments for the identification of C_2H_5CN ($M56^+$ in Figure 4) are the same as for CH_3CN. However, since we have only estimates for the rate constants of ethyl cyanide, we consider its identification to be only probable at this time.

From the results presented in Table IV and Figures 4 and 5, and from those given by Altwegg et al. (1999) and Crovisier and Bockelée-Morvan (1999), we obtain N/C $\leq$ 0.08 for the material released into the gas phase for $R \leq$ 8000 km. This ratio is quite well established since $\sim$ 60% of carbon is present as CO + H_2CO (cf. Eberhardt, 1999) and $\sim$70% of N as NH_3 (Altwegg et al., 1999, and this work), all of which are well studied.

N/C = 0.08 is 4 times lower than the protosolar value of 0.3 (Anders and Grevesse, 1989). In the dust phase beyond 8000 km, Jessberger and Kissel (1991) determined an average N/C ratio of 0.05. In spite of the rather large uncertainty of this determination, any combination of gas and dust gives an N/C ratio for Halley that is lower than the protosolar value. It is very unlikely that the low N/C value in the gas phase could be compensated by a high N/C ratio in the dust, considering that also C1 chondrites, even more depleted in volatiles than Halley dust, have N/C = 0.04. Thus, we confirm the earlier conclusion (Geiss, 1987; Jessberger and Kissel, 1991; Encrénaz and Knacke, 1991) that nitrogen is depleted in Comet Halley. This is probably due to the loss of N_2 either at the time of condensation or later in the history of the cometary material.

Acknowledgements

We want to acknowledge the teams at the University of Bern, the Max-Planck-Institut für Aeronomie, Lindau, the Jet Propulsion Laboratory, Pasadena, and the Lockheed Space Science Laboratory, Palo Alto, for their splendid work in building IMS Giotto. The design of the HIS sensor, the data from which are discussed in this paper is due to Helmut Rosenbauer. We are grateful to Tamas Gombosi and Kenneth Hansen for making available detailed data from their MHD model and Walter Huebner, Wesley T. Huntress and Elmar Jessberger for discussions. We thank Ursula Pfander and Silvia Wenger for their help with the manuscript.

References

Anders, E., and Grevesse, E.: 1989, *Cosmochim. Acta* **46**, 2363-2380.
Altwegg, K., Balsiger, H., Geiss, J., Goldstein, R., Ip, W.-H., Meier, A., Neugebauer, M., Rosenbauer, H., and Shelley, E.G.: 1993, *Astron. Astrophys.* **279**, 260-266.
Altwegg, K., Balsiger, H. and Geiss, J.: 1999, *Space Sci. Rev.*, this volume.

Anicich, V.G., 1993, *J. Phys. Chem. Ref. Data* **22 (6)**, p. 14-96.
Anicich, V.G. and Huntress, W.T. Jr.: 1986, *Astrophys.J. Suppl. Series* **62**, 553.
Balsiger, H. et al.: 1986, *Nature* **321**, 330-336.
Balsiger, H. et al.: 1987, *J. Phys. E. Sci. Instrum.* **20**, 759.
Banaszkiewicz, M.: 1999, personal communication
Bockelée-Morvan, D. and Crovisier, J.: 1985, *Astron. Astrophys.* **151**, 90-100.
Cremonese, G., Boehnhardt, H., Crovisier, J., Fitzsimmons, A., Fulle, M., Licandro, J., Pollacco, D., Rauer, H., Totzi, G.P., West, R.M.: 1997, *Astrophys. J.* **94**, L199
Cremonese, G.: 1999, *Space Sci. Rev.*, this volume.
Crovisier, J., and Bockelée-Morvan, D.: 1999, *Space Sci. Rev.*, this volume.
Eberhardt, P.: 1999, *Space Sci.Rev.*, this volume.
Encrénaz, T., and Knacke, R.:1991, in *Comets in the Post-Halley Era*, R. Newburn, M. Neugebauer, and J. Rahe (eds.), Kluwer Dordrecht, 107-137.
Geiss, J., Bühler, F., Cerutti, H., Eberhardt, P. and Fillieux, C.: 1972, *Apollo 16 Preliminary Science Report*, NASA SP-315, Section 14/1-10.
Geiss, J.:1987, *Astron. Astrophys.* **187**, 859-866.
Geiss, J., Bochsler, P., Ogilvie, K.W., and Coplan, M.A.: 1987, *Astron. Astrophys.* **166**, L1-L4.
Geiss, J., Altwegg, K., Anders, E., Balsiger, H., Ip, W.-H., Meier, A., Neugebauer, M., Rosenbauer, H., and Shelley, E.G.: 1991, *Astron. Astrophys.* **247**, 226-234.
Geiss, J.: 1998, *Space Sci. Rev.* **85**, 241-252, and *Space Science Series of ISSI* **5**, 241-252.
Giguere, P.T. and Huebner, W.F.: 1978, *Astrophys. J.* **223**, 638-654.
Gombosi, T.I., DeZeeuw, D.L., Häberli, R.M. and Powell, K.G.: 1996, *J. Geophys.Res.* **101**, 15233-15243.
Häberli, R., Altwegg, K., Balsiger, H., and Geiss, J.: 1995, *Astron. Astrophys.* **297**, 881.
Häberli, R., Altwegg, K., Balsiger, H., and Geiss, J.: 1996, *J. Geophys. Res.* **101, No.A7**,, 15, 579-15, 589.
Häberli, R., Combi, M.R., Gombosi, T.I., DeZeeuw, D.L. and Powell, K.G.: 1997, *Icarus* **130**, 373-386.
Herbst, E. and Leung, C.M.: 1989, *Astrophys J. Suppl. Ser.* **69**, 271-300.
Huebner, W.F., Boice, D.C., Schmidt, H.U., and Wegmann, R.: 1991, in *Comets in the Post-Halley Era*, R. Newburn, M. Neugebauer, and J. Rahe (eds.), Kluwer Dordrecht, 907-926.
Huebner, W.F. and Keady, J.J., Lyon, S.P.: 1992, *Astrophys. and Space Sci.* **195**, 1-294.
Ip, W.H. and Axford, I.: 1990, in *'Physics and Chemistry of comets'*, W.F. Huebner (ed.), Springer, Heidelberg, 177 - 234.
Irvine, W.: 1999, *Space Sci. Rev.*, this volume.
Jessberger, E. and Kissel, J.; 1991, in *Comets in the Post-Halley Era*, R. Newburn, M. Neugebauer, and J. Rahe (eds.), Kluwer Dordrecht, 1075-1092.
Kallenbach, R. et al.: 1997, P*J. Geophys. Res.* **102**, 26895-26904.
Korth A., Krueger, F.R., Mendis, D.A., and Mitchell, D.L.: 1989, in *Asteroids, comets, meteors III*, C.-I. Lagerkvist, H. Rickman, B.A. Lindblad, an M. Lindgreen (eds.), Uppsala Univ. Press, Uppsala.
Körösmezey, A., Cravens, T.E., Gombosi, T.I., Nagy, A.F., Mendis, D.A., Szegö, K., Gribov, B.E., Sagdeev, R.Z., Shapiro, V.D. and Shevchenko, V.I.: 1987, *J. Geophys. Res.* **92**, 7331.
Krasnopolsky, V.A., Mumma, M.A., Abbott, M., Flynn, B.C., Meech, K.J., Yeomans, D.K., Feldman, P.D., Cosmovici, C.B.: 1997, *Science* **277**, 1488.
Meier, A.: 1988, *PhD Thesis*, University of Bern, Bern, Switzerland.
Millar, T.J., Rawlings, J.M.C., Bennet, A., Brown, P.D., and Charnley, S.B.: 1991, *Astron. Astrophys. Suppl. Ser.* **87**, 585.
Neubauer, F. et al.: 1986, *Nature* **321**, 352-355.
Schmidt, H.U., R. Wegmann, W.F. Huebner, and Boice, D.C.: 1988, *Computer Phys. Comm.* **49**, 17-59.

Stern, S.A.: 1999, *Space Sci. Rev.*, this volume.
Wilson, J.K., Mendillo, M., Baumgardner, J.: 1998, *Geophys. Res. Let.* **25**, 225.
Winnewisser, G. and Kramer, C.: 1999, *Space Sci. Rev.*, this volume.

Address for correspondence: Johannes Geiss, International Space Science Institute, Hallerstr. 6, 3012 Bern, Switzerland, johannes.geiss@issi.unibe.ch

ION IRRADIATION AND THE ORIGIN OF COMETARY MATERIALS

G. STRAZZULLA
Osservatorio Astrofisico di Catania, Città Universitaria, I-95125 Catania, Italy

Abstract. For about 20 years laboratory research has been carried out on the effects induced by energetic ions on materials (ices, silicates, carbons) of cometary relevance. Here I present some recent results and outline the relevance such laboratory investigations might have for understanding the origin of cometary materials.

1. Introduction

One of the key objectives of modern research on cometary physics is to understand the origin of the materials that form cometary nuclei. Is it possible that cometary matter reflects the composition of pre-cometary (interstellar, IS) grains? To what extent are IS grains "reprocessed" in the solar nebula? To what extent does energetic processing (UV and/or energetic particle irradiation) modify pre-cometary grains and/or grains that are reformed in the solar nebula and are accreted to form comets, and/or the entire comet once stored in the (different) regions (e.g. Kuiper belt, Oort cloud)?

I believe it is not yet possible to answer the above questions. One of the main reasons is that we have only a rough knowledge of the actual composition of comet nuclei (are they all the same?). Again we have sparse information about cometary aging and the relative role of different processes such as thermally and/or radiation driven evolution.

It is then particularly relevant to approach the problems from several different points of view: ground and space based observations are obviously needed as well as theoretical and laboratory work.

Some groups have been involved for about 20 years in laboratory research on the effects induced by energetic ions on materials (ices, silicates, carbons) of cometary relevance (for a review see Strazzulla, 1997). Here I present some recent results and try to outline the relevance such laboratory investigations might have for understanding the origin of cometary materials.

2. The Organic Crust

UV photons and cosmic ions deposit a large amount of energy on grains in the interstellar medium. Thus, based on experimental results obtained in several laboratories, we can expect, starting from simple ices, the production of many new species including organic refractory materials (e.g. Jenniskens *et al.,* 1993). The irradiation history, during comet accretion, may be at least as important as that in the ISM. This is particularly true if we assume that comet accretion occurred during an active phase of the young Sun (T-Tauri phase). Thus even if one would believe that cometary materials are fully re-formed in the solar nebula, irradiation processes could drive the formation of compounds quite similar to those present in the interstellar medium.

Once formed, comets are exposed for about 4.6×10^9 yr to the flux of galactic cosmic rays. Strazzulla and Johnson (1991) presented a review of the effects expected on comets and cometary debris. In particular they suggested that a comet exposed to background particle radiation obtains an outer web of non-volatile material which will lead to the formation of a substantial "crust". Other experiments (Strazzulla and Baratta, 1992) demonstrated that the organic crust has already been formed during bombardment at low temperature. This gives credit to the hypothesis that a cometary crust can already be formed during the long stay far from the Sun and its development does not require a first passage (heating) in the inner Solar System. Experiments on targets much thicker than the penetration depth of irradiating ions showed that an organic crust remains intact after the sublimation of underlying ices, thus giving support to the hypothesis that the ion-produced cometary organic crust can "survive" gas ejection from deeper layers (i.e. passages near the Sun) and it can be as old as the comet itself (Strazzulla *et al.,* 1991).

The existing observational data on the colors of cometary nuclei, Halley's in particular, and of some small bodies in the outer Solar System are compatible both with the hypothesis that comets have a substantial organic crust and that only the very outer cometary skin is dark. In the latter case darkness might be due to an organic-rich layer or to a very porous silicate skin. Hopefully, future cometary missions will answer the question about the existence of a substantial organic crust. Suppose that the crust exists. How thick is it? Strazzulla and Johnson (1991) estimated that the external (0.1-0.5 m) layers of a comet were subjected to an irradiation dose of 600 eV/molecule. Deeper layers were subjected to a lower dose because the most abundant but less energetic ions are stopped by the external layers. That estimate is based on the experimentally measured cross section ($\simeq 10^{-17}$ cm^2) for the conversion of frozen methane to a refractory residue (Foti *et al.,* 1984; Strazzulla *et al.,* 1984). We have now measured the fraction of carbon incorporated into the refractory material for more cases. That fraction is given by an exponential function: $R = 1 - e^{-\sigma \times \phi}$. Where ϕ (ions/cm^2) is the ion fluence and σ (cm^2) the cross section for the process. The experimental results are shown in Fig. 1 where the fraction of converted carbon is shown vs. ion fluence for irradiated frozen methane (10 K), frozen benzene (77 K) and for polystyrene (9000 amu, 300 K). In the latter case, the remaining fraction means the fraction of insoluble residue left over after irradiation. In Fig. 1 points refer to the experimental data and

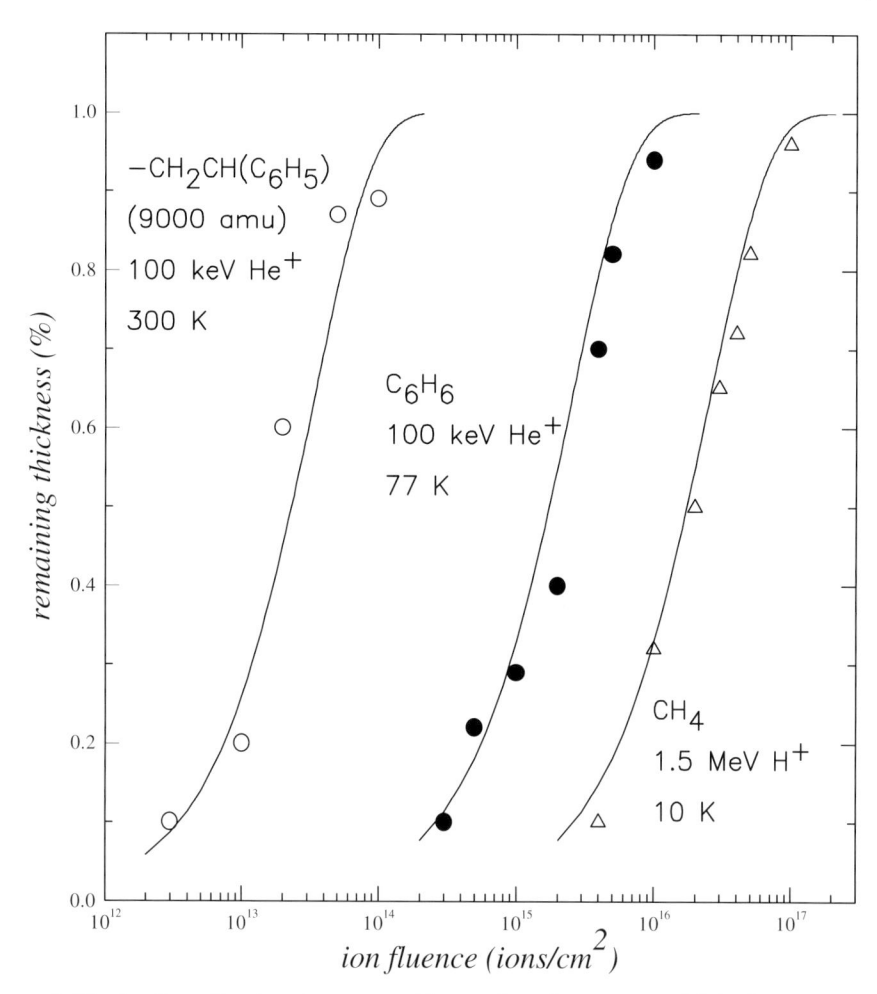

Figure 1. The fraction of carbon incorporated into a the refractory material is shown versus ion fluence for irradiated frozen methane (10 K), frozen benzene (77 K) and for polystyrene (9000 amu, 300 K). In the latter case the remaining fraction means the fraction of insoluble residue left over after irradiation. Data points have been fitted with an exponential function: $R = 1 - e^{-\sigma \times \phi}$, where ϕ (ions/cm^2) is the ion fluence and σ (cm^2) the cross section for the process. The measured cross section of the processes are 4×10^{-17} cm^2 for methane, 4×10^{-16} cm^2 for benzene and 3×10^{-14} cm^2 for polystyrene.

curves are the exponential fits. The cross section of the processes are 4×10^{-17} cm^2 for methane, 4×10^{-16} cm^2 for benzene and 3×10^{-14} cm^2 for polystyrene.

These results indicate that whatever the hydrocarbon is, it is converted to a refractory, insoluble residue. The cross section of the process depends on the molecular weight of the original material (it is about 10^3 times higher for polystyrene than for methane). This implies that the original material going into formation of comets already contains organics with high molecular weight (as could be the case for organics already present on interstellar pre-cometary grains). Therefore the crust formation process is much more efficient than in the case considered by Strazzulla and Johnson (1991). Thus the crust could be much thicker than the 0.1-0.5 m estimated by Strazzulla and Johnson (1991).

3. Specific Molecules

As stated above, we have only a rough knowledge of the actual composition of the cometary nuclei. Thus it is extremely difficult to say if there are specific molecules whose presence clearly indicate the origin of cometary materials. In particular, no specific molecule has yet been identified in the laboratory whose detection may clearly indicate that ion processing has been responsible for its formation. All of the parent molecules identified on comets are also observed on icy grains in dense molecular clouds, where they could be formed by direct condensation from the gas phase (e.g., CO), or by chemical reactions on grain surfaces (e.g. CO_2) or by energetic processing (e.g., CO and CO_2 produced by ion irradiation of water/methanol mixtures). Also, the so-called organic refractory residues presumed to be formed by energetic processing of simple ices containing hydrocarbons could in fact be produced in different ways (e.g., around carbon-rich evolved stars). In the latter case, however, one can expect that the organic material is the core on which ices can form in dense clouds. In the former case the organics should be formed as mantles on silicate cores (e.g. Greenberg, 1982). Although some PIA-PUMA data seem to indicate the latter might be the case (e.g., Kissel and Kruger, 1987), no firm conclusion can be drawn at present. I believe that only future cometary missions (e.g. the COSIMA instrument on board Rosetta) and/or sample return can answer to the correctness of the core-mantle model.

There is overwhelming evidence that the production rates of several volatile species (e.g., CO, CN, H_2CO, etc.) observed in comet comae cannot be explained by direct sublimation from the nucleus but require a major contribution from distributed sources (e.g., Mumma *et al.*, 1993; Crovisier and Bockelée-Morvan, 1999).

A large number of experiments indicate that ion irradiation, among other effects, induce the formation of different molecules. These latter include not only species more volatile than the irradiated ones (e.g., CO is produced by irradiation of CO_2) or very refractory (e.g., the organic material left over after irradiation of hydrocarbons) but also a large variety of species with intermediate volatility. These latter could help in explaining the distributed sources in the coma. In fact, if present in cometary nuclei, the species of intermediate volatility could be expelled with sublimating gas and/or dust and, when photodissociated far from the nucleus, produce the observed distributed sources.

In particular, Brucato *et al.* (1997) suggested some relatively stable compounds be responsible for the CO distributed sources: Namely carbon suboxides (C_3O_2 and/or C_3O) obtained by irradiating CO, CO:O_2, and CO:N_2 frozen mixtures

(Strazzulla *et al.*, 1997) and carbonic acid obtained by ion bombardment of frozen mixtures of $H_2O:CO_2$ (Moore and Khanna, 1991; Brucato *et al.*, 1996). The latter could also provide a source for formaldehyde that has been detected as well in the mass spectrum of the sublimating residue (Moore and Khanna, 1991).

Recently new experiments have been performed irradiating ternary mixtures $H_2O:NH_3$(or N_2)$:CH_4$ (Palumbo *et al.*, 1998): it has been shown that among the newly synthesized species there is an R-O-C≡N group (R is a group in the organic residue) which exhibits a band at about 4.62 μm. This band remains after warming up to room temperature and furnishes the best available fit to the so-called X-C≡N band observed towards some young stellar sources still embedded in their placental cloud (Pendleton *et al.*, 1999). Here I suggest that the sublimation of such a species could be responsible for the distributed CN emission in cometary comae.

In conclusion, I believe that ion irradiation could be an important process to be considered to understand the origin and evolution of cometary material. To fully understand its effective role it is important to establish, by future space missions, the nature of the cometary surfaces (organic vs. silicatic), the depth of the organic crust (if it exists) and the parent species responsible as distributed sources of some observed cometary species.

References

Brucato J.R., Palumbo M.E. and Strazzulla G.: 1996, 'Ion irradiation of frozen water-carbon dioxide mixtures', *Icarus* **125**, 135-144.

Brucato, J.R., Castorina, A.C., Palumbo, M.E., Strazzulla, G., Satorre, M.A.: 1997, 'Ion irradiation and extended CO emission in cometary comae', *Planet. Space Sci.* **45**, 835-840.

Cremonese, G.: 1999, 'Hale-Bopp and its sodium tails', *Space Sci. Rev.*, this volume.

Crovisier, J. and Bockelée-Morvan, D.: 1999, 'Remote observations of the composition of cometary volatiles', *Space Sci. Rev.*, this volume.

Foti, G., Calcagno, L., Sheng, K.L. and Strazzulla, G.: 1984, 'Micrometer-sized polymer layers synthesized by MeV ions impinging on frozen methane', *Nature* **310**, 126-128.

Greenberg, J. M.: 1982, 'What are comets made of? A model based on interstellar grains', in *Comets* L.L. Wilkening ed., University of Arizona Press, Tucson, p. 131-163.

Jenniskens, P., Baratta, G.A., Kouchi, G.A., de Groot, M., Greenberg, J.M., and Strazzulla, G.: 1993, 'Carbon dust formation on interstellar grains', *Astron. Astrophys.* **273**, 583-600.

Kissel, J. and Krueger, F.R.: 1987, *Nature* **326**, 755-760.

Moore, M.H. and Khanna, R.K.: 1991, 'Infrared and mass spectral studies of proton irradiated H_2O+CO_2 ice: evidence for carbonic acid', *Spectrochimica Acta* **47**, 255-262.

Mumma, M.J., Weissman, P.R. and Stern, S.A.: 1993, in *Planets and Protostars III* E.H. Levy, J.I. Lunine, M.S. Matthews eds., University of Arizona Press, Tucson, p. 1171-1252.

Palumbo, M.E., Strazzulla, G., Pendleton, Y.J. and Tielens, A.G.G.M.: 1999, 'R-O-C≡N species produced by ion irradiation of ice mixtures: comparison with astronomical observation', *Astrophys. J.* , submitted.

Pendleton, Y.J., Tielens, A.G.G.M., Tokunaga , A.T., and Bernstein, M.P.: 1999, 'The interstellar 4.62 μm band', *Astrophys J.*, in press.

Strazzulla, G.: 1997, 'Ion bombardment of comets', in *From Stardust to Planetesimals* Y.J. Pendleton, A.G.G.M. Tielens eds., ASP Conf Series Book, S. Francisco, p. 423-433.

Strazzulla, G. and Baratta, G.A.: 1992, 'Carbonaceous material by ion-irradiation in space', *Astron. Astrophys.* **266**, 434-438.
Strazzulla, G. and Johnson, R.E.: 1991, 'Irradiation effects on comets and cometary debris', in *Comets in the Post-Halley Era* R. Jr Newburn, M. Neugebauer, J. Rahe eds., Kluwer, Dordrecht, p. 243-275.
Strazzulla, G., Calcagno, L. and Foti, G.: 1984, 'Build up of carbonaceous material by fast protons on Pluto and Triton', *Astron. Astrophys.* **140**, 441-444.
Strazzulla, G., Baratta, G.A., Johnson, R.E. and Donn, B.: 1991, 'Primordial comet mantle: irradiation production of a stable', organic crust, *Icarus* **91**, 101-104.

Address for correspondence: Gianni Strazzulla, gianni@alpha4.ct.astro.it

FORMATION AND PROCESSING OF ORGANICS IN THE EARLY SOLAR SYSTEM

JOHN F. KERRIDGE
Department of Chemistry, University of California, San Diego, 9500 Gilman Drive, La Jolla, CA 92093-0424 USA

Abstract. Until pristine samples can be returned from cometary nuclei, primitive meteorites represent our best source of information about organic chemistry in the early solar system. However, this material has been affected by secondary processing on asteroidal parent bodies which probably did not affect the material now present in cometary nuclei. Production of meteoritic organic matter apparently involved the following sequence of events: Molecule formation by a variety of reaction pathways in dense interstellar clouds; Condensation of those molecules onto refractory interstellar grains; Irradiation of organic-rich interstellar-grain mantles producing a range of molecular fragments and free radicals; Inclusion of those interstellar grains into the protosolar nebula with probable heating of at least some grain mantles during passage through the shock wave bounding the solar accretion disc; Agglomeration of residual interstellar grains and locally produced nebular condensates into asteroid-sized planetesimals; Heating of planetesimals by decay of extinct radionuclides; Melting of ice to produce liquid water within asteroidal bodies; Reaction of interstellar molecules, fragments and radicals with each other and with the aqueous environment, possibly catalysed by mineral grains; Loss of water and other volatiles to space yielding a partially hydrated lithology containing a complex suite of organic molecules; Heating of some of this organic matter to generate a kerogen-like complex; Mixing of heated and unheated material to yield the meteoritic material now observed. Properties of meteoritic organic matter believed to be consistent with this scenario include: Systematic decrease of abundance with increasing C number in homologous series of characterisable molecules; Complete structural diversity within homologous series; Predominance of branched-chain isomers; Considerable isotopic variability among characterisable molecules and within kerogen-like material; Substantial deuterium enrichment in all organic fractions; Some fractions significantly enriched in nitrogen-15; Modest excesses of L-enantiomers in some racemisation-resistant molecules but no general enantiomeric preference. Despite much speculation about the possible role of Fischer-Tropsch catalytic hydrogenation of CO in production of organic molecules in the solar nebula, no convincing evidence for such material has been found in meteorites. A similarity between some meteoritic organics and those produced by Miller-Urey discharge synthesis may reflect involvement of common intermediates rather than the operation of electric discharges in the early solar system. Meteoritic organic matter constitutes a useful, but not exact, guide to what we shall find with in situ analytical and sample-return missions to cometary nuclei.

1. Introduction

Probably our best opportunity to investigate the origin and evolutionary history of organic matter in the early solar system will be through the analysis of comet nucleus samples. Until such samples are available in pristine condition, however, our next-best source of such information consists of primitive meteorites, particularly those known as carbonaceous chondrites. Although a cometary origin has occasionally been proposed for these meteorites, the present consensus is that they are derived from asteroids (Zolensky and McSween, 1988). This difference is significant because it is likely that asteroids were affected by secondary processes which did not occur on cometary nuclei. Interpreting the meteoritic record in terms of the nature of cometary, i.e., primitive, organic matter therefore requires one to correct for the effects of asteroidal, i.e., secondary, processing. This paper will consist of five sections: A brief review of the properties and genesis of carbonaceous chondrites; A description of the molecular and isotopic characteristics of meteoritic organic matter; An assessment of possible mechanisms that might have been responsible for synthesis of that organic matter; An outline of the likely scenario involved in the production of meteoritic organics; And a summary of implications for the origin and evolution of organic matter in comets.

2. Primitive Meteorites

Carbonaceous chondrites are extraterrestrial, organic-rich, sedimentary rocks. They contain up to about 4wt% carbon, mostly in the form of organic matter but also including some carbonates and minor amounts of elemental carbon (graphite and nanodiamonds) and refractory carbides. The organic matter occurs dispersed throughout a matrix lithology consisting of a disequilibrium mixture of high- and low-temperature minerals, e.g., mafic silicates such as olivine and pyroxene, and clays and water- soluble salts. Lack of equilibrium is also apparent in the range of coexisting oxidized and reduced species, and in the juxtaposition of hydrated and anhydrous minerals. Petrologic study reveals evidence for partial aqueous alteration of an initially anhydrous lithology, combined with impact brecciation and compaction (e.g. Kerridge and Bunch, 1979). Stable-isotope thermometry indicates that this alteration took place at about room temperature, though some data suggest a temperature somewhat over $100°C$ (Clayton and Mayeda, 1984). A record of solar-wind and solar-flare irradiation preserved within some minerals points towards a near-surface location on an airless solar-system body for this secondary processing (Woolum and Hohenberg, 1993). The number of parental asteroids responsible for the carbonaceous meteorites in our collections is not known but is certainly more than half a dozen. Also, it should be noted that relatively few such meteorites have been analyzed in depth; in fact, almost all rigorous studies to date have been carried out on just one meteorite, the Murchison carbonaceous chondrite which fell in 1969. Radiometric dating of the products of secondary mineralization in carbonaceous chondrites shows that they were formed within a few million years of the formation of the first solids in the solar nebula (Macdougall *et al.*, 1984). Alteration of the organic matter originally accreted by the parent asteroids would have taken place at the same time. Carbonaceous chondrites tend to be quite porous, so that organic analysis is hampered by prevalent terrestrial contamination.

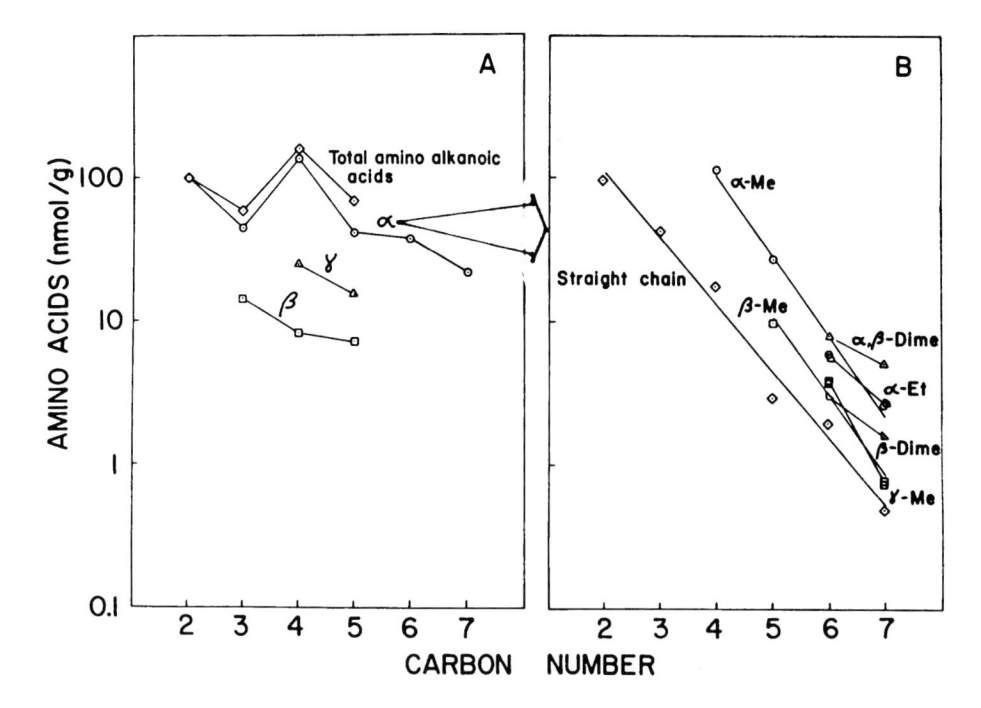

Figure 1. Abundance of amino alkanoic acids as a function of C number in the Murchison meteorite. Right-hand panel shows a-amino acids broken down into homologous series. Me = methyl substituted, Et = ethyl substituted. After Cronin *et al.* (1988).

3. Characteristics of Meteoritic Organic Matter

3.1. MOLECULAR FEATURES

Only about 10% of the organic matter in a meteorite exists in the form of soluble, and therefore characterizable, molecules; the rest consists of an insoluble macromolecular material superficially resembling terrestrial kerogen. The complexity of this material reflects a complex history, disentangling which has proved to be quite challenging. The kerogen-like fraction is therefore much less informative about primordial organic matter than are the soluble molecules, though even they, too, have been affected by secondary alteration.

Three striking characteristics of the soluble material are illustrated, for the amino acids, in Figure 1, which shows the abundance of individual amino alkanoic acids as a function of carbon number. Although the left-hand panel, giving total abundances, reveals no systematic pattern, the right-hand panel, in which the amino acids are broken down into homologous series, reveals the following systematic features (Cronin *et al.*, 1988): Each homologous series shows a regular decline in abundance with increasing carbon number; Complete structural diversity is apparent (every amino acid that could occur, does occur); Branched-chain compounds are more abundant than the corresponding straight-chain variety. These features will serve as effective constraints on possible synthesis scenarios. Despite occasional reports to the contrary, most of the chiral compounds occurring in meteorites are racemic. However, a recent study of 2-amino-2,3-dimethylpentanoic acid has found evidence for a $\sim$ 10% L-enantiomeric excess (Cronin and Pizzarello, 1997). This is significant for two reasons: This amino acid is not found in nature and is therefore most unlikely to be a terrestrial contaminant; Because it lacks an α-hydrogen, it is highly resistant to racemization, and is therefore most likely to have retained any primordial enantiomeric excess. The cause of the chiral excess is not known, though selective destruction by circularly polarized light emanating from neutron stars has been invoked (Bonner and Rubenstein, 1987). It is generally believed that establishment of the homochirality characteristic of life would have required some level of chirality in the prebiotic reservoir from which the first life-forms emerged. The magnitude of the excess observed by Cronin and Pizzarello may have been large enough to permit protobiotic amplification to achieve the required homochirality. The molecular characteristics of the amino acids described earlier (abundance decline, structural diversity, prevalence of branched-chains) are found also in many of the > 400 compounds identified in meteorites to date, though some apparent exceptions are found. How significant, or even real, those exceptions may be will require further work. Compared with the soluble fraction, little molecular information can be derived from the kerogen-like material. It consists of relatively small aromatic moieties interconnected by aliphatic cross links and side chains (Hayatsu *et al.*, 1983). By terrestrial kerogen standards, it is relatively mature, with a C/H ratio of about 2, most of the C being present in aromatic rings and most of the H in aliphatic functional groups. Estimates of the number of rings in each aromatic moiety range from four to 60; the mean value is probably somewhat greater than 10 (Kerridge *et al.*, 1987).

3.2. Isotopic Features

Meteoritic organic matter reveals considerable isotopic variability. It is important to bear in mind that variations in isotopic composition can result from either fractionation during physical or chemical processing, or admixture of material with a different presolar history. Distinguishing between these possibilities can be quite difficult. Furthermore, many meteoritic compounds have resulted from processing of one or more generations of precursor compounds. It is therefore necessary to distinguish diagnostic isotopic characteristics of the final product from those of its precursors, which were almost certainly formed under very different circumstances. For example, C in an alkyl functional group of an amino acid could have been derived from an aldehyde or ketone precursor, whereas carboxyl-group C could have been inherited from a cyanide precursor (Kerridge, 1991).

Isotopic studies have been carried out on many classes of organic compound and on some individual compounds. Agreement among different studies has been

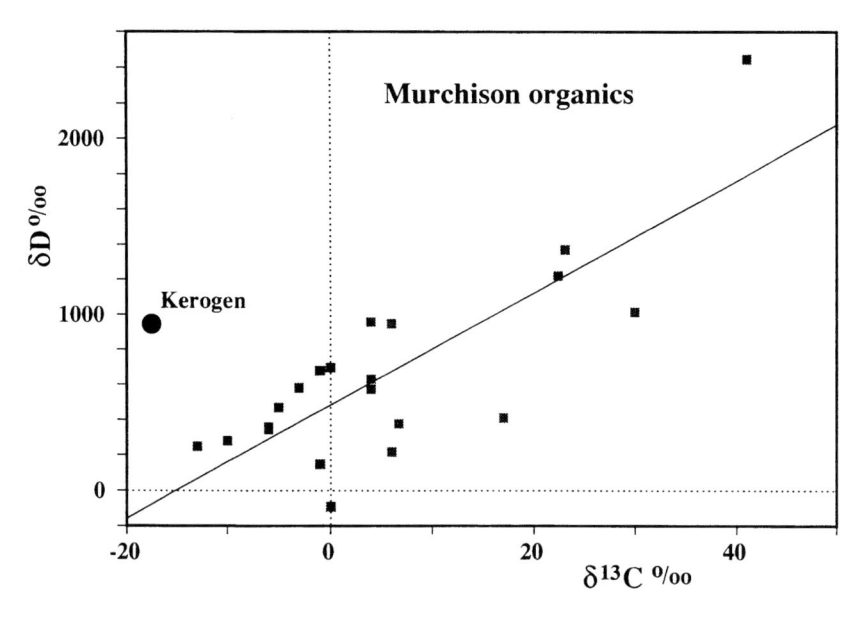

Figure 2. Carbon and H isotopic compositions of different organic species in Murchison. Note departure of kerogen point from general trend. Isotopic compositions given as deviations in parts per thousand ($^o/_{oo}$) from terrestrial standard values.

mixed, especially for studies of individual compounds. Also, few broad systematic patterns have emerged from this work, though some conclusions can be drawn. For different classes of soluble compounds, a positive trend exists between δD and $\delta^{13}C$, Figure 2. The significance of this trend is not clear at this time; large differences in C/H ratio among different compounds caution against interpreting it literally as a mixing line. However, the lower end of the trend line clearly passes close to what we believe to be "average solar-system", i.e., terrestrial, C and H. Conceivably, therefore, the trend may reflect mixing of indigenous solar-system organic matter with an exotic component enriched in D and ^{13}C. What is significant is that the trend line does not include the point for the kerogen-like material; apparently this was derived from a different isotopic reservoir(s) from the soluble fraction, and in particular, the aromatic moieties in the kerogen appear to be unrelated to the structurally very similar polycyclic aromatic hydrocarbons (PAH) found as free molecules in the meteorite, as they plot on the trend line (Krishnamurthy *et al.*, 1992; Gilmour and Pillinger, 1994). Substantial enrichment of ^{15}N is found in some soluble compounds and in some moieties in the kerogen-like fraction. A hint of a systematic relationship between δD and $\delta^{15}N$ may be seen in some studies, though its reality is controversial (Kerridge, 1980, 1985; Alexander *et al.*, 1998). There is a general tendency for relatively oxidized compounds, such as carboxylic acids, to be enriched in ^{13}C relative to more reduced species, such as hydrocarbons (Cronin and Chang, 1993). Given the very high $\delta^{13}C$ values measured for carbonates in

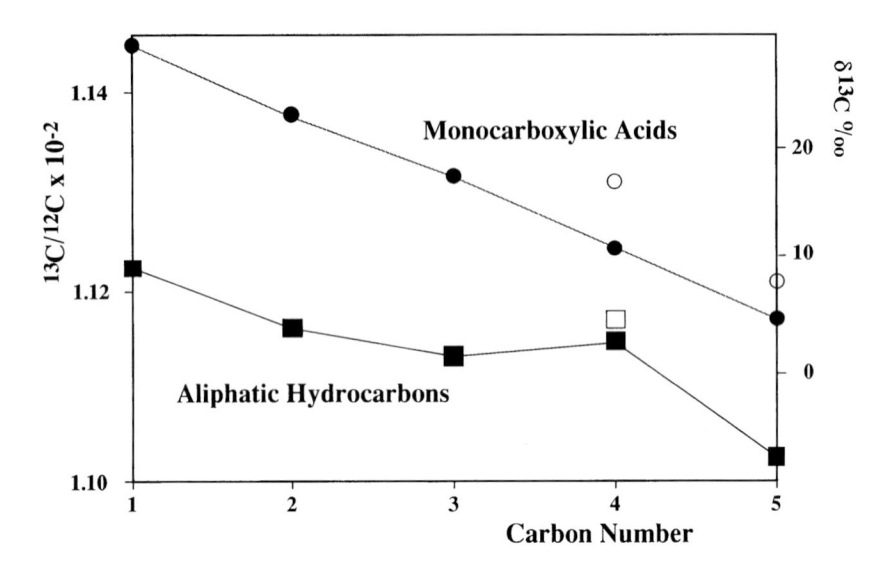

Figure 3. Isotopic composition of hydrocarbons and monocarboxylic acids in Murchison as a function of C number. Note systematic decrease in $\delta\ ^{13}C$ with increasing C number. After Yuen *et al.* (1984).

these meteorites (Grady *et al.*, 1988), this suggests that heavily ^{13}C-enriched CO_2 was involved in the synthesis.

One of the most significant isotopic findings about meteoritic organic matter is illustrated in Figure 3. The light hydrocarbons and the monocarboxylic acids both show a systematic decrease in $\delta^{13}C$ with increasing C number. This indicates that both classes of compound were generated by C-chain build-up involving addition of C, kinetically favoring the lighter isotope (Yuen *et al.*, 1984). Furthermore, the broad similarity between the two different compound classes suggests that they, or their precursors, enjoyed a common origin. Analysis of the S isotopes in methane sulfonate from the Murchison meteorite have revealed significant excesses in ^{33}S and ^{36}S (Cooper *et al.*, 1995). These excesses imply either preservation of a nucleosynthetically distinct S component or, as favored by Cooper *et al.*, mass-independent fractionation generated during gas-phase chemistry. The molecule in which the S isotope anomalies occur is also enriched in D, but whether there is a connection between the different isotope effects is not known.

Attempts to decipher the isotopic characteristics of the different moieties making up the kerogen-like fraction have been only partly successful. Figure 4 shows that the aliphatic and aromatic moieties differ in their distribution of C and H isotopes, but only the ^{13}C value of the aromatic material is well defined (Kerridge *et al.*, 1987). A feature of both Figures 2 and 4 is the remarkably high degree of D enrichment observed. This enrichment becomes even more dramatic when it is recognized that the starting material for production of primordial organic matter

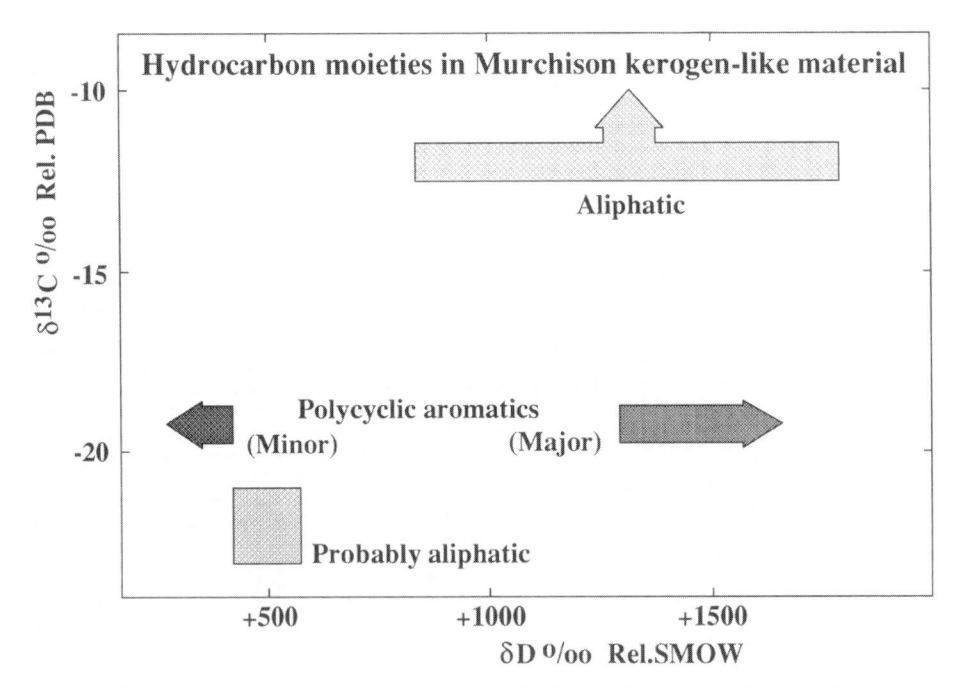

Figure 4. Carbon and H isotopic compositions of individual moieties within kerogen-like material extracted from Murchison. Note substantial isotopic inhomogeneity. The terms PDB and SMOW refer to the terrestrial standards employed. From Kerridge *et al.* (1987).

was characterized by a D content a factor of seven or so lower than the terrestrial value; D enrichment up to at least a factor of 20 was involved in production of the meteoritic organic matter. Such an enrichment so greatly exceeds the fractionating capability of neutral-molecule reactions in the solar nebula, Figure 5, that the meteoritic D content is generally recognized as a signature of presolar, interstellar-cloud material which survived incorporation into the forming solar system (Kolodny *et al.*, 1980). D enrichments of a factor of 10^3 are astronomically observed in interstellar molecular clouds, and are understood to result from synthesis by ion-molecule reactions (Geiss and Reeves, 1981), with a possible contribution from grain-surface reactions (Tielens, 1983). However, the exact nature of the interstellar material, how it was introduced into the meteorite, and the extent to which the other isotopic signatures reflect interstellar processes, remain unknown at this time.

4. Mechanisms for Organic Synthesis: An Assessment

Table I lists some of the sources and processes potentially involved in the production of the organic matter now found in meteorites. It is noteworthy that possible locations range from stellar outflows to asteroidal aquifers. For many

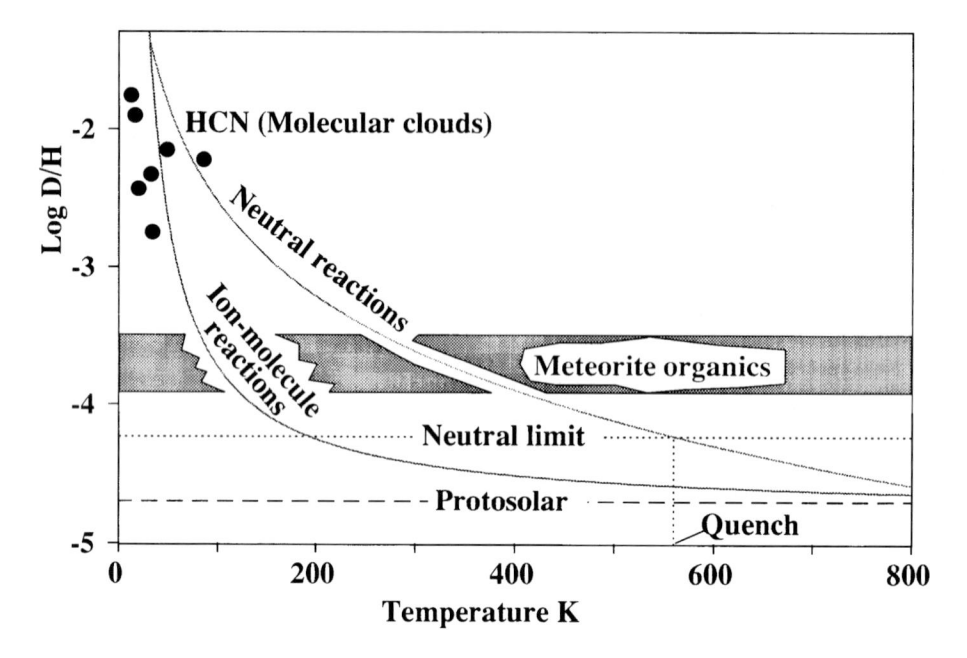

Figure 5. Fractionation of D/H between organic matter and H_2 reservoir as a function of temperature. The curves are the loci of the D/H ratios of organic matter in equilibrium with protosolar hydrogen at any temperature under neutral and ion-molecule conditions, respectively. "Quench" signifies the temperature at which reactions among neutral molecules would become kinetically inhibited in the solar nebula. The "neutral limit" is therefore the highest D/H ratio that can be achieved at equilibrium among neutral species in the nebula. Note that ion-molecule reactions are less efficient than neutral reactions at fractionating isotopes, but can continue to operate down to very low temperatures. After Geiss and Reeves (1981).

years, the favored interpretation was that the meteoritic material was synthesized by grain-catalyzed hydrogenation of CO in the solar nebula, the so-called Fischer-Tropsch-type synthesis (e.g., Hayatsu and Anders, 1981) . However, a feature of such catalyzed reactions is their structural selectivity, so that the discovery of complete structural diversity in homologous series in Murchison, Figure 1, dealt the Fischer-Tropsch hypothesis a fatal blow (Cronin and Pizzarello, 1990). Furthermore, isotopic fractionations accompanying Fischer-Tropsch reactions in the laboratory (Yuen *et al.*, 1991) do not explain the isotopic distributions observed in meteoritic material, and the Fischer- Tropsch hypothesis has trouble accounting for preservation of an interstellar D signature.

With the demise of the Fischer-Tropsch scenario has come the realization that no single mechanism can explain the full range of molecular and isotopic features observed in meteoritic organic matter, and that its production must have involved a concatenation of different processes taking place in a wide range of environments (Cronin and Chang, 1993). (It should be noted that the failure of the Fischer-

TABLE I
Possible sources and processes responsible for production of meteoritic organic matter

Ion-molecule reactions in interstellar clouds
Radiation chemistry in interstellar grain mantles
Condensation in stellar outflows
Equilibrium reactions in the solar nebula
Surface catalysis (Fischer-Tropsch) in the solar nebula
Kinetically controlled reactions in the solar nebula
Radiation chemistry (Miller-Urey) in the nebula
Photochemistry in nebular surface regions
Liquid-phase reactions on parent asteroid
Surface catalysis (Fischer-Tropsch) on asteroid
Radiation chemistry (Miller-Urey) in asteroid atmosphere
Thermal processing during asteroid metamorphism

Tropsch hypothesis to account for the properties of meteoritic organic matter does not rule out its possible operation elsewhere in the early solar system.) A nebular reaction has also been proposed for the insoluble fraction. Morgan *et al.* (1991) suggested that gas-phase synthesis of PAH in the nebula could lead to production of kerogen-like material. However, the isotopic inhomogeneity of that material rules out such an origin and points instead towards a mixture of material from a variety of sources. The petrographic evidence for an episode of aqueous activity in the history of carbonaceous chondrites and the isotopic evidence for interstellar organic material have led logically to the idea that the present population of organic molecules in these meteorites resulted from the action of liquid water on a suite of precursor molecules and/or radicals of at least partially interstellar origin (Cronin and Chang, 1993). This is supported by a rough correlation between total amino-acid abundance and extent of aqueous alteration (Cronin, 1989). Furthermore, the proportion of amino to hydroxy acids observed in carbonaceous chondrites broadly matches that predicted by the Strecker-cyanohydrin synthesis, a set of aqueous reactions in which cyanide, ammonia, and aldehydes and/or ketones are converted to amino acids and hydroxy acids (Peltzer *et al.*, 1984). However, the very low abundance of amino acids in meteorites apparently fails to match that predicted by the Strecker synthesis, suggesting that conditions in the aqueous-alteration episode were more complex than those assumed in the Strecker scenario (Cronin and Chang, 1993).

Table II summarizes those conclusions which can be drawn at this time about the origin and evolution of meteoritic organic matter. Note caveats in the text with regard to some simplified statements in the table.

TABLE II
Summary of key observations and related conclusions concerning origin and evolution of meteoritic organic matter

Structural diversity:	Random precursor combination; not Fischer-Tropsch
Branched-chain prevalence:	Control by thermodynamic stability; not Fischer-Tropsch
Amino/hydroxy acids:	Strecker-like synthesis; aqueous asteroidal environment
Chirality:	Physical, not biological, enantiomeric excess
D enrichment:	Involvement of interstellar processes
C,H isotopes in kerogen:	Disequilibrium mixture; not nebular gas-phase reaction
C isotopes in homologues:	Chain construction by addition of C atoms

5. Production Scenario for Meteoritic Organics

A complex sequence of events apparently led to production of the organic material now found in carbonaceous meteorites. At a minimum, the following stages must have been involved:

1. Molecule formation by a variety of reaction pathways in dense interstellar clouds. Ion-molecule reactions would have featured prominently at this stage.
2. Condensation of organic molecules as icy mantles on refractory interstellar grains. The composition of those mantles would have differed significantly from that of the ambient gas phase as volatility played a role.
3. Irradiation of icy grain mantles by UV and cosmic rays, leading to production of a wide range of molecular fragments and free radicals.
4. Repeated recycling of volatile material between icy mantles and the gas phase as grains experienced grain-grain collisions and interstellar shock waves, while moving between dense clouds and the diffuse interstellar medium.
5. Infall of interstellar grains into the protosolar nebula with probable heating of at least some grain mantles during passage through the shock front bounding the nebular accretion disk.
6. Agglomeration of locally produced nebular condensates and reaction products together with residual interstellar grains into progressively larger planetesimals, culminating — for our purposes — with formation of asteroid-sized objects, i.e., bodies with diameters in the range 10 to 10^3 km. At least the later stages of this process must have been accompanied by significant release of gravitational energy in the form of heat, though the extent of such heating is uncertain.
7. Heating of planetesimals by decay of short-lived radionuclides. Asteroid-sized objects which accreted within a few million years of nebula formation would have been heated, some to their melting point, by energy released during decay

of ^{26}Al, an unstable nucleus with a mean life of about 1 Myr which is known to have been present in the early solar system.
8. Generation of liquid water within at least some asteroidal bodies. Those asteroids which accreted ice, together with refractory solids, while significant ^{26}Al activity remained would have experienced melting of that ice to form a transient aqueous environment.
9. Reaction of interstellar molecules, fragments and radicals with each other and with the aqueous medium, possibly catalyzed by mineral grains. As temperature, pH, oxidation potential (Eh), and solute content changed, the last three factors as a result of mineral hydration, the nature of the reaction schemes involved, and their products, could have varied considerably.
10. Loss of water and other volatiles to space, yielding a partially hydrated lithology containing a complex suite of organic molecules. An object a few hundred km in diameter at the distance of the asteroid belt, could probably have maintained a liquid-water environment in its near-surface regions for a few million years.
11. Production of a kerogen-like macromolecular complex by continued heating of simpler organic molecules. Impact heating may have been involved in addition to radionuclide decay.
12. Mixing of heated and unheated material to yield the meteoritic complex now observed. Impact-induced gardening would have brought together material from different regions of an asteroid, and even from different asteroids, and impact-induced lithification, combined with aqueous mineralization, would have cemented the mixed products together into a rock.

6. Implications for Organic Matter in Comets

At least some of the processes described in the preceding section must have been involved also in production of protocometary material. Consequently, meteoritic organic matter can be a useful guide to what we will find in pristine samples from a comet nucleus, but with some important reservations.

— First, meteoritic material has clearly undergone aqueous alteration at around room temperature or higher. It is generally, though not universally, believed that cometary nuclei have not experienced aqueous activity because melting of ice would have resulted in differentiation of a "dirty snowball" nucleus, which is apparently contraindicated by astronomical observation (e.g. A'Hearn et al., 1995).
— Second, presolar organic precursors were probably not homogeneously distributed throughout the early solar system. This is suggested by detail differences in organic analyses of different types of meteorite. If true, this would

mean that the comet-formation locations would not necessarily have received the same inventory of organic matter as did the asteroid belt.
- Third, nebular processes not recorded in available meteorites may have been active elsewhere in the solar system. An example is the Fischer-Tropsch reaction, which can probably be ruled out for meteoritic organic matter but which in principle could have occurred wherever suitable catalytic minerals coexisted with CO and H_2 at appropriate temperatures and pressures.
- Fourth, since their fall to Earth, meteorites have been stored at room temperature so that compounds stable only at lower temperatures would have been lost, had they survived earlier aeons of meteorite history. A clear objective of spacecraft sampling of comet nucleus material will be to attempt to preserve a sub-surface sample at its ambient temperature within the nucleus, so that compounds unknown in meteorites would almost certainly be found.

However, despite these caveats, some clear lessons emerge from the meteorite experience.

- First, detailed analysis of homologous series will be of considerable value. This will put a premium on subnanomole sensitivity and high resolution of species.
- Second, a search for even a modest level of chiral selectivity will be highly desirable. This will require an approach capable of detecting a 10% enantiomeric excess in a compound quite possibly at a level of a few nanomoles g^{-1}.
- Third, isotopic analysis of C, H, and N, and possibly S, at least at the compound-class level and preferably at the species level, will be of utmost importance. This will require analytical precision at the per mil level for C and S, and at the level of a few per mil for H and N, on samples that could be in the nanomole range.
- These are challenging specifications but the payoff will be a unique insight into some of the earliest stages in development of organic matter in the solar system, and hence into the beginnings of solar-system biology.

References

A'Hearn, M.F., Millis, R.L., Schleicher, D.G., Osip, D.J., and Birch, P.V.: 1995, 'The ensemble properties of comets: Results from narrowband photometry of 85 comets', *Icarus* **118**, 223-270.

Alexander, C.M.O'D., Russell, S.S., Arden, J.W., Ash, R.D., Grady, M.M. and Pillinger, C.T.: 1998, 'The origin of chondritic macromolecular organic matter: A carbon and nitrogen isotope study', *Meteoritics* **33**, 603-622.

Bonner, W.A. and Rubenstein, E.: 1987, 'Supernovae, neutron stars and biomeolcular chirality', *Biosystems* **20**, 99-111.

Clayton, R.N. and Mayeda, T.K.: 1984, 'The oxygen isotope record in Murchison and other carbonaceous chondrites', *Earth Planet. Sci. Lett.* **67**, 151-161.

Cooper, G.W., Thiemens, M.H., Jackson, T. and Chang, S.: 1995, 'Sulfur and hydrogen isotopic anomalies in organic compounds from the Murchison meteorite', *Meteoritics* **30**, 500.
Cronin, J.R., Pizzarello, S. and Cruikshank, D.J.: 1988, 'Organic matter in carbonaceous chondrites, planetary satellites, asteroids and comets', In: *Meteorites and the Early Solar System.* Univ.Arizona. 819-857.
Cronin, J.R.: 1989, 'Origin of organic compounds in carbonaceous chondrites', *Adv. Space Res.* **9**, (2)59-(2)64.
Cronin, J.R. and Pizzarello, S.: 1990, 'Aliphatic hydrocarbons of the Murchison meteorite', *Geochim. Cosmochim. Acta* **54**, 2859-2868.
Cronin, J.R. and Chang, S.: 1993, 'Organic matter in meteorites: Molecular and isotopic analyses of the Murchison meteorite', In: *The Chemistry of Life's Origins*, Kluwer, 209-258.
Cronin, J.R. and Pizzarello, S.: 1997, 'Enantiomeric excesses in meteoritic amino acids', *Science* **275**, 951-955.
Geiss, J. and Reeves, H.: 1981, 'Deuterium in the early solar system', *A.&A.* **93**, 189-199.
Gilmour, I. and Pillinger, C.T.: 1994, 'Isotopic compositions of individual polycyclic aromatic hydrocarbons from the Murchison meteorite', *Mon. Not. R. Astron. Soc.* **268**, 235-24.
Grady, M.M., Wright, I.P., Swart, P.K. and Pillinger, C.T.: 1988, 'The carbon and oxygen isotopic composition of meteoritic carbonates', *Geochim. Cosmochim. Acta* **52**, 2855-2866.
Hayatsu R. and Anders E.: 1981, 'Organic compounds in meteorites and their origins', *Topics Curr. Chem.* **99**, 1-37.
Hayatsu, R., Scott, R.G. and Winans, R.E.: 1983, 'Comparative structural study of meteoritic polymer with terrestrial geopolymers, coal and kerogen', *Meteoritics* **18**, 310.
Kerridge, J.F. and Bunch, T.E.: 1979, 'Aqueous alteration on asteroids: Evidence from carbonaceous meteorites', In: *Asteroids* Univ. Arizona, 745-764.
Kerridge, J.F.: 1980, 'Isotopic clues to organic synthesis in the early solar system', *Lunar Planet. Sci.* **XI**, 538-540.
Kerridge, J.F.: 1985, 'Carbon, hydrogen and nitrogen in carbonaceous chondrites: Abundances and isotopic compositions in bulk samples', *Geochim. Cosmochim. Acta* **49**, 1707-1714.
Kerridge, J.F., Chang, S. and Shipp, R.: 1987, 'Isotopic characterization of kerogen-like material in the Murchison carbonaceous chondrite', *Geochim. Cosmochim. Acta* **51**, 2527-2540.
Kerridge, J.F.: 1991, 'A note on the prebiotic synthesis of organic acids in the early solar system', *Origins of Life & Evol. Biosphere* **21**, 19-29.
Kolodny, Y., Kerridge, J.F. and Kaplan, I.R.: 1980, 'Deuterium in carbonaceous chondrites', *Earth Planet. Sci. Lett.* **46**, 149-158.
Krishnamurthy, R.V., Epstein, S., Cronin, J.R., Pizzarello, S. and Yuen, G.U.: 1992, 'Isotopic and molecular analyses of hydrocarbons and monocarboxylic acids of the Murchison meteorite', *Geochim. Cosomochim. Acta* **56**, 4045-4058.
Macdougall, J.D., Lugmair, G.W. and Kerridge, J.F.: 1984, 'Early solar system aqueous activity: Sr isotopic evidence from the Orgueil CI meteorite', *Nature* **307**, 249-251.
Morgan, W.A., Feigelson, E.D., Wang, H. and Frenklach, M.: 1991, 'A new mechanism for the formation of meteoritic kerogen-like material', *Science* **251**, 109-112.
Peltzer, E.T. Bada, J.L., Schlesinger, G. and Miller, S.L.: 1984, 'The chemical conditions on the parent body of the Murchison meteorite: Some conclusions based on amino, hydroxy and dicarboxylic acids', *Adv.Space Res.* **4**, 69-74.
Tielens, A.G.G.M.: 1983, 'Surface chemistry of deuterated molecules', *Astron.Astrophys.* **119**, 177-184.
Woolum, D.S. and Hohenberg, C.: 1993, 'Energetic particle environment in the early solar system: Extremely long precompaction meteoritic ages or an enhanced early particle flux', In: *Protostars and Planets* **III**. Univ.Arizona. 903-919.

Yuen, G., Blair, N., DesMarais, D.J. and Chang, S.: 1984, 'Carbon isotopic composition of low molecular weight hydrocarbons and monocarboxylic acids from Murchison meteorite', *Nature* **307**, 252- 254.
Yuen, G., Pecore, J.A., Kerridge, J.F., Pinnavaia, T.J., Rightor, E.G., Flores, J., Wedeking, K.M., Mariner, R., DesMarais, D.J. and Chang, S.: 1991, 'Carbon isotopic fractionation in Fischer-Tropsch type reactions', *Lunar Planet.Sci.* **XXI**, 1367-1368.
Zolensky, M. and McSween, H.Y.: 1988, 'Aqueous alteration', In: *Meteorites and the Early Solar System.* Univ.Arizona. 114-143.

Address for correspondence: John F. Kerridge, jkerridg@ucsd.edu

PERSPECTIVES ON THE COMET-ASTEROID-METEORITE LINK

KATHARINA LODDERS and ROSE OSBORNE

Planetary Chemistry Laboratory, Department of Earth and Planetary Sciences, Washington University, Campus Box 1169, St. Louis,MO 63130-4899 USA

Abstract. We discuss the possibility that CI and CM carbonaceous chondrites are fragments of extinct cometary nuclei. Theoretical and observational work suggests that comets evolve into asteroids, and several extinct cometary nuclei are now suspected to be among the near Earth object population. This population is the most likely source of meteorites and consequently, we may expect that some meteorites are from extinct comets in this population. The mineralogy and chemistry of CI and CM chondrites is consistent with the view that they originate from asteroidal objects of carbonaceous spectral classes, and these objects in turn may have a cometary origin. We do not suggest that CI or CM chondrites are directly delivered by active comets during perihelion passage or that these chondrites come from cometary debris in meteor streams. Instead, we summarize arguments suggesting that CI and CM chondrites represent fragments of cometary nuclei which evolved into near Earth asteroids after losing their volatiles.

1. Introduction

It is a widely held view that meteorites are fragments of larger asteroidal bodies delivered via near Earth objects (NEOs). With estimated lifetimes of only 1-100 Ma due to collisional destruction, the NEO population must be continuously replenished to explain cratering rates on the terrestrial planets over geologic time and the current meteorite influx (e.g., Shoemaker *et al.*, 1979). The proposal of a cometary source for meteorites via Apollo asteroids has been discussed in the literature, but no firm conclusions have been reached about whether or not some carbonaceous chondrites could be cometary fragments (e.g., Öpik, 1963, 1966; Anders, 1971, 1975; Wetherill, 1971; Wasson and Wetherill, 1979).

Recently, Binzel *et al.* (1998) showed that S-type asteroids are the most likely source of ordinary chondrites, the most abundant meteorites, but we also need parent bodies for other chondrite and achondrite meteorites among the NEO population. As pointed out by Anders (1971), the mineralogy and chemistry of the major meteorite classes are consistent with an asteroidal parent body (e.g., metamorphic temperatures > 400 K). However, he also notes that comets may be sources of some micrometeorites and carbonaceous chondrites.

Orbits have been determined for four ordinary chondrites from camera network observations and their aphelia extend into the main asteroid belt. Such observations are not available for carbonaceous chondrites but would be useful because orbital characteristics may indicate either an asteroidal or cometary source.

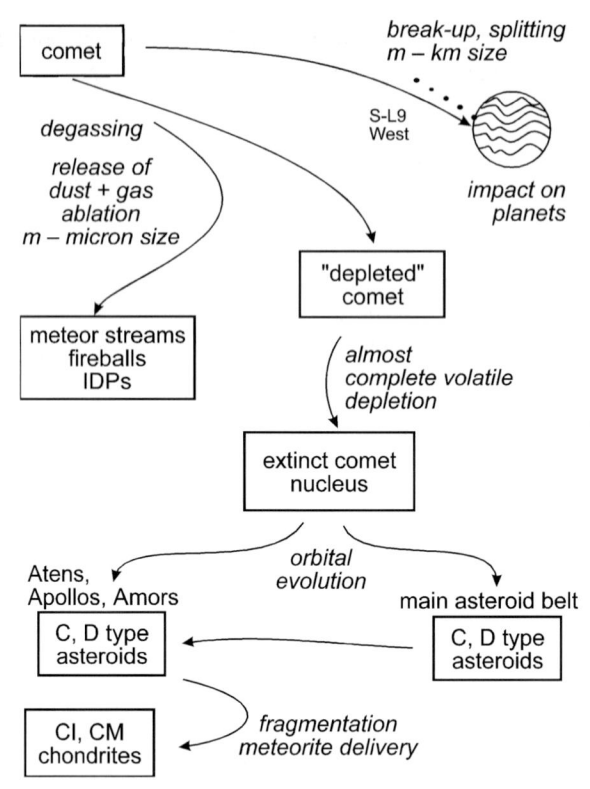

Figure 1. A schematic diagram showing how CI and CM chondrites may be delivered to Earth from extinct comet nuclei. During successive perihelion passages a comet becomes depleted in volatile ices and eventually cometary activity (tail, coma) ceases and only the more refractory cometary nucleus remains. The degassing process leads to ablation and the release of gas and fine-size dust, which leads to the formation of meteor streams. These streams are likely sources for interplanetary dust particles (IDPs) but are unlikely to produce recoverable meteorites. Degassing and cometary splitting (e.g., Comets West and Shoemaker-Levy 9) can also lead to changes in the comet's orbital path and to comet impacts on planets. Extinct cometary nuclei evolve into asteroids of carbonaceous spectral types (e.g., C, D type asteroids). If such cometary nuclei are among the near Earth asteroids (Apollos, Atens, Amors), further collisions and fragmentation may provide the CI and CM chondrites. See text for more details.

The ordinary, enstatite, and carbonaceous chondrites span a wide range in chemical and stable isotope composition and oxidation state, which is surprising if all chondrite parent bodies formed in the asteroid belt. The rare CI (5 falls) and CM (16 falls) carbonaceous chondrites with high contents of carbonaceous and volatile matter and aqueous alteration features may not have originated there.

In the next sections, we discuss the possibility that carbonaceous chondrites of type CI and CM are fragments of extinct cometary nuclei. Figure 1 schematically shows the model of the evolution of a comet to an asteroid delivering meteorites. Section 2 summarizes the arguments in support of comets evolving into asteroids. Section 3 describes the mineralogy and chemistry of CI and CM chondrites, which are consistent with their origin from asteroids of carbonaceous spectral classes, which may have a cometary origin. We are not suggesting that CI or CM chondrites are directly delivered by active comets during their perihelion passage or that these chondrites come from cometary debris in meteor streams. The CI and CM chondrites may represent fragments of cometary nuclei which evolved into near Earth asteroids after the parent comets lost their volatiles. We use the term "extinct" comets for the degassed cometary nucleus remaining after loss of volatile ices. Extinct comets appear as asteroids, if we adopt the observational definition that an asteroid is a body not displaying cometary activity, such as a coma or tail, during any time on its orbit. Orbital characteristics and surface properties derived from reflectance spectroscopy can be used to genetically distinguish between extinct comets and "real" asteroids.

2. The Comet-Asteroid Link

Weissman et al. (1979) review the possible evolution of comets into asteroids and we limit our discussion to a few examples. Orbital and dynamical studies suggest that both fragments of main belt asteroids and comets replenish the NEO population (e.g., Öpik, 1963; Shoemaker et al., 1979; Wetherill, 1971; 1988). Extinct comets may contribute up to half the NEO sample (Wetherill, 1988).

Observations of nearly extinct comets support these dynamical models. Comets displaying weak activity include 2P/Encke, 28P/Neujmin 1, and 49P/Arend-Rigaux (e.g., Degewij and Tedesco, 1982). One well known example is Comet 107P/Wilson-Harrington, which had a tail but no coma in 1949. This comet was then "lost" and rediscovered as asteroid 1979 VA.

Several asteroids have cometary orbital characteristics. Kresak (1979), Hahn and Rickman (1985) and Hartmann et al. (1987) use the Tisserant parameter (T) with respect to Jupiter to single out asteroids which may be extinct comets (see also Weissman et al., 1979). Asteroids with $T < 3$ among the NEOs include 3552 Don Quixote (T = 2.31) and 1984 BC (T = 2.78). Extinct cometary candidates outside the NEO population include 944 Hidalgo, 5335 Damocles, and 1996 PW (e.g., Weissman and Levison, 1997).

Hartmann et al. (1987) find that all asteroids with cometary orbital characteristics are of spectral type C, D, or P, indicating surfaces containing carbonaceous matter. Their comparison includes the D-type cometary nuclei of P/Neujmin 1 and Arend-Rigaux, which are among the candidates for the comet-asteroid transition.

Meteor streams related to asteroids may also indicate that these asteroids were once active comets. One well known example is 3200 Phaethon (spectral class F) associated with the Geminid meteors. Discussions of asteroids with associated meteor streams are given by Drummond (1981) and Olsson-Steel (1988).

3. The Asteroid-Meteorite Link

Meteorite cosmic ray exposure ages are within the dynamical lifetimes of the NEOs. Ordinary chondrites (H, L, LL) show exposure ages ranging from 10 to 30 Ma, while carbonaceous chondrites tend to have younger exposure ages; 0.2 to 8 Ma for CI and CM chondrites, and 2 to 40 Ma for CO and CV chondrites (Mazor et al., 1970; Crabb and Schultz, 1981). Lower exposure ages of CI and CM chondrites may indicate that resurfacing of their parent bodies is faster because they are more friable and less stable against collision than ordinary chondrites.

Asteroids of type C, B, F, and G with silicate hydration spectral features and low albedos indicating carbonaceous surfaces are believed to be parent bodies of CI and CM chondrites. In all carbonaceous asteroids (except the K type) the presence of ice is likely. The presence of ice may be important with respect to low exposure ages of carbonaceous chondrites, because ice coatings can prevent exposure of rocky material to cosmic rays. In that case, irradiation can only occur after sublimation of ice during perihelion passage.

Both comets and C-type asteroids appear to be very porous. The C-type asteroid Mathilde has a density of only 1.3 g cm^{-3} (Veverka et al., 1997), which is lower than the density of hydrous silicates ($\sim$ 2.5 g cm^{-3}). If the low density is due to porosity, about 40-45% of this asteroid is pore space. Similarly, CI chondrites have about 35% porosity and are stable enough to survive passage through the atmosphere. With respect to the stability of such matter passing the atmosphere, it should be recalled that sun-grazing comets survived even passage through the solar corona.

4. The Comet-Meteorite Link: Mineralogy and Chemistry

This section summarizes mineralogical and chemical arguments supporting the hypothesis that CI and CM chondrites are fragments of cometary nuclei. In Table I, we summarize some properties of CI and CM chondrites for reference. CI chondrites are the most chemically primitive meteorites, with elemental abundances resembling those of the solar photosphere. All elements forming compounds more refractory than H_2O have solar abundances in CI chondrites, indicating that CI chondrites were made from material formed in the solar nebula where temperatures were low enough to allow complete condensation of volatiles. The CM parent body also accreted low temperature condensates, but the volatile content in CM chondrites is somewhat lower than in CI chondrites.

Both CI and CM chondrites experienced aqueous alteration on their parent bodies. This means that water (ice) must have accreted to their parent bodies in a low temperature region. The accretion of ice along with rock to the CI and CM parent bodies indicates a similarity between them and comets, which are commonly thought to be assembled from rock and ice. The major unresolved questions are (1)

TABLE I
Selected Properties of CI- and CM-chondrites

Property	CI	CM
number of observed falls	5	16
bulk density, g cm^{-3}	1.6 - 2.2	2.6 - 2.8
porosity, vol%	10 - 35	2 - 23
fraction of fine grained, opaque matrix, vol%	95 - 100	50 - 90
aqueous alteration products and features	phyllosilicates (Mg & Fe-serpentines and chlorites) magnetite, elemental sulfur Mg-, Na-, Ca- sulfates Ca-, Mg-, Fe-carbonates NH_4Cl abundant salt-bearing veins	phyllosilicates (tochilinite, cronstedite, and Mg & Fe-serpentines) some elemental sulfur Mg-, Na-, Ca-sulfates Ca-, Mg-, Fe-carbonates few salt-bearing veins
other minerals and constituents	chondrules absent olivine very rare, with fo$\sim$ 100 pyrrhotite, pentlandite present CAI absent presolar diamonds 1000-1500 ppm	chondrules present olivine present with fo$\sim$ 98 $-$ 99 pyrrhotite, pentlandite present CAI present presolar diamonds $>$ 270 to $>$ 750 ppm
elemental abundances	refractory elements* = solar moderately volatile elements = solar highly volatile elements = solar	refractory elements = solar moderately volatile elements $\sim 0.4 - 0.6 \times$ solar highly volatile elements $\sim 0.4 \times$ solar
bulk water content, wt%	3 - 10	1 - 16
bulk carbon content, wt%	3 - 5	0.8 - 2.8
bulk sulfur content, wt%	5.4	2.7
D/H	$(18 - 20) \times 10^{-5}$	$(14 - 31) \times 10^{-5}$
aqueous alteration T, °C	$<$ 25, 50 - 150 (model dependent)	1-25 (model dependent)
time of aqueous alteration after CAI formation, Ma	50 - 100	not determined
percent of meteorites that are gas-rich	100	57
cosmic-ray exposure age, Ma	0.19 - 8	0.14 - 5.9

CAI = Ca, Al-rich inclusions. fo = mole percent forsterite in olivine.

* refractory elements = Ca, Al, Mg, Si, Fe...; moderately volatile elements = alkalis, halogens, S, Zn...; highly volatile elements = Bi, Cd, In, Tl... .

at which heliocentric distances did comets and the parent bodies of the CI and CM meteorites form, and (2) what is the rock to ice ratio in comets and what was this ratio in the CI and CM parent bodies?

The aqueous alteration conditions required to explain the presence of carbonates, sulfates, and hydrous silicates in CI and CM chondrites have been investigated by DuFresne and Anders (1962) and Boström and Fredriksson (1966). They infer that fluids containing H_2O, CO_2, SO_2, NH_3, H_2S, and organics are required. Clayton and Mayeda (1984) determine temperatures for aqueous alteration of 50 to 150°C for CI, and 1 to 25°C for CM chondrites, depending on the rock to ice ratio assumed. Currently the reactions forming the CO_2, SO_2, NH_3 etc. required for the alteration processes are not well known, but these reactants must have formed from organics, ices, and FeS originally accreted to the CI and CM parent bodies. All these required compounds are observed in comets.

Aqueous alteration after accretion of rock and ice requires a heat source sufficient to melt ice and produce a reactive fluid. Grimm and McSween (1989) suggest a model with ^{26}Al as a heat source. Aqueous alteration and salt formation seem to have occurred early in CI chondrites, supporting this model. Endress et al. (1996) determined from ^{53}Cr excesses that carbonates in CI chondrites formed within 20 Ma after Ca-Al-rich inclusions. Macdougall et al. (1984) used Rb-Sr systematics to find that aqueous alteration occurred within 100 Ma of formation of the CI parent body. Macdougall et al. also find that the Rb-Sr systematics are disturbed and that some of the Ca-sulfate deposits may be more recent. The disturbance of the Rb/Sr isotope systematics in CI and CM chondrites was already noted by Mittlefehldt and Wetherill (1979), who suggest the possibility of partial re-equilibration of Rb and/or Sr after chondrite formation during devolatilization of comets near perihelion. Devolatilization could cause element redistribution during removal of ice mantles and heating of ice plus rock as well as leaching and salt deposition, and formation of duricrust on the cometary core.

Enzian and Weissman (1998) developed thermal models for Comets 46/Wirtanen and 9P/Tempel 1. A cometary nucleus surrounded by a dust mantle may reach 320 K (47°C) at about 1 AU after five perihelion passages. These temperatures compare well to the aqueous alteration temperatures derived for CI and CM chondrites. Repeated heating and melting may cause repeated salt deposition in outgassing fractures or freezing cracks, so that on a small scale a cometary nucleus appears cemented together. This cementing can be compared to that illustrated for CI chondrites in Figure 1 of Richardson (1978) showing carbonate- and sulfate-filled veins and cracks. Due to space limitations, we refer to the detailed discussion by McSween and Weissman (1989) about possible aqueous alteration settings in cometary nuclei.

One open question is whether or not hydrous silicates, carbonates and sulfates are present in cometary nuclei as they are in CI and CM chondrites. The analyses of Comet Halley's dust particles by the PUMA mass spectrometer reveals the presence of Fe-poor, Mg-silicates; probably pyroxene and olivine (Schulze et *al.* 1997; Jessberger, this proceedings). Olivine and pyroxene are also detected in cometary spectra (Hanner, this proceedings). CI chondrites contain very few olivine and pyroxene grains and consist mainly of phyllosilicates, a situation not markedly different for CM chondrites (see Zolensky *et al.*, 1997, for the range in CM chondrite alteration). Hydrous silicates are not yet detected in comets, but this does not mean that they indeed are absent in comets. We should keep in mind that spectra sample the outer layer of cometary debris undergoing devolatilization, which may lead to dehydration of hydrous silicates, especially at close heliocentric distances where comets are most visible. It is also necessary to investigate whether or not the PUMA data clearly exclude the presence of hydrous silicates in Comet Halley.

Another connection of comets and carbonaceous meteorites may be derived from the study of interplanetary dust particles (IDPs), which form three major groups: olivine-rich, pyroxene-rich, and phyllosilicate-like IDPs. The last group is often compared to CI and CM chondrites. Sandford and Bradley (1989) argue that the olivine- and pyroxene-rich IDPs are from comets while the lattice silicate-rich IDPs are from asteroids. These conclusions do not necessarily conflict with the scenario that CI and CM chondrites (and hence, the phyllosilicate IDPs) are from extinct cometary nuclei because these nuclei may have evolved into asteroids. For example, such IDPs could originate from the hydrated F-type asteroid 3200 Phaeton, which is associated with the Geminid meteor stream.

We conclude this section with another argument linking CI and CM chondrites, as well as phyllosilicate IDPs, to comets. The CI and CM chondrites possess high D/H ratios of $(18 - 31) \times 10^{-5}$ (Zinner, 1988) which are comparable to the D/H ratios of $\sim 30 \times 10^{-5}$ seen in comets (Irvine *et al.*, 1998). Similarly, phyllosilicate IDPs are known to be strongly enriched in D (Zinner, 1988). These high D/H ratios cannot be explained by formation of the CI and CM chondrites in the asteroid belt because higher temperatures at these heliocentric distances in the solar nebula would decrease the D/H ratios. The high D/H ratios in the phyllosilicates of CI meteorites can only be caused by D-enriched fluids present during alteration. As discussed above, comets indeed possess all the ingredients as well as the high D/H ratios to explain the hydrous alteration features observed in CI and CM chondrites.

5. Outlook

We believe that the arguments above support the claim that CI and CM chondrites are fragments of cometary nuclei. Further work is needed to investigate whether the scenario outlined above also applies in modified form to other carbonaceous chondrites. In situ analyses and samples returned from cometary nuclei might provide proof that CI and CM chondrites originate from cometary nuclei.

Acknowledgements

Work supported by grant NAG5-6366 from the NASA Planetary Atmospheres Program. We thank Bruce Fegley for discussions, P. Ehrenfreund and an anonymous referee for useful comments, and the International Space Science Institute for the invitation to this workshop.

References

Anders, E.: 1971, 'Interrelations of meteorites, asteroids, and comets,' in *Physical studies of minor planets*, ed. T. Gehrels, NASA SP-**267**, 429-446.
Anders, E.: 1975, 'Do stony meteorites come from comets?' *Icarus* **24**, 363-371.
Binzel, R.P., Bus, S.J., and Burbine, T.H.: 1998, 'Relating S-asteroids and ordinary chondrite meteorites: The new big picture,' *Bull. Amer. Astron. Soc.* **30**, 1041.
Boström, K. and Fredriksson, K.: 1966, 'Surface conditions of the Orgueil meteorite parent body as indicated by mineral associations,' *Smithsonian Miscellaneous Collections* **151**, 1-39.
Clayton, R.N. and Mayeda, T.K.: 1984, 'The oxygen isotope record in Murchison and other carbonaceous chondrites,' *Earth Planet. Sci. Lett.* **67**, 151-161.
Crabb, J. and Schultz, L.: 1981, 'Cosmic-ray exposure ages for the ordinary chondrites and their significance for parent body stratigraphy,' *Geochim. Cosmochim. Acta* **45**, 2151-2160.
Degewij, J. and Tedesco, E.F.: 1982, 'Do comets evolve into asteroids? Evidence from physical studies,' in *Comets*, ed. L.L. Wilkening, Univ. of Arizona Press, 665-695.
Drummond, J. D.: 1981, 'A test of comet and meteor shower associations,' *Icarus* **45**, 545-533.
DuFresne and Anders: 1962, 'On the chemical evolution of the carbonaceous chondrites,' *Geochim. Cosmochim. Acta* **26**, 1085-1114.
Endress, M., Zinner, E., and Bischoff, A.: 1996, 'Early aqueous activity on primitive meteorite parent bodies,' *Nature* **279**, 701-703.
Enzian, A. and Weissman, P.R.: 1998, 'Thermal modeling of periodic comets 46P/Wirtanen and P9/Tempel 1,' *Bull. Amer. Astron. Soc.* **30**, 1095.
Grimm, R. F. and McSween, H. Y.: 1989, 'Water and the thermal evolution of carbonaceous chondrite parent bodies,' *Icarus* **82**, 244-280.
Hahn, G. and Rickman, H.: 1985, 'Asteroids in cometary orbits,' *Icarus* **61**, 417-442.
Hanner, M.: 1999, 'The silicate material in comets', *Space Sci. Rev.*, this volume.
Hartmann, W. K., Tholen, D. J., and Cruikshank, D. P.: 1987, 'The relationship of active comets, extinct comets, and dark asteroids,' *Icarus* **69**, 33-50.
Irvine, W.M, Schloerb, F.P., Crovisier, J., Fegley, B., and Mumma, M.J.: 1998, 'Comets: A link between interstellar and nebula chemistry,' in *Protostars and Planets* **IV**, in press.
Jessberger, E.K.: 1999, 'Rocky Cometary Particulates: their elemental, isotopic and mineralogical ingredients', *Space Sci. Rev.*, this volume.
Kresak, L.: 1979, 'Dynamical interrelations among comets and asteroids,' in *Asteroids*, ed. T. Gehrels, Univ. of Arizona Press, Tucson, 289-309.
Mazor, E., Heymann, D. and Anders, E.: 1970, 'Noble gases in carbonaceous chondrites,' *Geochim. Cosmochim. Acta* **34**, 781-824.
Macdougall, J.D., Lugmair, G.W., and Kerridge, J.F.: 1984, 'Early solar system aqueous activity: Sr isotope evidence from the Orgeuil CI meteorite,' *Nature* **307**, 249-251.
McSween, H.Y. and Weissman, P.R.: 1989, 'Cosmochemical implications of the physical processing of cometary nuclei,' *Geochim. Cosmochim. Acta* **53**, 3263-3271.

Mittlefehldt, D.W. and Wetherill, G.W.: 1979, 'Rb-Sr studies of CI and CM chondrites,' *Geochim. Cosmochim. Acta* **43**, 201-206.

Olsson-Steel, D.: 1988, 'Identification of meteoroid streams from Apollo asteroids in the Adelaide radar orbit surveys,' *Icarus* **75**, 64-96.

Öpik, E.J.: 1963, 'Survival of cometary nuclei and the asteroids,' *Adv. Astron. Astrophys.* **2**, 219-262.

Öpik, E.J.: 1966, 'The cometary origin of meteorites,' *Adv. Astron. Astrophys.* **4**, 301-336.

Richardson, S. M.: 1978, 'Vein formation in C1 carbonaceous chondrites,' *Meteoritics* **13**, 141-159.

Sandford, S.A. and Bradley, J.P.: 1989, 'Interplanetary dust particles collected in the stratosphere: Observations of atmospheric heating and constraints on their interrelationships and sources,' *Icarus* 82, 146-166.

Schulze, H., Kissel, J., and Jessberger, E.K.: 1997, 'Chemistry and mineralogy of comet Halley's dust,' in *From Stardust to Planetesimals, ASP Conf. Ser.* **122**, 397-414.

Shoemaker, E.M., Williams, J.G., Helin, E.F. and Wolfe, R.F.: 1979, 'Earth-crossing asteroids: Orbital classes, collision rates with Earth and origin,' in *Asteroids*, ed. T. Gehrels, Univ. of Arizona Press, 253-282.

Veverka, J., Thomas, P., Harch, A., Clark, B., Bell, J.F., III, Carcich, B., Joseph, J., Chapman, C., Merline, W., Robinson, M., Malin, M., McFadden, L.A., Murchi, S., Hawkins, S.E., III, Farquhar, R., Izenberg, N., and Cheng, A.: 1997, 'NEAR's flyby of 253 Mathilde: Images of a C asteroid,' *Science* **278**, 2109-2114.

Wasson, J. T., and Wetherill, G.W.: 1979, 'Dynamical, chemical and isotopic evidence regarding the formation locations of asteroids and meteorites,' in *Asteroids*, eds. T. Gehrels and M.S. Matthews, Univ. of Arizona Press, 926-974.

Weissman, P.R., A'Hearn, M.F., McFadden, L.A., and Rickman, H.: 1979, 'Evolution of comets into asteroids,' in *Asteroids II*, eds. R. P. Binzel, T. Gehrels, M.S. Matthews, Univ. of Arizona Press, Tucson, 880-920.

Weissman, P.R., and Levison, H.L.: 1997, 'Origin and evolution of the unusual object 1996 PW: Asteroids from the Oort cloud?' *Astrophys. J* **488**, L133-L136.

Wetherill, G.W.: 1971, 'Cometary versus asteroidal origin of chondritic meteorites', in *Physical studies of minor planets*, ed. T. Gehrels, NASA SP-267, 447-460.

Wetherill, G.W.: 1988, 'Where do the Apollo objects come from?' *Icarus* **76**, 1-18.

Wetherill, G. W.: 1989, 'Cratering of the terrestrial planets by Apollo objects,' *Meteoritics* **24**, 15-22.

Zolensky, M.E., Mittlefehldt, D.W., Lipschutz, M.E., Wang, M.-S., Clayton, R.N., Mayeda, T.K., Grady, M.N., Pillinger, C., and Barber, D.: 1997, 'CM chondrites exhibit the complete petrologic range from type 2 to 1,' *Geochim. Cosmochim. Acta* **61**, 5099-5115.

Zinner, E.: 1988, in *Meteorites and the Early Solar System*, eds. J.F. Kerridge and M.S. Matthews, Univ. of Arizona Press, 956-983.

Address for correspondence: Katharina Lodders, lodders@levee.wustl.edu

IV: CRITICAL MEASUREMENTS FOR THE FUTURE

DIVERSITY OF COMETS: FORMATION ZONES AND DYNAMICAL PATHS

PAUL R. WEISSMAN

Jet Propulsion Laboratory, Earth and Space Sciences Division, Mail stop 183-601, 4800 Oak Grove Drive, Pasadena, CA 91109 USA

Abstract. The past dozen years have produced a new paradigm with regard to the source regions of comets in the early solar system. It is now widely recognized that the likely source of the Jupiter-family short-period comets (those with Tisserand parameters, T > 2 and periods, P, generally < 20 years) is the Kuiper belt in the ecliptic plane beyond Neptune. In contrast, the source of the Halley-type and long-period comets (those with T < 2 and P > 20 years) appears to be the Oort cloud. However, the comets in the Oort cloud almost certainly originated elsewhere, since accretion is very inefficient at such large heliocentric distances. New dynamical studies now suggest that the source of the Oort cloud comets is the entire giant planets region from Jupiter to Neptune, rather than primarily the Uranus-Neptune region, as previously thought. Some fraction of the Oort cloud population may even be asteroidal bodies formed inside the orbit of Jupiter. These comets and asteroids underwent a complex dynamical random walk among the giant planets before they were ejected to distant orbits in the Oort cloud, with possible interesting consequences for their thermal and collisional histories. Observational evidence for diversity in cometary compositions is limited, at best.

Keywords: Oort cloud, Kuiper belt, formation zones, dynamical evolution, physical processing

1. Introduction

Our understanding of the nature and source of the long-period comets experienced a revolution in 1950 with the publication of two classic papers: Oort (1950), which described the vast spherical cometary cloud surrounding the solar system and extending to near-interstellar distances, and Whipple (1950), which proposed that cometary nuclei were icy-conglomerate bodies. Observational and theoretical evidence gathered over the past five decades has reinforced the conclusions of these landmark papers, refined the details of each hypothesis, and refuted competing explanations for the source and nature of the long-period comets (for recent reviews see Fernández, 1994 and Weissman, 1996).

Oort (1950) recognized that the long-period (LP) comets had likely not accreted *in situ* in their current distant orbits, and suggested that they were asteroids ejected from the planetary region. Kuiper (1951) correctly pointed out that the icy nature of comets required that their formation zones be further from the Sun, among the giant planets, where the solar nebula was cold enough for water and other volatile ices to condense. This idea was further refined by Safronov (1972) who noted that Jupiter and Saturn would likely eject most of the proto-comets in their zones to hyperbolic orbits, whereas the less violent gravitational perturbations of Uranus and Neptune were more likely to scatter objects to distant orbits, though still bound to the solar system. Safronov's work has been the prevailing explanation for the origin of the Oort cloud comets during the past three decades.

Another paradigm that developed during this period was that the short-period (SP) comets, those with orbital periods, $P < 200$ years, were simply long-period comets that had undergone a gravitational random walk to smaller semimajor axes due to planetary perturbations, primarily by Jupiter (e.g., Newton, 1893; Everhart, 1972). Some researchers questioned whether this mechanism could provide the observed number of SP comets but the uncertainty in the relevant parameters, in particular in the flux of LP comets, did not permit a firm conclusion either way.

The past dozen years have seen substantial changes in both of these paradigms. It is now generally accepted that the source of most of the Jupiter-family SP comets is a belt of comets in the ecliptic plane beyond the orbit of Neptune, known as the Kuiper belt. A much newer, and still to be confirmed result, is that the source of the Oort cloud comets is not primarily the Uranus-Neptune zone, but in fact encompasses the entire giant planets region, and possibly even the planetary zone interior to Jupiter's orbit. The implications of these new paradigms are quite substantial. They suggest that comets formed over a much wider range of heliocentric distances, and hence, a wider range of solar nebula environments, than previously thought. They also suggest that comets may have undergone varying degrees of thermal and collisional processing prior to their ejection to the Oort cloud, or while resident in the Kuiper belt.

An important part of this paradigm shift comes from a redefinition of short- and long-period comets. We now recognize two dynamically distinct groups within the traditional SP comet definition: 1) the Jupiter-family (JF) comets which have orbital periods generally < 20 years and are in low to moderate inclination orbits relative to the ecliptic, and 2) the Halley-type comets, which have orbital periods, $20 < P < 200$ years and whose orbital planes are generally randomly inclined to the ecliptic, much like the orbits of the LP comets. The different inclination distributions of the two groups are the key to understanding their origin.

Carusi et al. (1987) suggested a more formal dynamical definition of the differences between the two groups: that the Jupiter-family SP comets have Tisserand parameters, $T > 2$, whereas the Halley-type and long-period comets have $T < 2$. The Tisserand parameter is an approximate constant of the motion in the circularly restricted three-body problem, sun-planet-comet, and was first devised to recognize returning SP comets, even though their orbits had been perturbed by Jupiter. The Tisserand parameter is given by

$$T = a_J/a + 2\sqrt{(a/a_J)(1 - e^2)} \cos i \qquad (1)$$

where a, e, and i are the semimajor axis, eccentricity, and inclination of the comet's orbit and a_J is the semimajor axis of the perturbing planet's orbit, usually that of Jupiter.

This paper reviews the studies that led to the current paradigm shifts and discusses the possible implications of these new ideas for the nature of cometary nuclei. It also reviews the existing observational evidence on the diversity of comets in the light of these new ideas about the cometary formation zones.

2. The Source of the Short-Period Comets

The new understanding of the source of the short-period comets began with Fernández (1980) who proposed that the source of the low inclination, Jupiter-family comets might be a belt of comets beyond the orbit of Neptune, an idea originally proposed by Edgeworth (1949) and Kuiper (1951). Fernández argued that such a cometary belt would be dynamically more efficient for supplying SP comets because the comets would be energetically deeper within the Sun's gravitational potential well, and because the comets would already be in near-ecliptic orbits.

Fernández's suggestion was pursued by Duncan et al. (1988) and Quinn et al. (1990) who showed through dynamical simulations that a trans-Neptunian comet belt was the only way to obtain the low inclination distribution of the Jupiter-family comets. Comets evolved inward from the Oort cloud tended to preserve their random orbital inclinations, and were not able to recreate either the inclination or the semimajor axis distributions for the JF/SP comets. The dynamical simulations showed that the Oort cloud was still the likely source of the random inclination Halley-type comets, as planetary perturbations of trans-Neptunian objects will not generally drive comets to such high inclinations.

The discovery of the first object in a trans-Neptunian orbit, 1992 QB$_1$ (Jewitt and Luu, 1993) lent considerable credence to the idea of a comet belt beyond Neptune. Since then over 180 trans-Neptunian objects have been found in orbits between 30 and 50 AU from the Sun. Population estimates for this region are 7×10^4 bodies with radius, $r > 50$ km (Jewitt et al., 1996, assuming a cometary albedo of 0.04) and 6×10^9 objects with $r > 1$ km (Duncan et al., 1995), i.e, the typical size of JF/SP comet nuclei. The total mass of objects between 30 and 50 AU is estimated at about 0.1 Earth masses (Weissman and Levison, 1997a).

A related source of additional JF/SP comets may be the "scattered disk" (Duncan and Levison, 1997), comets in high eccentricity orbits in or near the plane of the ecliptic, which cross or closely approach the orbit of Neptune. These comets are presumably long-lived Uranus-Neptune zone and Kuiper belt planetesimals that were scattered to these orbits by Neptune, but with aphelia too small to permit their capture into the Oort cloud. Duncan and Levison (1997) showed that these objects could also contribute a substantial fraction of the JF/SP comets. The relative contributions of the Kuiper belt and the scattered disk are still to be determined. Observational searches have found several trans-Neptunian objects in scattered disk type orbits.

The recognition that the Kuiper belt (and perhaps the scattered disk) was the source of the JF/SP comets led to the interesting realization that the SP comets had likely formed farther from the Sun than the long-period Oort cloud comets. Since the temperature of the proto-solar nebula is expected to have declined with increasing heliocentric distance, this suggests that the Jupiter family comets may have originally incorporated more volatile ices in their interiors as compared with the Oort cloud comets. However, JF/SP comets experience considerable thermal processing during their long dynamical evolution to short-period orbits, and dynamical studies suggest that the average JF/SP comet has been in the inner planets region for $\sim 10^4$ years, equivalent to about 1500 orbits (Levison and Duncan, 1997). It is possible that this thermal processing removes any evidence of the colder formation zone for these comets. However, it may be that extremely volatile ices are still preserved deep in the interiors of the JF/SP comets, beneath the depths to which the diurnal and orbital thermal waves from solar insolation penetrate.

3. The Source of the Oort Cloud Comets

As noted above, the source of the long-period comets had been assumed to be primarily the Uranus-Neptune zone, since Jupiter and Saturn were expected to eject most of their icy planetesimals to interstellar space. This is illustrated in Table I, which shows the expected fraction of planetesimals from each giant planet zone placed in the Oort cloud, as estimated by several researchers. Using various analytical and dynamical techniques, these researchers showed that the relative efficiency of Uranus and Neptune in placing objects in the Oort cloud far exceeded that of Jupiter and Saturn, by factors up to 20 or more.

TABLE I
Fraction of comets in each giant planet zone ejected to the Oort cloud

	Safronov (1969)	Fernández and Ip (1981)	Duncan et al. (1987)
Jupiter	0.002	0.03	0.02
Saturn	0.005	0.14	0.06
Uranus	0.012	0.57	0.26
Neptune	0.020	0.72	0.40

However, several factors have begun to change this view, and suggest that Oort cloud comets originated over a much wider range of heliocentric distances in the early solar system. These include: 1) new integrations of icy planetesimals which show a larger fraction of Oort cloud comets coming from the Jupiter-Saturn zone; 2) a wider capture region in energy phase space for the Oort cloud, and 3) recognition of the fact that the spatial density of planetesimals was likely far higher in the

Jupiter-Saturn zone than in the Uranus-Neptune zone. The integrations also show that comets likely experienced a range of heliocentric distances and temperatures prior to being ejected to the Oort cloud. Lastly, it is likely that up to a few per cent of the Oort cloud population consists of asteroidal bodies formed within the orbit of Jupiter.

The new dynamical integrations have been performed by H. Levison, M. Duncan, L. Dones and P. Weissman (personal communication, hereafter Levison *et al.*) who simulated the dynamical evolution of planetesimals initially in low inclination and low to moderate eccentricity orbits between the giant planets. Levison *et al.* used a symplectic integrator, which provides a much more exact integration of the motion of the planetesimals as compared with the analytic treatment of Safronov (1969), the Öpik-type integrator of Fernández and Ip (1981), or the random-walk, diffusion approach of Duncan *et al.* (1987). Levison *et al.* followed 900 test particles for 10^9 years under the influence of the four giant planets, (assuming the current planetary masses) and galactic tides. Test particles were followed until they impacted the Sun or a planet, were ejected to interstellar space, or were captured into the Oort cloud. Oort cloud capture was defined as a test particle with a semimajor axis $< 10^6$ AU and a perihelion distance > 35 AU, i.e., detached from the planetary system.

Levison *et al.* found relative capture efficiencies to the Oort cloud of 0.03, 0.08, and 0.12 for the Jupiter-Saturn, Saturn-Uranus, and Uranus-Neptune zones (rather than identifying the ejecting planet, they denote the interplanetary zone from which the planetesimal originated). Note that the difference between the Jupiter-Saturn zone and the Uranus-Neptune zone is only a factor of four. Interestingly, most of the test particles were actually ejected to the Oort cloud by Jupiter or Saturn, despite having originated in other planetary zones. Levison *et al.* found that most Oort cloud comets had been to heliocentric distances < 10 AU at some time before they were ejected. They also found that there was a considerable dynamical exchange of planetesimals between the planetary zones, and that this exchange was necessary in order for planetesimals to be ejected.

This is readily seen if one considers the relative velocity between the planetesimals and the planets. Objects initially in low inclination, low eccentricity orbits near a planet encounter that planet with a relatively low velocity. Because of conservation of energy, a gravitational scattering event cannot change the magnitude of that relative velocity, only its direction However, the spacing between the giant planets is such that one planet can scatter the planetesimals in its zone to an encounter with another planet. These subsequent encounters are at higher relative velocities. After multiple scattering events a planetesimal's relative velocity is sufficiently high, close to 0.414 times the planet's orbital velocity (or more), that the encountered planet can eject the planetesimal from the planetary region. In most cases, Uranus and Neptune will pass their planetesimals to Jupiter and Saturn before they can be ejected, whereas Jupiter and Saturn can exchange the planetesimals in their zones to build up the necessary velocity.

Levison et al. also pointed out that the relative density of planetesimals was considerably higher near the orbits of Jupiter and Saturn than near Uranus and Neptune. Models of the solar nebula suggest that the surface density of the protosolar disk declined as $1/r^{1.5}$, where r is the heliocentric distance (Weidenschilling and Cuzzi, 1993). Thus, the surface density of primitive matter at the current orbit of Jupiter was $\sim$14 times higher than at the orbit of Neptune (this factor of 14 would be slightly different if the original planetary orbits migrated to their current locations following their formation, as a result of momentum exchange with the ejected comets; see, e.g., Fernández and Ip, 1984). Even if Jupiter and Saturn were less efficient in placing comets in the Oort cloud, the larger mass of icy planetesimals initially in their zones still allowed them to contribute a significant fraction of the Oort cloud's total mass. This is illustrated in Table II, which shows the fraction of the Oort cloud population originating in each planetary zone under a variety of assumptions, as estimated by Levison et al. Three cases are studied: 1) perturbation by the giant planets with their current masses and capture of comets into the Oort cloud due to galactic tides; 2) the same as (1) plus gas drag in the solar nebula for the first 10^7 years of the dynamical evolution; and 3) the same as (1) but with stellar perturbations due to the formation of the Sun and solar system in a young star cluster. The results in Table II show that a significant fraction of the initial Oort cloud population could have originated in the Saturn-Uranus zone, or even the Jupiter-Saturn zone. In the case of capture with gas drag, the capture efficiency of objects from the Jupiter-Saturn zone increased from 0.03 to 0.13; for formation in a star cluster, it increased from 0.03 to 0.12, while capture from the Saturn-Uranus zone increased from 0.08 to 0.16. The increased capture probability for Jupiter and Saturn zone comets assuming formation in a star cluster was also demonstrated analytically by Fernández (1997).

TABLE II

Fraction of comets initially captured into the Oort cloud originating in each planetary zone, based on the Levison et al. dynamical simulations

	Ejection by giant planets	Ejection plus gas drag from nebula	Formation of Sun in star cluster
Jupiter-Saturn	0.14	0.40	0.30
Saturn-Uranus	0.44	0.32	0.47
Uranus-Neptune	0.42	0.28	0.23

Note that these estimates by Levison et al. refer to the initial capture of objects into the Oort cloud, and do not necessarily reflect current-day statistics after 4 Gyr of dynamical evolution. Also, it should be pointed out that the Levison et al. integrations reported herein did not include the effects of random passing stars and only covered a time interval of 10^9 years. Inclusion of stars and longer duration integrations may alter the preliminary results discussed in this work. Those integrations are currently underway.

An important factor in increasing the efficiency of capture of icy planetesimals in the Oort cloud is the wider capture region afforded by the combination of galactic and stellar perturbations. Earlier studies, such as those by Safronov (1969) and Fernández and Ip (1981) assumed relatively narrow ranges of orbital energy for capture, typically semimajor axes between 10^4 and 10^6 AU. However, Duncan et al. (1987) showed that capture due to galactic tides could be effective at semimajor axes of 3 to 5×10^3 AU. This widened the capture region in energy phase space by a factor of two to three. In effect, it provided a larger "target" for the planets to scatter comets to. This effect was first pointed out by Weissman (1994) and Fernández (1997).

Lastly, new evidence suggests that the Oort cloud may even contain asteroidal bodies. An asteroid-like object, 1996 PW, was recently discovered in a typical long-period comet orbit (semimajor axis = 327 AU, eccentricity = 0.992). Weissman and Levison (1997b) examined the possible origin of this body and showed that it almost certainly came from the Oort cloud. They also showed that the object was more likely an asteroid than a dormant or extinct cometary nucleus. Weissman and Levison estimated that up to 3% of the Oort cloud population might be asteroidal bodies, formed inside the orbit of Jupiter and ejected by planetary perturbations.

4. Physical Processing of Comets in the Early Solar System

It would be expected that the wider range of formation zones for comets discussed in the previous sections would be reflected in the observed compositions of both long- and short-period comets. The maximum sub-solar, steady-state temperature, T (assuming no surface geometry effects), versus heliocentric distance is given by

$$T = [S_o(1 - A)/\epsilon \sigma r^2]^{1/4} \; K \tag{2}$$

where S_o is the solar constant, A is the albedo, ϵ is the surface emissivity, σ is the Stefan-Boltzmann constant, and r is the heliocentric distance. For unit emissivity and zero albedo, this is equal to

$$T = 396 \, r^{-1/2} \; K \tag{3}$$

where r is given in AU. Peak sub-solar, black-body temperatures currently range from 174 K at the orbit of Jupiter, sufficient to sublimate water and all other cometary ices, to 72 K at the orbit of Neptune and 56 K at 50 AU. At the latter two temperatures, water ice is a stable solid and can trap many other more volatile molecules.

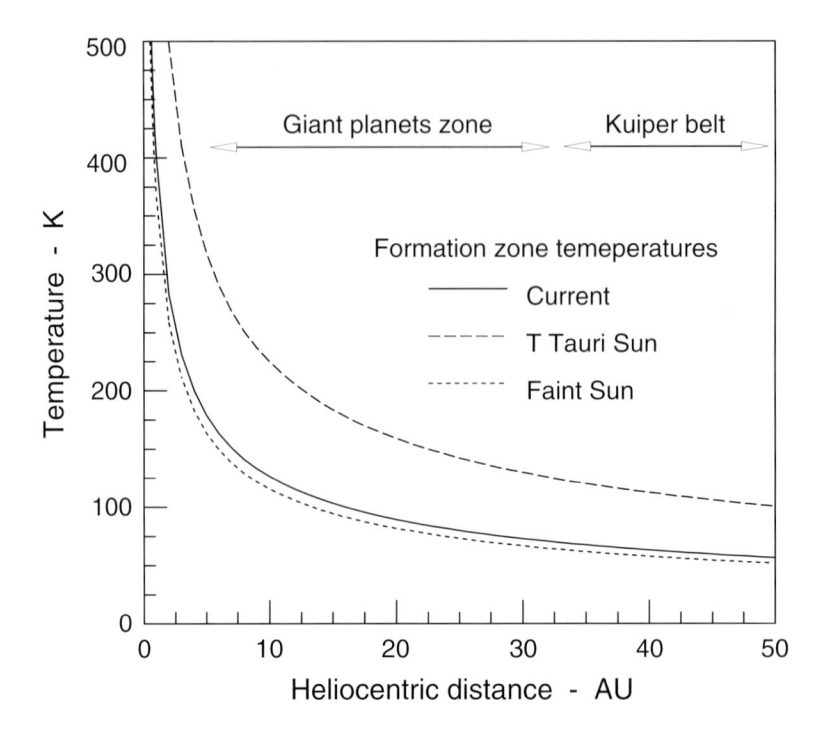

Figure 1. Sub-solar, steady-state, black-body temperatures versus heliocentric distance assuming the current solar luminosity, an early faint Sun (0.7 × current luminosity), and a bright T Tauri phase Sun (10 × current luminosity).

The expected sub-solar, black-body temperature versus heliocentric distance is shown in Figure 1. In addition to the current temperature range, curves are provided for an early faint Sun with 70% of the current solar luminosity (Gough, 1981), and for 10 times the current solar luminosity, reflecting a possible high activity T Tauri or FU Orionis stage (Stahler and Walter, 1993). These temperatures may have been mitigated in the early solar nebula by the opacity of dust in the central plane of the nebula.

Another important realization is that comets likely underwent considerable collisional processing prior to their ejection to the Oort cloud, or during their long residence in the Kuiper belt. Studies by Davis and Farinella (1997) and Stern and Colwell (1997) show that collisions over the history of the solar system have effectively "ground down" the population of the Kuiper belt between 30 and 50 AU. Davis and Farinella found that most objects which evolve to JF/SP comet orbits are fragments of primordial bodies that were collisionally disrupted. It is likely that similar collisional processing occurred to the icy planetesimals in the giant planets region. The results of this processing would have been to further increase the "rubble pile" nature of cometary nuclei (Weissman, 1986) as disrupted bodies re-accreted. Also, it would have increased the heterogeneity within individual cometary nuclei as planetesimal fragments from different interplanetary formation zones re-accreted.

This collisional processing, combined with the dynamical exchange of icy planetesimals in the giant planets zone, may provide an explanation for the lack of significant compositional diversity seen in cometary nuclei. It is possible that material from different formation zones was so thoroughly mixed that the composition of the comets was effectively "homogenized", each nucleus containing icy fragments from throughout the giant planets zone. This is offered as a highly speculative suggestion, but is certainly an area that would benefit from future investigation.

The only well documented evidence of compositional diversity in comets is the deficiency in long chain carbon molecules seen among some LP and Jupiter-family SP comets (A'Hearn et al., 1995). A'Hearn et al. found that about 30% of comets showed deficits in C_2 and C_3 relative to CN; almost all of the deficient comets were JF/SP comets with Tisserand parameters > 2. A'Hearn et al. suggested that the difference in the comet compositions was a signature of their formation zones and was not an evolutionary effect, because only a very few LP comets showed the same C_2 and C_3 deficiencies. They suggested that a compositional transition occurred in comets at some point in the Kuiper belt such that the comets from the giant planets zone and the inner portion of the Kuiper belt had normal C_2 and C_3 ratios, whereas the comets in the outer portion of the Kuiper belt were deficient in these long chain carbon molecules.

I do not find this argument compelling. The few LP comets that do show a deficiency are all dynamically older LP comets that have likely made more returns than most LP comets. The average LP comet is expected to have made only 5 passages through the planetary system (Weissman, 1979). The Jupiter-family SP comets have made far more returns on average, on the order of 1500 (Levison and Duncan, 1997). It may simply be that the process which destroys C_2 and C_3, and/or which prevents carbon chain molecules from escaping into the cometary comae, only acts after many years of thermal processing, requiring many perihelion passages. Only JF/SP comets and a few LP comets may achieve this large number of returns.

Conversely, it is interesting that P/Encke, which most cometary dynamicists would agree is a particularly old object (in terms of both total perihelion passages and perihelion passages close to the Sun), displays "normal" C_2 and C_3 abundances relative to CN. So I cannot say that I find my own arguments any more compelling.

What we have here is the age old question of "nature versus nurture." Are the recognizable compositional differences in comets due to the composition and physical conditions in their formation zones, or are the differences due to the different dynamical histories that comets may undergo, and their different evolutionary ages? The problem is an extremely complex one since we can only generalize about the dynamical history of any single comet. Because of the highly chaotic dynamics of comets in the solar system, as well as the unpredictable nongravitational forces from jetting of volatiles, we cannot extrapolate a comet's motion very far into the past or future. Thus, it is difficult to say whether a particular comet has been in its current orbit for very long, or what previous orbital evolution it may have experienced.

The answer may only come when we have the ability to directly sample a large number of comets, including Kuiper belt objects and dynamically new long-period comets. With that in mind, we look forward eagerly to the results of the Stardust, Contour, and Rosetta spacecraft missions, now in flight or under development.

Acknowledgements

I thank Hal Levison for the use of his unpublished dynamical simulations and for his helpful review of an earlier draft of this paper. This work was supported by the NASA Planetary Geology and Geophysics Program. It was performed at the Jet Propulsion Laboratory under a contract with the National Aeronautics and Space Administration.

References

A'Hearn, M. F., Millis, R. L., Schleicher, D. G., Osip, D. J., and Birch, P. V.: 1995, 'The ensemble properties of comets: Results from narrowband photometry of 85 comets', 1976-1992, *Icarus* **118**, 223–270.
Carusi, A., Kresák, L., Perozzi, E., and Valsecchi, G.: 1987, 'High-order librations of Halley-type comets', *Astron. Astrophys.* **187**, 899–905.
Davis, D. R., and Farinella, P.: 1997, 'Collisional evolution of Edgeworth-Kuiper belt objects', *Icarus* **125**, 50–60.
Duncan, M., Quinn, T., and Tremaine, S.: 1987, 'The formation and extent of the solar system comet cloud', *Astron. J.* **94**, 1330–1338.
Duncan, M., Quinn, T., and Tremaine, S.: 1988, 'The origin of short-period comets', *Astrophys. J.* **328**, L69–L73.
Duncan, M. J., Levison, H. F., and Budd, S. M.: 1995, 'The dynamical structure of the Kuiper belt', *Astron. J.* **110**, 3073–3081.
Duncan, M. J., and Levison, H. F.: 1997, 'A scattered comet disk and the origin of the Jupiter family comets', *Science* **276**, 1670–1672.
Edgeworth, K. E.: 1949, 'The origin and evolution of the solar system', *Mon. Not. Roy. Astron. Soc.* **109**, 600–609.
Everhart, E.: 1972, 'The origin of short-period comets', *Astrophys. Lett.* **10**, 131–135.
Fernández, J. A.: 1980, 'On the existence of a comet belt beyond Neptune', *Mon. Not. Roy. Astron. Soc.* **192**, 481–491.
Fernández, J. A.: 1994, 'Dynamics of comets: Recent developments and new challenges', In *Asteroids, Comets, Meteors 1993*, eds. A. Milani *et al.*, (Dordrecht: Kluwer), pp. 223–240.
Fernández, J. A.: 1997, 'The formation of the Oort cloud and the primitive galactic environment', *Icarus* **129**, 106–119.
Fernández, J. A., and Ip, W.-H.: 1981, 'Dynamical evolution of a cometary swarm in the outer planetary region', *Icarus* **47**, 470–479.

Fernández, J. A., and Ip, W.-H.: 1984, 'Some dynamical aspects of the accretion of Uranus and Neptune: The exchange of angular momentum with planetesimals', *Icarus* **58**, 109–120.

Gough, D. O.: 1981, 'Solar interior structure and luminosity variations', *Solar Phys.* **74**, 21–34.

Jewitt, D., and Luu, J.: 1993, 'Discovery of the candidate Kuiper belt object 1992 QB$_1$', *Nature* **362**, 730–732.

Jewitt, D., Luu, J., and Chen, J.: 1996, 'The Mauna Kea-Cerro-Tololo (MKCT) Kuiper belt and Centaur survey', *Astron. J.* **112**, 1225–1238.

Kuiper, G. P.: 1951, 'On the origin of the solar system'. In *Astrophysics*, ed. J. A. Hynek (New York: McGraw Hill), pp. 357–424.

Levison, H. F., and Duncan, M. J.: 1997, 'From the Kuiper belt to Jupiter-family comets: The spatial distribution of ecliptic comets', *Icarus* **127**, 13–32.

Newton, H. A.: 1893, 'On the capture of comets by planets, especially their capture by Jupiter', *Mem. Natl. Acad. Sci. USA* **6**, 7–23.

Oort, J. H.: 1950, 'The structure of the cloud of comets surrounding the solar system and a hypothesis concerning its origin', *Bull. Astron. Inst. Netherlands* **11**, 91–110.

Quinn, T., Tremaine, S., and Duncan, M.: 1990, 'Planetary perturbations and the origin of short-period comets', *Astrophys. J.* **355**, 667–679.

Safronov, V. S.: 1969, *Evolution of the Protoplanetary Cloud and the Formation of the Earth and Planets*, (Moscow: Nauka), NASA TTF-677 (1972).

Stahler, S. W., and Walter, F. M.: 1993, 'Pre-main sequence evolution and the birth population', In *Protostars and Planets* **III**, eds. E. H. Levy and J. I. Lunine, (Tucson: Univ. Arizona Press), pp. 405–428.

Stern, S. A., and Colwell, J. E.: 1997, 'Collisional erosion in the primordial Edgeworth-Kuiper belt and the generation of the 30–50 AU Kuiper gap', *Astrophys. J.* **490**, 879–882.

Weidenschilling, S. J., and Cuzzi, J. N.: 1993, 'Formation of planetesimals in the solar nebula', In *Protostars and Planets* **III**, eds. E. H. Levy and J. I. Lunine, (Tucson: Univ. Arizona Press), pp. 1031–1060.

Weissman, P. R.: 1979, 'Physical and dynamical evolution of long-period comets', In *Dynamics of the Solar System*, ed. R. L. Duncombe, (Dordrecht: D. Reidel), pp. 277-282.

Weissman, P. R.: 1986, 'Are cometary nuclei primordial rubble piles?' *Nature* **320**, 242–244.

Weissman, P. R.: 1994, 'Why are there no interstellar comets?' *Bull. Amer. Astron. Soc.* **26**, 1021 (abstract).

Weissman, P. R.: 1996, The Oort cloud, In *Completing the Inventory of the Solar System*, eds. T. W. Rettig and J. M. Hahn, ASP Conference Series **107**, pp. 265–288.

Weissman, P. R., and Levison, H. F.: 1997a, 'The population of the trans-Neptunian region'. In *Pluto-Charon*, eds. S. A. Stern and D. Tholen, (Tucson: Univ. Arizona Press), pp. 559–604.

Weissman, P. R., and Levison, H. F.: 1997b, 'Origin and evolution of the unusual object 1996 PW: Asteroids from the Oort cloud?' *Astrophys. J.* **488**, L133–L136.

Address for correspondence: Paul R. Weissman, pweissman@issac.jpl.nasa.gov

ROSETTA GOES TO COMET WIRTANEN

G. SCHWEHM and R. SCHULZ
ESA Space Science Department, ESTEC, Noordwijk, The Netherlands

Abstract. The International Rosetta Mission, approved by the Science Programme Committee of the European Space Agency as the Planetary Cornerstone Mission in ESA's long-term programme Horizon 2000, will rendezvous in 2011 with Comet 46P/Wirtanen close to its aphelion and will study the nucleus and the evolution of the coma for almost two years until it reaches perihelion. In addition to the investigations performed by the scientific instruments on board the orbiter, a Surface Science Package (Rosetta Lander) will be deployed onto the surface of the nucleus early during the near-nucleus study phase. On its way to Comet 46P/Wirtanen, Rosetta will fly by and study the two asteroids 4979 Otawara and 140 Siwa.

Keywords: ROSETTA

1. Introduction

The prime scientific objectives of the ROSETTA mission are to investigate the origin of the Solar System by studying the origin of comets, and to study the relationship between cometary and interstellar material and implications with regard to the origin of the Solar System. Comets are widely considered to contain the least processed material in the Solar System since its condensation from the proto-solar nebular. It is likely that even pre-solar grains have been preserved in these bodies. The physical and compositional properties of comets may therefore be a key to their formation and evolution, hence to the formation of the Solar System. Direct evidence on cometary material, in particular on the composition of cometary volatiles, is however extremely difficult to obtain. Very many physico-chemical processes, like sublimation, (photo)chemical reactions and interactions with the solar wind alter the material originally present in the nucleus. The species observable from Earth and even *in situ* during flyby missions are consequently not representing the actual molecular composition of the nucleus, although the currently available information already demonstrates the low level of evolution of cometary material. To retrieve information about the composition of a cometary nucleus and the processes altering it, the nucleus and near-nucleus environment of a comet have to be monitored *in situ* by very sensitive analytical investigations. The measurement goals of ROSETTA therefore include:

– Global characterization of the nucleus, determination of dynamic properties, surface morphology and composition

- Determination of chemical, mineralogical and isotopic compositions of volatiles and refractories in a cometary nucleus
- Determination of the physical properties and interrelation of volatiles and refractories in a cometary nucleus
- Study of the development of cometary activity and the processes in the surface layer of the nucleus and inner coma (dust/gas interaction).

To achieve these goals ROSETTA will study in detail the properties of the nucleus and the physico-chemical evolution of the cometary coma of Comet 46P/Wirtanen from onset of activity beyond 3 AU through perihelion at 1.06 AU. The target comet, 46P/Wirtanen, is described in Schulz and Schwehm (1999). After the rendezvous with the target comet at ≈ 4.5 AU, the spacecraft will go into orbit around the nucleus to accompany the comet on its pre-perihelion orbit. ROSETTA's payload comprises 12 scientific instruments on board the orbiter (Table II) and 9 on the Surface Science Package, *Rosetta Lander* (Table III). ROSETTA combines two strategies for characterizing the properties of a cometary nucleus. On one hand, the comet's evolution along the orbit with decreasing heliocentric distance, r_H, will be investigated with the orbiter instruments by monitoring the physical and chemical properties of the nucleus and in situ analysis of the near-nucleus environment. On the other hand, the Rosetta Lander will provide ground truth by analysing the nucleus material directly.

2. Mission Overview

ROSETTA is scheduled for launch by an Ariane-5 from Kourou, French Guiana, in January 2003. The spacecraft will employ three planetary gravity assist manoeuvres (Mars-Earth-Earth) to acquire sufficient energy to rendezvous with the comet. After each Earth gravity assist ROSETTA will fly by and study an asteroid. The selected asteroid targets are the S-type asteroid 4979 Otawara and the C-type asteroid 140 Siwa. Because of the long mission duration extended hibernation periods during cruise will be implemented. After a long cruise phase, the spacecraft will rendezvous with Comet 46P/Wirtanen in August 2011. The rendezvous manoeuvre at $r_H \approx 4.5$ AU is followed by a drift phase as the spacecraft slowly closes to the comet. Both will be performed on the basis of a ground-based determination of the orbit from dedicated astrometric observations, before the comet is detected by the on-board cameras. The final point of the near-comet drift phase, the Comet Acquisition Point (CAP), is reached at $r_H < 4.2$ AU. As soon as the solar arrays provide enough power to bring the spacecraft to full operational status, images of the comet will be acquired with the on-board navigation camera. The comet ephemeris will be updated from the on-board observations and ROSETTA will be placed in an orbit around the nucleus. The major mission events are summarized in Table I.

TABLE I
Major Events of the ROSETTA Mission

MAJOR MISSION EVENTS	NOMINAL DATE
Launch	21 January 2003
Mars gravity assist	26 August 2005
Earth gravity assist #1	26 November 2005
Otawara flyby	11 July 2006
Earth gravity assist #2	27 November 2007
Siwa flyby	25 July 2008
Rendezvous manoeuvre	24 August 2011
Start near- nucleus operations @ 3.25 AU (from Sun)	22 August 2012
Perihelion passage (end of mission)	10 July 2013

The preliminary survey of the surface of the nucleus is known as global mapping. Subject to solar illumination at least 80% of the surface is required to be mapped with high resolution (≤ 10 cm). During this phase the nucleus' shape, surface properties, kinematics and gravitational characteristics will be derived. Polar orbits around the comet at distances between 5 and 25 nucleus radii will be used for mapping of the surface. At the end of the global mapping phase, some five areas (500 x 500 m) will be selected for close observation (based on mapping and remote observation data). Ideally all the orbiter payload instruments are operating during this phase. Due to the large geocentric distance and the low downlink data rate of about 5 kbs on board pre-processing and intermediate storage of data in the on-board Solid State Mass Memory will be required. At the end of the close observation phase, the landing site for the Rosetta Lander will be selected on the basis of the collected data.

The Lander will be delivered from an eccentric orbit (pericenter altitude as low as possible, e.g. 1 km) with a pericenter passage near the desired landing site. An ejection mechanism will separate the Lander from the spacecraft with a maximum relative velocity of 1.5 m/s. The time and direction of the Rosetta Lander separation will be chosen such that the package arrives with minimum vertical and horizontal velocities relative to the local (rotating) surface. After delivery of the Lander, the spacecraft will be injected into an orbit which is optimized for receiving the data transmitted from the Lander and to relay them to the Earth. To adjust the payload operations sequences, the Lander can be commanded via the orbiter. The prime Lander activities will last about two weeks. After this the spacecraft will spend at least 200 days in orbit in close vicinity of the comet nucleus until perihelion passage, still collecting data from the Lander at regular intervals as long it remains operated. The objective of this phase is to monitor the nucleus (active regions), dust

TABLE II
Individual instruments on the Rosetta Orbiter

Short Name	Objective	Principal Investigator
	Remote Sensing	
OSIRIS	Multi-Colour Imaging (Narrow and Wide Angle Camera)	H. U. Keller, MPAe Katlenburg-Lindau, Germany
ALICE	UV-Spectroscopy (70 nm - 205 nm)	A. Stern, SRI Boulder, CO, USA
VIRTIS	VIS and IR Mapping Spectroscopy (0.25 μm - 5 μm)	A. Coradini IAS-CNR, Rome, Italy
MIRO	Microwave Spectroscopy (1.3 mm and .5 mm) (1.3 mm and 0.5 mm)	S. Gulkis, NASA-JPL Pasadena, CA, USA
	Composition Analysis	
ROSINA	Neutral Gas and Ion Mass Spectroscopy DFMS: 12-200 AMU, M/ΔM $\sim$ 3000 RTOF: 12-350 AMU, M/ΔM > 1000 incl. gas pressure sensor	H. Balsiger Univ. Bern, Switzerland
MODULUS	Isotopic Ratios of Light Elements by GCMS (D/H; ^{13}C/^{12}C; ^{18}O/^{16}O; ^{15}N/^{14}N)	C. Pillinger, Open Univ. Milton Keynes, UK
COSIMA	Dust Mass Spectrometer (SIMS, m/Δm $\sim$ 2000)	J. Kissel, MPE Garching, Germany
MIDAS	Grain Morphology (Atomic Force Microscope, nm Resolution)	W. Riedler, IWF Graz, Austria
	Nucleus Large Scale Structure	
CONSERT	Radio Sounding Nucleus Tomography	W. Kofman, CEPHAH Grenoble, France
	Dust Flux, Dust Mass Distribution	
GIADA	Dust Velocity and Impact Momentum Measurement, Contamination Monitor	E. Bussoletti Ist. Univ. Navale, Naples, Italy
	Comet Plasma Environment, Solar Wind Interaction	
RPC	Langmuir Probe (LP)	R. Boström, Swed. Inst. of Space Phys., Uppsala, Sweden
	Ion and Electron Sensor (IES)	J. Burch, SRI San Antonio, TX, USA
	Flux Gate Magnetometer (MAG)	K.-H. Glassmeier TU Braunschweig, Germany
	Ion Composition Analyser (ICA)	R. Lundin, Swed. Inst. of Space Phys., Kiruna, Sweden
	Mutual Impedance Probe (MIP)	J.G. Trotignon, LPCE/CNRS Orleans, France
RSI	Radio Science Experiment	M. Pätzold Univ. Köln, Germany

TABLE III
The Lander payload

APX	α-p-X-ray Spectrometer	R. Rieder, MPI Chem.
		Mainz, Germany
COSAC	Evolved Gas Analyser	H. Rosenbauer, MPAe
	elemental, molecular composition	Katlenburg-Lindau, Germany
MODULUS	Evolved Gas Analyser	C. Pillinger
	isotopic composition	Open University, UK
ÇIVA	Panoramic Camera	J.P. Bibring, IAS
	IR microscope	Orsay, France
ROLIS	Descent Camera	S. Mottola, DLR
		Berlin, Germany
SESAME	Comet Acoustic Surface Sounding	D. Möhlmann, DLR
	Experiment (CASSE)	Cologne, Germany
	Dust Impact Monitor (DIM)	I. Apathy,
		KFKI, Hungary
	Permittivity Probe (PP)	H. Laakso
		FMI, Finland
MUPUS	Multi-Purpose Sensor for Surface and	T. Spohn,
	Sub-Surface Science	Univ. Münster, Germany
ROMAP	RoLand Magnetometer and Plasma Monitor	U. Auster,
		DLR Berlin, Germany
CONSERT	Comet Nucleus Sounding	W. Kofman, CEPHAG
		Grenoble, France

and gas jets, and to analyse gas, dust and plasma in the inner coma from the onset to peak activity.

Orbital design will depend on safety considerations and scientific goals, taking into account that the communication relay service has to be provided for the long-life Rosetta Lander. Mission planning will depend on the results of previous observations, such as the activity pattern of the comet. Extended monitoring of different regions in the vicinity of the nucleus could be performed by successive quasi-hyperbolic flybys configuring petal-like trajectories.

3. Asteroid Science

The scientific objectives of the ROSETTA Mission also include the global characterization of asteroids. The spacecraft will, therefore, fly by two asteroids on its way to the comet, namely, 4979 Otawara and 140 Siwa. These asteroids have been selected after careful evaluation of the scientific significance of the reachable targets on one hand, and the available fuel budget of the spacecraft on the other hand. The target of highest interest is 140 Siwa, a C-type asteroid of diameter $\approx$ 110 km (Barucci *et al.*, 1998). C-type asteroids are believed to have undergone little or no heating, hence they should represent the most primitive type. Their meteorite analogs are carbonaceous chondrites. C-type asteroids are considered to be unaltered, volatile-rich bodies, darkened by opaque organic material, which causes their rather low albedo (6.7 % was derived for 140 Siwa).

The mission analysis showed that before the flyby at the preferred asteroid target 140 Siwa, another asteroid, 4979 Otawara, can be visited additionally within the limits of the fuel budget. So far, almost nothing is known of this asteroid. A first rough estimation of its size based on brightness measurements resulted in a diameter of about 5 km (Barucci, priv. com.). Due to the lack of spectroscopic data, no taxonomic classification is yet available for 4979 Otawara. However, observations of the target are planned for the beginning of 1999.

3.1. 4979 Otawara Flyby

Approximately three and a half years after launch, Rosetta will have its first asteroid flyby. The nominal date of closest approach to 4979 Otawara is 11 July 2006 with an uncertainty of one or two days, depending on the actual launch date and time. The flyby operations last from three months before to one month after the flyby. Parallel to the daily tracking with orbit determination and corrections, the scientific payload is to be checked out. Rosetta will pass the asteroid with a flyby speed of 10.6 km/s at a nominal distance of about 1500 km. The relative asteroid ephemeris will be determined with a cross track accuracy of 20 km by spacecraft optical navigation. During the few days immediately before the asteroid flyby, there will be the opportunity to improve the targeting and, if necessary or desired, manoeuvre the spacecraft. The cameras and scientific payload shall be pointing at the asteroid during the whole flyby, thus providing a nearly complete coverage and reconstruction of the 3 D shape.

3.2. 140 Siwa Flyby

Around 25 July 2008, Rosetta will encounter its main asteroid target, 140 Siwa. The operations are similar to the Otawara flyby. However, due to the higher flyby velocity (17.0 km/s) the nominal flyby distance will be 2500 km.

4. The Spacecraft and its Payload

The long mission time and the extreme environment in which the spacecraft has to operate (heliocentric distance from 1 AU to 5 AU with about 2 years very close to the comet nucleus) require a robust spacecraft with a high degree of on-board autonomy. To facilitate integration, the design of the ROSETTA spacecraft is based on a box-type central structure. The payload is mounted on one spacecraft wall, which during the close comet phase will continuously be pointed towards the nucleus. All instruments are body-mounted and the proper attitude is achieved by rotating the S/C with the high-gain antenna always pointing towards the Earth and the Solar Array pointing towards the Sun.

The orbiter payload has unprecedented capabilities to study the composition of both, the volatile and the refractory material released by the cometary nucleus, with very high resolution. Four instruments are dedicated to collect and analyse gas and dust samples from the inner coma as close as 1 km to the nucleus surface. The remote sensing suite of instruments will allow for a characterization of the nucleus surface in a wide wavelength range from UV to mm with high spatial resolution ($>$ 10 cm for the OSIRIS Narrow Angle Camera). For a complementary monitoring of the comet environment and its interaction with the solar wind, a Dust Flux Analyser and a Plasma Package have been selected.

The Rosetta Lander will focus on the in situ study of the composition and structure of the nucleus material. Measurement goals include the determination of the elemental, molecular, mineralogical, and isotopic composition of the cometary surface and subsurface material. Highest priority is given to the elemental and molecular determinations as it is believed that some mineralogical and isotopic measurements can be carried out adequately by orbiter science investigations. In addition properties like near-surface strength, density, texture, porosity, ice phases and thermal properties of the nucleus will be derived. Texture characterisation will include microscopic studies of individual grains.

CONSERT, with hardware on both the lander and orbiter will attempt to reveal the coarse structure of the nucleus through radio sounding.

References

Barucci, M.A., Doressoundiram, A., Fulchignoni, M., Florczak, M., Lazzarin, M. and Angeli, C.: 1998, 'Compositional type characterization of Rosetta asteroid candidates', *Plan. Sp. Sci.* **46**, 75-82.

Schulz, R. and Schwehm, G.: 1999, 'Coma Composition and Evolution of ROSETTA Target Comet 46P/Wirtanen', *Space Sci. Rev.*, this volume.

Address for correspondence: G. Schwehm, ESA Space Science Department, ESTEC, Postbus 299, NL-2200 AG Noordwijk, gschwehm@so.estec.esa.nl

COMA COMPOSITION AND EVOLUTION OF ROSETTA TARGET COMET 46P/WIRTANEN

R. SCHULZ and G. SCHWEHM
*ESA Space Science Department, ESTEC,
Noordwijk, The Netherlands*

Abstract. Comet 46P/Wirtanen is the target comet of the ROSETTA mission. Here, we give an overview of the information currently available on this comet from remote-sensing observations. Main emphasis is put on the description of the coma in terms of morphology, composition and evolution. We also summarize the current knowledge of the basic properties of the nucleus, in particular its size and rotational properties.

Keywords: ROSETTA, Space Mission Target Comets

1. Introduction

The European Space Agency is preparing ROSETTA, a mission to rendezvous with a Jupiter-family comet. ROSETTA will study the properties of the nucleus and the physico-chemical evolution of the cometary coma from onset of activity beyond 3 AU through perihelion (see: Schwehm and Schulz, (this issue) for a detailed description of the mission). Comet 46P/Wirtanen was selected as the target comet in 1994, mainly because it is basically the only target available for a ROSETTA type mission for the next decade. After its selection, it has been asked frequently whether 46P/Wirtanen is not only a reachable, but also an interesting target for a space mission. Providing an answer to this question is, therefore, one of the objectives of this paper.

2. Overview of Observations

Although 46P/Wirtanen was observed during nine of the ten apparitions it made since its discovery in January 1948 (not observed in 1980), very little was known about the comet before it approached perihelion in 1996/97. The available data from observations before its latest return were mainly restricted to visual and photographic magnitudes (see Jorda and Rickman (1995) for a summary). No spectroscopy and just two sets of observations with electronic detectors were obtained in 1991 (A'Hearn *et al.*, 1995; Cartwright *et al.*, 1997). Nevertheless, 46P/Wirtanen is now among the best observed short-period comets and one of the few comets that have been monitored along a major part of the orbit. In 1996/97 Comet 46P/Wirtanen was studied at different wavelengths from the ultraviolet to the radio from ground and from space. The observations from space were obtained with HST (Lamy et al., 1998; Stern et al., 1998), ISO (Colangeli et al., 1998) and SOHO (Bertaux et al., 1998). The monitoring from ground included visible imaging and spectrophotometry (Boehnhardt et al., 1997a,b; Fink et al., 1997; Meech et al., 1997; Farnham and Schleicher, 1998; Fink et al., 1998; Jockers et al., 1998; Schulz et al., 1998) and a search for the 18-cm radio lines of OH (Crovisier et al., 1999).

3. Size and Rotation of the Nucleus

Several attempts have been made to determine the size and rotational properties of the nucleus. The observations support a value of less than 1 km for the effective radius of the nucleus, assuming a geometric albedo of 4%. The smallest value results from the analysis of high-resolution images obtained with HST at a heliocentric distance of r_H = 2.45 AU. By separating the signal of the nucleus from that of the already present coma, Lamy et al. (1998) determined a value of 600±20 m for the average effective nucleus. The analysis of photometric observations of the presumedly inactive nucleus at large r_H also led to radii below 1 km. A maximum radius of 800 m was determined by Boehnhardt et al. (1997a) from a marginal detection of the comet as a point-like object at r_H = 4.6 AU. Later observations of the still point-like appearing nucleus with the ESO 3.5m NTT at r_H = 3.51 AU resulted in a radius of 900 m (Hainaut, priv. com.). Other estimates led to values between 1 km and 1.6 km (Fink et al., 1997; Jorda and Rickman, 1995). (All values apply to an assumed geometric albedo of 4%.) The rotational properties of Comet 46P/Wirtanen are still an open issue. Three different rotation periods have been derived from the three observational data sets examined so far: 6.0±0.3 h (Lamy et al., 1998), near 7.6 h (Meech et al., 1997), near 7 h or 14 h (Boehnhardt et al., 1997b). First modeling of the rotational properties of the nucleus, using the available data from apparitions prior to 1996/97, indicated that a complex rotation is very likely (Samarasinha et al., 1996). Further investigations led to a minimum value of the rotational period for pure spin states of 3.5 h (Samarasinha, priv. com.). The maximum value is still undefined and may be days.

TABLE I

Detected Species in Comet 46P/Wirtanen

Neutrals:	OH	CN	C_3	C_2	NH	NH_2	CS	H (Lyman-α)	[OI]
Ions:	H_2O^+	CO_2^+							

4. Coma Morphology, Composition and Evolution

The evolution of Comet 46P/Wirtanen in terms of brightness, coma morphology and composition was investigated as a function of heliocentric distance starting, at 4.6 AU (aphelion distance: 5.1 AU). At the time of recovery (June 23-26, 1995; r_H = 4.6 AU) the comet appeared as a point-source without detectable coma (Boehnhardt *et al.*, 1997a) and remained point-like in appearance also at its next observation at 3.51 AU (Hainaut and West, 1996). The first detection of a dust coma (radius ≈ 10000 km) was reported at r_H = 3.30 AU (Boehnhardt *et al.*, 1996). The brightness and coma diameter were monitored by several investigators using broad-band CCD photometry. First photometric light curves were produced by Meech *et al.* (1997) and Boehnhardt *et al.* (1997b). Except for its already developed dust tail, Comet 46P/Wirtanen did not show any prominent features in its coma. The coma isophotes were circular and symmetric to the tail axis (Boehnhardt *et al.*, 1997b). Table I gives a synopsis of the neutral and ionized gas species detected in Comet 46P/Wirtanen. All of them are typically present in comets. The first gaseous species (CN) was firmly detected at r_H = 2.34 AU (Schulz *et al.*, 1998). A possible marginal detection of OH at r_H = 2.47 AU was reported by Stern *et al.* (1998). Figure 1 shows the production rates of several gaseous species and the dust (in *Afρ*) as a function of heliocentric distance. Originally the gas production rates were derived and published by several individual groups. They used different models (vectorial or Haser) and different lifetimes/scalelengths which led to systematic, model-dependent differences between the values.

The gas production rates shown in Figure 1 and listed in Table II are therefore re-calculated from the original emission band fluxes, using the same model and parameters in all cases. This increases the accuracy of the curves for the individual species with respect to relative changes. The possibility of a systematic error affecting all absolute values in the same way remains, however. We used the vectorial model (Festou, 1981) with the lifetimes given in Table III. The parent velocity was varied as 0.85 km/s × $r_H^{-0.5}$ while the daughter velocity was arbitrarily set to 1 km/s. The water production rate in Figure 1 was determined from either OH or O[I]. We only re-calculated the values coming from OH, because O[I] was measured by just one author (Fink *et al.*, 1998). The values resulting from both species are remarkably consistent. The evolution of the gas and dust production can be well followed in Figure 1. After their first detection, the production rates of all gaseous species increase steadily with decreasing distance to the Sun, whereas the *Afρ* values of the dust remain more or less constant* at least until r_H = 1.8 AU. If the *Afρ* represents the dust production rate and its variations reliably, this would mean that the dust production at r_H > 1.8 AU does not follow the same pattern as the production rates of the five gas species shown in Figure 1. However, one has to keep in mind that *Afρ* is determined from the light reflected by the dust

---

* No phase correction has been applied to the *Afρ* values in Figure 1. The phase-corrected values show a slight increase with decreasing r_H (Jorda, priv. com.).

TABLE II
The production rates and $Af\rho$ values for Comet 46P/Wirtanen plotted in Figure 1

r_H (AU)	$Q(H_2O)$ 10^{26} s^{-1}	$Q(CN)$ 10^{24} s^{-1}	$Q(C_2)$ 10^{24} s^{-1}	$Q(C_3)$ 10^{23} s^{-1}	$Q(NH_2)$ 10^{24} s^{-1}	$Af\rho$ cm	Ref.
-2.81	-	<1.20	<0.26	<0.41	<0.72	<14	1
-2.72	<1.2	-	-	-	-	15±4	2
-2.47	2.6±0.5	-	-	-	-	18±3	2
-2.34	-	0.69±0.24	<0.16	0.15±0.12	<0.45	19±6	1
-2.04	-	1.92±0.36	0.48±0.18	0.60±0.17	0.45±0.25	16±3	1
-1.81	-	3.49±0.60	0.92±0.34	1.04±0.38	1.48±0.37	20±3	1
-1.60	-	10.17±2.63	5.87±1.10	3.56±0.85	6.12±0.46	76±17	1
-1.31	51.0±0.3	-	-	-	-	15±3	2
-1.22	(171)$^+$	19.5±0.71	33.3±1.63	-	50.1±0.10	85±1.4	3
-1.19	(187±14)	24.7±0.71	40.0±1.13	-	62.0±3.11	116.5±3.5	3
-1.18	(218±12)	26.8±2.88	43.6±2.69	-	58.4±3.18	110.8±3.3	3
-1.14	74.1±26	15.8±0.7	16.2±0.8	14.5±8	-	51±13	4
-1.12	77.6±14	17.8±0.8	19.1±0.9	28.8±5.9	-	50±4.5	4
-1.08	306.5±30.4	29.7±0.71	54.4±0.14	-	54.1±0.99	137	3
-1.07	259.0±11.3	32.3±0.78	59.3±6.08	-	56.6±1.41	125±1.4	3
-1.07	72.4±18.7	26.9±1.69	26.9±2.95	24.0±13.9	-	118±42	4
-1.07	309.0±9.9	32.0±1.23	54.2±0.14	-	51.6±0.28	140.5±2.1	3
1.08	175.5±9.2	27.2±2.90	50.7±0.64	-	48.2±1.70	114±1.4	3
1.09	283.5±24.7	39.2±1.91	64.8±5.16	-	66.2±6.79	129±4.2	3
1.23	(103±11)	15.4±0.21	27.9±1.56	-	29.5±1.27	80.5±2.1	3
1.23	(73±11)	18.7±0.78	24.2±4.88	-	26.8±0.57	69	3
1.24	(80±6)	17.7±2.19	20.3±3.32	-	18.8	75.5±16.3	3
1.24	(92±8)	17.7±2.19	20.3±3.32	-	18.8	75.5±16.3	3
1.25	(96±1)	14.5±1.13	21.6±2.40	-	25.9±0.42	67.5±3.5	3
1.47	(31)	12.9	12.7	-	14.2	32	3
1.49	-	7.1±1.6	7.9±2.6	-	-	30±14	4
1.72	-	4.0±2.0	5.6±2.5	-	-	41±41	4

/1/ Schulz et al. (1998)
/2/ Stern et al. (1998)
/3/ Fink et al. (1998)
/4/ Farnham and Schleicher, (1998)

$^+$Values in parentheses correspond to 40% of measured flux in OI+NH$_2$ complex and are not plotted in Fig. 1

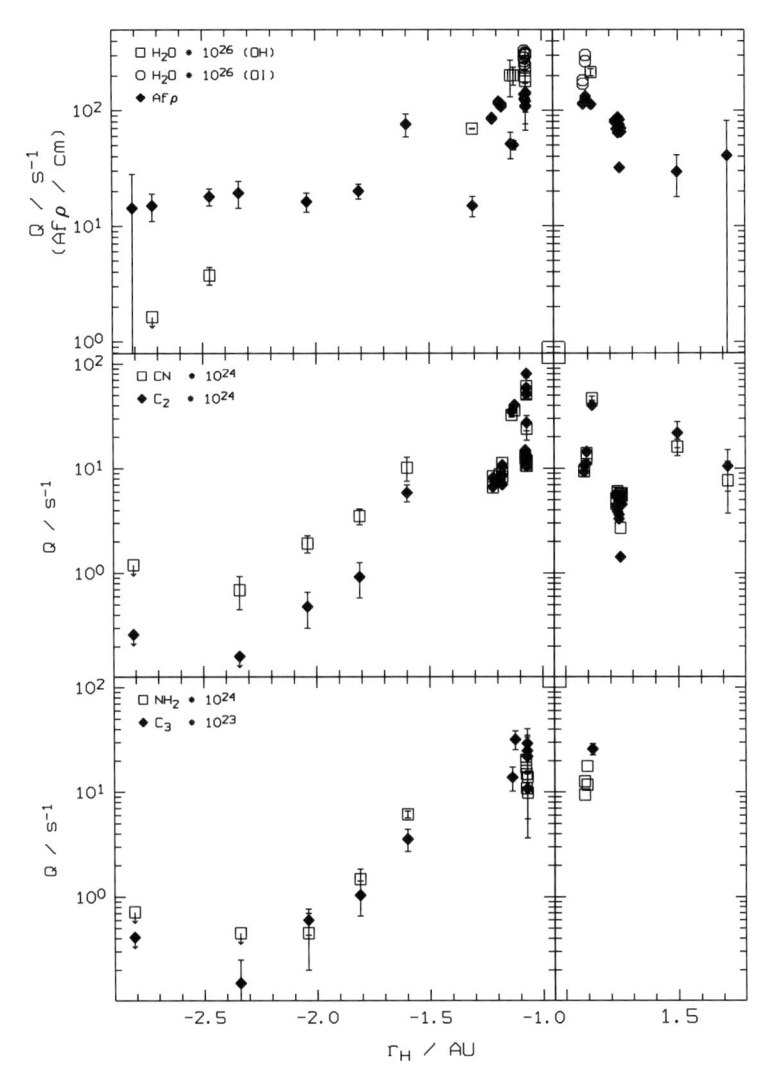

Figure 1. Production rate curves of Comet 46P/Wirtanen. The values were calculated from fluxes given in Farnham and Schleicher (1998), Fink *et al.* (1998), Schulz *et al.* (1998) and Stern *et al.* (1998).

particles in certain apertures. Hence, the $Af\rho$ value is also influenced by changes in the dust velocity and the dust size distribution. Below 1.8 AU the $Af\rho$ values show a significant dispersion of values. The discrepancy is particularly large for the values at r_H = 1.6 AU and r_H = 1.3 AU preperihelion, which were determined from spectrophotometric observations in two different spectral ranges (vis. at 1.6 AU, UV at 1.3 AU). This might indicate a different evolution of the $Af\rho$ versus r_H in different wavelengths. However, since variations in the $Af\rho$ value were also found in photometric observations taken in one single wavelength range with the same telescope/instrument combination (Farnham and Schleicher, 1998), they could also indicate the presence of a short-term variability in the dust production rate.

The production rate curves of the trace species (CN, C_2, C_3, NH_2) and of the dust show a steep increase between 1.8 AU and 1.6 AU (see Schulz et al. (1998) for the detailed analysis of this phenomenon). At about the same distance ($r_H \approx 1.9$ AU) the coma starts to increase strongly in R brightness (see Figure 2). Furthermore, Comet 46P/Wirtanen was previously reported to have displayed a similar rapid rise in brightness at its past two apparitions (Morris, 1994). However, these events were considered as ambiguous and needed confirmation. The production of C_2 continues to increase at higher level than the CN production which results in a significant increase in the C_2/CN ratio with decreasing r_H. During its post-perihelion phase the coma composition was monitored only until 1.7 AU where the values were already affected by very large error bars.

TABLE III

Dissociative parent- and daughter-lifetimes and fluorescence efficiencies ($r_H = 1$ AU) used for the re-calculation of the gas production rates in Figure 1.

	τ_p / s	τ_d / s	Reference	g / s^{-1}	Reference
OH	86000	159000	Stern et al. (1998)	variable	Schleicher and A'Hearn (1988)
CN	28500	400000	Schulz et al. (1993)	variable	Schleicher (1983)
C_3	4000	140000	Festou et al. (1990)	0.2960	A'Hearn et al. (1995)
C_2	37400	67900	Schulz et al. (1993)	0.1164	A'Hearn et al. (1995)
NH_2	5300	62000	Schulz et al. (1998)	0.0056	Schulz et al. (1998)

5. Discussion

Comet 46P/Wirtanen shows many peculiarities in its activity and compositional evolution. Its appears to have a very small radius which can barely account for its relatively high activity at perihelion. At heliocentric distances beyond about 1.8-1.9 AU it shows a rather constant R brightness and $Af\rho$ values, which indicates a fairly constant low dust production rate. The concomitant increase in (1) the production rates of gaseous trace species, (2) the dust production, (3) the R brightnesss and (4) the C_2/CN ratio between 1.8 AU and 1.6 AU indicate a distinct change in outgassing conditions in this part of the orbit. As the dust is dragged out of the nucleus by subliming volatiles, the observed increase of dust production requires a parallel increase of a gaseous species which is more abundant than trace species. It may be related to a parallel increase in the water production. Unfortunately, no data on the water production rate exist for this part of the orbit. However, since the overall shape of the water production rate curve is comparable to those of the

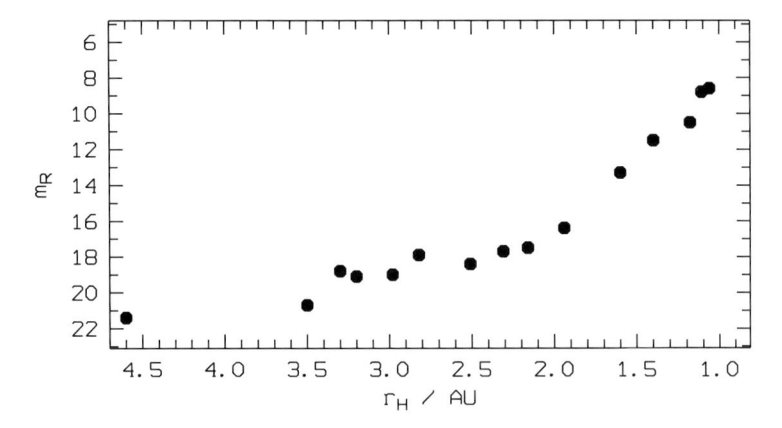

Figure 2. The lightcurve of 46P/Wirtanen in broad-band R (Boehnhardt *et al.*, 1997b).

trace species, it is likely that the water production rate also increases. The possible short-term variability in the $Af\rho$ values below 1.8 AU is very difficult to understand, as it does not seem to be connected with any variation in the gas production. The reason why the C_2/CN ratio increases with decreasing heliocentric distance is not known. It might indicate that a significant amount of the C_2 originates from the dust. Hence, the increase in the C_2 production would be related to the increased dust production.

In conclusion, the extensive study of Comet 46P/Wirtanen by remote-sensing techniques led to a large amount of new information about this comet. Many new questions were posed on the development and evolution of its activity as well as on the composition of the nucleus and the coma. Most of them will not be answerable by means of ground-based observations. Thus, this comet is indeed an interesting target for a space mission and ROSETTA will hopefully be able to provide the information we need to understand the evolution of a comet approaching perihelion.

References

A'Hearn, M.F., Millis, R.L., Schleicher, D.G., Osip, D.J. and Birch, P.V.: 1995, 'The Ensemble Properties of Comets: Results from Narrowband Photometry of 85 Comets', 1976-92, *Icarus* **118**, 223-270.
Bertaux, J.-L., Costa, J., Makkinen, T., Quémerais, E., Lellement, R., Kyrölä, E. and Schmidt, W.: 1998, 'Lyman-alpha Observations of Comet 46P/Wirtanen with SWAN on SOHO: H_2O Production Rate near 1997 Perihelion', *Plan. Space Sci.* **46**, in press.
Boehnhardt, H., Rauer, H., Motolla, S. and Nathus, A.: 1996, 'Comet P/Wirtanen', *IAU Circ. 6392*.
Boehnhardt, H., Babion, J. and West, R.M.: 1997a, *Astron. Astrophys.* **320**, 642-651.
Boehnhardt, H., Fiedler, A., Geffert, M. and Sanner, J.: 1997b, *ESA Contract Report*.
Cartwright, I.M., Fitzsimmons, A., Williams, I.P. and Kemp, S.N.: 1997, 'CCD photometry of comet 46P/Wirtanen', *Plan. Space Sci.* **45**, 821-826.

Colangeli, L., Bussoletti, E., Cecchi Pestellini, C., Fulle, M., Mennella, V., Palumbo, P. and Rotundi, A.: 1998, 'ISOCAM Imaging of Comets 65P/Gunn and 46P/Wirtanen', *Icarus* **134**, 35-46.

Crovisier, J., Biver, N., Bockelée-Morvan, Colom, P., Gérard, E. and Rauer, H.: 1999, 'Radio observations of the OH radical in comets 46P/Wirtanen and 81P/Wild 2', to be published.

Farnham, T.L. and Schleicher, D.G.: 1998, 'Narrowband photometric results for comet 46P/Wirtanen', *Astron. Astrophys.* **335**, L50-L55.

Festou, M.C.: 1981, 'The density distribution of neutral compounds in cometary atmospheres. I. Models and equations', *Astron. Astrophys.* **95**, 69-79.

Festou, M.C., Tozzi, G.P., Smaldone, A., Felenbok, P., Falciani, R. and Zucconi, J.-M.: 1990, 'Did an outburst occur on 4 December 1985 in Halley's comet?' *Astron. Astrophys.* **227**, 609-618.

Fink, U., Fevig, R.A. Tegler, S.C. and Romanishin, W.: 1997, 'CCD imaging and photometry of 46P/Wirtanen during October 1996', *Plan. Space Sci.* **45**, 1383-1387.

Fink, U., Hicks, M.D., Fevig, R.A. and Collins, J.: 1998, 'Spectroscopy of 46P/Wirtanen during its 1997 apparition', *Astron. Astrophys.* **335**, L37-L45.

Hainaut, O. and West, R.M.: 1996, *MPC 27351*.

Jockers, K., Credner, T. and Bonev, T.: 1998, 'Water ions, dust and CN in comet 46P/Wirtanen', *Astron. Astrophys.* **335**, L56-L59.

Jorda, L. and Rickman, H.: 1995, 'Comet P/Wirtanen, summary of observational data', *Plan. Space Sci.* **43**, 575-579.

Lamy, P.I., Toth, I., Jorda, L., Weaver, H.A. and A'Hearn, M.F.: 1998, 'The nucleus and the inner coma of Comet 46P/Wirtanen', *Astron. Astrophys.* **335**, L25-L29.

Meech, K.J., Bauer, , J.M. and Hainaut, O.R.: 1997, 'Rotation of Comet 46P/Wirtanen', *Astron. Astrophys.* **326**, 1268-1277.

Morris, C.: 1994, 'The light curve of comet 46P/Wirtanen', *Int. Comet Q* **92**, 178-180.

Samarasinha, N.H., Mueller, B.E.A. and Belton, M.J.S.: 1996, 'Comments on the Rotational State and Non-Gravitational Forces of Comet 46P/Wirtanen', *Plan. Space Sci.* **44**, 275-281.

Schleicher, D.G.: 1983, 'The Fluorescence of OH and CN', Thesis, Deparment of Physics and Astronomy, University of Maryland, pp. 108-110.

Schleicher, D.G. and A'Hearn, M.F.: 1988, 'The fluorescence of cometary OH', *Astrophys. J.* **331**, 1058-1077.

Schwehm, G. and Schulz, R.: 1999, 'ROSETTA goes to comet Wirtanen', *Space Sci. Rev.*, this volume.

Schulz, R., A'Hearn, M.F., Birch, P.V., Bowers, C., Kempin, M. and Martin, R.: 1993, 'CCD Imaging of Comet Wilson (1987VII): A Quantitative Coma Analysis', *Icarus* **104**, 206-225.

Schulz, R., Arpigny, C., Manfroid, J., Stüwe, J.A., Tozzi, G.P., Rembor, K., Cremonese, G. and Peschke, S.: 1998, 'Spectral evolution of ROSETTA target comet 46P/Wirtanen', *Astron. Astrophys.* **335**, L46-L49.

Stern, S.A., Parker, J.Wm., Festou, M.C., A'Hearn, M.F., Feldman, P.D., Schwehm, G., Schulz, R., Bertaux, J.-L. and Slater, D.C.: 1998, 'HST mid-ultraviolet spectroscopy of comet 46P/Wirtanen during its approach to perihelion in 1996-1997', *Astron. Astrophys.* **335**, L30-L36.

Address for correspondence: R. Schulz, ESA Space Science Department, ESTEC, Postbus 299, NL-2200 AG Noordwijk, rschulz@so.estec.esa.nl

MISSIONS TO COMETS AND ASTEROIDS

WESLEY T. HUNTRESS, JR.
Geophysical Laboratory, Carnegie Institution of Washington

Former NASA Associate Administrator for Space Science

1. History

For the greater part of the history of space exploration, now only just forty years in extent, the small bodies in our Solar System have been relatively ignored. The focus of attention in planetary exploration has been on the nearest planets to Earth, Mars and Venus, and on the large, dominant planets of the Outer Solar System, Jupiter, Saturn, Uranus and Neptune. Even the smallest planets Mercury and Pluto have been relatively ignored. There has been only one mission to Mercury, Mariner 10 in 1973, a multiple flyby mission which imaged only one hemisphere of the small planet. Pluto has yet to be explored by spacecraft and will not be visited until more than 50 years have past since the dawn of the space age. The Pluto mission, now scheduled by NASA for launch in 2003 and flyby in 2011, will mark the end of the reconnaissance phase of exploring the planetary region of the Solar System.

Until the set of missions sent to investigate Comet Halley in 1986, no dedicated flight mission to an asteroid or comet had been considered seriously. The public memory of Halley's spectacular apparition in 1910, and the anticipation of it's return in 1986, were the impetus for flyby missions to Halley conducted by the European Space Agency (Giotto), the Soviet Union (Vega 1 and 2), and Japan (Suisei and Sakigake).

Giotto was the only mission dedicated to Halley exploration. It carried a comprehensive set of 12 science instruments including a camera. Giotto flew within 596 km of Halley's nucleus on March 14, 1986, to give us our best pictures of an active cometary nucleus and in situ measurements of cometary dust, gas, and plasma in both the inner and outer coma.

The two Soviet missions were initiated as another in the series of highly successful Venera missions to Venus, specifically to carry balloon-carrying entry probes to the planet. After mission designers discovered the possibility to divert the carrier spacecraft to Halley using the gravity field of Venus, the missions were revised to include a Halley intercept and renamed. The carrier spacecraft were outfitted with an international array of instruments and flown to within 9000 km of the nucleus on March 6 and March 9, 1986.

Likewise, the two Japanese spacecraft were originally designed to orbit the sun and investigate the solar wind. Japanese mission designers also arranged for their solar spacecraft to fly near Halley (7 million km by Sakigake and 150,000 km by Suisei) where they returned in situ plasma data and images of the H-atom corona.

An international organization, the IACG, was set up to coordinate the observations by this armada of spacecraft and succeeded well in optimizing the science return from the missions through joint science campaigns. The U.S. participated only peripherally in the Halley mission enterprise, using the Deep Space Network to provide essential tracking and data acquisition in return for scientific participation. When it became clear that there would be no U.S. mission to Halley, the U.S. turned to the IACG for participation, and redirected the International Sun Earth Explorer (ISEE) from high earth orbit to fly through the distant tail of Giacobini-Zinner in 1985.

NASA was unable to capitalize on the popularity of Comet Halley to mount a mission of its own. After the Halley mission successes by other nations in 1986, NASA developed a mission to orbit a short period comet, the Comet Rendezvous Asteroid Flyby (CRAF), taking the next logical exploration step following the multiple flybys of Halley. This mission became part of a dual new start for the U.S. planetary program in 1990. The second of these missions is the joint NASA/ESA Cassini/Huygens mission to Saturn and Titan. Two years later, in January, 1991, the CRAF mission was cancelled by the Administration as part of a budget reduction exercise and Congress declined to resurrect it the following summer. The CRAF cancellation was yet another knockdown in a long series of blows in the quest for NASA missions to small bodies.

Some solace was provided for U.S. small body scientists by the expedient of flying missions with other objectives past main belt asteroids on their way through the asteroid belt. On October 29, 1991, history was made when the Galileo spacecraft made the first flyby of an asteroid, 951 Gaspra, on its first excursion through the asteroid belt. On its second pass through the asteroid belt, Galileo flew past the asteroid 243 Ida and discovered its tiny satellite Dactyl.

The European Space Agency developed a comet rendezvous mission of its own, Rosetta, which included a surface lander as well. Europe's success with Giotto, coupled with anticipation of an ambitious asteroid mission from Japan, and the public interest generated by the Gaspra and Ida images from Galileo, finally brought serious attention to small body missions in the U.S. In 1994, the first U.S. mission dedicated to small body exploration was approved-the Near Earth Asteroid Rendezvous (NEAR) mission to the earth- approaching asteroid Eros. Other attributes which made NEAR attractive were increasing public attention to the potential danger from earth-crossing asteroids and the fact that at the time NEAR was by far the most inexpensive planetary mission ever proposed. NEAR became the first in a new series of low-cost planetary missions called the Discovery program, which also included the Mars Pathfinder project.

In revising its strategy towards Space Science mission planning in the decade of the 1990's, NASA established three flight programs under which small body missions could be mounted. The first is the Discovery Program, which has produced three small body missions:

1. NEAR, which flew by the main belt asteroid 253 Mathilda in June, 1997, and which reaches Eros in February, 2000,
2. Stardust, which will launch in February, 1999, and return samples of comet refractory material from the coma of Wild 2 in January, 2006, and
3. Contour, which will fly by the three comets Encke (2003), Schwassmann-Wachmann-3 1 (2006) and d'Arrest (2008) to study the diversity of cometary nuclei.

The second mission line is the NASA New Millennium program. New Millennium is a spacecraft technology flight test program, but each mission also carries secondary science objectives. New Millennium has produced two small body missions:

1. DS-1, whose target is the asteroid 1992 KD but with an extended mission option to fly past Comets Wilson-Harrington and Borrelly in 2001, and
2. DS-4 carrying the Champollion lander to Comet Tempel 1 in December 2005.

The third program relevant to small body missions is the new NASA Outer Solar System mission line. A Comet Nucleus Sample Return mission is under study as a candidate to follow the first three missions in this line-Europa Orbiter, Pluto/Kuiper Belt Flyby and Solar Probe.

The 1990's have finally established comet and asteroid missions as primary objectives for scientific exploration of the Solar System on a par with missions to the planets. The number of dedicated small body missions approved today is astonishing considering the difficulty in establishing them prior to this decade. The only small body in the Solar System visited by spacecraft prior to 1990 was Comet Halley in 1986. As of this writing, spacecraft have flown past the asteroids Gaspra, Ida, and Mathilda. NASA has one mission nearing its target (NEAR), one recently launched and on its way (DS-1), another ready for launch (Stardust), and two others in preparation (Contour launch in 2002 and DS-4/Champollion launch in 2003). NASA has also committed to major participation in Japan's MUSES-C asteroid sample return mission (launch 2002), and ESA's Rosetta comet orbiter mission (launch 2003). The range of mission types include flybys (DS-1, Contour), rendezvous (NEAR, DS-4, Rosetta), landers (Rosetta, Champollion) and sample return (Stardust, MUSES-C, DS-4(?)). We can look forward to a rich return of science from these missions and a new era of small body exploration beginning with the DS-1 flyby of 1992KD in July, 1999, and extending through the next decade.

2. Science

Comets are believed to be the most primitive remnants of the primordial solar system. The newest of comets, those which appear to be making their initial pass by the Sun, are believed to be relatively unchanged since the time of the formation of the solar system. As long as comets show the outgassing activity which mark their passage by the Sun, they are believed to be expelling relatively pristine volatile material from their interior. The evidence is compelling that these comets have been stored since the beginning of the Solar System beyond Neptune in the Kuiper Belt or even farther beyond in the Oort Cloud.

While more similar than different, there is some variability among comets in their coma compositions and more so in their gas and dust production rates. The objectives in carrying out space-flight missions to comets are to understand:

- the degree of modification since formation
- the conditions and chemistry near the time of solar system formation
- the relationships among comets, asteroids and interplanetary dust
- the conditions and location of comet formation
- the content of comet reservoirs-Kuiper belt and Oort cloud
- the role of comets in early planetary bombardment and periodic mass extinctions, and
- the potential for delivery of water and organic material to the planets.

Small solar system bodies exhibiting cometary activity may represent one extreme of a continuum of small objects from light and volatile comets to dense and dry asteroids. Many comets and asteroids cross the Earth's orbit and have impacted the Earth over geologic time. Comets may also represent a reservoir of potential future interplanetary resources for space exploration or export to Earth. It is important therefore to devise a program that over time investigates a diverse set of comets and asteroids to:

- study their surface characteristics, sizes, shapes, bulk densities and internal properties
- establish their mineralogy and elemental compositions, and
- understand their relationship to remotely sensed asteroid spectral types, to known meteorite types and to interplanetary dust particles collected on Earth.

The search for the origin of life in the Solar System requires an understanding of the environments in the early Solar System that enabled the development of life, and an understanding of chemical and prebiological origins including the original chemical materials and processes leading to biology. This requires detailed analysis of the elemental, isotopic, molecular and mineralogical compositions of primitive objects in the Solar System including meteorites, comets, asteroids, icy satellites and Kuiper belt objects all of which are rich in organics, water and other crucial volatiles and prebiological organic materials.

TABLE I
Mission Objectives

Small Body Missions	NEAR	DS-1	Muses-C	Star-dust	Contour	DS-4/Champ.	Rosetta
Bulk Properties							
Size	x	x	x	x	x	x	x
Shape	x	x	x	x	x	x	x
Mass	x	x	x	x	x	x	x
Density	x		x			x	x
Dynamics, spin state	x	x	x	x	x	x	x
Internal Properties							
Structure	x		x			x	x
Magnetic field	x		x				x
Gravity field	x		x			x	x
Surface Properties							
Geology	x	x	x	x	x	x	x
Morphology	x	x	x	x	x	x	x
Elemental composition	x		x			x	x
Isotopic comp.						x	x
Chemical comp.						x	x
Mineralogical comp.	x	x	x		x	x	x
Regolith properties*			x			x	x
Volatile Evolution		x		x	x	x	x
Coma Properties							
Dust flux				x	x		x
Dust composition				x	x		x
Gas flux					x		x
Gas composition					x		x
Gas/dust chemistry					x		x
Plasma dynamics		x					x
Plasma chemistry		x					x
Sample Return							
Rocks and soil			x				
Refractory Grains				x			
Volatiles						x(?)	

* Strength, porosity, thermal, dielectric

APPROVED INT'L MISSIONS TO SMALL BODIES
(Launches 1996-2005)

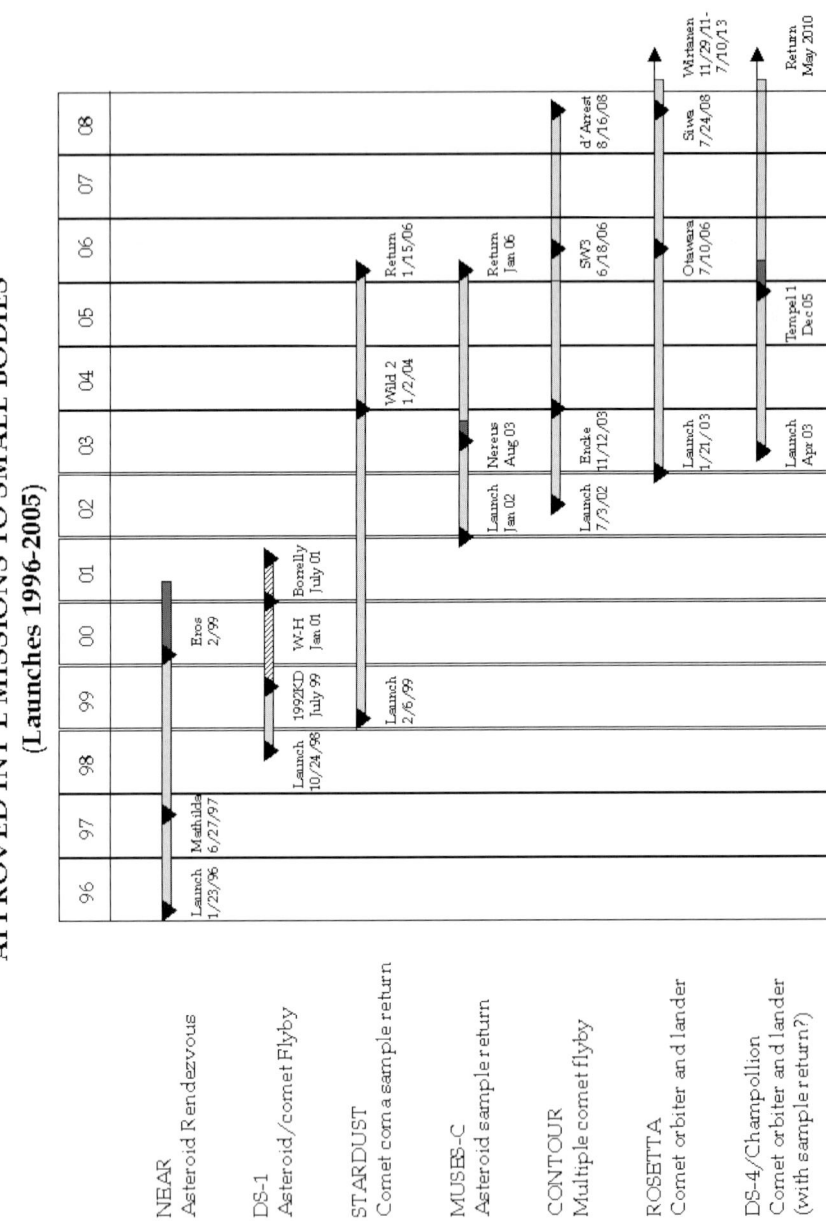

	96	97	98	99	00	01	02	03	04	05	06	07	08
NEAR — Asteroid Rendezvous	Launch 1/23/96	Mathilda 6/27/97			Eros 2/99								
DS-1 — Asteroid/comet Flyby			Launch 10/24/98	1992KD July 99	W-H Jan 01	Borrelly July 01							
STARDUST — Comet coma sample return				Launch 2/6/99					Wild 2 1/2/04		Return 1/15/06		
MUSES-C — Asteroid sample return							Launch Jan 02	Nereus Aug 03			Return Jan 06		
CONTOUR — Multiple comet flyby							Launch 7/3/02	Encke 11/12/03			SW3 6/18/06		d'Arrest 8/16/08
ROSETTA — Comet orbiter and lander								Launch 1/21/03			Otawara 7/10/06		Siwa 7/24/08 — Wirtanen 11/29/11-7/10/13
DS-4/Champollion — Comet orbiter and lander (with sample return?)								Launch Apr 03		Tempel 1 Dec 05			Return May 2010

3. Missions

3.1. NEAR

The Near Earth Asteroid Rendezvous mission was launched on January 23, 1996, and flew by the C-type asteroid Mathilda on June 27, 1997. NEAR will reach its primary target, the earth-approaching asteroid Eros, for a rendezvous encounter in February, 2000. NEAR will orbit the asteroid at altitudes from 200 km to 35 km for approximately one year.

NEAR carries a payload of six instruments; a multispectral imager, a near-IR spectrograph from 0.8 to 2.6 microns, an X-ray spectrometer covering 1 to 10 keV, a gamma-ray spectrometer covering 0.3 to 10 MeV, a three-axis fluxgate magnetometer and a laser rangefinder. There is also a radio science investigation using the spacecraft communication system.

The science objectives to be addressed by NEAR include surface topography and properties, elemental and mineralogical composition, compositional and structural heterogeneity, evolutionary history, cometary activity, relationship to meteorite types, magnetic field and potential satellites. The measurements to be made to support these objectives include:

1. Bulk properties-size, shape, mass, density, gravity field and spin state,
2. Surface properties-elemental and mineralogical composition, geology, morphology, and texture, and
3. Internal properties-heterogeneity and magnetic field.

3.2. DS-1

The Deep Space 1 mission is a spacecraft technology flight test mission whose primary function is to test advanced technologies for spacecraft and science instruments with secondary science objectives at the mission targets. The key spacecraft technology to be tested on DS-1 is solar electric propulsion; an enabling technology for future solar system exploration missions. Other technologies to be tested include a solar concentrator array, integrated microelectronics and spacecraft structure, an autonomous navigation system, an autonomous commanding system, and new telecommunication devices and spacecraft operations systems.

The science instrument technologies to be tested on DS-1 are a miniature multiwavelength camera and combined imaging spectrometer system, and a miniature ion and electron spectrometer package. These technologies will be tested by navigation and scientific targeting to at least one small body, the asteroid 1992 KD, with options to continue to comets Wilson-Harrington and Borrelly.

DS-1 was successfully launched on October 24, 1998, and the ion drive successfully started two weeks later. The expected science return from the comets will include imagery and spectroscopic mapping of both the nucleus and coma from the UV through near infra-red, and in situ space plasma data.

3.3. STARDUST

Stardust is a mission to return samples of the refractory dust component of a comet's coma. Scheduled for launch on February, 6, 1999, Stardust will encounter Comet Wild-2 on January 2, 2004, at a relatively low velocity, 6.1 km/s, collecting samples of coma dust and trapped gas during the active phase of the comet. The spacecraft will also image the nucleus and coma, measure dust fluence, and conduct in- situ analysis of cometary dust. During cruise the spacecraft will also collect separate samples of interstellar material. Return of these samples to earth is scheduled for January 15, 2006.

The spacecraft will carry four science investigations. The prime investigation is an aerogel capture device, one side of which will be exposed during comet flyby to trap cometary dust and trapped gas, and the other side of which will be exposed during cruise to collect particles from the interstellar stream. Stardust will also carry out in situ measurements with a dust flux monitor and with a replica of the Giotto particle composition analyzer. A camera is on-board for remote imaging of the coma and nucleus.

The science goals are:
1. Collect more than 1000 cometary particles larger than 15 microns,
2. Collect more than 100 interstellar particles larger than 0.1 microns,
3. Provide in situ particle analysis of the abundant elements for cometary, interstellar and interplanetary dust particles,
4. Provide dust flux measurements, and
5. Provide more than 65 images of Wild 2 with a resolution of at least 67 microrad/pxl taken within 2000 km from the nucleus.

3.4. MUSES-C

MUSES-C is an asteroid sample return mission recently approved in Japan. The NASA role in the mission is to provide communication and navigational support with the Deep Space Network, provide for the landing range and recovery, provide a micro-rover for deployment on the asteroid, and to participate in scientific analysis of the returned samples. The mission will be launched in January 2002 for arrival at the asteroid Nereus in August 2003.

The spacecraft will characterize the asteroid from orbit, then descend to within a meter or so of the surface to collect samples while hovering, and deploy the NASA microrover on the surface. Following sample retrieval and microrover deployment, the spacecraft will retreat from the asteroid for microrover operations and then enter an earth return trajectory. Arrival of the sample return capsule at the U.S. range in Utah is scheduled for January, 2006, the same month as the Stardust sample return.

The microrover will carry out four types of investigations on the asteroid surface:

1. High-resolution (1 mrad/pxl) panoramic imaging in 8 spectral bands in the visible and near-IR,
2. Microscopic imaging at better than 10 micrometers/pxl in the same bands,
3. Near-IR spectrometry from 1.2 to 2.2 microns at about 20 nm spectral resolution, and
4. Alpha-Xray Spectrometry of surface materials.

The key science objectives of MUSES-C are to:
1. Determine the asteroid's bulk properties-size, shape, volume, density and spin state,
2. Measure the elemental and mineralogical composition of the surface to enable comparison with meteorite types,
3. Geology and morphology of the surface,
4. Regolith and texture properties of the surface, and
5. Returned samples.

3.5. CONTOUR

The Contour mission is a multiple flyby mission to obtain detailed information on three diverse comets so that a detailed comparison can be made between the five to eight comets that will have been approached by spacecraft at that time: Encke, Schwassmann-Wachmann-3 and d'Arrest, by Contour, Halley by the 1986 Halley armada, Wild-2 by Stardust, and perhaps Wilson-Harrington and Borrelly by DS-1 and Tempel-1 by DS-4. There is also a possibility to retarget Contour to a new comet, should one be discovered during the mission.

Contour carries four science instruments:
1. A camera to image the cometary nucleus down to a resolution of 4 m/pxl,
2. A mapping spectrometer to measure the signatures of nucleus components to scales of 100 m or better,
3. A neutral and ion mass spectrometer to determine the composition of volatile cometary materials in the coma, and
4. A comet impact dust analyzer to determine the elemental composition, mass and density of dust particles in the coma.

Contour is designed to address six fundamental questions about comets:
1. Nucleus structure and bulk properties, including shape, spin state, construction and internal heterogeneity.
2. Nucleus surface processes, including geologic types, distributions, age relations, jet formation and evolution,
3. Volatile composition and dust/ice ratio, including relative abundance of volatiles, isotope chemistry and variations in ice chemistry both within and between cometary nuclei,
4. Dust composition, including scale and structure for individual comets, and bulk compositional differences between diverse comets,

5. Coma processes, including coma chemistry and the relative roles of dust and gas in forming coma species, and
6. Comparative comet science using Halley, Stardust, Contour, DS-1 and DS-4/Champollion data.

3.6. ROSETTA

The Rosetta mission is the planetary exploration cornerstone mission in the long-term "Space Science: Horizon 2000" strategic plan of the European Space Agency. The mission objectives are to rendezvous with an active short-period comet near aphelion, follow the comet through perihelion to observe the full cycle of cometary activity from cometary orbit, and to land a complimentary scientific package on the surface of the comet. Rosetta is the most scientifically comprehensive of the planned cometary missions and will spend the most time in the vicinity of the target comet.

Rosetta is planned for launch on an Ariane vehicle in January, 2003. After three planetary gravity assists-Mars in August, 2005, and Earth in November of 2005 and 2007-and two asteroid flybys-Otawara in July, 2006, and Siwa in July, 2008, Rosetta achieves rendezvous with Comet 46P/Wirtanen in November, 2011. The German-built lander will be deployed almost a year later in August, 2012, and orbital operations will continue as close to perihelion as possible in July, 2013. The U.S. is participating by providing three orbiter instruments, hardware for another, an interdisciplinary scientist, co-investigators and potential DSN support.

The Rosetta Orbiter carries 12 investigations. Instruments for remote sensing of the surface and coma include an imager (0.25 to 1.0 microns), vis/IR spectral mapper (0.25 to 5.0 microns), UV spectrometer (0.7 to 205 nm), and a millimeter spectrometer (0.5 and 1.3 mm). Instruments for in situ compositional measurements of gas and dust include a neutral/ion mass spectrometer, dust impact mass spectrometer, gas chromatograph, and an atomic force microscope. Dust flux and mass distribution will be measured by a dust flux monitor. A plasma package is included as well as a radio science investigation. There is also an innovative instrument on the orbiter for radio tomographic sounding of the nucleus in concert with a similar instrument on the lander.

The Rosetta Lander carries 7 instruments in addition to the sample acquisition system. These are an imaging system, an alpha-proton-Xray spectrometer, an evolved gas analyzer, a magnetometer and plasma monitor, a surface electrical and acoustic monitor, a multi-purpose surface/subsurface science sensor, and the complementary instrument to support radio tomographic sounding with the orbiter. The principle science objectives of the lander are:
1. Elemental, molecular, mineralogical and isotopic compositions of the surface and subsurface material, and
2. Near-surface strength, density, texture, porosity, ice phases and thermal properties.

The overall Rosetta science objectives are:
1. Global characterization of the nucleus, its dynamic properties, surface morphology and composition,
2. Chemical, mineralogical, and isotopic composition of volatiles and refractories in the nucleus,
3. Physical properties and interrelation of volatiles and refractories in the nucleus,
4. Study of the development of cometary activity and processes in the surface layer of the nucleus and in the inner coma,
5. Origin of comets and their relationship with interstellar material and origin of the solar system, and
6. Global characteristics of asteroids, their dynamic properties, surface morphology and composition.

3.7. DEEP SPACE 4/CHAMPOLLION

The Deep Space 4 mission is a combined spacecraft technology flight test mission (DS-4) and comet science lander (Champollion). DS-4 was designed as a mission to test technologies and operations for surface exploration of small bodies including lander deployment from orbit, lander anchoring for in situ operations, drilling, sampling and sample handling at the surface, return from landing and on-orbit rendez-vous, sample transfer and earth return. When schedules could not be reconciled between ESA's Rosetta mission and the proposed U.S./French Rosetta Champollion comet lander, Champollion was incorporated into the mission objectives of DS-4.

DS-4/Champollion is scheduled for an April, 2003, launch with arrival at Comet Tempel I in December, 2005. Stay time is approximately 3 months with an earth return in May 2010. The Champollion lander will return a cold sample from the surface to the DS-4 orbiter and transfer the sample to a refrigerated container on the orbiter. NASA has not yet made a decision to add resources for an earth-entry return capsule to the DS-4 spacecraft. The alternative would be to conduct in situ analysis of the Champollion sample on the DS-4 spacecraft.

The Champollion lander carries a set of four international science instruments in addition to the sampling experiments (including a drill to be supplied by Italy):
1. A surface panoramic camera to be supplied by the French (pending formal agreement),
2. A near-field camera and microscope including an IR spectrometer,
3. A gas chromatograph/mass spectrometer, and
4. A physical properties probe.

The science goals of Champollion are:
1. Large scale properties including surface morphology, texture and internal comet structure,
2. Physical properties including phases, texture, porosity, density, material strength, thermal and dielectric characteristics,

3. Chemical properties including elemental, molecular, mineralogical and isotopic composition, and
4. Bulk properties including nucleus mass, morphology, density and gravity field.

4. Future

The New Millennium will begin with a harvest of scientific results from missions approved and started in the last decade of this century. These missions are certainly not the last. There will be a continuing scientific requirement to study the diversity of comets and asteroids, and to understand the chemistry, physical structure and dynamics of the most primitive objects in the Solar System. The next decade should take the next steps with main-belt asteroid orbiters, new comet encounters, and continuing flybys, surface exploration, and sample return missions from comets and asteroids including return to earth of a refrigerated pristine sample of cometary nuclear material.

Address for correspondence: Wesley T. Huntress, huntress@gl.ciw.edu

CHARACTERIZATION OF COSMIC MATERIALS IN THE LABORATORY

L. COLANGELI, V. MENNELLA and J.R. BRUCATO
Osservatorio Astronomico di Capodimonte, Via Moiariello 16, 80131 Napoli, Italy

P. PALUMBO and A. ROTUNDI
Istituto Universitario Navale, Via A. de Gasperi 5, 80133 Napoli, Italy

Abstract. One of the main objectives of modern astrophysics is the characterisation of properties and evolution of materials present in space. Production, processing and analysis of cosmic dust analogues in the laboratory represents a powerful tool to interpret astronomical observations and to contribute to the solution of puzzling problems which are so far unsolved. In the present paper we summarize recent results obtained in our laboratory on carbon-based and silicate materials able to simulate various types of cosmic grains. The laboratory data are applied to discuss the nature of spectral features observed in the interstellar medium and in comets.

Keywords: Cosmic Dust, Laboratory Analogues

1. Introduction

The chemical and structural properties of comets are directly linked to the composition of the materials from which they originated and to the formation mechanisms. Remote infrared spectroscopy is a powerful tool to investigate materials present in comets. Since the first detection in Comet 1P/Halley (Combes *et al.*, 1988), a complex band falling at 3.4 μm due to organic compounds has been observed in several comets (e.g. Brooke *et al.*, 1991). Its attribution is consistent with a molecular origin (Disanti *et al.*, 1995; Mumma *et al.*, 1996), even if the contribution from carbon grains cannot be ruled out. Bands falling in the 3.3 − 3.4 μm range are due to C-H stretching modes and are observed in the interstellar medium and around sources rich in UV flux. These evidence suggests that similar features may have significantly different origin, depending on space environment and material processing. As far as silicates are concerned, for several years the amorphous form has been considered as the "only" possible state in space, except in comets where crystalline silicate features were observed (e.g. Campins and Rayan, 1989; Hanner *et al.*, 1994). The Infrared Space Observatory (ISO) spectra have shown that crystalline silicates are not negligible in comets and in the circumstellar medium (Waelkens *et al.*, 1996; Waters *et al.*, 1996; Crovisier *et al.*, 1997; Malfait *et al.*, 1998). These observations pose new constraints on models aimed at explaining the evolution of morphology, structure and chemistry of grains in different space environments.

Support to the identification of cometary materials comes from in situ analyses performed on 1P/Halley (Jessberger and Kissel, 1991; Jessberger, 1999). The two identified chemical phases (organic particles and Mg-rich silicates) are interdispersed and could occur in core-mantle or fluffy structures. The interstellar grain models would favor the first hypothesis, while the laboratory analyses on interplanetary dust particles are consistent with a random, fluffy aggregation. According to available data (Greenberg, 1998), the overall composition of comets seems compatible with the presence of about 26% in mass of silicates, 23% of complex organic refractory materials dominated by carbon and about 9% of very small carbonaceous grains and PAHs; the rest should be mainly water.

Moreover, in the context of the material evolution in space, the problem of cosmic elemental abundance budget must be accounted for (e.g. Snow and Witt, 1996) interstellar dust models must be compatible with observational evidences (Mathis, 1996; Dwek, 1997).

In this framework a substantial contribution to the clarification of dust composition comes from laboratory work on analogues. Experiments on grains condensed under carefully controlled conditions offer several keys for a quantitative interpretation of observations. Moreover, by tracing the sample evolution as a function of processes, similar to those expected in space, it is possible to interpret the modifications of spectra, morphology and structure of materials in different space environments. In an evolutionary reference frame also cometary observations may find a systematic interpretation. In the present paper, the most recent results obtained in the cosmic physics laboratory of Naples by producing and processing carbon and silicate cosmic dust analogues are presented and discussed. Their comparative analysis allows us to link intrinsic properties (such as chemical composition and structure) to the optical behaviour of grains. This approach is useful to interpret observations concerning interstellar and cometary dust.

2. Production and Processing of Cosmic Dust Analogues in Laboratory

The condensation of sub-micron disordered carbon grains is performed by striking an arc discharge between amorphous carbon electrodes in a 10 mbar atmosphere of pure Ar (ACAR sample), pure H_2 (ACH2 sample) or Ar-H_2 mixed gases. Other carbon grains are obtained by burning benzene in air (BE sample) at normal conditions (Colangeli *et al.*, 1995; Mennella *et al.*, 1997a). Sub-micron crystalline silicate grains are obtained by grinding minerals in an agate mill, followed by washing and sedimentation in ethanol. The amorphous counterparts are produced by bombarding the minerals by a high power ($\sim 10^8$ Wcm^{-2}) laser (Nd-YAG @ 1.064 μm) in a 10 mbar O_2 atmosphere. We have considered forsterite and fayalite, which are the Mg-rich and Fe-rich members of the olivines [$(Mg_xFe_{1-x})_2SiO_4$,

$0 \leq x \leq 1$], and enstatite, the Mg-rich pyroxene [$(Mg_xFe_{1-x})SiO_3$, $0 \leq x \leq 1$]. Size distribution, grain shapes and elemental composition of samples are characterized by a field emission scanning electron microscope (FESEM) equipped with an energy dispersive X-ray (EDX) device. Transmission electron microscopy (TEM) allows us to investigate the intimate structure of particles at the nanometer scale (Rotundi *et al.*, 1998). The spectroscopic analyses are performed in transmission from the far UV to the mm range (see Colangeli *et al.* (1995) for details about sample preparation and measurement procedures). Such a wide spectral range allows us to derive information on the structure of materials, as the solid state properties determine the spectral behaviour in the UV-Vis and far infrared, and to identify typical vibration modes, mostly active in the near and medium infrared.

Processing of the analogue samples is performed in order to reproduce the mechanisms most effective in space. Analyses before and after treatment evidence the impact of different processes on the physical and chemical characteristics of the analogues. A synthesis of the applied methods is given in Table I, where the doses are compared with those expected in space. We notice that the contribution of different processes may vary significantly, in function of the astrophysical environment, and it is not an easy task to detail their balance. Actually, processing by UV photons dominates in the diffuse interstellar medium (Greenberg, 1996), even if the role of cosmic-ray ions is certainly not negligible (Strazzulla and Johnson, 1991). In dense molecular clouds, cosmic rays penetrate while UV photons are efficiently attenuated (Tielens and Allamandola, 1987). The UV field induced by cosmic rays in the dense medium corresponds to 10^3 photons cm^{-2} s^{-1} (Sternberg *et al.*, 1987). During the formation of the solar system, fresh particles, condensed in the optically thick pre-solar disk, may have been incorporated in accreting objects together with processed interstellar matter. Once the Sun entered the T-Tau phase, a strong processing occurred in the inner solar system.

3. Results on Carbon Samples

Amorphous carbon smokes produced by arc discharge present a complex morphology and structure. The dominant component is spherical grains (diameter = $7 \div 15$ nm) occurring in chain-like aggregates. Less abundant forms of poorly graphitised carbon, bucky-structures, and rare graphitic structures are observed (Rotundi *et al.*, 1998).

The mass extinction spectra of carbon grains show a linear trend in the visible and two strong features in the UV range. The bump at 80 - 90 nm is due to $\sigma - \sigma^*$ electronic transitions, while the 200 - 260 nm feature is a $\pi - \pi^*$ electronic transition (Mennella *et al.*, 1995a; 1995b). In the medium IR ($3 - 20$ μm) several C-H and C-C stretching and bending vibrations fall. At longer wavelengths the spectral trend follows a power law, with spectral index $1.1 \div 0.5$, which depends on the specific material properties (Colangeli *et al.*, 1995). The behaviour of the

TABLE I
Processing of materials in laboratory and in space

Process	Doses	Reference

Laboratory

Process	Doses	Reference
Thermal annealing	up to 1300 K, $< 10^{-5}$ mbar	Mennella *et al.* 1995a, 1995b
		Brucato *et al.*, 1999
UV irradiation	up to 4×10^{22} eV cm^{-2}	Mennella *et al.*, 1996a
Ion bombardment	up to 660 eV/C-atom	Mennella *et al.*, 1997b

Space

Process	Doses	Reference
Thermal annealing		
stellar outflow and		
primitive solar nebula	< 1000 K	
Oort cloud		
(comet surface)	< 60 K	
UV irradiation	3×10^{23} eV cm^{-2} in 3×10^7 yr	Jenniskens, 1993
(diffuse interstellar	(carbonaceous grains)	
medium)		
Ion irradiation		Strazzulla and Johnson, 1991
interstellar medium	3×10^3 eV mol^{-1} in 10^9 yr	
pre-comet phase	10^6 eV mol^{-1} in 10^6 yr	
comet	6×10^2 eV mol^{-1} in 4.6×10^9 yr	
interplanetary dust	10-100 eV mol^{-1} (E~100 keV)	
	$10^5 - 10^6$ eV mol^{-1} (E~1 keV)	
	in $10^4 - 10^5$ yr	

band, due to π electron resonance, is influenced by the hydrogenation degree (Figure 1). These results indicate that the production conditions significantly affect the structure and optical properties of carbon-based materials and we can interpret the observed spectral differences in terms of structural properties. Grains produced by arc discharge have a dominant aliphatic character, as confirmed by the dominance of the (aliphatic) sp^3 C-H stretch at 3.4 μm over the (aromatic) sp^2 C-H stretch at 3.3 μm. Production in hydrogen-rich atmosphere inhibits the formation of large assemblies of C atoms, as testified by the weak band around 200 nm, with consequent localization of electronic carbon bonds and dominance of short range order.

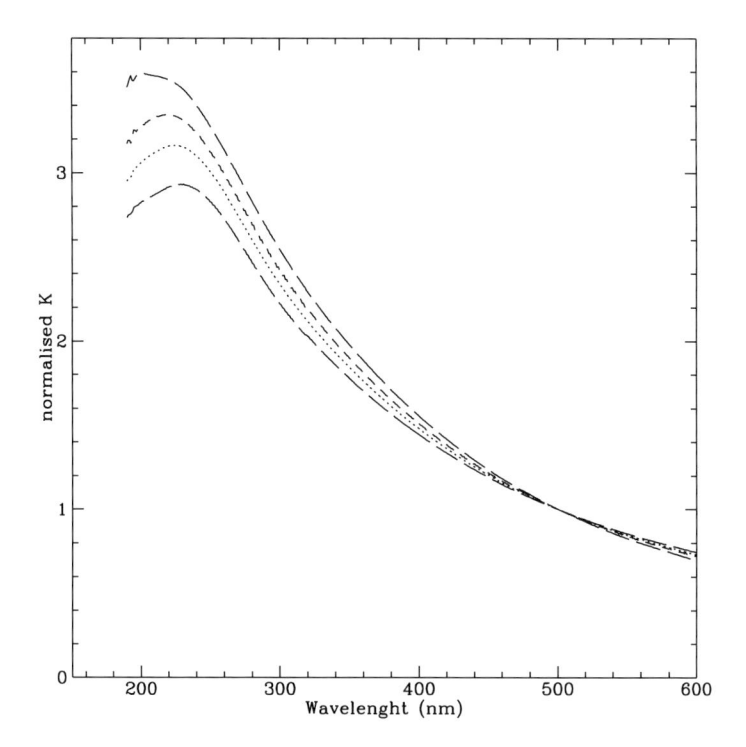

Figure 1. Extinction of carbon grains produced in Ar-H$_2$ mixed atmosphere. From bottom to top the curves refer to 0.4, 0.6, 0.8 and 1.0 % of H$_2$ in Ar, respectively (Mennella *et al.*, 1997a).

The H/C ratio determined by micro-combustion calorimetry for samples produced in pure H$_2$ atmosphere is about 0.62 (Mennella *et al.*, 1995a).

The comparison with other results, available in the literature, confirms that our interpretation may have a general validity for carbon grains. The experiments performed by Schnaiter *et al.* (1998) on carbon grains are in line with the correlation we find between structure properties and spectral behaviour. The same authors have experimentally evidenced the impact of particle clustering onto the extinction profile of nano-sized hydrogenated carbon grains. As far as the C-H stretching spectral region is concerned, samples produced by laser pyrolysis of hydrocarbons (Herlin *et al.*, 1998) present a sharp and very intense 3.3 μm feature, suggesting a marked aromatic character and, at the same time, a strong hydrogenation of the structure.

The effects of processing are evident for hydrogen-rich (ACH2) grains. Thermal annealing up to about 1100 K produces the release of hydrogen, with consequent "aromatisation". This evolution is testified by the spectral changes in the UV bump, which becomes more evident as thermal annealing proceeds, as well as in the C-H stretching bands (Figure 2). This behaviour is consistent with the increase in

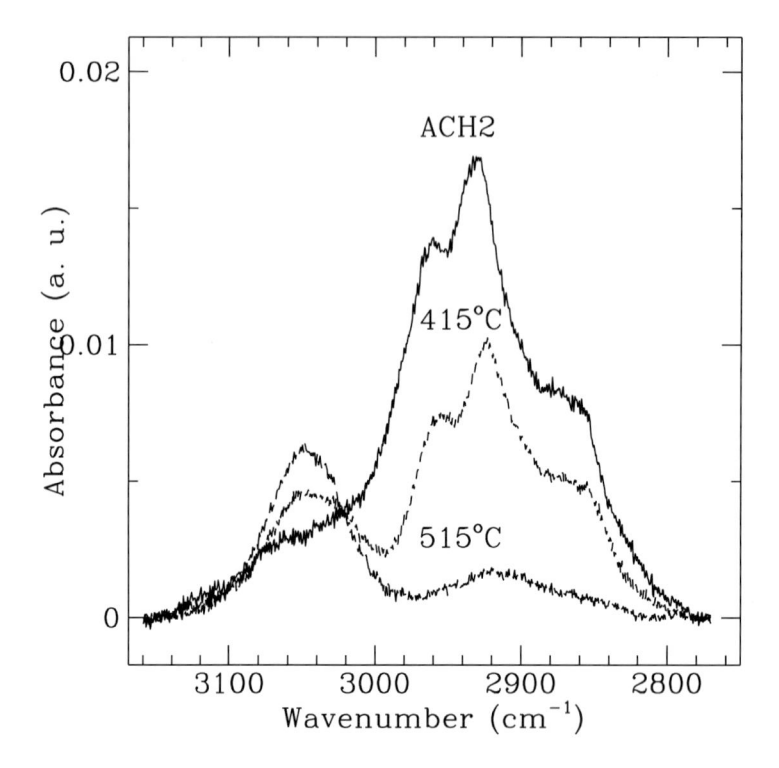

Figure 2. Thermal evolution of the C-H stretching bands. The overall band intensity decreases as hydrogen is removed and the C-H aromatic band at 3.3 μm overcomes the aliphatic stretch at 3.4 μm (Mennella *et al.*, 1996b). At temperatures higher than about 515°C most of hydrogen is removed and the spectrum in this spectral range becomes flat.

number and growth in size of sp^2 clusters, forming the intimate structure of grains. Aromatisation of ACH2 is produced also by UV irradiation, as shown in Figure 3 (Mennella *et al.*, 1996a), and by ion bombardment (Mennella *et al.*, 1997b). We conclude that various processes active in space tend to promote the formation of ordered (large sp^2) structures starting from disordered carbons, but with different efficiency.

4. Results on Silicate Samples

Crystalline silicate grains present irregular shapes, have a power law size distribution $n(d) \propto d^{-2.1}$ and average diameters $0.3 \div 0.9$ μm. Silicates condensed by laser bombardment occur as isolated or aggregated spheroids, have a power index p $\sim$ 2.6 of the size distribution and average diameters $0.01 \div 0.03$ μm. EDX analyses demonstrate that the relative elemental abundance does not vary after laser bombardment. In the view of interpreting the ISO cometary spectra, we

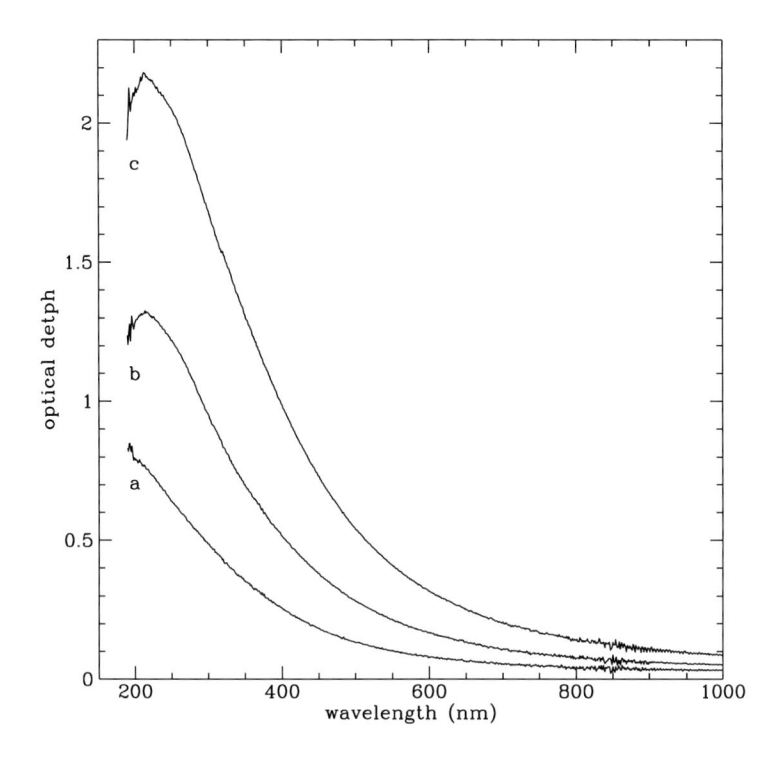

Figure 3. Evolution of the UV band after UV irradiation (Mennella *et al.*, 1996a). The curves refer to not irradiated carbon grains (a) produced in H$_2$ atmosphere and to samples processed by 1 x 10^{22} eV cm^{-2} (b) and 4 x 10^{22} eV cm^{-2} (c).

focus here on the bands falling around 10 μm (Si-O stretching) and 20 μm (O-Si-O bending). Crystalline materials present sharp features, while our condensed grains produce broad bands, as typical of amorphous silicates. The peak positions of the two olivines are rather close in the 10 μm region, while differing in the 20 μm bands. Olivines and enstatite have well distinct 10 μm bands. Examples of the mass extinction coefficient for enstatite are shown in Figures 4 and 5. The main peak positions in our spectra are consistent with those reported by Koike *et al.* (1993) and by Jäger *et al.* (1998) on olivine and pyroxene grains. Differences in band intensities are mainly due to the grain size distribution and average size obtained for the samples in different experiments.

By annealing (up to 1100 K) condensed enstatite, the broad bands evolve towards a structured profile, similar to that of the crystalline sample (Figure 5). This behaviour is in line with the results by Hallenbeck *et al.* (1998) for amorphous Mg silicate smokes subjected to thermal processing.

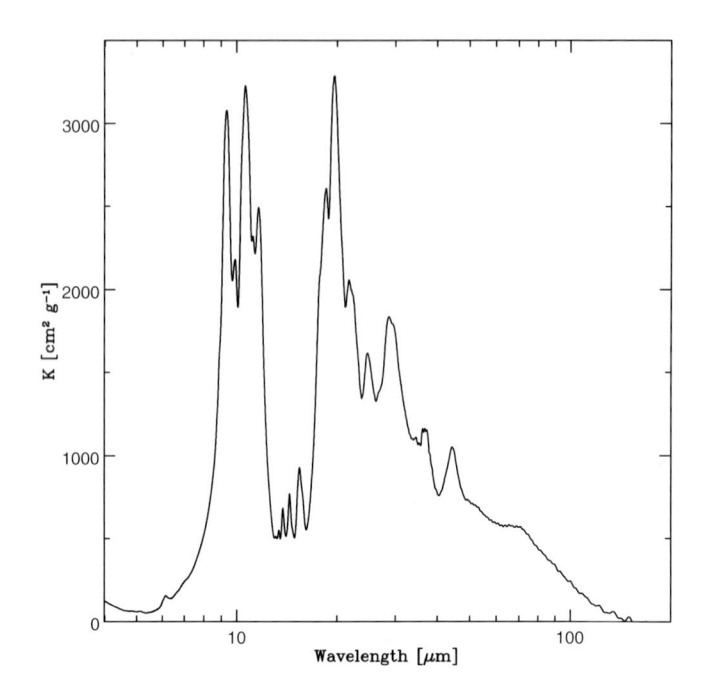

Figure 4. Mass extinction coefficient of crystalline enstatite grains. The sample has been prepared by grinding enstatite mineral fragments in an agata mill for time lags appropriate to obtain dominant sub-micron particles.

5. Astrophysical Applications

The results obtained from laboratory analyses on cosmic dust analogues form a set of reference data, useful to interpret astronomical observations. Here we report some applications to astrophysical problems. The measured dependence of the UV spectrum in carbon grains vs. UV processing can be applied to interpret the origin of the so-called UV interstellar extinction "bump" (e.g. Fitzpatrick and Massa, 1986). To this aim, the laboratory spectra (Figure 3) can be simulated by modeling the electronic properties of the carbon grains. Then, the behaviour of the dielectric function vs. UV processing can be derived and extended to UV doses compatible with the interstellar medium conditions (Table I). As shown in Figure 6, a linear combination of the extinction profiles derived for carbon grains at various degrees of UV processing can be fitted to the interstellar extinction bump (Mennella *et al.*, 1998). This is consistent with extinction observations, which are integrated along the line of sight and collect the contribution of regions containing carbon-based grains which have experienced various degrees of UV processing. This result is

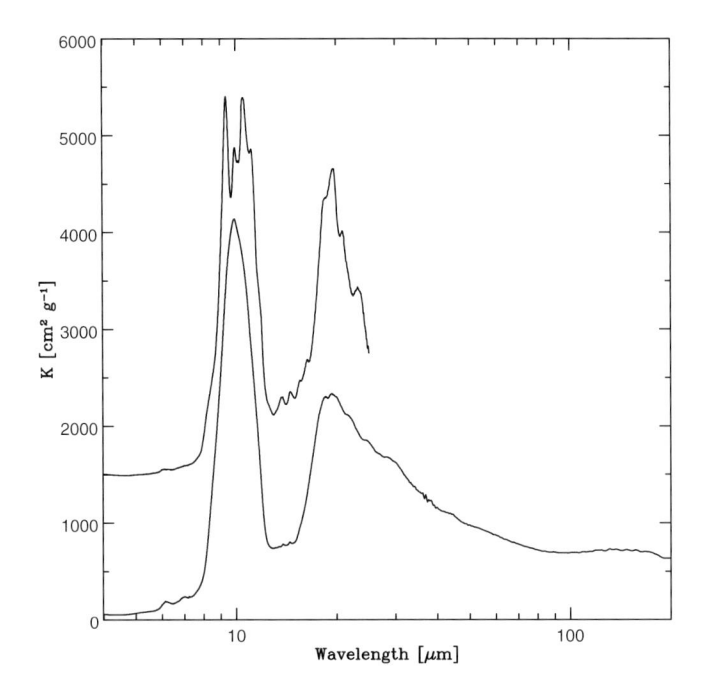

Figure 5. Mass extinction coefficient of enstatite condensed by the laser technique (lower curve) and after annealing in vacuum at about 1100 K for 300 hrs (upper curve). The upper curve has been shifted in ordinate arbitrarily for the sake of clarity.

rather important as it demonstrates that various forms of processed carbon do exist in space and may be incorporated in comet bodies, too.

A direct evidence of the latter consideration comes from the comparison of laboratory spectra with cometary observations. Actually, as already reported by Colangeli *et al.* (1990) the carbon bands around 2950 cm^{-1} (see Figure 2) are compatible with similar features observed in comets. However, as mentioned in the Introduction, recent observations seem to suggest a dominance of molecular species, even if a perfect match of observations has not been yet achieved and carbon grains may well contribute to the observed emission. The interpretation of ISO data for Comet Hale-Bopp (Crovisier *et al.*, 1997) is a tool to identify the silicate component of cometary dust, by comparison with laboratory data. We have simulated the observed flux by considering the contribution of two distinct populations of grains: "small" particles, with different composition (AC = amorphous carbon, ForC = crystalline forsterite, ForA = amorphous forsterite, EnsA = amor-

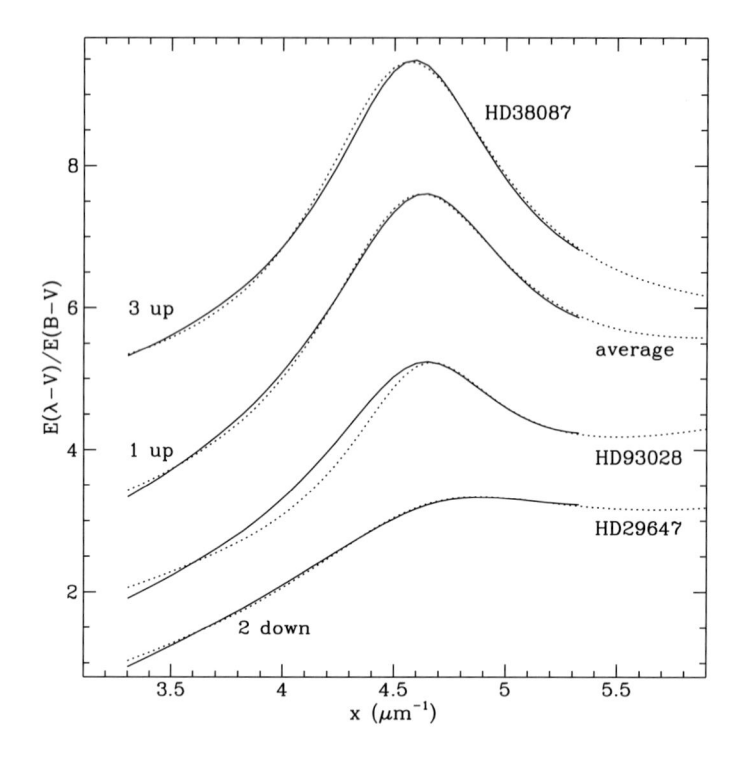

Figure 6. Comparison of interstellar curves (Cardelli and Savage, 1988; Fitzpatrick and Massa, 1990) — dashed lines — with profiles computed by using the optical properties of grains after different UV process degrees (Mennella *et al.*, 1998).

phous enstatite), and "large" grains (LG), behaving as black bodies (absorption efficiency = 1). Thus, the observed flux is:

$$F(\lambda) = \sum \beta_i K_i \cdot B(\lambda, T_1) + \delta_{LG} B(\lambda, T_2) \qquad (1)$$

where

$$\sum \beta_i K_i = \beta_{AC} K_{AC} + \beta_{ForC} K_{ForC} + \beta_{ForA} K_{ForA} + \beta_{EnsA} K_{EnsA} \qquad (2)$$

K is the mass absorption coefficient, $B(\lambda T)$ is the Planck function, δ is a dimensionless factor and $\beta = M/\Delta^2$ is the mass column density, with M = mass of each dust component within the telescope beam and Δ = Earth-comet distance at the time of observation. The result of the fit to observations is reported in Figure 7 and appears rather satisfactory (Brucato *et al.*, 1999a). The values of the free parameters in equation (1), derived from the best fit to observations, are reported in Table II.

The fit obtained by using amorphous and crystalline forsterite plus amorphous enstatite suggests that the contribution of crystalline enstatite to reproduce the

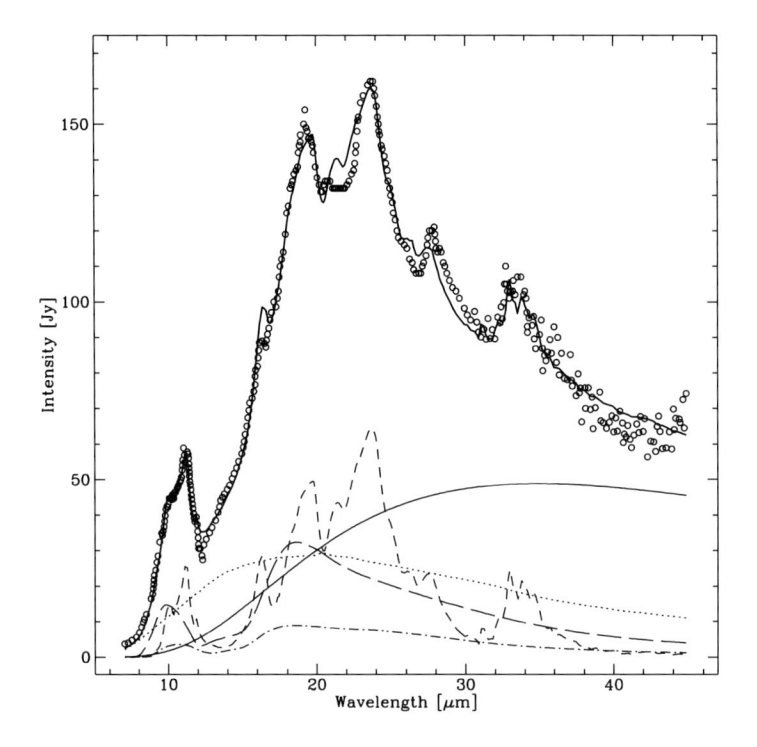

Figure 7. Fit of laboratory data (continuous thick line) to the emission spectrum of Comet Hale-Bopp (Crovisier *et al.*, 1997) - points. The contribution of the different components is reported: LG (continuous thin line), AC (dotted line), EnsA (long dashed line), ForA (dash-dotted line) and ForC (dashed line).

ISO spectrum is negligible. Actually, more recent observations of Hale-Bopp, performed at various heliocentric distances (Wooden *et al.*, 1999), indicate the presence of a further peak at 9.3 μm, consistent with Mg-rich crystalline pyroxene (i.e. enstatite), in the spectra taken around the perihelion. This result has been

TABLE II

Results of the fit to the Hale-Bopp ISO Spectrum (Crovisier *et al.*, 1997)

Grains	Mass fraction				T (K)
	AC	ForC	ForA	EnsA	
Sub-micron	0.18	0.41	0.08	0.33	201
Large	—	—	—	—	146

explained in terms of thermal properties of the materials: the Mg-rich crystalline pyroxene is less absorbing in the UV than the other silicates, so that its IR emission could be very low far from perihelion. Our results on thermal annealing (section 4) suggest that the presence of crystalline silicates in comets may be explained by two different approaches: a) the crystalline material was already present as interstellar dust at the time of comet formation; b) materials in the parent molecular cloud were annealed at temperatures as high as 1000 K or more. Similar temperatures, if applied to the carbon component (section 3), may produce dehydrogenation and the evolution towards a more aromatic carbon. The examples shown above demonstrate the importance of laboratory analyses on materials able to simulate species present in space, and in comets in particular. On the other hand, we are still far from a definitive solution of several problems concerning dust composition in space. The answer to open questions can only be derived from a combined effort of observations, systematic laboratory experiments and advanced modeling.

Acknowledgements

We warmly thank the ISSI institution for supporting this work. The research activity is carried on under ASI, MURST and CNR contracts. We thank S. Inarta, N. Staiano and E. Zona for technical assistance during experiment execution.

References

Brooke, T.Y., Tokunaga, A.T., and Knacke, R.F.: 1991, 'Detection of the 3.4 μm emission feature in comets P/Brorsen-Metcalf and Okazaki-Levy-Rudenko (1989R) and an observational summary', *Astron. J.* **101**, 268-278.
Brucato, J.R., Colangeli, L., Mennella, V., Palumbo, P., and Bussoletti, E.: 1999a, 'Silicates in Hale-Bopp: hints from laboratory studies', *Planet Spa. Sci.* **47**, 773-779.
Brucato, J.R., Colangeli, L., Mennella, V., Palumbo, P., Rotundi, A., and Bussoletti, E.: 1999, 'Mid-infrared spectral evolution of thermally annealed amorphous pyroxene', *Astron. Astrophys.*, in press.
Campins, H., and Rayan, E.V.: 1989, 'The identification of crystalline olivine in cometary silicates', *Astrophys. J.* **341**, 1059-1066.
Cardelli, J.A., and Savage, B.D.: 1988, 'Two lines of sight with exceedingly anomalous ultraviolet interstellar extinction', *Astrophys. J.* **325**, 864-879.
Colangeli, L., Schwehm, G., Bussoletti, E., Blanco, A., Fonti, S., and Orofino, V.: 1990, 'Hydrogenated amorphous carbon grains in comet Halley?', *Astrophys. J.* **348**, 718-724.
Colangeli, L., Mennella, V., Palumbo, P., Rotundi, A., and Bussoletti E.: 1995, 'Mass extinction coefficients of various submicron amorphous carbon grains: tabulated values from 40 nm to 2 mm', *Astron. Astrophys. Suppl. Ser.* **113**, 561-577.
Combes, M., Crovisier, J., Encrenaz, T., Moroz, V.I., and Bibring J. P.: 1988, 'The 2.5-12 micron spectrum of comet Halley from the IKS-VEGA experiment', *Icarus* **76**, 404-436.
Crovisier, J., Leech, K., Bockele-Morvan, D., Brooke, T.Y., Hanner, M.S., Altieri, B., Keller, H.U., and Lellouch, E.: 1997, 'The spectrum of comet Hale-Bopp (C/1995 O1) observed with the infrared space observatory at 2.9 AU from the Sun', *Science* **275**, 1904-1907.

Disanti, M.A., Mumma, M.J., Geballe, T.R., and Davies, J.K.: 1995, 'Systematic observations of methanol and other organics in comet P/Swift-Tuttle: Discovery of new spectral structure at 3.42 micron', *Icarus* **116**, 1-17.

Dwek, E.: 1997, 'Can composite fluffy dust particles solve the interstellar carbon crisis?', *Astrophys. J.* **484**, 779-784.

Fitzpatrick, E.L., and Massa, D.: 1986, 'An analysis of the shapes of ultraviolet extinction curves. I - The 2175 Å bump', *Astrophys. J.* **307**, 286-294.

Fitzpatrick, E.L., and Massa, D.: 1990, 'An analysis of the shapes of ultraviolet extinction curves. III - An atlas of ultraviolet extinction curves', *Astrophys. J. Supp.* **72**, 163-189.

Greenberg, J.M.: 1996, 'Chirality in interstellar medium and in comets: life from dead stars', in *Physical origin of homochirality in life*, edited by D.B. Cline, *AIP Proc.* **379**, pp.185-210.

Greenberg, J.M.: 1998, 'Making a comet nucleus', *Astron. Astrophys.* **330**, 375- 380.

Hallenbeck, S.L., Nuth III, J.A., and Daukantas, P.L.: 1998, 'Mid-infrared spectral evolution of amorphous magnesium silicate smokes annealed in vacuum: comparison to cometary spectra', *Icarus* **131**, 198-209.

Hanner, M.S., David, K.L., and Russell, R.W.: 1994, 'The 8-13 micron spectra of comets and the composition of silicate grains', *Astrophys. J.* **425**, 274-285.

Herlin, N., Bohn, I., Reynaud, C., Cauchetier, M., Golvez, A., and Rouzaud, J.N.: 1998, 'Nanoparticles produced by laser pyrolysis of hydrocarbons: analogy with carbon cosmic dust', *Astron. Astrophys.* **330**, 1127-1135.

Jäger, C., Molster, F.J., Dorschner, J., Henning, Th., Mutschke, H., and Waters L.B.F.M.: 1998, 'Steps toward interstellar silicate mineralogy. IV. The crystalline revolution', *Astron. Astrophys.* **339**, 904-916.

Jenniskens, P.: 1993, 'Optical constants of organic refractory residue', *Astron. Astrophys.* **274**, 653-661.

Jessberger, E.K., and Kissel, J.: 1991, 'Chemical properties of cometary dust and a note on carbon isotopes', in *Comets in the Post-Halley Era*, edited by R.L. Newburn, M. Neugebauer and J. Rahe, J., pp. 1075-1092, Kluwer, Dordrecht.

Jessberger, E.K.: 1999, 'Rocky Cometary Particulates: Their Elemental, Isotopic and Mineralogical Ingredients', *Space Sci. Rev.*, this volume.

Koike, C., Shibai, H., and Tuchiyama, A.: 1993, 'Extinction of olivine and pyroxene in the mid- and far-infrared', *Mon. Not. R. Astron. Soc.* **264**, 654-658.

Malfait, K., Waelkens, C., Waters, L.B.F.M., Vandenbussche, B., Huygen, E., and De Graauw, M.S.: 1998, 'The spectrum of young star HD 100546 observed with the Infrared Space Observatory', *Astron. Astrophys.* **332**, L25-L28.

Mathis, J.: 1996, 'Dust models with tight abundance constraints', *Astrophys. J.* **472**, 643-655.

Mennella, V., Colangeli, L., Blanco, A., Bussoletti, E., Fonti, S., Palumbo, P., and Mertins, H.C.: 1995a, 'A dehydrogenation study of cosmic carbon analogue grains', *Astrophys. J.* **444**, 288-292.

Mennella, V., Colangeli, L., Bussoletti, E., Monaco, G., Palumbo, P., and Rotundi, A.: 1995b, 'On the electronic structure of small carbon grains of astrophysical interest'; *Astrophys. J. Suppl. Ser.* **100**, 149-157.

Mennella, V., Colangeli, L., Palumbo, P., Rotundi, A., Schutte, W., and Bussoletti, E.: 1996a, 'Activation of a UV resonance in hydrogenated amorphous carbon grains by exposure to UV radiation', *Astrophys. J.* **464**, L191-L194.

Mennella, V., Colangeli, L., Cecchi Pestellini, C., Palomba, E., Palumbo, P., Rotundi, A., and Bussoletti, E.: 1996b, 'The role of hydrogen in small amorphous carbon grains: the IR spectrum', in *From stardust to Planetisimals*, edited by M.E. Kress, A.G.G.M. Tielens and Y.J. Pendleton, pp. 109-112, NASA CP-3343.

Mennella, V., Brucato, J.R., Bussoletti, E., Cecchi Pestellini, C., Colangeli, L., Palomba, E., Palumbo, P., Robinson, M., and Rotundi, A.: 1997a, 'Laboratory studies of cosmic dust analogues grains', in *Cosmic Physics in the Year 2000*, edited by S. Aiello, N. Iucci, G. Sironi, A. Treves and U. Villante, pp. 225-232, Ed. Compositiori, Bologna.

Mennella, V., Baratta, G.A., Colangeli, L., Palumbo, P., Rotundi, A., Bussoletti E., and Strazzulla, G.: 1997b, 'Ultraviolet spectral changes in amorphous carbon grains induced by ion irradiation', *Astrophys. J.* **481**, 545-549.

Mennella, V., Colangeli, L., Bussoletti, E., Palumbo, P., and Rotundi, A.: 1998, 'A new approach to the puzzle of the ultraviolet interstellar extinction bump', *Astrophys. J. Lett.* **507**, L177-L180.

Mumma, M.J., Disanti, M.A., Dello Russo, N., Fomenkova, M., Magee-Sauer, K., Kaminski, C.D., and Xie, D.X.: 1996, 'Detection of abundant ethane and methane, along with carbon monoxide and water, in comet C/1996 B2 Hyakutake: Evidence for interstellar origin', *Science* **272**, 1310-1314.

Rotundi, A., Rietmaijer, F.J.M., Colangeli, L., Mennella, V., Palumbo, P., and Bussoletti, E.: 1998, 'Identification of carbon forms in soot materials of astrophysical interest', *Astron. Astrophys.* **329**, 1087-1096.

Schnaiter, M., Mutschke, H., Dorschner, J., Henning, Th., and Salama, F.: 1998 'Matrix-isolated nano-sized carbon grains as an analog for the 217.5 nanometer feature carrier', *Astrophys. J.* **498**, 486-496.

Snow, T.P., and Witt, A.N.: 1996, 'Interstellar depletions update: where all the atoms went', *Astrophys. J.* **468**, L65-L68

Sternberg, A., Dalgarno, A., and Lepp, S.: 1987 'Cosmic-ray-induced photodestruction of interstellar molecules in dense clouds', *Astrophys. J.* **320**, 676-682.

Strazzula, G., and Johnson, R.E.: 1991, 'Irradiation effects on comets and cometary debris', in *Comets in the Post-Halley Era*, edited by R.L. Newburn, M. Neugebauer and J. Rahe, Vol. **1**, pp.243-275, Kluwer, Dordrecht.

Tielens, A.G.G.M., and Allamandola, L.J.: 1987, 'Composition, structure and chemistry of interstellar dust', in *Interstellar Processes*, edited by D.J. Hollenbach and H.A. Thronson, Jr, pp.397-469, Reidel, Dordrecht.

Waelkens, C., *et al.*: 1996, 'SWS observations of young main-sequence stars with dusty circumstellar disks', *Astron. Astrophys.* **315**, L245-L248.

Waters, L.B.F.M., *et al.*: 1996, 'Mineralogy of oxygen-rich dust shells', *Astron. Astrophys.* **315**, L361-L364.

Wooden, D.H., Harker, D.E., Woodward, C.E., Butner, H.M., Koike, C., Witteborn, F.C., and McMurtry, C.W.: 1999, 'Silicate mineralogy of the dust in the inner coma of comet C/1995 01 (Hale-Bopp) pre- and post-perihelion', *Astrophys. J.* in press.

Address for correspondence: Luigi Colangeli, Osservatorio Astronomico di Capodimonte, Via Moiariello 16, 80131 Napoli, Italy, colangeli@na.astro.it

STUDIES OF COMETS IN THE ULTRAVIOLET: THE PAST AND THE FUTURE

S.A. STERN
Southwest Research Institute, Department of Space Studies,
1050 Walnut St., No. 426, Boulder, CO 80302

Abstract. The remote sensing of comets in the ultraviolet bandpass has been a valuable tool for studying the structure, composition, variability, and physical processes at work in cometary comae. By extension, these studies of comae have revealed key insights into the composition of cometary nuclei. Here we briefly review the ultraviolet studies of comets, and then take a look toward the future of such work as anticipated by the advent of several key new instruments.

Keywords: comets, ultraviolet instrumentation, Rosetta

1. Introduction

The history of ultraviolet (UV) studies of comets stretches back to 1970, being dominated by results from sounding rocket missions (which opened up all of the main UV bandpass regions between 250 Å and 3000 Å), and spacecraft orbiting the Earth, including OGO-5, OAO-2, Skylab, IUE, HST, EUVE, SOHO, USSR/Astron, and various payloads aboard the Space Shuttle. Among these, the IUE workhorse was *the* greatest source of contributions (e.g., Festou and Feldman, 1987). To date, no UV instrument except the low-resolution TKS spectrometer aboard the USSR's Vega mission to Halley has been flown to a comet, but that situation is changing (see § 4, below).

The ultraviolet bandpass, like the radio, millimeter-wave, and IR, is a powerful tool for studying astrophysical objects and has been applied with dramatic success to the study of comets. In this brief review, we hope to give some perspective on the value of ultraviolet remote sensing of comets, the contributions UV work has made in the past, and the prospects for future advances. This review cannot, within its space limitations, cover this subject in detail, and the reader is referred to previous expository reviews on UV studies of comets, most particularly, P.D. Feldman's (1982) review and his (1991) update and extension of that review, Bowyer and Malina's excellent (1991) exposition on the value of EUV studies, the intensive, detailed discussion relevant to our topic in Festou, Rickman and West (1993) and Meier (1991), and the historically valuable early perspective provided by Keller (1976).

2. The Value of the Ultraviolet

The UV region of the electromagnetic spectrum is loosely defined as that region between $\approx$250 and 4000 Å, with the blue end constrained by the (relatively arbitrarily defined) soft X-rays, and the red edge defined by the approximate blue boundary of human eyesight. Within the UV bandpass, commonly-used subdivisions include the near-UV (NUV; 3000–4000 Å), mid-UV (MUV; 2000–3000 Å), far-UV (FUV; 900–2000 Å), and extreme-UV (EUV; $\approx$200–900 Å).

NUV observations can be carried out from the ground and high-altitude aircraft, but the increasing opacity of the atmosphere due to ozone absorption makes observations difficult below $\approx$3500 Å, and impossible below $\approx$3050 Å. As such, space observations are required in the MUV, FUV, and EUV. A key characteristic of all solar system UV observations, including in the NUV, is the inherent weakness of UV signals compared to visible-bandpass signals (this is largely due to the UV being on the Wien side of the Sun's G5V blackbody). Therefore UV detectors are particularly sensitive to visible light leaks (so-called "red-leaks"), which can easily pollute desired signals with an undesirable and potentially fatal source of noise. Red leaks can be controlled in various ways, most notably by suitable filtering, or more decisively, by employing detector photocathodes which are blind to long wavelength radiation due to the quantum mechanically-assured protection of photon work functions via the photoelectric effect.

Once water production dominates the nuclear activity, UV spectra of comets are, to first order, similar, after adjustments are made for absolute water production rates and the removal of the dust continuum. (At greater heliocentric distances, emissions from CO and the high fluorescence-efficiency trace molecule CN's 3883 Å 0-0 band emission dominate). Further, once water production dominates, a combination of the high relative abundance *and* the large resonance fluorescence efficiencies of H and OH, emissions from these two species completely dominate cometary UV spectra (although, owing to its high g-factor, CN emission can rival these two at times). Near 1 AU, for example, all of the other numerous (i.e., >50) UV features detected to date in comets are typically two or more orders of magnitude fainter than the main H (Lyα) and OH (0–0) emissions. Despite this, however, the contributions to our understanding of coma processes and composition that these "minor species" have made far outweigh their contribution to the ultraviolet luminosity of comets.

We now illustrate this point by identifying some of the key utilities of the various UV regions for cometary studies:
- *The NUV*: Trace metal species derived from the refractory and organic solids in the nucleus (e.g., Fe, Mn, Ni, Cu, etc. seen in sungrazers); daughter radicals OH, NH, and CN; CO^+, CO_2^+, OH^+, N_2^+ (and C_3, CH, etc., but the brightest emissions from these latter two are longward of 4000 Å); the small-particle

dust continuum; and the dust/ice ratio of both grains and the cometary surface itself.
- *The MUV:* Daughter molecules, most particularly including the neutrals C_2, CS, S_2, CO, and OH (again), and the ions CO^+ and CO_2^+; SO_2 and SO (not yet detected in the UV); and the small-particle dust continuum. We note that the CO Cameron bands in this region are produced by the prompt emission from CO_2 dissociation or electron-excitation of CO, and therefore can be used as a neutral CO_2 tracer.
- *The FUV:* The parent molecules CO and CO_2; the neutral atomic daughter products of molecular dissociation: C, H, O, N, and S; H/D ratios; the C^+ ion; the Kr and Xe noble gases (for which only upper limits have as yet been obtained); the H_2 daughter (for which, again, only upper limits are available); absorption studies of the parent molecule H_2O in both its gaseous and solid-state phases (which requires very high spatial resolution indeed, and has as yet only been planned for spacecraft missions to comets); and the small-particle dust continuum which can be studied using the light source at Lyman-α. The lack of detected signals from H_2, and N and N_2 constitutes a pair of major objectives for future research.
- *The EUV:* Noble gas emissions including He, He^+, Ne, and Ar; the H Lyman series below Ly-α (which avoids the optical depth problems often encountered at Ly-α); numerous ions including C^+, S^+, O^{++}, S^{++}, and O^{++}; absorption studies involving the parent molecules CO, CO_2, as well as N_2; and the so-called X-ray emission of comets.

Though certainly not exclusively true, most of the emissions referred to above are due to solar-stimulated fluorescence. Key secondary mechanisms include electron impact excitation, dissociative recombination, and photodissociative excitation.

Ultraviolet studies of comets have been dominated by spectroscopic investigations, but UV imaging have also been successfully employed. In the next section we briefly review some of the major contributions that such UV observations have made to our understanding of comets.

3. Notable Results from the UV

UV observations of cometary comae have revealed a rich variety of features and processes in cometary comae. Here we touch only on a few items of notable historic import:
- *OH and H: The Chemical Clue to Water's Dominant Role in Generating Cometary Activity Near 1 AU.* Perhaps the single most fundamental contribution of UV work to cometary science was the almost inescapable chemical evidence that the detections of large OH (Swings *et al.*, 1941) and H (Code *et al.*, 1972; Bertaux *et al.*, 1973) abundances in comets secured the case for H_2O as the dominant cometary volatile and the source of activity in comets

inside 2.0–2.5 AU. Over the past ∼30 years, a considerable database of H and OH abundance measurements have been built up (see Feldman, 1991), which have yielded a database on the production rate, Q, of H_2O for many comets as a function of heliocentric distance. These data have revealed that individual comets display a considerable variety of H_2O production curves, and have (along with dust continuum and direct H_2O measurements in the IR and additional OH data in the radio) allowed researchers to discover that comets actually display a wide variation of active surface fractions (e.g., A'Hearn et al., 1995).

– *The Lack of Sulfur and Nitrogen Molecule Detections.* Although sulfur is cosmogonically abundant, only S, CS, and S_2 have been detected in UV spectra of comets. (And it remains true that, as Feldman (1991) and Festou et al. (1993) also point out, that other sulfur-bearing species, notably including SO and SO_2 have never been detected in the UV, no doubt owing in part to a recent downward revision in oscillator strengths.) S_2 is difficult to detect because it is short-lived against photodissociation, and has only been seen in the Earth-approaching comets IRAS-Iraki-Alcock (A'Hearn et al., 1983; and at abundance levels near that of CS (i.e., $\sim 10^{-3}$ H_2O's abundance), and then much later, in 1996, in Hyakutake (Weaver, 1998; see also A'Hearn and Livengood, 1996). Regarding nitrogen: The difficulty of detecting N_2, a homonuclear species with forbidden electric dipole transitions, is well known. However, the complete lack of N_2 molecules (and N atoms themselves) remains puzzling on cosmogonic grounds (e.g., Mumma et al., 1993).

– *The Unexpected Discovery of EUV/X-ray Emission.* Without a doubt, one of the most unexpected, and most intriguing cometary discoveries in recent years has been the discovery of unexpectedly bright cometary X-ray emission. Cometary X-rays were first detected in Hyakutake in early 1996, using the ROSAT observatory in Earth orbit (Lisse et al., 1996). Rapidly thereafter, comet C/Tabur 1996 Q1 and then four ROSAT all-sky survey comets (Dennerl et al., 1997) were found to also emit X-rays, indicating that the X-ray emission phenomenon is common. The observed X-ray emission is quite soft ($kT \sim 0.2 \pm 0.1$ KeV). Although it is not fully clear as yet whether the emission is dominated by a continuum process such as thick-target bremsstrahlung, or is simply a dense forest of line emission due to solar wind charge exchange processes, the charge exchange mechanism seems the early favorite. The detection of EUV emissions apparently related to the X-ray emission (e.g., Krasnopolsky et al., 1997) provides additional clues to and leverage on the problems posed in understanding the root cause of this emission.

– *Ne and Ar : Witnesses to Cometary Thermal History.* Noble gases, owing to their disaffinity for chemical reactions and their (related) high volatility, have long been known to provide powerful probes of the thermal history of comets. Although the interpretation of noble gas abundances in cometary coma is complicated by the details of their trapping and release efficiencies (Owen et *al.*, 1991), their detection has nonetheless been highly anticipated. Significant upper limits revealing depletions of order 10^4 in He (Stern *et al.*, 1992) and of order 25 in Ne (Krasnopolsky *et al.*, 1997) have been reported. Very recently, Stern *et al.* (1999) have reported the detection of Ar at near-cosmogonic levels in Hale-Bopp. Combining this detection with the Ne depletion reported in this same comet (Krasnopolsky *et al.*, 1997), it is possible to conclude that the cometary ices formed at temperatures between 20 and perhaps 35 K, indicating an origin for Hale-Bopp in the Uranus-Neptune region.

4. Key Upcoming UV Opportunities for Comet Science

Many future advances in UV remote sensing will be driven by the advent of new observational capabilities. Major new observing opportunities include:

- *The Far Ultraviolet Spectroscopic Explorer (FUSE) mission set to launch in 1999.* FUSE (Moos, *et al.*, 1997) is a general purpose, Earth orbital observatory with high-resolution ($\lambda/\Delta\lambda$=24,000–30,000) spectroscopic capabilities between 905 Å and 1195 Å. The instrument will achieve an effective area of 20–80 cm^2 (varying across the bandpass) and a 1.5 arcsecond resolution for a Strehl ratio of 0.90.

- *The MICAS UV Spectrometer Comet flybys aboard New Millennium/Deep Space 1 (NMP/DS1) set for 2000–2001.* NMP/DS1 (Nelson, 1998) was launched in October 1998, and is en route to a flyby with near-Earth asteroid 1992KD on 29 July 1999. If all goes well, DS1 will also execute a flyby with either the comet/asteroid transition object Wilson-Harrington (October 2000), or comet 19P/Borrelly (January 2001). The DS1 UVS offers a characteristic resolution of 12 Å across its 800–1850 Å bandpass, a 0.1×6.0 deg slit, and a characteristic effective area of $\sim$0.5 cm^2— all vastly superior to the 70 Å resolution TKS instrument aboard Vega. DS1 also includes a visible imager, a low-resolution IR spectrograph, and the PEPE plasma sensor package.

- *The installation of the Advanced Camera for Surveys (ACS) aboard HST in 2000.* The ACS (http://www.stsci.edu/instruments/acs/index2.html) offers to provide HST with a photon-counting, high-resolution (0.05"), high-throughput, near-UV (>2000 Å) imaging capabilities with a FOV about twice that of the present Wide Field (WF) Camera 2. ACS also includes coronagraphic and polarization capabilities. The UV sensitivity of the ACS is comparable to HST/STIS and about 10×that of WFPC2, and is also expected to contribute to imaging surveys of cometary nuclei (H.A. Weaver, pers. comm.).

- *The installation of the Cosmic Origins Spectrograph (COS) aboard HST in 2003.* COS (Morse *et al.*, 1998) is a major new, high-throughput UV facility planned for installation aboard the Hubble Space Telescope. COS will cover the 1150–3200 Å bandpass, achieve spectral resolutions up to 30,000, and offer effective areas exceeding 1000 cm^2 at peak performance (2600 Å). As a figure of merit, COS is expected to achieve S/N=10 at R=20,000 for flux levels of $1-2\times10^{-15}$ ergs cm^{-2} s^{-1} Å^{-1} in a 10,000 second integration across much of its 1150–3200 Å bandpass.

- *The rendezvous of the Alice UV Spectrometer, aboard Rosetta, at comet 46P/Wirtanen in 2011.* The Rosetta/Comet Wirtanen rendezvous is the most ambitious comet mission now in development. It will include a lander and some 12 orbiter investigations. Alice (Stern *et al.*, 1998) will obtain 700–2050 Å spectra with a characteristic resolution of 6 Å, and a characteristic effective area of 0.1–0.5 cm^2 through its 6 deg long slit. Resolutions as high as 1 Å are possible in occultation mode. Rosetta will make detailed studies of Wirtanen beginning in November 2011, when the comet is near 3.3 AU, until the comet reaches perihelion in 2013.

5. Future Needs

Despite the enormous potential afforded by the FUSE, NMP/DS1, HST/COS, and Rosetta/Alice opportunities, a number of important desires remain unfulfilled. These include:

- *Extremely High UV Spectral Resolution ($R>3\times10^5$)* to directly probe the velocity fields of species revealed through UV emissions, to study hyperfine structure, and to measure OH/OD abundances.
- *Low Solar Elongation Angle (SEA) Observing Capability* to permit the routine observation of perihelion passages substantially inside 1 AU, as well as those comets with poor apparition geometry. This capability will, for example, allow many trace species that either have low g-factors, or are still bound in grains at 1 AU, to reveal themselves.
- *Flyby Reconnaissance of Multiple Comets with UV Spectroscopic Capabilities* to sample the diversity of noble gas and other species abundances seen only in the UV with high spatial resolution, which is particularly important both for studies of nuclear heterogeneity and those species with short scale lengths.

One looks forward to these and other aspirations being achieved as well in the years to come.

Acknowledgements

I thank my colleagues Will Colwell, Joel Parker, and Dave Slater for their comments on the initial draft of this manuscript, and Mike A'Hearn, Dan Boice, Michel Festou, Hal Weaver, and the ISSI reviewer for comments on the subsequent draft.

References

A'Hearn, M.F., Feldman, P.D., and D.G. Schleicher: 1983, *ApJL* **274**, L99.
A'Hearn, M.F., Schleicher, D.G., Feldman, P.D., Millis, R.L., and Thompson, D.T.: 1984, *AJ* **89**, 589.
A'Hearn, M.F., and Livengood, T.A.: 1998, In *Ultraviolet Astrophysics Beyond the IUE Final Archive* (W. Wamstekker and R. González- Riestra, eds.), ESA SP-413, Noordwijk, ESA Pub. Div., 625.
Bertaux, J.-L., Blamont, J.E., and Festou, M.C.: 1973, *A&A* **25**, 415.
Bowyer, S. and Malina, R.: 1991, EUV Astronomy, *Adv. Space Res.* **11** (11), 205.
Code, A.D., Houch, T.E., and Lillie, C.F.: 1972, In The Scientific Results from Orbiting Astronomical Observatory (OAO-2) (A.D. Code, ed.), NASA SP-**310**, 109.
Dennerl, K., Englhauser, J., and J. Trümper: 1997, *Science* **277**, 1625.
Feldman, P.D.: 1982, In *Comets* (L.L. Wilkening, ed.) U. Az. Press, Tucson, 461.
Feldman, P.D.: 1991, In *Comets in the Post-Halley Era* (R. Newburn, M. Neugebauer, and J. Rahe, eds). Kluwer, Dordrecht, 339.
Festou, M.C., and Feldman, P.D.: 1987, In The IUE Satellite (Y. Konda, ed.) Reidel, 101–118.
Festou, M.C., Rickman, H., and West, R.: 1993, *A & A Revs.* **4**, 363, and **5**, 37.
Keller, H.U.: 1976, *Sp. Sci. Revs.* **18**, 641.
Krasnopolsky, V.A., *et al.*: 1997, *Science* **277**, 1488.
Lisse, C.M., *et al.*: 1996, *Science* **274**, 205.
Meier, R.R.: 1991, *Space Sci. Revs.* **58**, 1.
Moos, W., Sembach, K., and Bianchi, L.: 1997, In *Origins* (C.E. Woodward, J.M. Shull, and H.A. Thronson, eds.), ASP Conf. Series **148**, 304.
Morse, J.A., Green, J.C., Ebbets, D., Andrews, J., Heap, S.R., Leitherer, C., Linsky, J., Savage, B.D., Shull, J.M., Snow, T.P., Stern, S.A., Stocke, J.T. and Wilkinson, E.: 1998, Proc. SPIE **3356**, 361.
Nelson, R.M.: 1998, *EOS* **79**, 493.
Owen, T., Bar-Nun, A., and Kleinfeld, J.: 1991, In *Comets in the Post-Halley Era* (R. Newburn and J. Rahe., eds.) Kluwer Press, Dordrecht, 429.
Stern, S.A., Green, J.C., Cash, W., and Cook, T.A.: 1992, *Icarus* **95**: 157.
Stern, S.A., Slater, D.C., Gibson, W., Scherrer, J., A'Hearn, M.F., Bertaux, J.-L., Feldman, P.D., and Festou, M.C.: 1998, *Adv. Sp. Res.* **21**, No. 11, 1517.
Stern, S.A., Slater, D.C., Festou, M.C., Parker, J.P., Gladstone, R., and A'Hearn, M.F.: 1999, *BAAS* **30**, No. 4, .
Weaver, H.A.: 1998, In *The Scientific Impact of the Goddard High Resolution Spectrograph* (J.C. Brandt, T.B. Ake, and C.C. Peterson, eds.), ASP Conf. Series **143**, 213.

Address for correspondence: Alan Stern, Southwest Research Institute, Space Studies Dept., 1050 Walnut St., No. 426, Boulder, CO 80302, USA, alan@everest.boulder.swri.edu

CRITICAL QUESTIONS AND FUTURE MEASUREMENTS — COLLATED VIEWS OF THE WORKSHOP PARTICIPANTS

A. J. BALL
International Space Science Institute, Hallerstrasse 6, 3012 Bern, Switzerland
(now at the Institut für Planetologie, Westfälische Wilhelms-Universität Münster, Germany)

H. U. KELLER
Max-Planck-Institut für Aeronomie, D-37191 Katlenburg-Lindau, Germany

R. SCHULZ
Solar System Division, ESA ESTEC, Noordwijk, Netherlands

Abstract. At the ISSI Workshop 'The Origin and Composition of Cometary Material' a short questionnaire was devised by the 'Critical Measurements for the Future' Working Group and distributed to the attendees. The aim was to find out what they thought were the 'critical questions' and the key measurements needed to find answers. Results from the 15 respondents are collated and summarized.

1. Questionnaire and Results

The Workshop's fourth Working Group (charged with addressing the topic of 'Critical Measurements for the Future') decided that a short questionnaire (Table I) would be a suitable way to gauge the opinions of those present at the Workshop. Although feedback on 'Critical Measurements' was sought, the Working Group felt it fundamentally important to identify 'Critical Questions' in addition to the required observations or measurements. These two categories were then subdivided into three fields: composition, origin and physical properties. Responses were received from 15 of the ~ 45 participants. Many replies were quite detailed and specific, and almost inevitably related to the respondents' own fields. Nevertheless it was encouraging to see some degree of overlap, i.e. agreement. Two-thirds of the respondents supplied separate answers for the two columns, as intended. The remainder gave combined answers, perhaps indicating some confusion between 'Questions' and the suggested means to obtain answers — an impression reinforced by the fact that some of the separate suggestions in the questions column (e.g. 'What is x?') resembled entries in the measurements column (i.e. 'Measure x.').

The questionnaire responses were collated to remove repetition and reorganized into a more logical structure. Table II shows a synthesis of the 'composition questions' or, if the reply was written across the page, the combined response. Tables III and IV relate to physical properties and origins in a similar way. In Table II one can see that composition questions may be formulated quite naturally in terms of

TABLE I
Format of the questionnaire

What do you think are the most critical measurements and observations of cometary science that should be made? (Consider both short- and long-term timescales.)	
Critical Questions	Accompanying Observation/Measurement
A: relating to composition	
B: relating to origin(s)	
C: relating to physical properties	

TABLE II
Synthesis of the 'composition questions' questionnaire responses

Four basic categories of composition	• Isotopic • Elemental • Chemical (incl. ortho/para ratio) • Mineralogical
Spatial heterogeneity	• Within and between nuclei (both Oort & Kuiper types) • Within and between grains and ices • Between the nucleus and coma • Over the surface • In the coma (radial and transverse)
Temporal variations	• Diurnal • Orbital • Lifetime • Time since emission from surface
Specific targets	• D/H, $^{12}C/^{13}C$, $^{16}O/^{17}O/^{18}O$ • Sulfur • Sodium • Noble gases • H_2O (ortho/para ratio) • Abundance of CO_2 and other volatiles • Isomers, e.g. HCN/HNC • Nitriles • Homologous species • PAHs • Reduced vs. oxidised molecules • Radicals • Tracer molecules • Dust : Gas ratio • Crust material

TABLE III
Synthesis of the 'physical properties questions' questionnaire responses

- To what depths do the modifying processes penetrate into the nucleus?
- What is the nucleus structure and morphology from small to global scales, e.g. distribution of dust and volatiles in 3D?
- What is the surface strength?
- What is the diurnal variation of surface temperature?
- What is the nature of the active regions?
- How are jets formed?
- What are the dynamics and physical properties of dust vs. distance from the nucleus?
- What are the critical parameters necessary for modeling coma observations?
- Was liquid water ever present in comet nuclei?
- What is the source of sodium emission?
- How diverse are comet nuclei?

TABLE IV
Synthesis of the 'origins questions' questionnaire responses

- Separate origin and evolution effects
- Distinguish parent molecules from secondary species
- Do PAHs originate from photochemically modified interstellar ice?
- What is the aggregation state of nucleus material compared to IDPs?
- How does the internal structure of the nucleus relate to its origin?
- Are there interstellar grains in comet nuclei?
- Are there radicals in cometary ice?
- Can the rôle of sodium as a tracer in the ISM be applied to cometary material?
- Where is sulfur in the ISM, as compared to sulfur-bearing species in comets?
- What is the inventory of organics and PAHs in cometary material?
- What is the composition and mineralogy of dust and refractory organics?
- How does isotopic composition vary between molecules, between grains and between comets?
- What are the relative importances of the various sources, reservoirs and loss processes for comets?
- Assuming dead comet nuclei exist, can they be distinguished from the most primitive asteroids?

'composition in space and time'. The respondents cited many specific targets for study, ranging from particular isotopes and elements to various classes of molecule or mineral mixture. The corresponding results for physical properties (Table III) are less easy to categorize. Most are concerned with the nucleus surface and interior and many relate to thermal, mechanical or structural features in space and/or time. The collated 'origin questions' (Table IV) show that there are currently many quite

TABLE V
Suggested observations / measurements from the questionnaire responses

- *In situ* analysis at and sample return from multiple sites on nucleus surface & the coma, and from several comets, to determine composition. [IS,SR]
- Mass spectrometry of nucleus material and coma gas and dust [IS,SR]
  - with high sensitivity at M = 50 → 300
  - Resolve critical doublets like N_2/CO and ^{13}C/CH
  - Inhomogeneity of isotope ratios
  - Isotope ratios in individual dust particles and sub-units of dust particles, and of volatiles from different comets
- Measure and map (vs. heliocentric distance, pre- and post-perihelion) [RG,RS]
  - Chemical composition
  - Flux of H_2O and minor volatiles
- Identify PAHs [IS,SR]
- Dissociation products in nucleus and coma [RG,RS,IS]
- CO Cameron bands from space [RS]
- High spatial resolution IR observations [RS,IS]
- Radio observations of radicals in coma [RG]
- Measurement of structural properties of refractories; aggregation state of dust & nucleus material, from small to large scales [IS]
- Composition and mineralogy of dust and refractory organics [IS]
- Can more useful (e.g. statistical) information be extracted from the SOHO comets? [RS]
- Observations of a dynamically new comet from *before* the onset of activity [RG,RS,IS]
- Ice modification processes by UV, X-rays, ions [IS,SR]
- Nucleus tomography, seismic sounding and imaging; diversity of structure between comets [IS]
- Noble gas abundances [IS]
- Observations of many comets over a wide range of wavelengths [RG,RS]
- Study of KBOs for comparison with comets
  - Space mission [IS]
  - IR spectroscopic observations [RS]
- Coma measurements close to the nucleus [IS]
- Volatile measurements and temperature mapping of nucleus surface and sub-surface, during a nucleus rotation [IS]
- Look for evidence of liquid water [IS,SR]
- Look for sodium- UV observations of ions[RS]

RG: Ground-based Remote Sensing; RS: Satellite Remote Sensing;
IS: *In Situ* Measurements; SR: Sample Return.

to be continued on next page

TABLE V
Suggested observations / measurements from the questionnaire responses (continued)

Modeling and lab work:
- D/H exchange kinetics
- Gas-grain chemistry
- Lab measurements of dust emission and light scattering
- Lab data for spectroscopic parameters
- Lab simulations of structural properties of refractory materials
- Find critical chemical rates (photo-rates)
- Study ice modification by UV, X-rays and ions
- Identify species which are unique tracers of specific processes

fundamental issues to be addressed in cometary science. Looking at the points raised, one can say that the key steps required are as follows:
1. to characterize the comets we see today,
2. to understand how comets evolve with time, and
3. to link the primordial features of cometary material to the environment and processes of its formation.

The suggested measurements — for the three rows of the questionnaire combined — are brought together in Table V (Part 1 and 2). It is evident that space missions (sample returns as well as *in situ* analyses of comet nuclei and comas), Earth-based telescope observations, laboratory experiments and computer modeling are all important and complementary. It seems clear that critical measurements will have to focus on the nucleus material, see for example the proceedings of the recent IAU Colloquium 'Cometary Nuclei in Space and Time' (A'Hearn, 1999).

The results from the questionnaires, the contributions to, and the discussions during the Workshop show that our understanding of comets is being consolidated by observational results (the outstanding Comet Hale-Bopp being of particular importance) in many areas but also that comets may be more diverse, having a wide range of origins and hence diverse histories, rendering them less pristine or even collisionally evolved (e.g. Kuiper Belt comets). The anticipation (or hope) that the detailed analysis of a single comet by one sample return mission will provide all the answers has dwindled.

The recently reopened questions of how and where comets were formed within the early Solar System and whether (and how) comets could be classified according to their place of formation (heliocentric distance), together with the indications of nuclear heterogeneity, will require results from any one comet to be related to the full ensemble. Compositional classification (taxonomy) of comets has failed in the past or is controversial. Relating a present day comet to its origin and evolutionary path will be essential for further understanding of the physics of comets and their relation to the formation of our Solar System. The situation may be even more complex if single comets are formed from material of different origin. How much of the nucleus material is pristine (of interstellar origin)? How strongly mixed in is the material processed in the inner Solar System at different heliocentric distances? Cometary material (in parts) may turn out to resemble primitive meteorites with strong compositional variation on minute scales.

The diversity and likely inhomogeneity of comets, together with the nature of their dynamical origins and, therefore, individual evolutions demand that a large number of nuclei be investigated, and at several locations. Sample retrieval is of course not the final goal but the beginning of a lengthy process of analysis and interpretation (a situation analogous to the study of meteorites). Information on a large number of bodies (perhaps 10^4) is also important for dynamical studies of the Oort cloud and Kuiper belt objects. The detection of Oort clouds and KBOs around other stars would help improve our understanding of those of our own Solar System.

2. Conclusion

The survey conducted by Working Group 4 proved to be a useful way to gather concise opinions from a broad cross-section of workers. It was encouraging to see some degree of overlap — in other words, agreement — between many of the responses. With only 15 respondents, however, there are no doubt some important points which were missed. Perhaps a questionnaire approach could be adopted at other well-focused workshops, with the stated aim of receiving answers from the majority of participants. It may also be worth looking back at these results in perhaps 15-20 years, by which time most of the forthcoming cometary investigations by spacecraft and other means will have reached fruition. To what extent will the fundamental questions have been answered? What new questions will have emerged? When discussing the long-term future of cometary science, it is certainly important to concentrate on fundamental science questions. These could be summarized as follows:

— From where does cometary material originate?
— Under what conditions and by what processes did comets form?
— How have comets evolved?
— What makes comets comets, how does their activity work?

These most basic questions give rise to secondary questions relating to specific issues of composition, physical properties and possible primordial features. The array of key questions has remained broadly similar for at least the last decade. For example, the results presented here match closely the points summarized already during planning for the *Rosetta* mission (ESA, 1993).

Acknowledgements

Thanks are due to all those who completed the questionnaire, and to those who participated in the discussions of Working Group 4. A. J. Ball wishes to thank the Swiss Academy of Sciences for financial support during a visit to the Institut für Weltraumforschung, Graz in Nov-Dec 1998, where much of the text of this paper was completed.

References

A'Hearn, M. F. (editor): 1999, *Cometary Nuclei in Space and Time*, Proc. IAU Colloquium 168, Nanjing, China, 18-22 May 1998: Astron. Soc. Pacific.
European Space Agency: 1993, *Rosetta Comet Rendezvous Mission*, ESA SCI(93)7: European Space Agency.

Address for correspondence: Andrew J. Ball, balla@uni-muenster.de

V: SUMMARY AND INDEXES

COMETARY MATERIALS: PROGRESS TOWARD UNDERSTANDING THE COMPOSITION IN THE OUTER SOLAR NEBULA

K. ALTWEGG
Physikalisches Institut, Universität Bern, CH-3012 Bern, Switzerland

P. EHRENFREUND
Leiden Observatory, 2300 RA Leiden, The Netherlands

J. GEISS
International Space Science Institute, CH-3012 Bern, Switzerland

W. F. HUEBNER
Southwest Research Institute, San Antonio, TX 78228-0510, USA

A.-C. LEVASSEUR-REGOURD
Université Paris VI & Service d'Aéronomie CNRS, F-91371 Verrières, France

Abstract. A major objective of the workshop was to learn about the chemical composition, physical structure, and thermodynamic conditions of the outer parts of the solar nebula where comets formed. Here we sum up what we have learned from years of research about the molecular constituents of comet comae primarily from *in situ* measurements of Comet 1P/Halley and remote sensing of Comets 1P/Halley, Hale-Bopp (C/1995 O1), and Hyakutake (C/1996 B2). These three bright comets are presumably captured Oort cloud comets. We summarize the analyses of these data to predict the composition of comet nuclei and project them further to the composition, structure, and thermodynamic conditions in the nebula.

Near-future comet missions are directed toward less active short-period Jupiter-family comets. Thus, future analyses will afford a better understanding of the diversity of these two major groups of comets and their respective regions of origin in the solar or presolar nebula.

We conclude with recommendations for determining critical data needed to aid in further analyses. Results of the workshop provide new guidelines and constraints for modeling the solar nebula.

1. Gas and Dust in Comet Comae

1.1. GAS

Since the March 1986 encounter of an international fleet of spacecraft (Giotto, Vega 1, Vega 2, Suisei, Sakigake, and ICE) with Comet 1P/Halley, the search for mother molecules in the coma of comets has made an enormous step forward. At the time of these encounters, models to deduce abundances of mother molecules from *in situ* data or from remote sensing of radicals were in their infancy. Now, that the Giotto ion mass spectrometer data have been analyzed, many more mother molecules have been identified, and with new technology it has also been possible to observe some mother molecules directly using radio and infrared spectroscopic techniques. At least twenty-four mother molecules have been identified (Crovisier and Bockelée-Morvan, this volume). New results from *in situ* mass spectrometry of Comet 1P/Halley and remote sensing converge and are in most cases compatible (Altwegg *et al.*, this volume; Crovisier and Bockelée-Morvan, this volume; Eberhardt, this volume). The two big comets of this decade, Hale-Bopp (C/1995 O1) and Hyakutake (C/1996 B2), were sufficiently active to observe molecules of minor abundances such as OCS, CH_4, C_2H_2, and C_2H_6 by remote sensing. While the detection of Na in the coma and in the dust tail was not surprising, its detection in separate tails was unexpected (Cremonese, this volume). If we compare the Comets Halley, Hale-Bopp, and Hyakutake the agreement of molecular abundances is quite remarkable. This may not be too surprising because all three comets are thought to be originally Oort cloud comets. One of the major differences for the more abundant species is thought to be ammonia. The value for ammonia determined from *in situ* mass spectrometry in the coma of Comet Halley at $r \approx 0.9$ AU heliocentric distance is 1.5% relative to water. This is about double the amount in the comae of Comets Hyakutake and Hale-Bopp (Altwegg *et al.*, 1993; Meier *et al.*, 1994). However, Bird *et al.* (1997) report an abundance of 1% to 1.8% of NH_3 in Comet Hale-Bopp, bringing it in line with the value for Comet Halley. From this comet we know that nitrogen is depleted compared to solar values (Geiss *et al.*, this volume). Most of the nitrogen may be in ammonia. This means that for Comet Hyakutake the nitrogen depletion in the gas phase may be even more severe, or more of the nitrogen may still be in the form of N_2. However, the abundance for N_2 is not known; the bands of N_2^+, which are an indirect indicator of N_2, have not been reported. For Comet Halley the optically observed value for N_2 is 0.2% (Wyckoff *et al.*, 1991). N_2 is hard to detect by remote sensing and by *in situ* mass spectrometry (overlap of molecular mass with that of CO). It has been suggested that comets of dynamically different classes may be chemically different (A'Hearn *et al.*, 1995; Feldman *et al.*, 1997). The statistics are still poor and an answer to this question has to be deferred to further remote sensing and space missions (e.g., Rosetta).

Most of the volatile material detected in comets comes from the nucleus directly: Water from the surface of the nucleus and gases from species more volatile than water from the interior. These gases may be stored as a separate frozen phase or trapped in amorphous water ice. Ice and refractory organic materials (CHON particles or hydrocarbon polycondensates) in the dust contribute additionally to the coma gas (Eberhardt, this volume; Huebner and Benkhoff, this volume). However, the unusual heliocentric distance behavior of most molecular species in the coma (in particular H_2O) may be related to seasonal effects because of the orientation of the spin axis of the nucleus (Farnham *et al.*, 1999; Kührt, this volume). The organic material in dust particles is a distributed source for coma gas (radicals as well as mother molecules). Among them is the distributed source for H_2CO and CO (Boice *et al.*, 1990). Eberhardt *et al.* (1995) conclude that in Comet Halley about two-thirds of the CO observed at $r \approx 0.9$ AU can be explained by the dissociation of H_2CO, which itself seems to come primarily from the dust (Eberhardt, this volume). However, this is contested for other comets (see Crovisier and Bockelée-Morvan, this volume).

It seems that nucleus sources of coma gas dominate at large heliocentric distance r, while nuclear and distributed sources coexist at small r. DiSanti et al. (1999) indicates that the distributed source is effective for r less than about 2 AU. Festou (this volume) suggests that the observed density profile of CO can be explained by outgassing of inhomogeneous regions of the comet — e.g., jet-like features ("jets") from CO-rich ice. The nature of the distributed sources of H_2CO and CO and of the release mechanisms from dust remain unexplained in the models by Greenberg and Li (1998). A similar situation exists for the observed CN. There is no doubt that some of the CN is related to dissociation of HCN and CH_3CN. These two molecules may be the only parents of CN at large heliocentric distances. However, an additional source seems to be required and it has been suggested that it is related to dust (A'Hearn et al., 1995).

Remote sensing and in situ measurements have identified a large number of radicals in the comae of comets. Most of these radicals are daughter products of stable molecules. However, the cometocentric distance profile of CH_2, observed in Comet Halley, is not compatible with any one suspected parent molecule. It seems to require several mother molecules (perhaps evaporation of hydrocarbon polycondensates) or be released directly from the nucleus (Altwegg et al., this volume). Also observed in Comet Halley are the radicals C_2H (Altwegg et al., this volume) and C_4H (Geiss et al., this volume). These radicals have not been studied as fully as CH_2, but the C_4H data point to the nucleus as the source.

Also important for the origin of cometary material and the history of comet formation are isotopes. From Comet Halley in situ measurements, the D/H ratio in water has been determined accurately and is about twice the value of seawater (Balsiger et al., 1995; Eberhardt et al., 1995). The values determined in Comets Hale-Bopp and Hyakutake are compatible with the Halley data (Bockelée-Morvan et al., 1998; Meyer and Owen, this volume). Since Comets Halley, Hyakutake, and Hale-Bopp may come from the giant planet region of the original Edgewood-Kuiper belt (via the Oort cloud), it will be important to also measure the D/H ratio in Jupiter family comets whose origin is thought to be at larger distances from the Sun (cf. Levison, 1996). This may help to distinguish between processes that led to enrichment of deuterated molecules: Short-time, high density ion–molecule reactions during comet formation (e.g., caused by x-rays from the neighborhood of the proto-Sun (cf. Lécluse and Robert, 1994; Shue et al., 1996; Benz et al., 2000) or long-time low-density reactions before comet formation in molecular clouds.

The isotope ratios of heavier atoms (C, O, N, and S) seem to be compatible with cosmic values (Meier and Owen, this volume). No significant deviations from normal solar isotope abundances were found either in the dust or the ice, with the notable exception of carbon in some dust particles (Jessberger, this volume).

In both techniques of remote sensing (global and long-term) and of *in situ* measurements (local and instantaneous), many results depend on laboratory reaction rates. Unfortunately, even for well-known species not all of the required rates are known. For minor species and radicals, the situation is even worse. This can easily lead to large discrepancies in determining molecular abundances. There are still many unidentified lines in the spectra of comet comae that may reveal interesting species and states of excitation. For analysis of coma data it is important to know photo rate coefficients, ion–molecule reaction rates, dissociative electron recombination rates, and all the relevant branching ratios to be able to derive densities of mother molecules. Hopefully some of the missing information can be supplied from laboratory measurements or theoretical calculations before the next space mission to a comet.

In addition, to derive the nucleus composition from observed molecular abundances in the coma, more modeling has to be performed. A significant first step has been initiated at ISSI by setting up a team (Huebner *et al.*, 1999) to work on this problem.

1.2. DUST

In the field of coma observations, Fomenkova (this volume) summarized the dust component and compared it with the CI-chondrites and the carbonaceous component in dense molecular clouds. Some major dust groups have been identified in comets, namely "rocks," which contain only very small fractions of organic materials, and the heavier "CHON" and "mixed" particles, which contain a large fraction of organic materials (Clark *et al.*, 1987). The majority of such grains contain a mixture of complex organic molecules, including alcohols, ketones, etc. (Grün and Jessberger, 1990). Jessberger (this volume) summarized the major findings of the PUMA experiment on board Vega, with emphasis on the rock forming elements. PUMA 1 and PUMA 2 data revealed an overall mass ratio of siliceous to organic materials in Comet Halley dust between 2 and 1 (Fomenkova and Chang, 1993).

Hanner (this volume) presented silicate spectral features in comets, and discussed their possible pre-solar origin; she stressed the first direct observational link between these silicates, the dust properties inferred from *in situ* measurements, and some stratospheric interplanetary dust particles. Amorphous and crystalline silicate particles in comets do not necessarily share a common origin. The glassy component of particles (GEMS — glass with embedded metals and sulfides) appear to constitute a major fraction of noncrystalline silicates of interstellar origin. On the other hand, crystalline particles can only form by sublimation (direct condensation from the vapor phase) at high temperatures (1200 to 1400 K) with very slow cooling or by annealing. The crystalline grains could not have formed by short-term heating. Thus, their origin could be the mid-plane of the inner solar nebula. However, the transport from the inner to the outer solar nebula (where comets form) has not been demonstrated in model calculations.

Levasseur-Regourd (this volume) analyzed the polarization data obtained for various comets; her results agree with the Greenberg model of fluffy dust aggregates, and point towards the existence of different regions of formation or stages of evolution. She finds differences in polarization both with the regions of formation of the comet nuclei and with age of the dust particles in the coma.

2. Laboratory Work

Laboratory astrophysics is a field that has recently come into its own right for characterizing interstellar molecules and cosmic dust. Physical and chemical processes occurring in the ISM and the solar system can be well simulated in the laboratory. One of the major techniques used is spectroscopy in the UV, optical, and infrared ranges. The preparation of laboratory spectra under simulated space conditions is crucial for the identification and abundance determination of molecular species in space. The comparison of laboratory spectra with astronomical data from ground- and space-based observations allows us to define parameters of the environment, such as temperature and radiation fields, in which the species reside. In the laboratory we can simulate vacuum conditions ($< 10^{-10}$ mbar), low temperature (with liquid He we can reach 4 K), as well as UV irradiation and ion bombardment relevant for simulating the interstellar medium and the solar system (cf. Allamandola, this volume; Strazzulla, this volume).

Other powerful methods are currently used to characterize the structure and composition of comet dust analogs. Among these methods are electron microscopy and x-ray diffraction (see, e.g., Colangeli *et al.*, this volume). Large-scale experiments simulating cometary environments have been carried out (the KOSI experiments, Sears *et al.*, 1999). Laboratory experiments to simulate pre-planetary dust collision and aggregation with micrometer-sized dust particles are currently successfully performed. The CODAG experiment to simulate pre-planetary dust collisions has recently flown on the ESA MASER 8 mission. A significant temporal evolution in light scattering properties (cross sections and physical properties of the particles) has been noticed. A recent book on solid state matter summarizes laboratory work in support to the ISO mission (d'Hendecourt *et al.*, 1999). For the characterization of extraterrestrial matter, particularly for the analysis of meteorite probes, many additional techniques such as laser spectroscopy, mass-spectrometry, and chromatography are in use.

Laboratory science is important for preparing future space missions and for analyzing samples from solar system bodies to be returned in the future (Ehrenfreund *et al.*, 1999).

3. From Coma Abundances to Nucleus Composition

3.1. Heat and Gas Diffusion Algorithms

Gas flow through the porous comet nucleus from subsurface ices more volatile than water ice depends on the heat flow into the interior. Physico-chemical modeling to describe this process is becoming increasingly more sophisticated. Several numerical algorithms are being developed to model the heat and gas flow based on multi-dimensional or multi-species calculations. Some models take into account compositions that include amorphous water ice with trapped gases, dust mantling of the nucleus, dust flow through pores, changing pore sizes caused by evaporation and recondensation, or dust entrainment from the surface of the nucleus. No model includes all of these effects yet, but as these models evolve, it becomes increasingly more important to identify which of these effects and processes are relevant for data analyses and interpretations.

A small ISSI team of comet nucleus modelers began comparisons of existing algorithms of numerical models based on mutually agreed upon parameters and in the simplest possible way. Only algorithms are compared that were either developed independently or evolved independently of each other from a common source. The aim is to compare the algorithms and to develop a reference model. Thus, all models are one-dimensional and make very simple assumptions: The orbit of the comet nucleus is that of Comet 46P/Wirtanen, the spin axis is normal to the orbital plane, the spin period is an even multiple of the orbital period, and the composition is restricted to intimate mixtures of crystalline H_2O ice, CO ice, and dust. All models use the same physical parameters. Since there are many different ways of manipulating dust, it was agreed not to include dust dynamics. As ice evaporates, the dust simply accumulates on the surface without compaction. Attention is focused on one spot on the equator of the spinning nucleus. This, however, does not mean that the nucleus has to be spherical.

Although agreement of results is good for some models, there are still very significant discrepancies in several models (Huebner *et al.*, 1999). At present, the conclusion of the study is that results from nucleus models diverge too much to convey more than a very limited credibility. Discrepancies must be resolved and eliminated or reduced in further investigations. These investigations include: (1) power balance, (2) temperature profiles into the nucleus (3) determination of the effective thermal conductivity, (4) energy flow profiles in the nucleus, (5) gas flux profiles out of the nucleus, (6) porosity profiles in the nucleus, and (7) density profiles in the nucleus. All team members have agreed to carry out these additional tests and tasks.

3.2. ANALYSIS OF COMET HALE-BOPP AND PREVIEW OF APPLICATIONS TO COMET WIRTANEN

The wealth of observations of recent Comets Hale-Bopp (C/1995 O1) and Hyakutake (C/1996 B2) (see, e.g., Crovisier and Bockelée-Morvan, this volume) are most suitable for comet modeling to gain insight into the structure and composition of the nucleus (Benkhoff, this volume; Enzian, this volume; Huebner and Benkhoff, this volume). Comet Hale-Bopp has provided a continuous flow of observational data for molecules from instruments ranging from the UV to the radio range and from heliocentric distances of about 7 AU pre-perihelion through perihelion to large distances post-perihelion. Comet Hyakutake, on the other hand, was so close to the Earth that it provided high spatial resolution in the inner coma. Since these two comets derived from the Oort cloud of comets, similar to 1P/Halley, it is tempting to assume similar compositions and reactions and compare results from all three comets. Attempts to carry out such analyses are under way but will take considerably more time to complete (see, e.g., Proceedings of the First International Conference on Comet Hale-Bopp, 1999).

Since Comet 46P/Wirtanen (the target of the Rosetta mission) is a short-period, Jupiter family comet, we may expect to see molecular differences in it when compared to the above three comets. At the very least, some compounds more volatile than water may be depleted. Dust particles derived from the mantled surface of the nucleus may be more processed by UV radiation during repeated passages through the inner solar system. This comet is (1) relatively inactive and (2) its activity profile around perihelion is considerably shorter than that of Comets Halley, Hale-Bopp, and Hyakutake. Measuring the D/H ratio in water from Comet Wirtanen will be of particular interest.

4. Origin of Cometary Constituents

4.1. GAS

One of the challenges we face is the interpretation of remote sensing and *in situ* data from comets. Many of the molecular species identified in comets closely resemble species detected in interstellar clouds (Allamandola *et al.*, 1997; Geiss and Altwegg, 1998; Huebner and Benkhoff, 1999; Ehrenfreund and Schutte, 1999; Irvine *et al.*, 1999; Irvine, this volume). However, there are still many unanswered questions about the similarity of relative abundances of various species and their physical state in the ice in both comets and interstellar clouds. These questions have to be resolved before we can draw firm conclusions. The D/H ratio (in water, hydrogen cyanide, and other molecules) strongly suggests ion–molecule reactions in the gas phase. However, do these reactions occur at low densities over long time spans in the interstellar medium or at higher densities over short time spans in an x-ray and EUV environment in the solar nebula? Observational evidence indicates that crystalline ice exists in accretion disks in which comets may form around young stars (Waelkens, 1999). Consistent with this, theoretical considerations indicate that amorphous ice can form in dense interstellar clouds (Ehrenfreund, this volume), but not in the solar nebula (Kouchi et al., 1994). Yet, it has been suggested that water ice in the interior of comets is amorphous (see, e.g., Enzian, this volume). How did amorphous water ice from the ISM survive heating in the accretion shock of the solar nebula? How "hot" was the shock?

Star formation is an ongoing process in collapsing interstellar molecular clouds (IMCs). It is generally accepted that 4.6 Gy ago our Sun formed similarly, by the collapse of a fragment of an interstellar cloud. Thus, observations of contemporary IMCs can serve as a model and guide for the initial processes of comet formation. However, for chemical composition only proto-stellar disks in which a spectral type G2 star (similar to the Sun) has formed are admissible for comparison. Even then, the initial material must have undergone alterations by heating during the collapse phase, particularly in the accretion shock (see, e.g., Winnewisser and Kramer, this volume). During this epoch, the more volatile components (e.g., CHON) of the grains may have evaporated, molecular bonds may have been broken, and chemical reactions may have taken place.

Further modifications probably occurred after the solar nebula had formed. Judging from observations of contemporary proto-planetary disks, the solar nebula was not a quiet place (Shue et al., 1996; Benz et al., 2000). UV-photochemistry and ion–molecule reactions in the outer nebula may have changed molecular abundances in the gas phase, and the grain composition may have been modified before comet accretion (Fegley, this volume). Moreover, radial transport has probably mixed thoroughly processed material from the inner parts of the disk into the regions at which comets formed. Thus, we expect a mixture of severely and less severely processed interstellar material in comet ices.

Finally, the molecular composition may have changed during and after accretion. For example, if comet nuclei form very early in the solar nebula, portions of not-yet-decayed short-lived ^{26}Al may be incorporated, raising the temperature in the inner regions of comet nuclei (Wallis, 1980; Prialnik and Podolak, this volume).

There is a wealth of data on abundances of molecules in IMCs, and modeling these results has led to a consistent picture of IMC gas-phase chemistry with ion–molecule reactions playing a major role (Huntress and Mitchell, 1979; Leung et al., 1984; Winnewisser and Herbst, 1993; Herbst, 1995). Molecular abundance data in contemporary proto-planetary disks are just beginning to become available. On the other hand, model predictions of proto-stellar disks depend on unknown parameters, such as the degree of ionization or the temperatures to which the gas or grains are heated by the accretion shock. Thus, we cannot yet fully exclude the possibility that a general correspondence can also be obtained for the products of nebula chemistry and comet molecular abundances.

An inventory of interstellar ices is being assembled from ISO and VLT data for many proto-stellar regions to enhance our knowledge of interstellar ice compositions. Another approach is the search for molecules that are uniquely produced under IMC conditions. At this time, two types of species are considered as specific indicators for IMC production: deuterated molecules and certain, unusual radicals.

Deuterium is highly enriched in condensable interstellar molecules (Irvine, this volume; Winnewisser and Kramer, this volume). Similarly, D is enriched – above the proto-solar abundance – by an order of magnitude in the water of P/Halley and two orders of magnitude in the hydrogen-cyanide of C/Hale-Bopp (Meier and Owen, this volume). Such strong enrichment can be achieved at very low temperatures in ion–molecule reactions (Geiss and Reeves, 1981) and, assuming reaction equilibrium, Meier and Owen (this volume) derive a formation temperature of 30 K, which fits very well the temperature conditions in IMCs. In fact, the D/H ratios measured in the water and the hydrogen-cyanide of 1P/Halley falls into the range of values found in the hot cores of IMCs (Irvine, this volume). This suggests that significant fractions of H_2O and HCN molecules of P/Halley were synthesized in the ancient IMC from which the solar system formed.

Significant D enrichments could also have been produced in the outer solar nebula by UV-photochemistry and ion–molecule reactions (Fegley, this volume). So far, modeling of the solar nebula has explained D enrichment of a factor of three at most (Lécluse and Robert, 1994; Meier and Owen, this volume). The constraints on temperature, time, and degree of ionization for producing high deuterium enrichments are so severe that it is exceedingly difficult to produce highly enriched species in the solar nebula (Geiss and Reeves, 1981; Meier and Owen, this volume). Thus the high D/H ratios in cometary molecules can still be considered the most direct and reliable indicator of an IMC origin.

Deuterated molecules reflect the temperature of their formation and not so much the temperature history afterward. Thus, once deuterated molecules are formed at low temperatures, their abundances can only be changed at higher temperatures by neutral–neutral reactions or by solid state processes.

Free atoms or radicals of IMC origin, if found in comet ice, could be used as a sensitive monitor of the chemical and thermal environment that the ice has experienced during the past 4.6 Gy. C_4H is such a radical, it has a remarkably high abundance in molecular clouds. CH_2 (Altwegg et al., this volume) and C_4H (Geiss et al., this volume), which were detected in the inner coma of 1P/Halley, appear to be primary species that may have been sublimated from the ice of the nucleus. If so, these radicals are likely to be of IMC origin, and the fact that they survived may turn out to place important constraints on the thermal history of comet ices. This is discussed by the above-mentioned authors.

Dust particles of 1P/Halley contain sodium in chondritic abundance (cf. Jessberger, this volume). Thus, like the other materials, its abundance is consistent with solar composition. The abundance of sodium in the coma of Comet Halley is so low (cf. Geiss et al., this volume) that sputtering from dust particles alone may account for the Na source in this coma. Cremonese (this volume) also finds that the broad

Na-tail of Comet Hale-Bopp is consistent with release of Na, possibly by sputtering or heat desorption, from particles in the dust tail.

On the other hand, Cremonese (this volume) finds that a second, narrow sodium tail of Comet Hale-Bopp probably has its origin close to the nucleus. Thus, it could come from the ice or from slowly moving large dust particles rapidly fragmenting near the nucleus. Combi et al. (1996) suggested that a major source of sodium is a distributed source of Na-bearing ions (NaX^+) that produce Na by the process of electron dissociative recombination.

Further modeling of the formation of the narrow Na-tail and additional studies of the chemical implications are needed before the origin of this tail is fully understood.

Abundances of inert gases place further constraints on the history of cometary ice. The apparent depletion of N_2 and the low abundance of neon in the coma of 1P/Halley (Geiss et al., this volume) constrain the temperature of ice formation and its subsequent warming. If, in addition, a low enough upper limit for argon could be established and compared with N_2, it would be a useful indicator for the thermal history of comet ice.

4.2. Dust

Dust grains are interspersed in the interstellar gas, which fills the space between the stars. The composition of dust grains changes according to the environment in which they are formed and processed (Dorschner and Henning, 1995). The bulk of stardust originates in the circumstellar environment of oxygen-rich M giants, radio-luminous OH/IR stars, supergiants, in carbon stars, and probably super novae. In oxygen-rich environments (regions with O/C > 1) oxygen is first incorporated into silicates and the remaining oxygen is mainly locked into CO. Oxygen-rich environments are therefore dominated by silicates and metal oxide grains. In carbon-rich environments (regions with O/C < 1) carbon is mostly incorporated in organic molecules and carbonaceous grains. Dust grains made of silicates, amorphous carbon, PAHs, graphite, organic refractories, SiC, and metallic oxides have been identified in the diffuse interstellar medium. In dense cold clouds dust grains accrete an icy mantle of simple molecules, such as H_2O, CO, CO_2, CH_3OH, CH_4 etc. (Whittet et al., 1996; Schutte, 1999). Aggregation of dust grains into fluffy particles in the dense medium is a rapid and efficient process.

Interstellar grains act as an important catalyst in the interstellar medium. Dust models based on astronomical observations in the UV, visual, and IR allowed to infer the dust size distribution, which provides a good fit to the mean interstellar extinction curve (Mathis et al., 1977; Dwek et al., 1997). Desert et al. (1990) proposed a three component-model of interstellar dust, where big grains, silicates with organic refractory mantles, very small grains (VSG — carbonaceous), and polycyclic aromatic hydrocarbons (PAHs) coexist in the interstellar medium. Recently Li and Greenberg (1997) proposed a unified model of interstellar dust. Interstellar dust strongly influences the physical and chemical processes in the interstellar medium. The cold and dense environment in quiescent molecular clouds provides an ideal basis for the accretion of icy grain mantles and aggregation of particles. At the densities of dense interstellar clouds ($10^3 - 10^5$ H atoms cm^{-3}) the timescale for the accretion of molecules on interstellar dust grains is short ($\sim 10^5$ yr; e.g., Schutte, 1996) and all species except H_2, He, and probably Ne can be indefinitely retained at the temperature of dense cloud grains ($\sim 10 - 30$ K).

Molecular clouds, however, evolve from an initial cold, low-density and quiescent phase (10 K, 10^3 cm^{-3}) to warm, dense and active proto-stellar regions (100 K, 10^6 cm^{-3}). Energetic proto-stellar outflows create shocks that can raise the temperatures locally to more than 2000 K. Processes such as ultraviolet irradiation, cosmic ray bombardment, and temperature variations become important in the evolutionary stages of molecular clouds and strongly influence the grain mantle growth and chemical evolution of the ice mantles (Allamandola *et al.*, 1997; this volume). Grains are exposed to a considerable metamorphism during their lifetime while they cycle back and forth between diffuse and dense clouds. Energetic ultraviolet photoprocessing of icy grain mantles results in a variety of new molecules and radicals (Bernstein, 1995, 1999), which subsequently can re-enter the gas phase. Simple molecules that accreted in the dense clouds are converted into complex organic materials by the UV irradiation in diffuse clouds. More than 120 interstellar molecules are known in the literature with diethylether as the largest gaseous species recently reported (cf. Winnewisser and Kramer, this volume; Irvine, this volume). About a third of those interstellar molecules have been identified in comet comae. For a recent comparison of interstellar and cometary molecules see Huebner and Benkhoff (1999) and Irvine (this volume). The incorporation of interstellar matter in meteorites and comets provides the basis of the cosmic dust connection. While it is likely that comets also contain processed solar system materials (Fegley, this volume), interstellar grains also may have been incorporated into comets (Hanner, this volume). Observations of Comets Hale-Bopp and Hyakutake show remarkable resemblance between the composition of interstellar and cometary dust, strengthening the interstellar – solar system connection.

5. Diversity of Comets

5.1. Diversity Based on Orbital Evolution

It is now thought (cf. Levison, 1996; Weissman, this volume) that comets formed in the region of the outer planets and beyond in the Edgewood-Kuiper (EK) belt. Interaction of comets with the giant planets threw some comets into the inner solar system, some into the outer solar system (into the Oort cloud) and some completely out of the solar system. Comets of the latter type have become extra solar system comets. However, it should be noted that we have not observed any extra solar comets originating from other stellar regions. Presently observed long-period (LP) comets visit the inner solar system because of perturbations they experienced in the Oort cloud. Those presently observed short-period (SP) comets with orbits that are close to the ecliptic, appear to have their origin in the trans-Neptunian region of the EK belt. Thus, if LP comets had their origin in the inner part of the comet-forming region, then we might expect them to be compositionally different from SP comets, perhaps even reflecting the composition of the proto-planetary nebulae of the giant planets.

SP comets fall into two groups, the Jupiter family of comets that have orbits close to the ecliptic and had their origin in the trans-Neptunian region, and the Halley family of comets, which may be captured LP comets. Halley family comets come from the Oort cloud to which they were ejected from their region of origin between the giant planets. Thus, we expect Comet Halley to be compositionally more closely related to Comets Hale-Bopp and Hyakutake than to comets in the Jupiter family.

5.2. DIVERSITY BASED ON CHEMICAL COMPOSITION

As mentioned above, the original composition of comets may differ with their place of origin, i.e., with distance from the Sun. The strength and temperature of the accretion shock and the temperature of the solar nebula will be lower with increasing distance from the Sun. However, there are also mitigating circumstances that must be considered: There may have been large-scale radial mixing in the solar nebula and collisional disruption, mixing, and re-accretion of comet nuclei in the EK belt (see Weissman, this volume).

Compositional diversity has been studied in only one sample that has a statistically significant number of comets, all coming from the same database and acquired by the same observers (A'Hearn *et al.*, 1995). They showed that about 30% of comets were deficient in C_2 and C_3 relative to CN. Most of the deficient comets were Jupiter family comets. A'Hearn *et al.* suggest that this deficiency is associated with the region of formation of these comets. Since we do not know the mother molecules of C_2, C_3, and CN (and particularly since C_2 may be associated with more than one mother molecule), it is difficult to say whether this deficiency arose during comet formation or was caused by evolution.

6. Critical Measurements for the Future

During the last decade, two bright comets (Comets Hyakutake and Hale-Bopp) have been investigated extensively with ground- and space-based telescopes using virtually the entire electromagnetic spectrum. On the other hand, Comet Halley remains the only comet that has been investigated extensively *in situ*. Since space agencies recognize the significance of comet research for understanding the origin of the solar system, they have planned for several missions (Huntress, this volume). This year, NASA launched the Stardust mission to Comet 81P/Wild 2 to return a sample of comet dust for laboratory examination. Missions to other comets are planned: CONTOUR will fly by several comets, Deep Space 1 (DS-1) may have an extended mission to fly by a comet, and ESA's Rosetta mission is scheduled to rendezvous with Comet Wirtanen using an orbiter and lander (Schwehm et al., this volume; Schulz et al., this volume). The Rosetta mission will use advanced instrumentation to study in situ many aspects of comet science from physico-chemical properties of the nucleus, outgassing activity, to dust and gas composition. All of these missions have a common goal: Study the least changed bodies of our solar system to deduce the physical and chemical parameters that prevailed in the solar nebula.

Comets are complex objects. It is not straightforward to deduce from measured parameters in the coma of one comet the complete history and composition of comets in general. Many comets have to be studied using a broad range of methods. New techniques need to be developed (e.g., Stern, this volume). It also requires a substantial effort in accompanying laboratory effort and theoretical modeling (Huebner and Benkhoff, this volume).

A summary of proposed measurements for future comet missions is provided by Ball et al. (this volume). Their list includes measurements ranging from chemical composition to physical properties of the nucleus and from isotope ratios to the ortho/para ratio of water.

The goal is to characterize and classify comets as we see them today, to distinguish between original and evolved properties, to compare comets with other solar system bodies, with proto-stellar clouds and accretion disks, and with interstellar dust grains. This needs a dedicated interdisciplinary effort from astronomers, experimental physicists and chemists, and theoreticians and modelers to obtain the interlocking pieces of the puzzle for a deep insight in the history of the solar system.

7. General Discussion on the Outcome of the Workshop

7.1. DATA ANALYSIS AND MODELING

From the foregoing discussions, it is clear that data analysis and interpretation cannot make full use of observations because some of the basic atomic and molecular data are lacking. In particular photo rate coefficients, ion–molecule reaction rates, protonization reactions, dissociative electron recombination rates, and all the associated branching ratios are needed for radicals and for the more complex species. Also processes of desorption from dust particles are poorly known. These processes, which entail heat desorption and sputtering by electron impacts, ion impacts, impacts by energetic neutrals, and photon impacts, are important to obtain quantitative molecular release rates of species from dust.

Understanding the diversity and similarities of comets is a major goal of comet research. A start in this direction was made by A'Hearn et al. (1995). It was based completely on remote sensing. While remote sensing has the advantage of long-term, time-averaged, and global perspectives, it does not have the sensitivity that mass spectrometry provides with instantaneous and *in situ* measurements. Both types of measurements, closely coupled with *global and local* modeling are needed to decipher the unpredictably variable nature of comets. The distinction between *in situ* measurements and global remote sensing were more extreme during the Comet Halley encounters then they will be during future rendezvous missions. Many comets need to be investigated and related to their respective orbital histories to statistically differentiate between inherent differences related to their origin and acquired differences related to their evolution. Lodders and Osborne (this volume) suggest an extreme case of evolution in which comets lost nearly all their volatiles and became near-Earth asteroids. Their arguments are based on the possibility that CI and CM carbonaceous chondrites are fragments of such NEOs. Kerridge (this volume) discusses production of meteoritic organic materials and the origin and evolution of organic matter in comets.

Many comet coma analyses have been semi-empirical, i.e., they depend in part on measured data such as electron temperature profiles. Global, internally self-consistent models are needed that combine physics and chemistry for a complete understanding of the nature of comets (Huebner et al., 1991). The importance of internal consistent modeling is not always fully appreciated. For example, the inner coma of an active comet is optically thick in the UV. This means that the photodissociation and photoionization rates are smaller than in the optically thin case. It also means that the products of the photo processes receive less of the excess energy (the energy of a photon above the threshold of the respective photo process). With a smaller excess energy the dissociation products are less efficient in heating and accelerating the coma gas, and less energetic electrons are more efficient in dissociative recombination. Using measured values for the gas velocity or for the electron temperature masks these effects. Confidence in a model prediction will only be attained when all modeled quantities (including gas velocity, electron temperature, electron temperature profile, etc.) are in agreement with observable, measured quantities.

Gas dynamics, dust entrainment, and chemical reactions are coupled processes. All must be modeled together.

Progress has been made on the main objective of comet research: To link the coma observations to properties of the nucleus (in particular composition) and then link the properties of the nucleus to the physical and chemical conditions at the time and place of comet formation and beyond that to the proto-solar nebula.

Several questions have been raised: How does amorphous ice survive the accretion shock? Is the accretion shock so "hot" that it vaporizes ice or even dust? How much radial mixing of processed solar nebula materials with interstellar material occurs in the solar nebula before comets form? Present evidence indicates that

— aided by radial transport – comets contain a mixture of unaltered interstellar molecules and grains and material thoroughly processed in the solar nebula. However, the presence of amorphous ice cannot be explained in such simple terms. The temperatures in the outer regions of the accretion shock must have been sufficiently high ($T > 135$ K) to convert amorphous ice to crystalline ice. Our interpretation of comet data may not be correct. On the other hand, if they are correct, then the solar nebula models may not be correct.

The chemistry in interstellar molecular clouds has been modeled for more than twenty years. However, new models for the formation of the solar nebula including collapse of the pre-solar nebula, disk formation, strength of accretion shock, etc. are needed. Observations of proto-planetary disks around young stars and modern data of physical properties must be included. The rapidly emerging observations and data on present-day circumstellar and proto-planetary disks should help to validate such models.

Comparison between the abundances in interstellar clouds (specifically in hot cores – Irvine, this volume), with abundances in the ice-phase of comet nuclei needs to be expanded and put on a firmer base. Laboratory experiments are of great importance in this area.

By including chemical reactions in simple models for cloud collapse and formation of the pre-solar disk (in particular its outer parts), one should be able to develop the sequences that lead to the origin of cometary volatiles. The constituents are likely to include molecular cloud materials, products of ion–molecule reactions in the early solar nebula, and thoroughly processed constituents from the inner regions of the solar nebula that were transported to the source region of comet nuclei by radial mixing.

7.2. Recommendations

Three recommendations emerge from the workshop:
1. Comet data acquisition, analysis, and evaluation.
   - Acquire data by remote sensing over large heliocentric distance ranges with modern techniques in all wavelength regions.
   - Acquire data by *in situ* measurements.
   - Acquire data from sample returns.
   - Acquire data from comet simulation experiments.
   - Develop global models of comet activity.
2. Establish a comet source book for data on material properties on a web site for easy access.
   - Establish a working group to deal with basic material properties.
   - Atomic and molecular cross sections and rate coefficients.
   - Sputtering and heat desorption from solids.
   - Contact laboratories to make measurements and theoretical groups to make calculations.

3. Establish an interdisciplinary working group to model the collapse of an interstellar cloud. Include formation of the accretion shock in the outer regions of the solar nebula where comets form.
   – Use comet data as a guide to constrain the models.
   – Investigate the proto-planetary disks.
   – Pursue the emerging field of Edgeworth-Kuiper belt objects.

Acknowledgements

WFH gratefully acknowledges support from ISSI and NASA grant NAG5-6785 and thanks the ISSI staff for their hospitality.

References

A'Hearn, M. F., Millis, R. L., Schleicher, D. G., Osip, D. J., Birch, P. V.: 1995, *Icarus* **118**, 223.
Allamandola, L. J., Bernstein, M. P., Sandford, S. A.: 1997, in *Astronomical and Biochemical Origins and the Search for Life in the Universe*, eds. C. B. Cosmovici, S. Bowyer, D. Werthimer, Editrice Compositori, Bologna, p. 23.
Altwegg, K., Balsiger, H., Geiss, J., Goldstein, R., Ip, W.-H., Meier, A., Neugebauer, M., Rosenbauer, N., Shelley, E.: 1993, *Astron. Astrophys.* **279**, 260.
Balsiger, H., Altwegg, K., Geiss, J.: 1995, *J. Geophys. Res.* **100**, 5827.
Benz, W., Kallenbach, R., Lugmair, G., Podosek, F.: 2000, (eds.) *From Dust to Terrestrial Planets*, Proceedings of the 9th ISSI Workshop, Kluwer.
Bernstein, M. P., Sandford, S. A., Allamandola, L. J., Chang, S., Scharberg, M. A.: 1995, *Astrophys. J.* **454**, 327.
Bernstein, M. P., Sandford, S. A., Allamandola, L. J., Gillette, J. S. B., Clemett, S. J., Zare, R. N.: 1999, *Science* **283**, 1135.
Bird, M. K., Huchtmeier, W. K., Gensheimer, P., Wilson, T. J., Janardhan, P., Lemme, C.: 1997, *Astron. Astrophys. Lett.* **325**, L5.
Bockelée-Morvan, D., Gautier, D., Lis, D. C., Young, K., Keene, J., Phillips, T., Owen, T., Crovisier, J., Goldsmith, P. F., Bergin, E. A., Despois, D., Wootten, A.: 1998, *Icarus* **133**, 147 (1998).
Boice, D. C., Huebner, W. F., Sablik, M. J., Konno, I.: 1990, *Geophys. Res. Lett.* **17**, 1813.
Combi, M. R., Di Santi, M. A., Fink, U.: 1996, in *Asteroids, Comets, Meteors*.
Clark, B. C., Mason, L. W., Kissel, J.: 1987, *Astron. Astrophys.* **187**, 779.
Desert, X., Boulanger, F., Puget, J. L.: 1990, *Astron. Astrophys.* **237**, 215.
d'Hendecourt, L., Joblin, C., Jones, A.: 1999, *Solid Interstellar Matter: The ISO Revolution*, Springer-Verlag, EDP Sciences.
DiSanti, M. A., Mumma, M. J., Dello Russo, N., Magee-Sauer, K., Novak, R., Rettig, T. W.: 1999, *Nature* **339**, 662.
Dorschner, J., Henning, T.: 1995, *Astron. Astrophys. Rev.* **6** 271.
Dwek, E.: 1997, *Astrophys. J.* **475**, 565.
Eberhardt, P., Reber, M., Krankowsky, D., Hodges, R. R.: 1995, *Astron. Astrophys.* **302**, 301.
Ehrenfreund, P., Krafft, C., Kochan, H., Pirronello, V.: 1999, (eds.) *Laboratory Astrophysics and Space Research*, Kluwer Academic Publishers.
Ehrenfreund, P., Schutte, W. A.: 1999, in *Adv. Space Res.* in press.
Engel, S., Lunine, J. I., Lewis, J. S.: 1990, *Icarus* **85** 380.

Farnham, T. L., Schleicher, D. G., Blount, E. A., Ford, E.: 1999, *Earth, Moon, Planets*, in press.
Feldman, P. D., Festou, M. C., Tozzi, G. P., Weaver, H. A.: 1997, *Astrophys. J.* **475**, 829.
Fomenkova, M., Chang, S.: 1993, *Lunar. Plan. Sci. Conf.* **24**, 501.
Geiss, J.: 1987, *Astron. Astrophys.* **187**, 859.
Geiss, J., Altwegg, K.: 1998, *ESA SPC-431*, 103.
Geiss, J., Reeves, H.: 1981, *Astron. Astrophys.* **93**, 189-199.
Greenberg, J. M.: 1982, in *Comets*, ed. L. L. Wilkening, University of Arizona Press, Tucson, p. 131.
Greenberg, J. M., Li, A.: 1998, *Astron. Astrophys.* **332**, 374.
Grün, E., Jessberger, E. K.: 1990, in *Physics and Chemistry of Comets*, ed. W. F. Huebner, Springer-Verlag, p. 113.
Herbst, E.: 1995, *Ann. Ref. Phys. Chem.* **46**, 27.
Huebner, W. F., Benkhoff, J.: 1999, in *Proc. IAU Colloq. 168, Cometary Nuclei in Space and Time*, ed. M. A'Hearn, Astron. Soc. Pacific Conf. Series, in press.
Huebner, W. F., Boice, D. C., Schmidt, H. U., Wegmann, R.: 1991, in *Comets in the Post-Halley Era*, eds. R. L. Newburn, Jr., M. Neugebauer, J. Rahe, Kluwer, p. 907.
Huebner, W. F., Benkhoff, J., Capria, M. T., Coradini, A., De Sanctis, M. C., Enzian, A., Orosei, R., Prialnik, D.: 1999, *Adv. Space Res.* **23**, 1823.
Huntress, W. T., Jr., Mitchell, G. F.: 1979, *Astrophys. J.* **231**, 456.
Irvine, W. M., Schloerb, F. P., Crovisier, J., Mumma, M.: 1999, in *Protostars and Planets* **IV**.
Kouchi, A., Yamamoto, T., Kozasa, T., Kuroda, T., Greenberg, J. M.: 1994, *Astron. Astrophys.* **290**, 1009.
Lécluse, C., Robert, F.: 1994, *Geochim. Cosmochim. Acta* **58**, 2927.
Leung, C. M., Herbst, E., Huebner, W. F.: 1984, *Astrophys. J. Suppl.* **56**, 231.
Levison, H.: 1996, in *Completing the Inventory of the Solar System*, eds. T. W. Rettig, J. M. Hahn, Astron. Soc. Pac., San Francisco, p. 173.
Li, A., Greenberg, J. M.: 1997, *Astron. Astrophys.* **323**, 566.
Lunine, J. I., Engel, S., Rizk, R., Horanyi, M.: 1991, *Icarus* **94**, 333.
Mathis, J. S., Rumpl, W., Nordsieck, K. H.: 1977, *Astrophys. J.* **217**, 425.
Meier R., Eberhardt, P., Krankowsky, D., Hodges, R. R.: 1994, *Astron. Astrophys.* **287**, 268.
Proceedings of the First International Conference on Comet Hale-Bopp: 1999, *Earth, Moon, Planets* in press.
Schutte, W. A.: 1996, in *The Cosmic Dust Connection*, ed. J. M. Greenberg, Kluwer, Dordrecht, p. 1.
Schutte, W. A.: 1999, in *Laboratory Astrophysics and Space Research*, eds. P. Ehrenfreund, C. Krafft, H. Kochan, V. Pirronello, Kluwer Academic Publishers, p. 69.
Sears, D.W.G., Kochan, H., Huebner, W. F.: 1999, *Meteoritics Planet. Sci.* in press.
Shue, F. H., Shang, H., Lee, T.: 1996, *Science* **271**, 1545.
Vandenbussche, B., Ehrenfreund, P., Boogert, A. C. A., Van Dishoeck, E. F., Schutte, W. A., Gerakines, P. A., Chiar, J., Tielens, A. G. G. M., Keane, J., Whittet, D. C. B., Breitfellner, M., Burgdorf, M.: 1999, *Astron. Astrophys. Lett.* **346**, L57.
Waelkens, C.: 1999, *Earth, Moon, Planets*, in press.
Wallis, M. K.: 1980, radiogenic heating of primordial comet interiors, *Nature* **284**, 431.
Whittet, D. C. B.: 1996, *Astron. Astrophys. Lett.* **315**, L357.
Winnewisser, G., Herbst, E.: 1993, *Rep. Prog. Phys.* **56**, 1209.
Wyckoff S., Tegler, S. C., Engel, L.: 1991, *Astrophys. J.* **367**, 641.

SUBJECT INDEX

Abundances, 5, 7–9, 11–14, 16–19, 23, 26, 28–31, 33, 37–41, 45–47, 50, 54–56, 58, 64, 72, 73, 82, 91–94, 102, 103, 109, 111–114, 117, 119, 121, 128, 131, 138, 141, 143, 149, 151, 153, 156, 169, 172, 176, 182, 187, 203, 204, 206, 207, 209, 210, 212–215, 219, 221, 223–225, 228, 229, 233–237, 240–242, 251, 253–257, 260–266, 275, 277, 278, 283, 287, 292, 309, 337, 353, 356–358, 360, 364, 366, 373, 374, 377–382, 387

  comparative, 208

  elemental, 5, 6, 14, 96, 118, 204, 292, 293, 342, 346

  isotopic, 14, 15, 26, 93, 375

  molecular, 8, 14, 23, 25, 26, 36, 46, 47, 119, 131, 141, 153, 203, 204, 221, 226, 374, 376, 380

Accretion, 36, 97, 106, 109, 120, 131, 152, 166, 178, 203, 206, 219, 225, 243–245, 250, 270, 292, 294, 301, 311, 380, 383, 384

  disk, 36, 201, 203, 206, 243, 275, 284, 380, 385

  shock, 105, 119, 121, 239, 246, 380, 384, 386–388

Aggregation, 48, 119, 166, 342, 365, 366, 377, 382, 383

Albedo, 77, 122–124, 133, 134, 136, 137, 143, 163, 165, 172, 292, 303, 307, 318, 322

Analytical techniques, *see also* Laboratory techniques

  electron microscopy, 343, 377

  gas chromatography, 338, 339

  IR spectroscopy, 58, 95, 316, 335, 337–339, 359, 366

  mass spectrometry, 3–5, 8, 9, 11, 15, 17, 25, 34, 45, 46, 92, 97, 103, 208, 213, 214, 256, 259, 294, 337–339, 373, 374, 386

  UV photolysis, 219, 220, 227, 228

Asteroid, 18, 104, 118, 167, 184, 203, 244, 250, 252, 275, 276, 281, 283–287, 289–292, 295–297, 301, 305, 307, 313, 314, 317–319, 329–332, 335–340, 359, 365, 386

Chemical differentiation, 131, 132, 138, 141

CHON, 48, 55, 61, 64, 91, 93, 94, 97, 103, 111, 151, 214, 374, 376, 380, *see also* Hydrocarbon polycondensate, Organic refractory

Comet, *see* Comet Index

  coma, 3–5, 9–14, 16–19, 22, 24, 25, 29, 30, 33, 38, 40, 45, 46, 48–50, 53–65, 69–71, 73, 75, 76, 78, 83, 84, 86–89, 92, 97, 105, 108, 110, 111, 117, 121–123, 132, 135, 136, 138, 141, 143, 149–151, 153, 154, 156, 157, 163–167, 209, 224, 228, 235, 242, 253, 254, 256–258, 260, 263–265, 272, 290, 291, 309, 313, 314, 317,

319, 321–323, 326–329, 331–333, 335, 336, 338, 354–356, 358, 364–367, 373–376, 379, 381, 382, 385, 386
dust, 11, 14, 22, 24, 25, 36, 45, 48, 50, 51, 55, 57, 58, 61, 69–73, 83, 84, 86–88, 91, 92, 94–97, 99, 103, 109–114, 119, 121–123, 125, 131, 132, 134–136, 138, 141–145, 148–151, 153–159, 163, 165–169, 171, 174, 181, 182, 187, 193, 196, 203, 206, 210, 219, 221, 233, 245, 249, 260, 266, 272, 273, 294, 308, 315, 317, 319, 323, 325–329, 332, 336–338, 342, 344, 350, 352–354, 356–358, 365–367, 374–379, 381, 382, 385, 386
    dust to gas mass ratio, 71, 73, 111, 150, 154, 364
    dust to gas production rate, 71, 73
    fragmentation, 61, 73, 121, 156, 159
    long-period, 83, 100, 131, 156, 301, 302, 304, 307, 309–311, 384
    nucleus, 4–6, 9, 11–14, 19, 22–26, 30, 32, 38–40, 46–48, 51, 53–65, 69, 70, 73, 75–78, 82–86, 88, 89, 93, 103, 105, 109–111, 114, 117–125, 127, 129, 131–139, 141–143, 147–151, 153–155, 157, 159, 165, 166, 169–172, 175, 178, 203, 209, 214, 254, 258, 260–265, 269, 270, 272, 275, 285, 286, 289–292, 294–297, 301, 303, 307–309, 311, 313–315, 317, 319, 321, 322, 326–329, 331, 335–339, 355, 356, 359, 364–368, 373–382, 384–387

density, 3, 11, 12, 56, 141, 154, 155, 319, 338–340, 378
differentiation, 131, 138, 141
dust to ice ratio, 94, 337, 357
heterogeneity, 309, 337, 360, 364, 367
model, 58, 82, 118, 131, 132, 134–136, 138, 139, 142, 148, 170, 353, 378
morphology, 60, 69, 149, 313, 316, 335, 339, 340
porosity, 123, 125, 152, 169, 170, 177, 178, 319, 338, 339, 378
splitting, 178, 290
temperature, 24, 144, 169, 170, 173, 174, 176, 177, 378, 380
texture, 319, 335, 337–339
short-period, 156, 302–304, 308, 309, 384
Halley family, 384
Jupiter family, 26, 28, 37, 40, 137, 143, 146, 148, 150, 155, 156, 159, 301–304, 308, 309, 321, 373, 375, 379, 384
tail, 53, 56, 60, 83–87, 89, 149, 163, 263, 273, 290, 291, 323, 330, 382
    dust, 60, 83–86, 88, 135, 137, 323, 374, 382
    plasma, 32, 122
    sodium, 60, 83–89, 263, 381, 382
Condensation, 38, 60, 104, 105, 123, 131, 132, 141, 142, 170, 196, 225, 246, 266, 272, 275, 283, 284, 292, 313, 342, 376, 378
Conduction, *see* Heat conduction
Cores
    cold, 39
    dense, 204, 206

hot, 36, 38, 181, 187, 200, 203, 204, 207–209, 212–215, 240, 242, 381, 387
icy, 33, 41
warm, 200
Cosmic dust, 341, 342, 348, 353, 354, 377, 383
Cosmic rays, 36, 103, 157, 206, 219, 226, 244, 249, 270, 284, 292, 293, 296, 343, 354
Crystallization, 131, 132, 135, 137, 157, 170–174
front, 123, 135, 178

D/H ratio, 7, 8, 14, 17, 18, 29, 30, 33–41, 206, 211–213, 240, 242, 247–251, 282, 293, 295, 316, 364, 365, 367, 375, 379, 381
Dark cloud, 38, 203, 204, 207, 211, 213, 214
Diffuse cloud, 152, 204, 383
Diffusion, 254, 305
gas, 117, 123, 125, 129, 131, 132, 135, 378
heat, 76, 117, 123, 125, 129, 131, 378
radial, 106
Distributed source, 17, 24, 25, 53, 54, 57, 58, 60, 62–65, 117, 123, 127–129, 135, 272, 273, 374, 375, 382, *see also* Extended source
Dust, *see* Comet dust, Particle
interplanetary, *see* Interplanetary dust
interstellar, *see* Interstellar dust

Edgewood-Kuiper belt, *see* Kuiper belt
EK, *see* Kuiper belt
Extended source, 8, 9, 11, 14, 18, 23, 26, 30, 32, 45–51, 53, 61, 63, 69, 73, 80, 86, 88, 111, 242, 254, 260–263, *see also* Distributed source

Extinction, 108, 233, 332, 343, 345, 347–349, 352–354, 382

Giant Planet, *see* Planet
Grain
carbon, 113, 155, 156, 341–343, 345, 347–349, 352–354
mantle, 3, 70, 71, 73, 113, 131, 149, 151–155, 157, 159, 182, 203, 206, 207, 209, 210, 212, 214, 215, 219, 221, 225, 233, 235–237, 272, 275, 283, 284, 342, 382, 383
organic, 70, 73, 110, 111, 113, 152, 154, 284
silicate, 99, 104, 111, 152, 154, 156, 167, 222, 272, 342, 353, 382

Heat conduction, 75, 76, 122, 123, 135, 142, 171, 173, 174, 176
Hot core, *see* Cores
Hydrocarbon, 8, 22, 33, 38, 62, 109, 229, 242, 247, 248, 251, 271, 272, 279, 280, 287, 288, 345, 353
polycondensate, 117, 123, 135, 374, 375, *see also* CHON, Organic refractory

Ice, 11, 19, 25, 26, 30, 31, 33, 34, 39, 41, 53, 54, 58, 59, 70, 81, 94, 105, 109, 112, 114, 117, 119, 120, 122, 123, 131, 132, 134–136, 138, 139, 141–143, 147–154, 159, 169–178, 182, 187, 207, 209, 214, 219–230, 233–237, 239, 244–247, 249–251, 262–264, 270, 272, 273, 275, 285, 290–292, 294, 301, 304, 307, 319, 337, 338, 357, 359, 364–367, 378–383, 386
amorphous, 123, 128, 131, 132, 134, 135, 137, 139, 142, 148,

169–171, 173, 174, 177, 178, 374, 380, 386, 387
crystalline, 119, 123, 136, 170, 173, 380, 387
non-polar, 223–225, 227, 230, 237
polar, 224, 228, 230, 237
IDP, 95, 96, 99, 101–105, 111, 166, 229, 290, 295, 365, 366, *see also* Interplanetary dust
IMC, *see* Molecular cloud
Infrared, 20, 22–26, 29, 30, 75, 82, 91, 100, 103, 106–108, 125, 126, 132, 133, 158, 181, 182, 187, 190, 191, 193, 203, 206, 208, 210, 225, 233, 234, 236, 237, 242, 316, 317, 343, 353, 355, 358, 366, 377, 382
emission, 76, 79, 102, 122, 143, 150, 154, 157, 159, 352
radiation, 150, 157
spectra, 19, 21, 22, 28–30, 32, 59, 82, 99, 104, 107, 108, 113, 121, 127, 149, 221, 224, 233, 273, 337, 341, 352, 353, 374
Insolation, 122, 124, 125, 157, 178, 304
Interplanetary dust, 91, 96, 97, 99, 103, 107–109, 114, 118, 166, 290, 295, 297, 332, 336, 342, 344, 376, *see also* Particles
Interstellar dust, 30, 69, 70, 73, 96, 106, 107, 113, 114, 149–155, 157, 159, 183, 189, 191, 217, 233, 283, 342, 352, 354, 382, 383, 385
Interstellar molecular cloud, *see* Molecular cloud
IR, *see* Infrared
Irradiation, 104, 220, 224, 228, 270–276, 284, 292, 354
energetic particle irradiation, 269
ion, 269, 272–274, 344, 354

photon, 51, 110, 155, 156, 206, 214, 221, 223, 226, 269, 344, 346, 347, 377, 383
ISO, *see* Infrared
Isomer, 10, 184, 203, 204, 208, 209, 275, 364, *see also* Molecule indexes
Isomerization, *see* Isomer
Isotope, 7, 8, 14–16, 19, 26, 27, 30, 31, 33, 34, 39–41, 91, 92, 94–97, 110, 113, 114, 118, 169, 171, 172, 178, 203, 204, 206, 209, 211–215, 223, 239, 240, 246–248, 250–252, 275, 276, 278–284, 286–288, 290, 294, 296, 297, 314, 316, 319, 332, 333, 337–340, 353, 364–366, 375, 385, *see also* Molecule indexes
Isotope ratios, *see* Isotope
Isotopic abundances, *see* Abundances
Isotopic composition, *see* Isotope
Isotopic fractionation, *see* Isotope

Jupiter, *see* Planet

Kuiper belt, 14, 16, 33, 37, 40, 119, 121, 137, 244, 245, 250, 269, 301–304, 308–311, 331, 332, 364, 365, 367, 368, 375, 383, 384, 388
Kuiper belt comets, *see* Kuiper belt
Kuiper belt objects, *see* Kuiper belt

Laboratory data, *see* Laboratory techniques
Laboratory measurements, *see* Laboratory techniques
Laboratory techniques, 21, 24, 35, 39, 46, 51, 60, 70, 73, 92, 95, 99, 101–103, 112, 153, 166, 177, 181, 182, 186–188, 190, 191, 198, 200, 214, 215, 219, 221–224, 227–229, 233, 234,

SUBJECT INDEX

236, 240, 251, 257, 269, 272, 282, 341, 342, 344, 348, 349, 351, 352, 354, 367, 376, 377, 385, 387
LP (long-period) comet, *see* Comet

Meteorites, 33, 37, 39, 95–97, 103, 105, 108, 112, 113, 118, 170, 184, 204, 215, 219, 229, 230, 243, 247, 251, 252, 275–290, 292–297, 318, 332, 335, 337, 368, 377, 383, 386
Micrometeorites, 289
Molecular
    cloud, 7, 8, 18, 70, 106, 109, 118, 119, 152, 153, 181, 182, 184, 187, 191–196, 198, 200, 201, 203, 207, 214, 215, 219–222, 225, 226, 230, 239, 240, 243, 249–251, 253, 264, 265, 272, 281, 343, 352, 375, 376, 380, 381, 383, 387, *see also* Dark cloud, Diffuse cloud
    dissociation, 5, 24, 34, 45, 56, 63, 64, 119, 121, 213, 249, 253, 254, 263, 264, 357, 358, 366, 374, 375, 386
    dissociative electron recombination, 35, 40, 57, 253, 357, 376, 382, 386
    ionization, 5, 36, 46, 60, 85, 111, 194, 249, 250, 253, 254, 380, 381, 386
    reaction
        ion-molecule, 7, 17, 33, 35, 36, 38, 182, 204, 209, 214, 239, 240, 243, 247–251, 253–255, 281–284, 375, 376, 379–381, 385, 387
        protonization, 256, 265, 385
        surface, 33, 36, 206, 220, 281, 287
Molecule, *see* Molecule indexes

Neptune, *see* Planet
Noble gas, 14, 33, 39, 118, 296, 357, 358, 360, 364, 366
Nucleosynthesis, 33, 39, 186, 214, 280

Oort cloud, 14, 16, 26, 28, 33, 36–38, 91, 103, 119, 121, 131, 178, 269, 297, 301–308, 310, 311, 332, 344, 364, 368, 373–375, 379, 383, 384
Oort cloud comets, *see* Oort cloud
Oort cloud objects, *see* Oort cloud
Organic refractory, 69–71, 91, 93, 94, 106, 109, 149–155, 159, 219, 247, 270, 272, 342, 353, 365, 366, 374, 382, *see also* CHON, Hydrocarbon polycondensate

PAH, *see* Polycyclic aromatic hydrocarbon
Particle, 39, 50, 55, 56, 58–61, 71, 83–88, 91–97, 99–103, 105, 106, 109–111, 113, 114, 119, 123, 125, 135, 149, 151, 152, 154, 155, 157–159, 163, 165–168, 182, 191, 192, 270, 287, 290, 295, 297, 305, 325, 332, 336, 337, 342, 343, 345, 348, 349, 353, 357, 366, 374–377, 379, 381–383, 385
Photodissociation, *see* Molecular dissociation
Photoionizaton, *see* Molecular ionization
Photoprocessing, 151, 249, 343, 383, 386
Planet, 33, 37, 38, 41, 109, 118, 119, 158, 184, 201, 203, 222, 230, 243, 246, 250, 289, 290, 297, 301, 302, 304–307, 311, 329, 331, 332, 383
    giant planets, 41, 119, 170, 178, 301, 302, 304–306, 308, 309, 375, 383, 384

Jupiter, 33, 37, 41, 118, 156, 184, 243–245, 291, 301, 302, 304–307, 311, 329
Neptune, 37, 38, 40, 41, 301–307, 310, 311, 329, 332, 359, *see also* trans-Neptunian
Pluto, 244, 250, 274, 329, 331
Saturn, 41, 243, 247, 249, 301, 304–306, 329, 330
Uranus, 37, 40, 41, 301–306, 311, 329, 359
Planetesimal, 33, 37, 39, 41, 109, 119, 170, 275, 284, 303–309
Plutino, 119
Pluto, *see* Planet
Point source, 253, 254, 261–264
Polarization, 163–167, 359, 377
Polycyclic aromatic hydrocarbon (PAH), 22, 62, 149, 154, 181–183, 190, 204, 219, 225, 228–230, 279, 283, 287, 342, 365, 366, 382
Polyoxymethylene (POM), 48, 50, 51, 62, 73, 123, 219, 228, 230
Porosity, 70–73, 141–143, 154–156, 159, 169–171, 173–177, 292, 293, 319, 333, 338, 339, 378
Proto-solar nebula, 109, 275, 284, 304, 313, 386
Protoplanetary disk, 8, 36, 106, 184, 240, 243, 251, 252, 380, 387, 388
Protoplanetary subnebula, 239, 242, 245–250
Protostar, 193, 206, 207, 210, 221, 222, 225, 233–237, 251

Radical, 3, 5, 8, 9, 11, 12, 14, 18, 21, 24–26, 28, 30, 53, 55, 59, 61, 62, 64, 119, 132, 181, 186, 200, 206, 224, 245, 246, 249, 253, 256, 257, 264, 265, 275, 283–285, 328, 356, 364–366, 373–376, 381, 383, 385
Radio observation, 19–21, 23–27, 29, 30, 48, 59, 63–65, 75, 76, 125–127, 135, 138, 146, 147, 225, 316, 328, 366, 374
Radioactive, 169, 171, 172, 176, 178
Radiogenic, 131, 169–175, 177, 178, 389
Re-accretion, *see* Accretion
Recondensation, *see* Condensation
Regolith, 79, 88, 333, 337

Saturn, *see* Planet
Shocks, 99, 119, 181, 192, 207, 275, 284, 380, 383
Silicates, *see* Molecule Indexes
Solar nebula, 8, 14, 31, 33, 36–41, 62, 99, 104–106, 117–120, 170, 184, 203, 204, 206, 213, 215, 239, 240, 242–251, 269, 270, 276, 281–283, 292, 295, 301, 302, 306, 308, 311, 344, 373, 376, 379–381, 384–388
Solar wind, 4, 5, 13, 37, 51, 55, 56, 88, 121, 256, 257, 313, 316, 319, 329, 358
SP (short-period) comet, *see* Comet
Sublimation, 25, 38, 48, 50, 51, 76–78, 80, 81, 142, 143, 145, 148, 170–176, 178, 184, 219, 270, 272, 273, 292, 313, 376, *see also* Vaporization, Condensation
front, 26, 145–148
SW, *see* Solar wind

Trans-Neptunian, 119, 303, 311, 384

Ultraviolet, 19, 20, 22, 23, 26, 34, 43, 51, 62, 75, 117, 121, 125, 139, 158, 192, 193, 226, 227, 235, 239, 240, 243, 248–251, 269, 316, 319, 322, 325, 335,

338, 341, 343–345, 347, 348, 350, 352–357, 359, 360, 366, 367, 377, 379–382, 386
 emission, 59, 150
 radiation, 119, 194, 200, 219, 221, 237, 243, 244, 249, 353, 379
 spectra, 17, 22, 29, 126, 193, 328, 348, 356–358, 360
Uranus, *see* Planet
UV, *see* Ultraviolet
UV irradiation, *see* Irradiation (photon)
UV photochemistry, *see* Ultraviolet
UV photolysis, *see* Analytical techniques
UV photoprocessing, *see* Photoprocessing
UV processing, *see* Ultraviolet

Vaporization, 64, 121, 123, 125, 142, 226

COMET INDEX

49P/Arend-Rigaux, 291
47P/Ashbrook-Jackson, 165, 168
Austin (C/1982 M1), 34
Austin (C/1989 X1), 20, 29, 34
Bennett (C/1969 Y1), 84, 89
19P/Borrelly, 103, 107, 149, 150, 155, 159, 331, 335, 337, 359
Bradfield (C/1979 Y1), 34
Bradfield (C/1987 P1), 100
23P/Brorsen-Metcalf, 29, 103, 108, 113, 352
6P/d'Arrest, 331, 337
2P/Encke, 34, 57, 291, 309, 331, 337
4P/Faye, 103, 107
21P/Giacobini-Zinner, 3, 34, 330
24P/Grigg-Skjellerup, 3, 165, 167
Hale-Bopp (C/1995 O1), 12, 17, 19–33, 35, 38, 39, 46, 47, 53–60, 63–65, 75–77, 79–84, 87–89, 91, 95, 99–108, 117, 121, 125, 131–139, 141, 142, 145–150, 156–159, 164–167, 209, 212–215, 224, 235, 240, 242, 247, 250, 261, 263, 266, 273, 349–354, 359, 367, 373–375, 379–384, 389
1P/Halley, 3, 5–11, 16–20, 22, 25–35, 38–40, 45–48, 50–52, 54–64, 69, 71, 73, 84, 91–97, 99–101, 103–114, 118, 129, 142, 149–151, 153–156, 158, 159, 164–168, 212–215, 220, 224, 228, 235, 240, 242, 247, 252–254, 256, 259–261, 263–266, 270, 294, 295, 297, 301–303, 328–331, 337, 338, 341, 342, 352, 355, 373–376, 379, 381–384, 386
103P/Hartley 2 , 28, 32, 103, 240
Hyakutake (C/1996 B2), 12, 17, 19–23, 25, 26, 28–35, 38, 39, 46, 47, 59, 61, 88, 100, 184, 185, 209, 212–215, 224, 240, 242, 247, 266, 354, 358, 373–375, 379, 383, 384
Ikeya-Seki (C/1965 S1), 84, 100
IRAS-Araki-Alcock (C/1983 H1), 20, 29, 34, 38, 242, 358
Kohoutek (C/1973 E1), 20, 30, 84
Levy (C/1990 K1), 20, 29, 34, 47, 100, 108, 165, 168
124P/Mrkos (1957 d), 83, 89
28P/Neujmin 1 , 291
29P/Schwassmann-Wachmann 1 , 30, 138
73P/Schwassmann-Wachmann 3 , 331, 337
Seargent (C/1978 T1), 34
Shoemaker-Levy 9 (D/1993 F2), 290
109P/Swift-Tuttle, 113, 353
Tabur (C/1996 Q1), 358
9P/Tempel 1 , 294, 296, 331, 337, 339
West (C/1975 V1), 20, 30, 32, 84, 89, 290
81P/Wild 2 , 328, 331, 336, 337, 385
Wilson (C/1986 P1), 28, 38, 103, 240
107P/Wilson-Harrington, 291, 331, 335, 337, 359
46P/Wirtanen, 3, 4, 14, 16, 18, 30, 32, 131–133, 136–139, 141–143, 145, 146, 148, 260, 294, 296, 313, 314, 319, 321–328, 338, 360, 378, 379, 385

Space Science Reviews 90: 398, 1999.

MOLECULE INDEXES

S. GRAF and U. PFANDER
International Space Science Institute, Hallerstr. 6, 3012 Bern, Switzerland

There are two molecule indexes given below, a name index and a formula index. For the latter, the sequence of entries is determined by the number of atoms in the chemical compounds. Entries with the same number of atoms are ordered using the Hill convention. In this convention, the molecular formula for carbon-containing compounds is written with C first, H second, and then all other elements in alphabetical order of their chemical symbols. For non-carbon compounds, the elements are arranged in alphabetical order.

Note that this convention does not differentiate between isomers (e.g., hydrogen cyanide HCN and hydrogen isocyanide HNC or acetic acid CH_3COOH and methyl formate $HCOOCH_3$). The assigned names in the index are those given in the paper. If there was no name given by the authors, we have chosen the most common isomer or we wrote the formula only.

Isotopically specified compounds are listed after the unspecified entries. For deuterated species we used the ^2H symbol except for atomic Deuterium.

Since some species occur on a large number of pages, we have included only references with quantitative or comparative information, preferably with additional references for already published results. Therefore, the article titles in the reference sections are also indexed.

Note that the entries in the two large tables on pages 183 (Winnewisser and Kramer) and 205 (Irvine) are not listed and the ionized or protonized species are omitted as well.

MOLECULE INDEX: NAMES

Acetaldehyde, C_2H_4O, 208
Acetamide, C_2H_5NO, 227
Acetonitrile, *see* Methyl cyanide
Acetylene, C_2H_2, 9, 12, 20, 23, 29, 31, 47, 146–148, 208, 242
Aliphatic hydrocarbons, 287
Aluminum, Al, 95
Aluminum-26, ^{26}Al, 169, 171, 178, 285, 294
Amino acid, 277, 287
Amino alkanoic acid, 277
Amino radical, H_2N, 21, 30, 58, 324, 326
2-Amino-2,3-dimethylpentanoic acid, $C_7H_{15}O_2$, 278
Ammonia, H_3N, 9, 17, 18, 20, 23, 29, 31, 47, 206, 208, 224, 226, 233, 237, 242, 247, 266, 294, 374
($^2H_1$)Ammonia, $^2HH_2N$, 212
Argon, Ar, 358, 382
Benzene, C_6H_6, 270, 271
Bromine, Br, 96
1,3-Butadiyne, *see* Diacetylene
Butadiynyl, C_4H, 253, 255, 262–265, 375, 381
Caesium, Cs, 186
Calcium, Ca, 95
Carbon, C, 5, 30–32, 91–97, 109–114, 150–156, 182, 189, 190, 204, 213, 240, 266, 276, 287, 342–345, 351–354
Carbon-13, ^{13}C, 8, 31, 39, 91, 94, 113, 212–215, 279–281, 286–288
Carbon dimer, C_2, 57, 58, 61, 309, 324, 326, 384

Carbon dioxide, CO_2, 9, 20, 23, 30, 47, 63, 73, 146–148, 154, 172, 175, 206, 208, 223, 224, 226, 227, 233–237, 241, 273, 294, 382
Carbon disulfide, CS_2, 10, 23, 47, 242
Carbon monosulfide, CS, 10, 20, 32, 197, 208, 242, 358
Carbon (^{34}S)monosulfide, $C^{34}S$, 27
Carbon monoxide, CO, 9, 11, 12, 17, 20, 23–32, 39, 45–51, 62–65, 69–73, 82, 113, 117, 126–128, 131–137, 145–148, 154, 172, 175, 187–198, 206, 208, 222–226, 233, 235, 237, 241, 242, 247, 252, 266, 272, 273, 282, 354, 374, 375, 382
(^{12}C)Carbon monoxide, ^{12}CO, 197, 213
(^{13}C)Carbon monoxide, ^{13}CO, 194, 196, 197, 213
Carbon suboxide, C_3O_2, 73, 272
Carbon trimer, C_3, 58, 182, 309, 324, 326, 384
Carbon trioxide, CO_3, 224, 227
Carbonates, 95
Carbonic acid, $C_3H_4O_3$, 273
Carbonyl sulfide, COS, 10, 20, 23, 30, 32, 47, 64, 208, 224, 233, 242
Chromium-53, ^{53}Cr, 294
Coronene, $C_{24}H_{12}$, 182
Cyanide (XCN), 222–224
Cyanide radical, CN, 25, 30, 48, 57–64, 147, 156, 272, 309, 323, 324, 326, 328, 375, 384
(^{13}C)Cyanide radical, ^{13}CN, 27
(^{15}N)Cyanide radical, $C^{15}N$, 27

Cyano radical, *see* Cyanide radical
Cyanoacetylene, *see* 2-Propynenitrile
Cyanoethynyl, C_3N, 255, 262, 263
Cyanogen, 113
Cyanopolyynes, 206
Deuterium, D, 7, 18, 26–31, 33–41, 211–213, 247, 279–287, 295, 381
Deuterium cyanide, *see* (^{2}H)Hydrogen cyanide
Diacetylene, C_4H_2, 211, 255, 262, 264, 265
Diamond, 96
Dicarbon monoxide, C_2O, 12
Dihydrogen sulfide, *see* Hydrogen sulfide
Enstatite, 95, 101, 102, 104, 105, 343, 348, 349, 351
Ethanal, *see* Acetaldehyde
Ethane, C_2H_6, 9, 20, 23, 31, 39, 47, 208, 236, 242, 354
Ethanol, C_2H_6O, 208, 227
Ethene, *see* Ethylene
Ethyl cyanide, C_3H_5N, 10, 208, 253, 255, 262–264, 266
Ethylene, C_2H_4, 9, 47
Ethylene carbonate, *see* Carbonic acid
Ethyne, *see* Acetylene
Ethynyl, C_2H, 12, 265, 375
Fayalite, 342
Formaldehyde, CH_2O, 9, 11, 18, 20, 23–26, 29, 32, 45–48, 50, 51, 62, 73, 154, 208, 210–214, 224, 227, 233, 241, 242, 264, 266, 272, 273, 374
(^{2}H$_1$)Formaldehyde, C^2HHO, 35, 207, 212
(^{2}H$_2$)Formaldehyde, C^2H_2O, 207, 212
Formamide, CH_3NO, 20, 23, 48, 208, 227
Formic acid, CH_2O_2, 20–23, 208, 233, 236
Formyl, CHO, 224, 227

Formyl radical, *see* Formyl
Forsterite, 95, 102, 104–106, 342, 351
Graphite, 96, 97
Helium, He, 186, 359
Helium-3, ^{3}He, 18, 37
Hexamethylenetetramine, $C_6H_{12}N_4$, 227, 228
HMT, *see* Hexamethylenetetramine
Hydrogen, H, 7, 91, 109–112, 152, 153, 211, 287, 345, 353, 357
Hydrogen, H_2, 28, 190, 192, 220, 223, 224, 247
(^{2}H$_1$)Hydrogen, ^{2}HH, 37
Hydrogen cyanide, CHN, 9, 20, 23, 25, 30–32, 40, 47, 64, 145–148, 184, 185, 208, 209, 214, 241
(^{2}H)Hydrogen cyanide, C^2HN, 27, 31, 35, 37, 207, 212, 215, 242
Hydrogen (^{13}C)cyanide, ^{13}CHN, 27, 212, 242
Hydrogen (^{15}N)cyanide, $CH^{15}N$, 27, 242
Hydrogen isocyanate, *see* Isocyanic acid
Hydrogen isocyanide, CHN, 20, 23, 25, 30, 32, 40, 47, 153, 208, 209, 214, 242
(^{2}H)Hydrogen isocyanide, C^2HN, 35, 212
Hydrogen sulfide, H_2S, 9, 17, 20, 23, 29, 30, 47, 146–148, 206, 208, 241, 242, 294
Hydroxyl radical, HO, 56–59, 76, 82, 132, 134, 135, 146, 323–328, 357
(^{2}H)Hydroxyl radical, ^{2}HO, 34
Imidogen, HN, 199, 200
(^{2}H)Imidogen, ^{2}HN, 27, 35, 200
Iron, Fe, 91, 95–97, 103, 150
Iron sulfide, FeS, 104
Iron-oxides, 95
Iron-sulfides, 91, 95

Isocyanic acid, CHNO, 20, 23, 48, 208
Ketenyl radical, C_2HO, 265
Magnesium, Mg, 94, 97, 150
Magnesium-25, ^{25}Mg, 39
Mercapto radical, HS, 199, 200
(^2H)Mercapto radical, ^2HS, 200
Methane, CH_4, 8, 10, 17, 20, 23, 31, 39, 47, 146–148, 153, 206, 208, 224, 226, 233–236, 241, 242, 247, 252, 270–273, 354, 382
(^2H)Methane, (^2H)CH_4, 41
($^2H_1$)Methane, C^2HH_3, 248, 252
Methanoic acid, see Formic acid
Methanol, CH_4O, 9, 17, 20–24, 29, 30, 47, 62, 73, 110, 113, 154, 208, 221–226, 233, 237, 241, 242, 251, 353, 382
(^2H)Methanol, (^2H)CH_4O, 27, 35, 242, 247
(1-$^2H_1$)Methanol, C^2HH_3O, 212
(O-^2H)Methanol, C^2HH_3O, 207, 212
Methyl cyanide, C_2H_3N, 9, 20, 23, 25, 48, 208, 211, 214, 253, 255, 262, 264, 265
($^2H_1$)Methyl cyanide, $C_2^2HH_2N$, 212
Methyl formate, $C_2H_4O_2$, 20, 21, 23, 208
Methylamine, CH_5N, 10, 208, 255, 262
Methylene, CH_2, 3, 9–12, 375, 381
Methylidyne, CH, 8, 17
(^2H)Methylidyne, C^2H, 27, 35
Naphthalene, $C_{10}H_8$, 182
Neon, Ne, 97, 253–262, 358, 382
Neon-22, ^{22}Ne, 261
Nitric oxide, NO, 10, 227
Nitrogen, N, 7, 27, 91, 97, 109–112, 150–153, 182, 204, 237, 266, 287, 374
Nitrogen, N_2, 10, 17, 28, 113, 223, 224, 237, 242, 358, 374, 382
Nitrogen-15, ^{15}N, 31, 213, 279, 286

Nitrogen dioxide, NO_2, 227
Nitrous oxide, N_2O, 227
Olivine, 101–103, 105, 107, 108, 276, 294, 342, 347, 352, 353
Oxygen, O, 5, 109–113, 150–153, 182, 204, 235, 296, 354
Oxygen, O_2, 10, 28, 223, 224, 235–237
Oxygen-18, ^{18}O, 7, 8, 29, 30, 213, 240, 286, 287
Ozone, O_3, 18, 227
Phenanthrene, $C_{14}H_{10}$, 22
Phosphinidene, HP, 199, 200
Phyllosilicate, 295
Polycyclic aromatic hydrocarbon (PAH), see Subject Index
Polyoxymethylene (POM), see Subject Index
Polystyrene, 270, 271
Potassium, K, 89
Potassium-40, ^{40}K, 171
Propanenitrile, see Ethyl cyanide
Propiolnitrile, see 2-Propynenitrile
Propynal, C_3H_2O, 265
Propyne, C_3H_4, 211, 214
(3-$^2H_1$)Propyne, $C_3^2HH_3$, 212
2-Propynenitrile, C_3HN, 10, 20, 23, 25, 48, 208, 213, 255, 262
(^2H)2-Propynenitrile, C_3^2HN, 212
Pyroxene, 101, 102, 105, 107, 108, 276, 294, 343, 347, 351–353
Rubidium, Rb, 186, 294, 297
Silicate, 91, 95, 99, 100, 103–108, 149, 153–158, 165, 167, 222, 346, 347, 352–354
Silicon, Si, 5, 94, 97, 150
Silicon carbide, CSi, 96
Sodium, Na, 60, 61, 83–89, 255–262, 381, 382
Sodium cyanide, CNNa, 263
Strontium, Sr, 287, 294, 297
Sulfates, 95
Sulfur, S, 16, 32, 94

Sulfur-33, ^{33}S, 280
Sulfur-34, ^{34}S, 8, 27, 31, 213, 287
Sulfur-36, ^{36}S, 280
Sulfur dimer, S_2, 20, 23, 29, 38, 48, 54, 208, 224, 242, 358
Sulfur dioxide, O_2S, 20, 23, 24, 31, 187–190, 208, 242, 294, 358
Sulfur monoxide, OS, 10, 20, 23, 24, 31, 47, 208, 242, 358
Thioformaldehyde, CH_2S, 20, 21, 23, 208, 210, 211, 214, 242
Thorium-232, ^{232}Th, 171
Uranium-235, ^{235}U, 171
Uranium-238, ^{238}U, 171
Water, H_2O, 9, 17, 18, 20, 23–26, 29–31, 38, 40, 45, 47, 49, 53, 58, 59, 70, 76, 79–82, 119, 120, 124–126, 134, 135, 139, 145–149, 153, 154, 172, 178, 206–211, 214, 221–226, 233, 235–237, 240, 251, 252, 255, 273, 285, 294, 296, 324, 328, 354, 356, 374, 382
($^2H_1$)Water, ^{2}HHO, 27, 29, 31, 34, 35, 37, 40, 207, 212, 215, 240, 247, 248, 252, 375

MOLECULE INDEX: FORMULAS

1 atomic

Al, Aluminum, 95
^{26}Al, Aluminum-26, 169, 171, 178, 285, 294
Ar, Argon, 358, 382
Br, Bromine, 96
C, Carbon, 5, 30–32, 91–97, 109–114, 150–156, 182, 189, 190, 204, 213, 240, 266, 276, 287, 342–345, 351–354
^{13}C, Carbon-13, 8, 31, 39, 91, 94, 113, 212–215, 279–281, 286–288
Ca, Calcium, 95
^{53}Cr, Chromium-53, 294
Cs, Caesium, 186
D, Deuterium, 7, 18, 26–31, 33–41, 211–213, 247, 279–287, 295, 381
Fe, Iron, 91, 95–97, 103, 150
H, Hydrogen, 7, 91, 109–112, 152, 153, 211, 287, 345, 353, 357
He, Helium, 186, 359
^{3}He, Helium-3, 18, 37
K, Potassium, 89
^{40}K, Potassium-40, 171
Mg, Magnesium, 94, 97, 150
^{25}Mg, Magnesium-25, 39
N, Nitrogen, 7, 27, 91, 97, 109–112, 150–153, 182, 204, 237, 266, 287, 374
^{15}N, Nitrogen-15, 31, 213, 279, 286
Na, Sodium, 60, 61, 83–89, 255–262, 381, 382
Ne, Neon, 97, 253–262, 358, 382
^{22}Ne, Neon-22, 261
O, Oxygen, 5, 109–113, 150–153, 182, 204, 235, 296, 354

^{18}O, Oxygen-18, 7, 8, 29, 30, 213, 240, 286, 287
Rb, Rubidium, 186, 294, 297
S, Sulfur, 16, 32, 94
^{33}S, Sulfur-33, 280
^{34}S, Sulfur-34, 8, 27, 31, 213, 287
^{36}S, Sulfur-36, 280
Si, Silicon, 5, 94, 97, 150
Sr, Strontium, 287, 294, 297
^{232}Th, Thorium-232, 171
^{235}U, Uranium-235, 171
^{238}U, Uranium-238, 171

2 atomic

CH, Methylidyne, 8, 17
C^2H, (^{2}H)Methylidyne, 27, 35
CN, Cyanide radical, 25, 30, 48, 57–64, 147, 156, 272, 309, 323, 324, 326, 328, 375, 384
$C^{15}N$, (^{15}N)Cyanide radical, 27
^{13}CN, (^{13}C)Cyanide radical, 27
CO, Carbon monoxide, 9, 11, 12, 17, 20, 23–32, 39, 45–51, 62–65, 69–73, 82, 113, 117, 126–128, 131–137, 145–148, 154, 172, 175, 187–198, 206, 208, 222–226, 233, 235, 237, 241, 242, 247, 252, 266, 272, 273, 282, 354, 374, 375, 382
^{12}CO, (^{12}C)Carbon monoxide, 197, 213
^{13}CO, (^{13}C)Carbon monoxide, 194, 196, 197, 213
CS, Carbon monosulfide, 10, 20, 32, 197, 208, 242, 358
$C^{34}S$, Carbon (^{34}S)monosulfide, 27
CSi, Silicon carbide, 96

C_2, Carbon dimer, 57, 58, 61, 309, 324, 326, 384
FeNi, 104
FeS, Iron sulfide, 104
HN, Imidogen, 199, 200
^{2}HN, (^{2}H)Imidogen, 27, 35, 200
HO, Hydroxyl radical, 56–59, 76, 82, 132, 134, 135, 146, 323–328, 357
^{2}HO, (^{2}H)Hydroxyl radical, 34
HP, Phosphinidene, 199, 200
HS, Mercapto radical, 199, 200
^{2}HS, (^{2}H)Mercapto radical, 200
H_2, Hydrogen, 28, 190, 192, 220, 223, 224, 247
^{2}HH, (^{2}H$_1$)Hydrogen, 37
NO, Nitric oxide, 10, 227
N_2, Nitrogen, 10, 17, 28, 113, 223, 224, 237, 242, 358, 374, 382
OS, Sulfur monoxide, 10, 20, 23, 24, 31, 47, 208, 242, 358
O_2, Oxygen, 10, 28, 223, 224, 235–237
S_2, Sulfur dimer, 20, 23, 29, 38, 48, 54, 208, 224, 242, 358

3 atomic

CHN, Hydrogen cyanide, 9, 20, 23, 25, 30–32, 40, 47, 64, 145–148, 184, 185, 208, 209, 214, 241
CHN, Hydrogen isocyanide, 20, 23, 25, 30, 32, 40, 47, 153, 208, 209, 214, 242
CH^{15}N, Hydrogen (^{15}N)cyanide, 27, 242
C^{2}HN, (^{2}H)Hydrogen cyanide, 27, 31, 35, 37, 207, 212, 215, 242
C^{2}HN, (^{2}H)Hydrogen isocyanide, 35, 212
^{13}CHN, Hydrogen (^{13}C)cyanide, 27, 212, 242
CHO, Formyl, 224, 227

CH_2, Methylene, 3, 9–12, 375, 381
CNNa, Sodium cyanide, 263
COS, Carbonyl sulfide, 10, 20, 23, 30, 32, 47, 64, 208, 224, 233, 242
CO_2, Carbon dioxide, 9, 20, 23, 30, 47, 63, 73, 146–148, 154, 172, 175, 206, 208, 223, 224, 226, 227, 233–237, 241, 273, 294, 382
CS_2, Carbon disulfide, 10, 23, 47, 242
C_2H, Ethynyl, 12, 265, 375
C_2O, Dicarbon monoxide, 12
C_3, Carbon trimer, 58, 182, 309, 324, 326, 384
H_2N, Amino radical, 21, 30, 58, 324, 326
H_2O, Water, 9, 17, 18, 20, 23–26, 29–31, 38, 40, 45, 47, 49, 53, 58, 59, 70, 75, 79–82, 119, 120, 124–126, 134, 135, 139, 145–149, 153, 154, 172, 178, 206–211, 214, 221–226, 233, 235–237, 240, 251, 252, 255, 273, 285, 294, 296, 324, 328, 354, 356, 374, 382
^{2}HHO, (^{2}H$_1$)Water, 27, 29, 31, 34, 35, 37, 40, 207, 212, 215, 240, 247, 248, 252, 375
H_2S, Hydrogen sulfide, 9, 17, 20, 23, 29, 30, 47, 146–148, 206, 208, 241, 242, 294
NO_2, Nitrogen dioxide, 227
N_2O, Nitrous oxide, 227
O_2S, Sulfur dioxide, 20, 23, 24, 31, 187–190, 208, 242, 294, 358
O_3, Ozone, 18, 227

4 atomic

CHNO, Isocyanic acid, 20, 23, 48, 208
CH_2O, Formaldehyde, 9, 11, 18, 20, 23–26, 29, 32, 45–48, 50, 51, 62, 73, 154, 208, 210–214,

224, 227, 233, 241, 242, 264, 266, 272, 273, 374

C^2HHO, $(^2H_1)$Formaldehyde, 35, 207, 212

C^2H_2O, $(^2H_2)$Formaldehyde, 207, 212

CH_2S, Thioformaldehyde, 20, 21, 23, 208, 210, 211, 214, 242

CO_3, Carbon trioxide, 224, 227

C_2HO, Ketenyl radical, 265

C_2H_2, Acetylene, 9, 12, 20, 23, 29, 31, 47, 146–148, 208, 242

C_2N_2, 30

C_3H, 253, 262, 264, 265

C_3N, Cyanoethynyl, 255, 262, 263

C_3O, 272

C_4, 255, 262

H_3N, Ammonia, 9, 17, 18, 20, 23, 29, 31, 47, 206, 208, 224, 226, 233, 237, 242, 247, 266, 294, 374

$^2HH_2N$, $(^2H_1)$Ammonia, 212

5 atomic

CH_2O_2, Formic acid, 20–23, 208, 233, 236

CH_4, Methane, 8, 10, 17, 20, 23, 31, 39, 47, 146–148, 153, 206, 208, 224, 226, 233–236, 241, 242, 247, 252, 270–273, 354, 382

$(^2H)CH_4$, (^2H)Methane, 41

C^2HH_3, $(^2H_1)$Methane, 248, 252

C_2H_2O, 211, 265

C_3HN, 2-Propynenitrile, 10, 20, 23, 25, 48, 208, 213, 255, 262

C_3^2HN, (^2H)2-Propynenitrile, 212

C_3H_2, 10, 12, 208, 211, 265

C_3^2HH, 212

C_3O_2, Carbon suboxide, 73, 272

C_4H, Butadiynyl, 253, 255, 262–265, 375, 381

C_5, 182

6 atomic

CH_3NO, Formamide, 20, 23, 48, 208, 227

CH_4O, Methanol, 9, 17, 20–24, 29, 30, 47, 62, 73, 110, 113, 154, 208, 221–226, 233, 237, 241, 242, 251, 353, 382

$(^2H)CH_4O$, (^2H)Methanol, 27, 35, 242, 247

C^2HH_3O, $(1-^2H_1)$Methanol, 212

C^2HH_3O, $(O-^2H)$Methanol, 207, 212

C_2H_3N, Methyl cyanide, 9, 20, 23, 25, 48, 208, 211, 214, 253, 255, 262, 264, 265

$C_2^2HH_2N$, $(^2H_1)$Methyl cyanide, 212

C_2H_4, Ethylene, 9, 47

C_3H_2O, Propynal, 265

C_4H_2, Diacetylene, 211, 255, 262, 264, 265

7 atomic

CH_5N, Methylamine, 10, 208, 255, 262

C_2H_4O, 10

C_2H_4O, Acetaldehyde, 208

C_3H_4, 265

C_3H_4, Propyne, 211, 214

$C_3^2HH_3$, $(3-^2H_1)$Propyne, 212

8 atomic

$C_2H_4O_2$, Methyl formate, 20, 21, 23, 208

C_2H_6, Ethane, 9, 20, 23, 31, 39, 47, 208, 236, 242, 354

C_3H_5, 265

C_4H_4, 265

9 + atomic

C_2H_5NO, Acetamide, 227

C_2H_6O, Ethanol, 208, 227

$C_3H_4O_3$, Carbonic acid, 273

MOLECULE INDEX: FORMULAS

C_3H_5N, Ethyl cyanide, 10, 208, 253, 255, 262–264, 266
C_3H_6, 265
C_6H_6, Benzene, 270, 271
$C_6H_{12}N_4$, Hexamethylenetetramine, 227, 228
$C_7H_{15}O_2$, 2-Amino-2,3-dimethylpentanoic acid, 278
$C_{10}H_8$, Naphthalene, 182
$C_{11}HN$, 206
$C_{14}H_{10}$, Phenanthrene, 22
$C_{24}H_{12}$, Coronene, 182

ABBREVIATIONS

AC	Amorphous carbon
ACAr	Amorphous carbon - Ar
ACH_2	Amorphous carbon - H_2
AU	Astronomical unit
BIMA	Berkeley-Illinois-Maryland Association
CHON	Carbon-hydrogen-oxygen-nitrogen
CODAG	Cosmic dust aggregation
CONTOUR	Comet Nucleus Tour
COS	Cosmic Origin Spectrograph
CRAF	Comet Rendezvous Asteroid Flyby
DFMS	Double Focusing Magnetic Spectrometer
DIB	Diffuse interstellar band
DS-1	New Millennium Deep Space 1 mission
DS-4	New Millennium Deep Space 4 mission (see also STM-4)
EDX	Energy dispersive x-ray
EK belt	Edgeworth-Kuiper belt (see also KBO)
EnsA	Amorphous enstatite
EUV	Extreme-ultraviolet
FESEM	Field emission scanning electron microscope
ForA	Amorphous forsterite
ForC	Crystalline forsterite
FOV	Field of view
FUSE	Far Ultraviolet Spectroscopic Explorer
FUV	Far-ultraviolet
FWHM	Full width half maximum
GEMS	Glass with embedded metals and sulfides
GMC	Giant molecular cloud
HEB	Hot Electron Bolometer
HERS	High Energy Range Spectrometer
HIS	High Intensity-range Spectrometer
HMT	Hexamethylenetetramine
HST	Hubble Space Telescope
IAU	International Astronomical Union
ICE	International Comet Explorer
IDP	Interplanetary dust particle

IF	Intermediate frequency
IMS	Ion Mass Spectrometer
IR	Infrared
IRAM	Institut de Radioastronomie Millimètrique
IRTF	Infrared Telescope Facility
IS	*In situ* measurements
IS	Interstellar
ISEE	International Sun-Earth Explorer
ISM	Interstellar medium
ISO	Infrared Space Observatory
IUE	International Ultraviolet Explorer
JF	Jupiter-family
KAO	Kuiper Airborne Observatory
KBO	Kuiper belt object
KOSI	Comet Simulation (from German)
LG	Large grain
LO	Low oscillator
LP	Long-period
MUV	Mid-ultraviolet
NEAR	Near-Earth Asteroid Rendezvous
NEO	Near-Earth object
NGE	Neutral Gas Experiment
NMS	Neutral Mass Spectrometer
NUV	Near-ultraviolet
OPR	Ortho-to-para ratio
OSIRIS	Optical, Spectroscopic, and Infrared Remote Imaging System
PAH	Polycyclic aromatic hydrocarbon
PDR	Photon-dominated region
PHO	Potentially hazardous object
PIA	Particulate Impact Analyser
PICCA	Positive Ion Cluster Composition Analyser
POM	Polyoxymethylene
PUMA	Dust Mass Impact Analyzer (from Russian)
RG	Ground-based remote sensing
ROSINA	Rosetta Orbiter Sensor for Ions and Neutral Analysis
RPA	Rème Plasma Analyzer
RTOF	Reflectron Time Of Flight
RS	Satellite remote sensing
SEA	Solar elongation angle
SMOW	Standard mean ocean water
SOFIA	Stratospheric Observatory for Far-Infrared Astronomy
SOHO	Solar and Heliospheric Observatory
SP	Short-period

SR	Sample return
STM-4	Space Technology Mission 4 (formerly DS-4)
TEM	Transmission electron microscopy
TKS	Three-Channel Spectrometer
TMC	Taurus molecular cloud
TNO	Trans-Neptunian object
UV	Ultraviolet
Vis	Visible
WF	Wide field

AUTHOR INDEX

Allamandola, L.J., 219
Altwegg, K., 3, 253, 373
Ball, A.J., 363
Balsiger, H., 3, 253
Benkhoff, J., 117, 141
Bernstein, M.P., 219
Bockelée-Morvan, D., 19
Brucato, J.R., 341
Colangeli, L., 341
Cremonese, G., 83
Crovisier, J., 19
Eberhardt, P., 45
Ehrenfreund, P., 233, 373
Enzian, A., 131
Fegley, B., 239
Festou, M.C., 53
Fomenkova, M.N., 109
Geiss, J., 3, 253, 373
Graf, S., 253, 399
Greenberg, J.M., 69, 149
Hanner, M.S., 99
Huebner, W.F., 117, 373
Huntress, W.T., 329
Irvine, W.M., 203
Jessberger, E.K., 91
Keller, H.U., 363
Kerridge, J.F., 275
Kramer, C., 181
Kührt, E., 75
Levasseur-Regourd, A.C., 163, 373
Li, A., 69, 149
Lodders, K., 289
Meier, R., 33
Mennella, V., 341
Osborne, R., 289
Owen, T., 33
Palumbo, P., 341

Pfander, U., 399
Podolak, M., 169
Prialnik, D., 169
Rotundi, A., 341
Sandford, S.A., 219
Schulz, R., 313, 321, 363
Schwehm, G., 313, 321
Stern, S.A., 355
Strazzulla, G., 269
Walker, R.L., 219
Weissman, P.R., 301
Winnewisser, G., 181

LIST OF PARTICIPANTS

Louis J. Allamandola, *NASA/Ames Research Center*, lallamandola@mail.arc.nasa.gov
Kathrin Altwegg, *University of Bern*, altwegg@phim.unibe.ch
Andrew J. Ball, *ISSI*, balla@uni-muenster.de
Hans Balsiger, *University of Bern*, balsiger@phim.unibe.ch
Johannes Benkhoff, *DRL Institut für Planetenerkundung*, Johannes.Benkhoff@dlr.de
Willy Benz, *University of Bern*, benz@phim.unibe.ch
Luigi Colangeli, *Osservatorio Astronomico di Capodimonte*, colangeli@na.astro.it
Jean-François Crifo, *Service Aéronomie CNRS*, Jean-Francois.Crifo@aerov.jussieu.fr
Jacques Crovisier, *Observatoire de Paris*, crovisie@obspm.fr
Gabriele Cremonese, *Osservatorio Astronomico di Padova,*, cremonese@pd.astro.it
Peter Eberhardt, *University of Bern*, eberhardt@phim.unibe.ch
Pascale Ehrenfreund, *Leiden Observatory*, pascale@strw.leidenuniv.nl
Achim Enzian, *Jet Propulsion Laboratory*, achim.enzian@jpl.nasa.gov
Bruce Fegley Jr., *Washington University*, bfegley@levee.wustl.edu
Michel C. Festou, *Southwest Research Institute*, festou@obs-mip.fr
Marina N. Fomenkova, *University of California*, mfomenkova@ucsd.edu
Johannes Geiss, *ISSI*, johannes.geiss@issi.unibe.ch
Stephan Graf, *ISSI*, stephan.graf@issi.unibe.ch
J. Mayo Greenberg, *Leiden Observatory*, greenber@strw.leidenuniv.nl
Martha S. Hanner, *Jet Propulsion Laboratory*, msh@scn1.jpl.nasa.gov
Walter F. Huebner, *Southwest Research Institute*, whuebner@swri.edu
Wesley T. Huntress Jr., *Carnegie Institution of Washington*, huntress@gl.ciw.edu
William M. Irvine, *University of Massachusetts*, Irvine@fcrao1.phast.umass.edu
Elmar K. Jessberger, *University of Münster*, ekj@uni-muenster.de
Horst-U. Keller, *MPI Aeronomie*, keller@linax1.mpae.gwdg.de
John F. Kerridge, *University of California San Diego*, jkerridg@ucsd.edu
Ekkehardt Kührt, *DLR Institut für Weltraumsensorik*, ekkehard.kuehrt@dlr.de
L. Maria Lara, *ESA ESTEC SSD*, lara@estec.esa.nl
AnnChantal Levasseur-Regourd, *Service Aéronomie CNRS*, aclr@aerov.jussieu.fr
Katharina Lodders, it *Washington University* lodders@levee.wustl.edu
Roland Meier, *University of Hawaii*, rmeier@zrh.che.xerox.com
Michael J. Mumma, *Goddard Space Flight Center*, mmumma@kuiper.gsfc.nasa.gov
Dina Prialnik, *University of Tel Aviv*, dina@planet.tau.ac.il
Hans Rickman, *University of Uppsala*, hans@astro.uu.se
Rita Schulz, *ESA ESTEC SSD*, rschulz@so.estec.esa.nl
S. Alan Stern, *Southwest Research Institute*, alan@everest.boulder.swri.edu
Gianni Strazzulla, *Osservatorio Astrofisico di Catania*, gianni@alpha4.ct.astro.it
Stefano Verani, *ISSI*, stefano.verani@issi.unibe.ch
Paul R. Weissman, *Jet Propulsion Laboratory*, pweissman@issac.jpl.nasa.gov
Richard West, *ESO*, rwest@eso.org
Gisbert Winnewisser, *Universität zu Köln*, winnewisser@ph1.uni-koeln.de